B. Liščić · H. M. Tensi · W. Luty
(Eds.)

Theory and Technology of Quenching

A Handbook

With 379 Figures and 43 Tables

Springer Science+Business Media, LLC

Professor
Dr.-Ing. Božidar Liščić
University of Zagreb
Faculty of Mech. Engineering
Dept. for Material Science,
Djure Salaja 1

HR-41001 Zagreb

Professor
Dr.-Ing. Hans M. Tensi
Technical University of Munich
Institute for Materials
and Processing Sciences
Arcisstr. 21

W-8000 München 2 (FRG)

Professor
Dr.-Ing. Wacław Luty
Institute of Precision Mechanics
ul. Duchnicka 3

ISBN 978-3-662-01598-8 Springer Science+Business Media New York
ISBN 978-3-662-01598-8 ISBN 978-3-662-01596-4 (eBook)
DOI 10.1007/978-3-662-01596-4

Library of Congress Cataloging-in-Publication Data
Theory and technology of quenching / B. Liščić, H. M. Tensi, W. Luty
(eds.). p. cm.
ISBN 3-540-52040-6 (Berlin : acid-free paper)
ISBN 0-387-52040-6 (New York : acid-free paper)
1. Metals--Quenching. I. Liščić, B. (Božidar) II. Tensi, H. M. III. Luty, W.
TN672.T48 1991 671.3'6--dc20 91-25653 CIP

61/3020-5 4 3 2 1 0 – Printed on acid-free paper

Preface

Heat treatment of metallic alloys constitutes an important step within the production process. The heat treatment process itself is considered as a cycle of heating the workpieces to a predetermined temperature, keeping them at this temperature for the time period required, and cooling them to room temperature in an appropriate way.

The process of heating and keeping workpieces at the required temperature is nowadays well mastered and mostly automatized. The process of cooling or quenching which determines actually the resulting properties, is handicapped with many physical and technical uncertainties. Good results can already be obtained predominantly by using empirically based practice. But increased demands on the properties of the products as well as demands on safety and environment conditions of the quenching media require efforts to investigate the details of the quenching process and to transfer the results of the research to practical application.

Advances in the knowledge about quenching processes have been achieved by modern applied thermodynamics especially by the heat and mass transfer researches; further the application of computer technology was helpful to new approaches in quenching processes.

Special emphases has been given to:

- The theory of heat transfer and heat exchange intensification during quenching
- Wetting kinematics
- Residual stresses after quenching
- Determination of the quenching intensity
- Prediction of microstructural transformation and hardness distribution after quenching, the latter with some limitations.

The idea to write this book originated with the Technical Committee: "Scientific and Technological Aspects of Quenching" of the INTERNATIONAL FEDERATION FOR HEAT TREATMENT AND SURFACE ENGINEERING (IFHT).

While the development of quenching media is pushed on by the chemical industry, the development of quenching techniques lies with heat treatment equipment manufacturers. The above named Committee deals primarily with standardization of methods for testing the quenching intensity (cooling power) of different quenchants in laboratory and in practical conditions, as well as with the upgrading of the theoretical explanations of different quenching phenomena.

As a consequence of the multidisciplinary approach of the very complicated process of quenching, 17 authors from 6 different countries, have contributed to this book. Only in this way we have been able to deal with this specific matter from many different aspects.

Being aware of the very intensive research going on during the recent years in the field of quenching we would be grateful for every further comments or suggestions.

Zagreb
München
Warszawa, February 1992

B. Liščić
H. M. Tensi
W. Luty

Table of Contents

Contributors

Dr. P. Archambault
Université de Nancy I, Ecole des Mines de Nancy,
Laboratoire de Science et Génie
des Matériaux Métalliques,
Parc de Saurupt, F-54042 Nancy Cedex, France

H. E. Boyer (deceased)
Formerly at Consulting Service Materials, Manufacturing Processes,
7935 Chagrin Road, Chagrin Falls, OH 44022, USA

Prof. Dr. G. Beck
Université de Nancy I, Ecole des Mines de Nancy,
Laboratoire de Science et Génie des Matériaux Métalliques,
Parc de Saurupt, F-54042 Nancy Cedex, France

Dr. T. Filetin
University of Zagreb, Faculty of Mechanical Engineering and Naval Architecture,
Dept. of Materials Science,
Djure Salaja 1, 41001 Zagreb, Croatia

Prof. Dr. H. P. Hougardy
Max-Planck-Institut für Eisenforschung GmbH,
Max-Planck-Straße 1, W-4000 Düsseldorf 1,
Federal Republic of Germany

Prof. Dr. R. Jeschar
Technische Universität Clausthal, Institut für Energieverfahrenstechnik,
Agricolastraße 4, W-3392 Clausthal-Zellerfeld,
Federal Republic of Germany

Dr. N. T. Kobasko
Head of Laboratory, Ukrainian Academy of Sciences,
Institute of Engineering Thermophysics,
Zelyaborstreet 2 A, 252057 Kiev, CIS

Prof. Dr. Božidar Liščić
University of Zagreb, Faculty of Mechanical Engineering and Naval Architecture,
Dept. of Materials Science,
Djure Salaja 1, 41001 Zagreb, Croatia

Prof. Dr. W. Luty
Instytut Mechaniki Precyzyjnej,
ul. Duchnicka 3, skr. poczt. 11, PL-00967 Warszawa, Poland

Prof. Dr. E. Macherauch
Universität Karlsruhe, Institut für Werkstoffkunde I,
Kaiserstraße 12, W-7500 Karlsruhe,
Federal Republik of Germany

Dr. F. Moreaux
Université de Nancy I, Ecole des Mines de Nancy,
Laboratoire de Science et Génie des Matériaux Métalliques,
Parc de Saurupt, F-54042 Nancy Cedex

Prof. Dr. E.-J. Spies
Bergakademie Freiberg, Institut für Werkstofftechnik,
O-9200 Freiberg/Sachsen,
Federal Republic of Germany

Prof. Dr. H. M. Tensi
Technische Universität München, Institut für Werkstoffwissenschaften,
Arcisstraße 21, W-8000 München 2,
Federal Republic of Germany

Prof. Dr. O. Vöhringer
Universität Karlsruhe, Institut für Werkstoffkunde I,
Kaiserstraße 12, W-7500 Karlsruhe 1,
Federal Republic of Germany

1 Transformation of Steels During Cooling

Hans P. Hougardy

1.1 Introduction

The properties of steels can be influenced in a wide range by changing the thermodynamical properties (e.g. the constitution) by alloying elements or by suppression of the equilibrium states during cooling.

For steels with an austenite–ferrite transformation just by changing the cooling rate from extremely slow to extremely fast the yield strength can be changed from 200 MPa (microstructure of ferrite and carbide) to 2500 MPa (microstructure martensite). Therefore, to get a sufficient predictability and reproducibility of the service performance of steel components, the correct handling of the cooling rate during heat treatment is decisive. A survey on these correlations has been published recently [1].

1.2 Constitution of Iron Alloys

Figure 1.1 shows the diagram of the metastable iron–cementite equilibrium [2–5]. Cubic face centred gamma iron, called austenite, is the stable phase at temperatures between about 900 °C and 1100 °C for carbon concentrations greater than 0.02 wt.-%. At room temperature the cubic body-centred alpha iron, called ferrite is the stable phase, besides the metastable phase Fe_3C, called cementite. The volume fraction of cementite increases with the carbon concentration. In the stable system iron–carbon ferrite and graphite are the stable phases at room temperature. This system is not of interest for steels.

During a very slow cooling of an iron–carbon alloy with 0.5 wt.-% carbon from 1000 °C, the transformation starts with a precipitation of proeutectoide ferrite followed by a simultaneous precipitation of ferrite and cementite at 723 °C, the temperature of the eutectoid metastable equilibrium, line P–S–K in Fig. 1.1. In an alloy with 1.3 wt.-% carbon, see Fig. 1.1, under the same conditions transformation starts with the proeutectoid precipitation of cementite followed by a simultaneous precipitation of ferrite and cementite at 723 °C.

In an alloy with 0.01 wt.-% C, at 700 °C there is a solid solution of iron and carbon. During very slow cooling from 700 °C to room temperature cementite precipitates. At room temperature there is a metastable equilibrium between ferrite

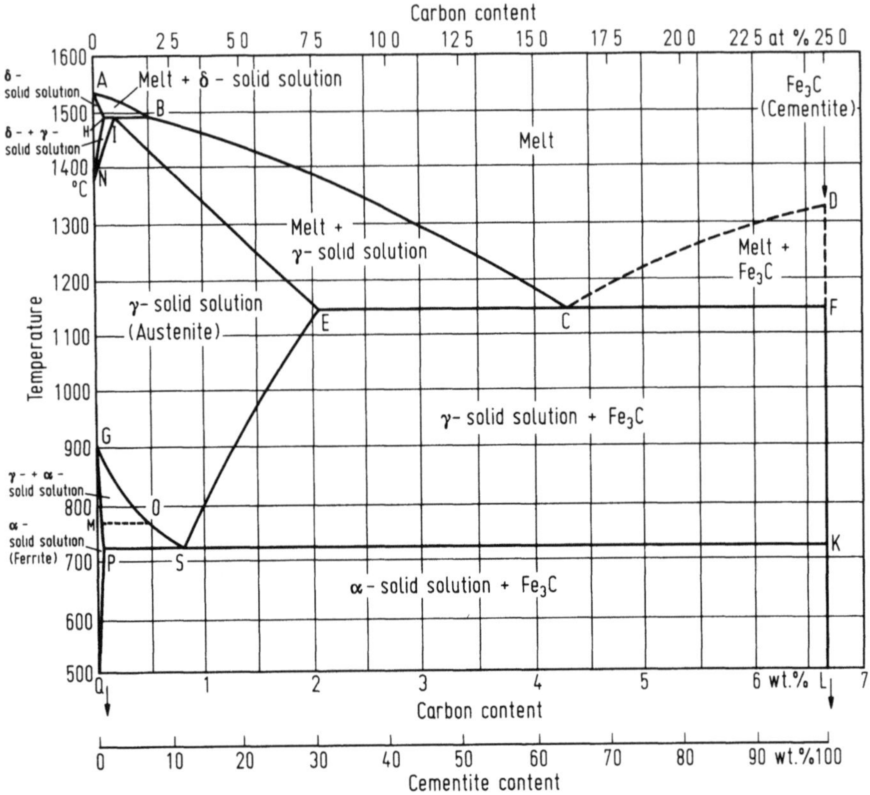

Fig. 1.1. The iron cementite phase diagram, representing the metastable equilibrium [2]

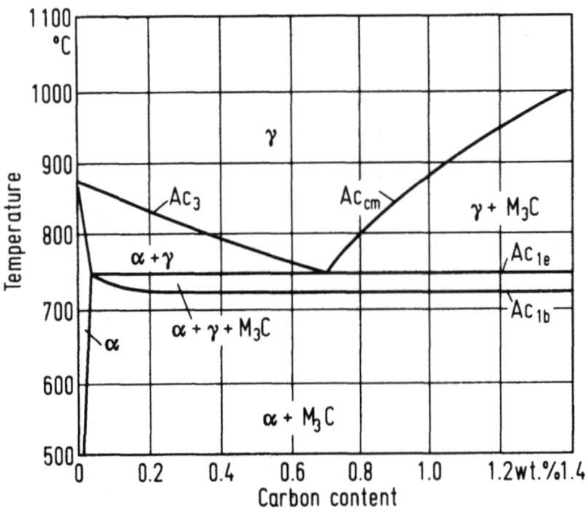

Fig. 1.2. Section through a three compotent equilibrium system Fe–M–C at a low content of *M*, with the definition of the Ac-temperatures [2]

and cementite precipitations. To minimize total energy, all phases in equilibrium as well as in metastable equilibrium build up large grains and particles to reduce surface and interface boundary energy. The description of the transformation given above is correctly valid only for infinite slow cooling rates.

The phase diagram iron–cementite can be changed remarkably by alloying elements. Figure 1.2 shows, that in alloyed steels instead of cementite a carbide $(Fe, X)_3C$ is the stable phase (X represents alloying elements). Instead of $(Fe, X)_3C$ only M_3C (M for metal) is written. At room temperature a carbide M_3C with only 0.7 wt.-% Cr is a stable phase, while Fe_3C is metastable. With increasing alloy content carbides as M_7C_3, $M_{23}C_6$, MC or M_2C may precipitate. In Fig. 1.2 the definitions of Ac temperatures are given, used in TTT-diagrams, see Chaps. 1.3.4 and 2.

Alloying elements change the stability range of austenite. In iron–silicon alloys with more than 3 wt.-% silicon no austenite is formed, the ferritic phase is stable from room temperature up to the melting point. A similar effect has chromium, the opposite effect have manganese and nickel or combinations of chromium and nickel. In alloys with about 18 wt.-% Cr and 10 wt.-% Ni, the "austenitic steels", austenite is stable down to room temperature. By special combinations of alloying elements a wide variety of all ferritic and austenitic steels and all transitions between them are possible. So called duplex steels e.g. have a microstructure containing about 50 wt.-% austenite and 50 wt.-% ferrite.

1.3 Kinetics of Transformation

1.3.1 Principles

If high cooling rates are applied, the transformation must be described including the time, instead of the phase diagrams. Transformations and precipitations as described in Chap. 1.2 are controlled by the diffusion of iron, carbon and alloying elements and therefore time-dependent. Cooling the solid solution of austenite with increased cooling rate gives no chance to finish diffusion processes. Therefore the transformation- and precipitation temperatures are lowered, supersaturated solid solutions arise or the austenite transforms diffusionless. Metastable phases, not present in the equilibrium state, may occur [6]. During fast cooling, opposite to the equilibrium state, small grains, small precipitations and arrangements of phases with high inferfacial energy as pearlite or bainite arise. During industrial production of steel components these effects are used to produce by increased cooling rates non-equilibrium states.

Details of the kinetic of the formation of these microstructures are given in Chap. 1.3.4.

1.3.2 Microstructures of Steels

1.3.2.1 Types of Microstructure

During very slow cooling—about 100 K per week—in unalloyed steels austenite transforms to ferrite and spheroidized cementite, a microstructure that can also be produced by long tempering below Ac_1, Fig. 1.3. This microstructure is close to the equilibrium. During technical heat treatment with higher cooling rates than 100 K per week three groups of microstructures can be differentiated, see Fig. 1.4. The microstructures of the pearlite range grow mainly diffusion-controlled, the microstructures of the bainite range grow partly diffusion-controlled and the microstructures of the martensite range mainly diffusionless. The temperature range of the formation of the three groups depends on the chemical composition of the steel and the cooling rate, but also for one steel there is an overlapping. An excellent summary of microstructures in steel is given in [7].

For the description below, after austenitization an isothermal transformation of a low alloyed steel is assumed. In high alloyed steels other carbides than M_3C or nonmetallic and intermetallic phases as well as other eutectoids than pearlite may arise, not discussed here.

1.3.2.2 Ferrite and Pearlite

In unalloyed steels with eutectoid composition (carbon concentration with the lowest temperature of the austenite range, Fig. 1.2, corresponding to the concentration of point S in Fig. 1.1) austenite transforms between about 700 °C and 600 °C by a simultaneous precipitation of ferrite and cementite in a metastable, lamellar-like arrangement, called pearlite (Figs. 1.5 and 1.6) [6, 8].

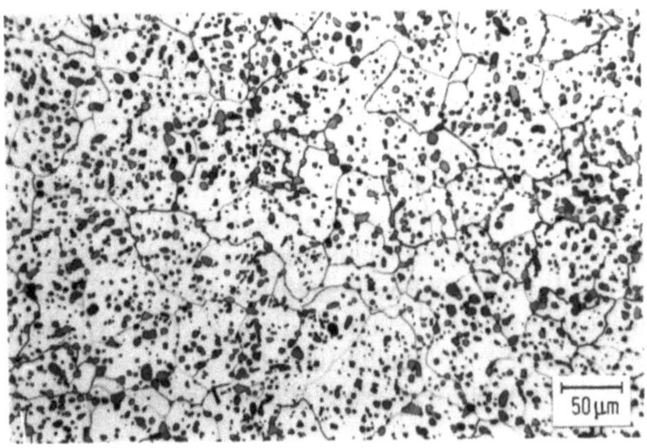

Fig. 1.3. Ferrite (white) and cementite (grey or black) in a Fe–C-alloy with 1.02 wt.% C. The dark lines are ferrite grain boundaries. Heat treatment: 700 °C 1000 hr. Magnification 200:1

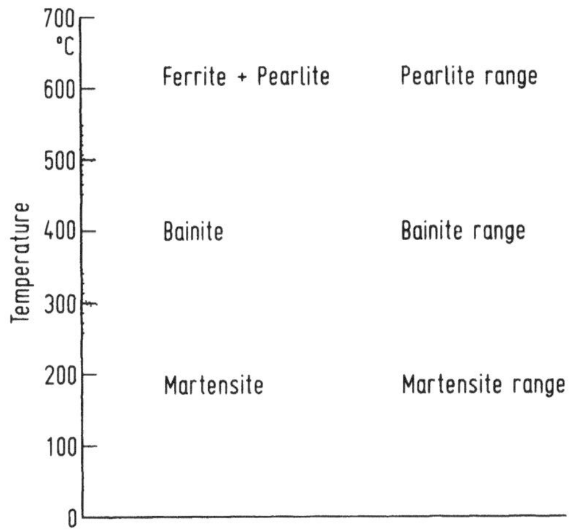

Fig. 1.4. Temperature ranges of the formation of microstructures in unalloyed steels [8]

The interlamellar distance of pearlite decreases with decreasing transformation temperature. In most technical steels the pearlite has such a small interlamellar distance, that ferrite and cementite plates cannot be resolved with light optical microscopy, Fig. 1.7. An area with parallel cementite lamella or lamella grown from one central point is called colony.

In hypoeutectoid steels, in general, a transformation to pro-eutectoide ferrite starts at the austenite grain boundaries. The thickening rate of this grain boundary ferrite decreases with temperature. Therefore, the volume fraction of ferrite transformed at one temperature below Ac_1 increases with decreasing grain size of austentite. At high volume fractions of ferrite the nucleation at austenite grain boundaries is not visible after complete transformation, Fig. 1.7. Steels with low carbon concentration may build too ferrite in Widmannstätten-arrangement, see Fig. 1.8. After precipitation of ferrite, with increasing transformation time, the retained austenite transforms into pearlite, Fig. 1.7.

1.3.2.3 Martensite

If, by very high cooling rates, the solid solution of austenite maintains down to temperatures of about 300 °C, a diffusionless transformation to martensite occurs [6]. In steels with carbon concentrations greater 0.6 wt.-%, this microstructure has a mainly plate-like (Fig. 1.9) in low carbon steels a lath-like (Fig. 1.10) substructure. In medium carbon steels both the substructures of martensite arise. In specimens with 100% martensite it is difficult to make the microstructure visible by etching and to differentiate between plates and laths, Fig. 1.11. The formation of martensite starts at a temperature M_S. At a temperature T less than M_S austenite transforms to a defined volume fraction of martensite, independent on the time. If the M_S temperature is above about 250 °C or after tempering of martensite, carbides precipitate, see Chap. 1.3.2.6.

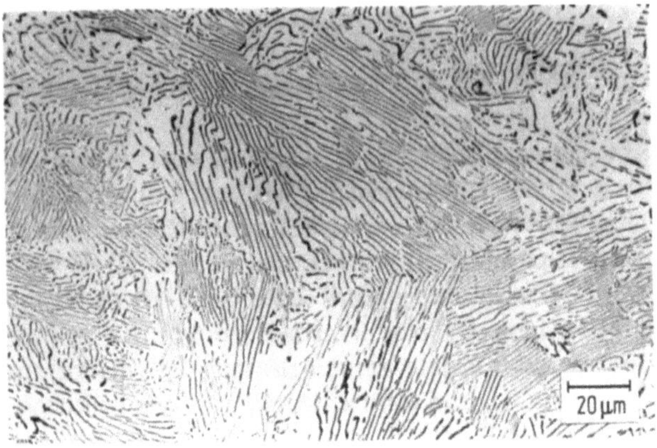

Fig. 1.5. Pearlite: ferrite (white) and cementite (grey or black) in an arrangement of parallel plates. Carbon steel with 1.0 wt.-% C. Heat treatment: 850 °C 15 min/cooling rate about 100 °C/hr. Magnification 500:1

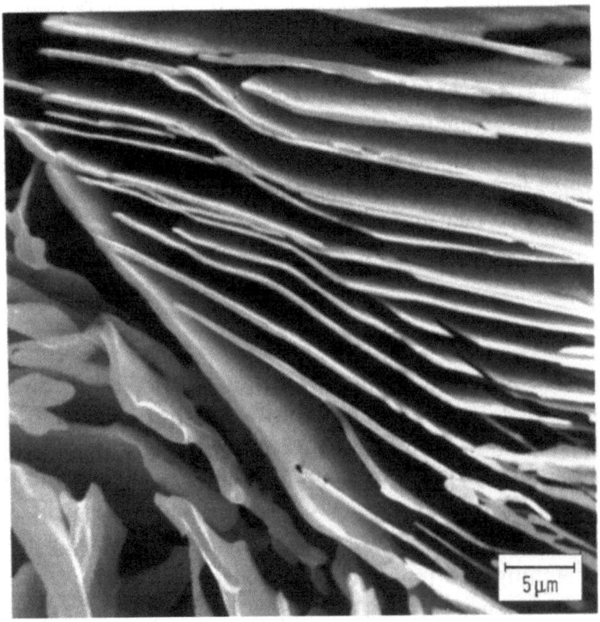

Fig. 1.6. Pearlite after a deep etching 4 min in 5% nital. The micrograph from a scanning electron microscope shows the grey cementite plates. The ferrite in between is dissolved by etching. Magnification 2400:1

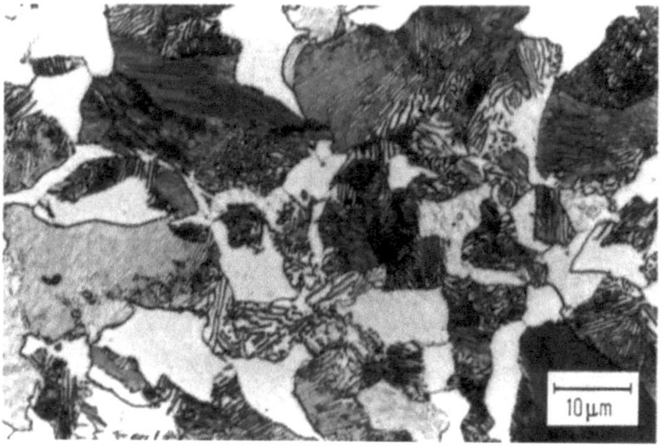

Fig. 1.7. Proeutectoid ferrite (white) and pearlite (grey). Most of the pearlite colonies have a lammellar distance below resolution. Steel with 0.45 wt.-% C and 0.5 wt.-% Cr. Heat treatment: 850 °C 20 min continuous cooling from 800 to 500 °C in 200 s. Magnification 1250:1

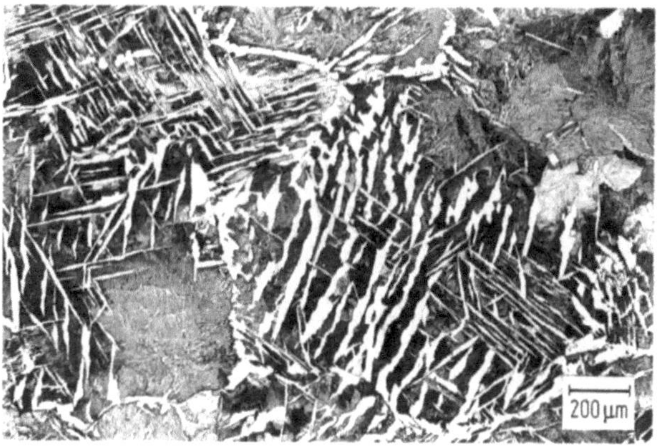

Fig. 1.8. Ferrite in Widmannstätten arrangement and pearlite. Steel with 0.45 wt.-% C. Heat treatment: 1000 °C 15 min continuous cooling from 800 °C to 500 °C in 100 s. Magnification 50:1

1.3.2.4 Bainite

In a temperature range between martensite and pearlite, see Fig. 1.4 austenite transforms to bainite by a mechanism, partly diffusion-controlled and partly diffusionless [6, 9]. In the upper bainite, Fig. 1.12, transformation starts with formation of bainitic ferrite, which has a microstructure, similar to that of lath martensite. The austenite retaining between these ferrite lathes is enriched with carbon. During further transformation from that carbon enriched austenite, cementite or other carbides precipitate. The final microstructure is ferrite

Fig. 1.9. Plate martensite (dark) and retained austenite (white). Carbon steel with 1.6 wt.-% C. Heat treatment: 1060 °C 20 min/brine. Magnification 1500:1

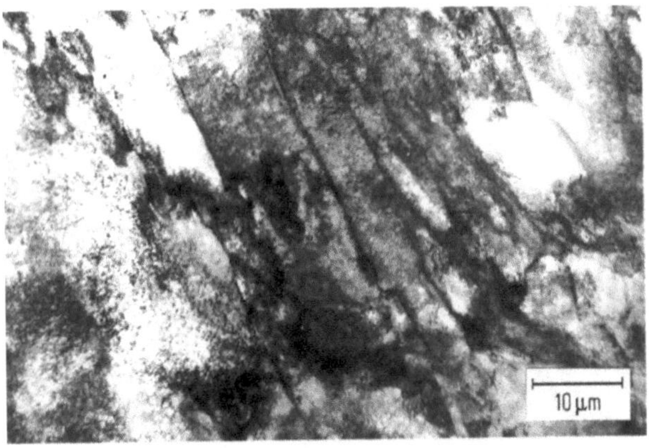

Fig. 1.10. Lath martensite in a steel with 4 wt.-% Mn, 3 wt.-% Ni, 1 wt.-% Cr, 2 wt.-% Si and 1 wt.-% Ti. Heat treatment: 1100 °C 30 min/brine. Micrograph from transmission electron microscope. Magnification 45000:1

surrounded by carbides. At transformation temperatures at the lower end of the bainite range, Fig. 1.4, ferrite and carbides are very fine and hardly to resolve by light optical microscopy, Fig. 1.13. Areas with parallel lathes of bainite are called packets.

In steels with carbon contents above 0.3 wt.-%, autenite transforms to lower bainite. In that case austenite transforms partly diffusionless to a high supersaturated ferrite which is similar to plate martensite. Within that ferrite with subsequent transformation carbides precipitate. To differentiate between upper and lower bainite is nearly impossible by light optical microscopy. Therefore, bainite should be characterized by 'fine' (Fig. 1.13) or 'coarse' (Figs. 1.12 and

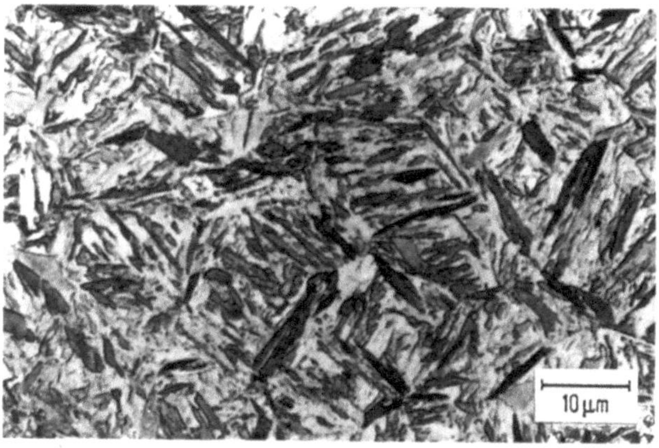

Fig. 1.11. Martensite. Carbon steel with 0.45 wt.-% C. Heat treatment: 950 °C 10 min/brine. Magnification 1500:1 [8]

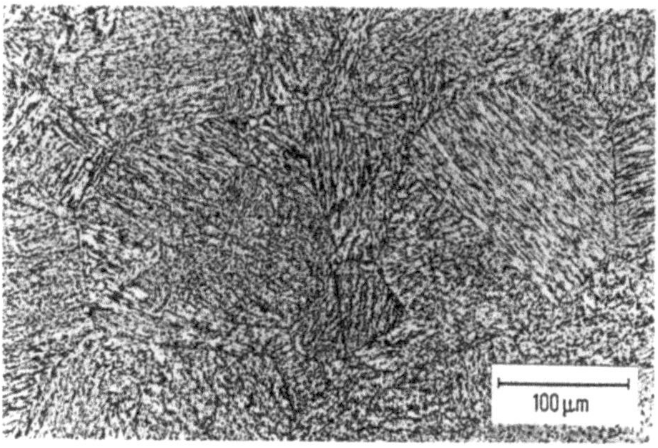

Fig. 1.12. Upper bainite aggregate of ferrite (white) and carbide (black). Steel with 0.17 wt.-% C. Heat treatment: 1300 °C continuous cooling from 800 °C to 500 °C in 15 min. Magnification 1500:1

1.15) [10] corresponding to the mean distance of bainitic carbides, a parameter that can be correlated to mechanical properties [12].

During continuous cooling also in low alloyed steels it may happen, that during formation of bainite the carbide precipitation from the austenite is not finished, the austenite transforms to martensite, Fig. 1.14, or retained austenite remains. In general the phases precipitated besides the bainitic ferrite have dimensions less than 1 μm. It is very difficult to differentiate between carbide, martensite and austenite. Therefore, for industrial applications, the whole arrangement is just called bainite [10].

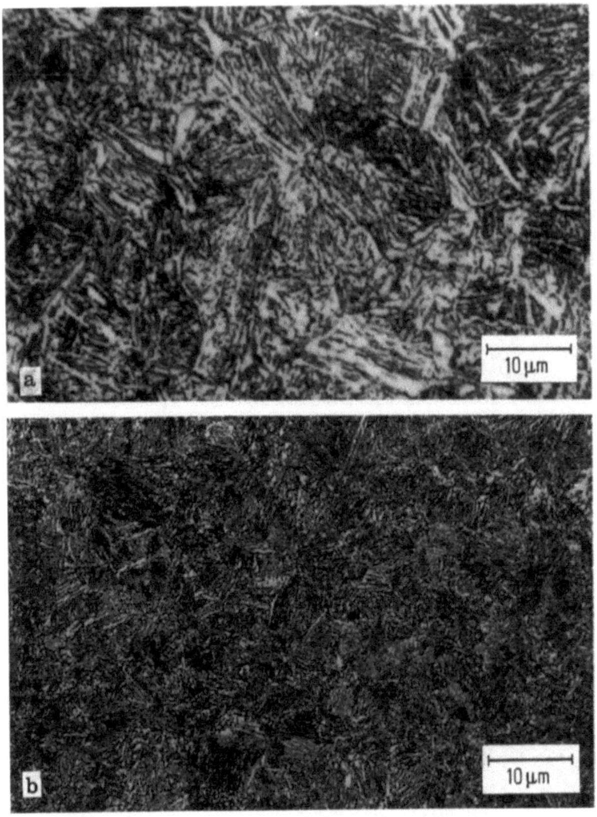

Fig. 1.13 a, b. Fine bainite in a steel with 0.5 wt.-% C, 1 wt.-% Cr and 0.25 wt.-% Mo. Heat treatment: 850 °C 20 min/360 °C 1700 s/brine. **a** Magnification 1500:1; **b** Magnification 500:1

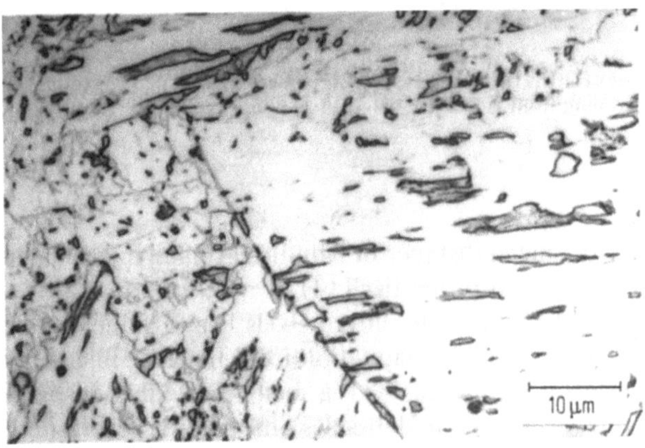

Fig. 1.14. Bainitic in ferrite (white) and martensitic islands (grey or black). Steel with 0.2 wt.-% C. Heat treatment 1300 °C 5 min/continuous cooling from 800 °C to 500 °C in 250 s. Magnification 1500:1

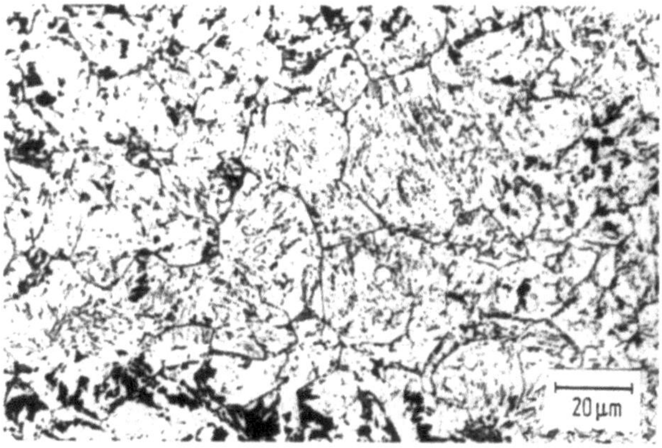

Fig. 1.15. Pearlite (large black areas) and bainite (ferrite, white, and small black carbides). Steel with 0.35 wt.-% C, and 1 wt.-% Cr. Heat treatment 850 °C 10 min/490 °C 9000 s/brine. Magnification 1500:1

1.3.2.5 Influence of Transformation Temperature

With decreasing transformation temperature the number of nuclei for pearlite and bainite increases for a constant austenite grain size. The size of colonies of pearlite and packets of bainite, the interlamellar distance of pearlite and the distance between carbides in bainite decrease remarkably. Pearlite formed at the same temperature as bainite has a much finer structure than bainite, see Fig. 1.15. The size of martensite lathes or plates depends mainly on the size of the austenite grains, the martensite is grown from [6, 8].

During continuous cooling the transformation is spread over a wide range of temperatures and ferrite, pearlite, bainite and martensite may be observed in one specimen. For details see Chap. 1.3.4. Also in one specimen there are pearlite colonies grown at different temperatures having different lamellar distances, see Fig. 1.7.

1.3.2.6 Tempering

If a specimen or component is annealed at a temperature higher than the temperature, the microstructure was formed, by diffusion processes type, shape, size and arrangement of phases change towards the equilibrium e.g. a microstructure presented in Fig. 1.3 [6]. Martensite annealed at 500 °C for example has a very fine distribution of carbides within a ferritic matrix. Such a microstructure, got by quenching and tempering, has a very good combination of mechanical properties. Annealing of supersaturated solid solutions is also used to increase strength by precipitation hardening.

1.3.3 Mechanical Properties of Microstructures

By industrial heat treatments different microstructures are produced to get a sufficient toughness at a defined level of yield strength. Details are given in Chap. 2. Martensite has the highest yield-strength, a steel of given chemical composition can get. The best relation between toughness and yield strength can be obtained by a homogeneous distribution of the alloying elements, the dislocation density, the precipitations and a small grain size. Only with decreasing grain size of the microstructure the yield strength *and* the toughness to yield strength ratio increase [11]. By all other strengthening mechanisms with increasing yield strength the toughness to strength ratio is lowered.

From all microstructures described above, tempered martensite has the most homogeneous distribution of the strenghtening elements and the highest toughness to strength ratio. The next best microstructure is tempered bainite, formed at low temperatures. The worst values have ferritic pearlitic microstructures. Ferrite is nearly pure iron with a very low yield strength and pearlite a mixture of cementite and ferrite with a high yield strength, therefore, the yield strength is inhomogeneous across the volume. Toughness to strength ratios between the values of ferrite/pearlite and tempered martensite have bainite, transformed at temperatures at the upper limit of the bainite range, and microstructures containing ferrite in Widmannstätten arrangement. From this point of view the optimal microstructure is tempered martensite, but to get martensite in a large component may be impossible, see below. Fortunately, for many applications ferritic pearlitic microstructures, which can be produced with minimum effort, have a sufficient strength and sufficient toughness [12].

1.3.4 Transformation Diagrams

The miocrostructures to be realized in a component can be derived from the desired mechanical properties. An important tool to realize that are the transformation diagrams, that give the correlation between alloying content, cooling rate and microstructure. But before the transformation, the material must be austenitized, as described by the time temperature austenitization diagrams.

1.3.4.1 Austenitization

Figure 1.16 shows a diagram for isothermal austenitization [13]. The diagram is valid for a heating rate of 130 K/s to the austenitizing temperature and subsequent isothermal annealing. At 800 °C e.g. transformation starts immediately with heating to the transformation temperature. After 8 seconds the transformation is complete, the austenite has a grain size of about ASTM 11. With increas ing time only the grain size increases. But at this temperature just above Ac_3 or at 850 °C and short austenitizing time the austenite is inhomogeneous, (see

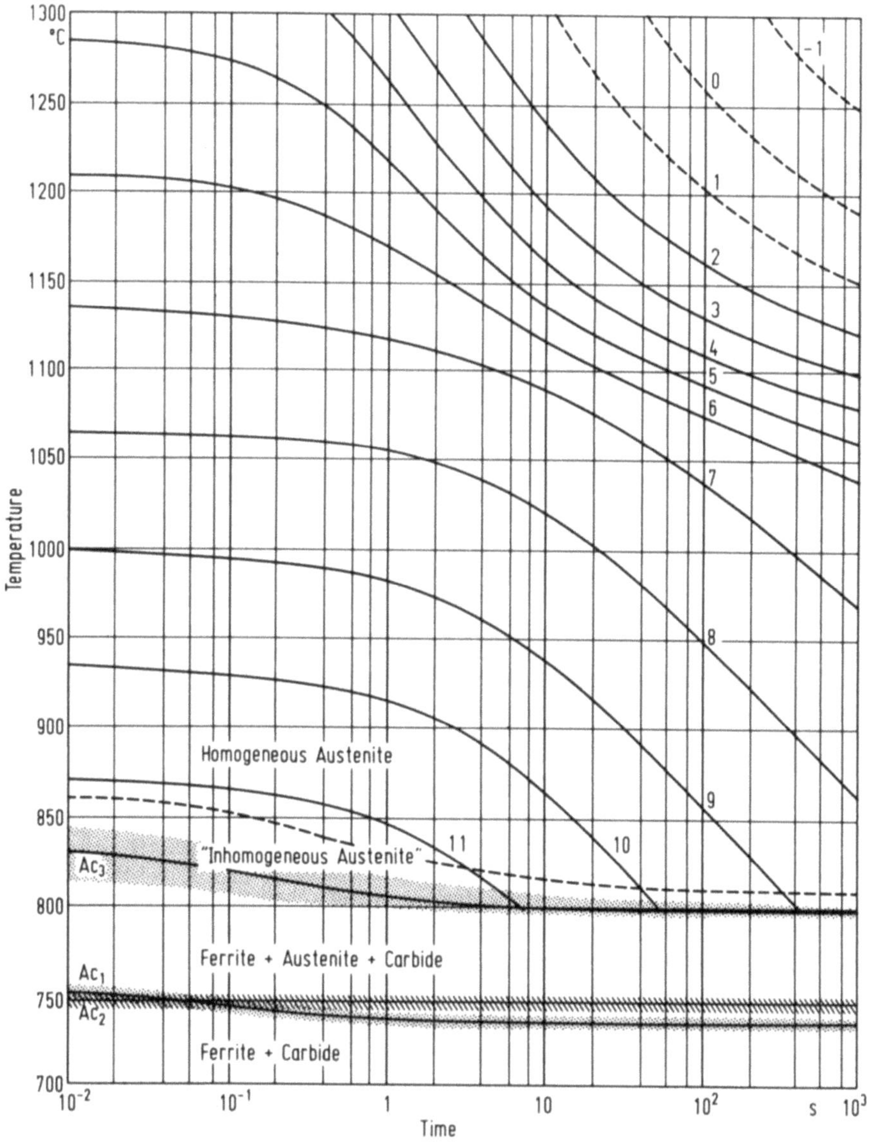

Fig. 1.16. Time temperature austenitization diagram for isothermal austenitization. Steel with 0.40 wt.-% C, 0.22 wt.-% Si, 0.57 wt.-% Mn, 0.95 wt.-% Cr, 0.02 wt.-% Al, 0.0098 wt.-% N. Microstructure before austenitization: tempered martensite. Heating rate 130 K/s [13]

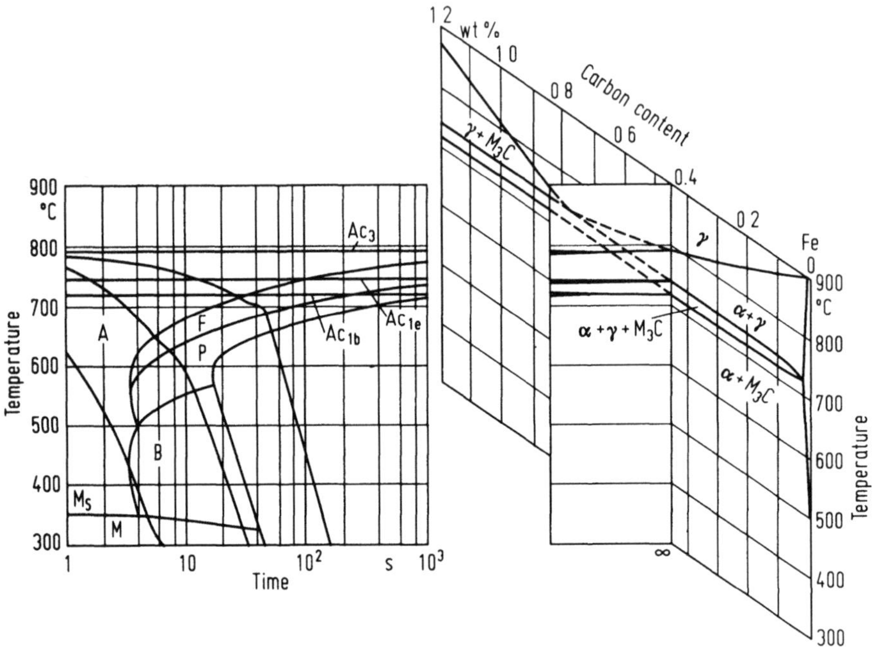

Fig. 1.17. The equilibrium-(phase-) diagram Fe–M_3C as limitation of a CCT diagram for infinite slow cooling [8]

Fig. 1.16) and transforms very fast during cooling, and in general, gives insufficient hardness after quenching. The diagram shows clearly, that with increasing austenitizing temperature or at one temperature with increasing austenitizing time, the austenite becomes homogeneous, the austenite grains grow [2, 8, 13, 14]. Therefore, the standard austenitizing temperature is 30–50 °C above Ac_3 for practical application, and the austenitizing time about 30 min, to austenitize the component in the region of homogeneous austenite. In industrial heat treatment there are a lot of other conditions, especially for tool steels [15].

1.3.4.2 Isothermal Transformation

The transformation of austenite at one temperature into the microstructures described in Chap. 1.3.2 can be described by the isothermal transformation diagrams, Fig. 2.12 in Chap. 2. For these diagrams is assumed, that after a complete austenitization the temperature is changed very fast to the transformation temperature [2, 8, 16]. As a function of time the development of transformation will be observed. Details are given in Chap. 2.1.

1.3.4.3 Transformation During Continuous Cooling

Figure 2.13 in Chap. 2 shows an example for a continuous cooling transformation (CCT)-diagram. In these diagrams the transformation is presented for the time temperature cycles plotted in the diagram. In the lower part of the diagram the volume fraction of microstructures arising during the different cooling rates are presented [2, 8, 16]. It can be seen clearly, that the volume fraction of ferrite and pearlite depends on the cooling rate.

With decreasing cooling rates the transformation follows more and more the equilibrium. For infinite slow cooling rate the transformation gives microstructures which are identical with that, described by the equilibrium diagram, Fig. 1.1. The equilibrium diagram is the time temperature transformation diagram for infinite slow cooling rate. This is valid for the austenitization, isothermal, and continuous cooling, Fig. 1.17.

1.4 Factors Influencing the Transformation

1.4.1 Austenitizing Conditions

After complete austenitization the grain size changes with austenitizing time and temperature, see Fig. 1.16. In general, with increasing grain size the transformation rate decreases. But on the other hand, with increasing grain size of austenite, for a given cooling rate the grain size of ferrite, pearlite, bainite and especially martensite increases with some exceptions only [17].

1.4.2 Cooling Rate

According to Fig. 2.13 in Chap. 2, the cooling rate has decisive influence on the microstructure produced. At high cooling rates austenite transforms only to martensite, at low cooling rates ferrite and pearlite are produced, at medium cooling rates microstructures of ferrite, pearlite, bainite, martensite and probably retained austenite arise.

Not only the cooling time $t_{8/5}$ from 800 to 500 °C defines the transformation of austenite, but also the shape of the curves in the transformation diagrams. For the same time $t_{8/5}$ the microstructure produced during cooling may be quite different if a constant cooling rate or an exponential time-temperature cycle are applied [18]. Therefore, transformation diagrams are only valid for the time-temperature cycle used for the measurement.

1.4.3 Alloying Elements

Compared with pure iron–carbon alloys, the transformation rate can be reduced drastically by alloying elements. In Fig. 2.13, Chap. 2, after a cooling according to the cooling line crossing the temperature of 500 °C at 300 sec the austenite transforms to 18% ferrite and 82% pearlite. After cooling with the same time temperature cycle a steel with 0.4% C, 1.3% Cr, 0.25% Mo and 4% Ni would transform only to martensite [16]. This influence of alloying elements is used to produce steels which have transformation characteristics adopted to the desired microstructure and to the diameters of components to be manufactured, see Chap. 1.5. A list of collections of transformation diagrams is given in [2].

1.5 Description of Transformation in Components

Figure 2.13 in Chap. 2 shows the time temperature cycle of cylinders with different diameters during quenching in oil. It is clearly seen, that the cooling rate changes across the section and for the same position within the section with the diameter. At one time, the temperature difference between surface and core increases from air cooling to quenching in brine.

The time temperature cycles used to establish the CCT-diagram and measured within the cylinders are different. That means that the transformation described by the continuous cooling transformation diagrams can be used only as a first approximation for the transformation in a component manufactured in industries. According to the transformation characteristic and the cooling rate only small differences in the time temperature cycle can result in large differences in microstructure and from that in mechanical properties [18].

As example in Fig. 1.18 the hardness of cylinders manufactured from different steels is plotted as calculated from the corresponding CCT diagrams taking the time $t_{8/5}$ as a basis for the description of the cooling cycle. At least the differences

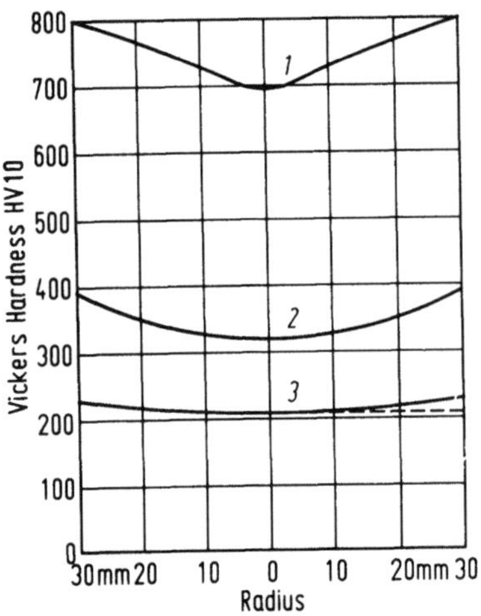

Fig. 1.18. Hardness at different positions on the cross section of cylinders with 60 mm diameters quenched in oil for three different steels. Steel 1: 0.5 wt.-% C, 1 wt.-% Cr, 0.25 wt.-% Mo. Steel 2: 34 wt.-% C, 1 wt.-% Cr, 0.25 wt.-% Mo. Steel 3: 0.35 wt.-% C (8)

between different steel grades become evident. To get a better prediction of microstructure and properties of heat treated components the time temperature cycle at each position must be known. There are methods developed for a calculation of these time temperature distributions [19].

1.6 Calculation of Transformation and Properties

To estimate the transformation of austenite in a complex part as a crankshaft, considering all influences discussed in Chap. 1.4 is not possible. Therefore calculation methods have been worked out with simplifying assumptions. Input are measured diagrams for isothermal transformation. Calculated is the transformation for any time temperature cycle [19–21]. If the heat transfer coefficient for each point of the surface of a part is known during heating and cooling, for every point of the cross section the time—temperature cycle and the transformation can be calculated. The result is the microstructure at each position across the working piece [21]. These calculations can be extended to start with the chemical composition instead of a measured isothermal transformation diagram if the influence of alloying elements on the isothermal transformation characteristic is known. At time this can be realized only for some steels grades [22]. If the mechanical properties of the microstructure are known, the mechanical properties of the part can be calculated by use of an addition rule [23]. Such calculations show, that the heat transfer coefficient has a great influence on the transformation during cooling of a component [24]. But on the other hand, it is difficult to determine these coefficients for industrial equipments, see Chaps. 8 and 9. The calculation of the time temperature cycle and the transformation are the basis for a calculation of residual stresses, see Chap. 5.

References

1. Steel, A handbook for material research and engeneering, vol. 1 and vol. 2 (1991, 1992) Springer-Verlag, Berlin Heidelberg New York Tokyo; Verlag Stahleisen mbH Düsseldorf
2. Hougardy HP, Description and Control of Transformation in Technical Application. In: (1), vol. 1
3. Horstmann D (1985) Das Zustandsschaubild Eisen-Kohlenstoff, 6. Aufl. In: Bericht des Werkstoffausschusses des Vereins Deutscher Eisenhüttenleute Nr. 180. Verlag Stahleisen
4. Kubaschewski O (1982) Iron—Binary phase diagrams. Springer Verlag, Berlin Heidelberg New York
5. Schürmann E, Sdmidt R (1979) In: Arch. Eisenhüttenwes. 50: 185–186
6. Pitsch W, Sauthof S, Microstructure of Steels. In: (1), vol. 1
7. Rose A, Schrader A (1966) De Ferri Metallographia, vol. 2. Verlag Stahleisen mbH Düsseldorf (The text is written in English, French and German)
8. Hougardy H (1990) Umwandlung und Gefüge unlegierter Stähle, eine Einführung. Verlag Stahleisen mbH Düsseldorf
9. Badeshia HKDH, Christian JW, Metall. Trans., vol. 21A, pp. 767–798. Other authors Metall. Trans., vol. 21A, pp. 799–894 and 1343–1553

10. Lotter U, Hougardy HP (1992) To the definition of bainite. In: Practical Metallography
11. Dahl W, Mechanical Properties. In: (1), vol. 1
12. Hougardy HP, Heat treatability, Heat Treatment, and Properties. In: (1), vol. 1
13. Orlich J, Rose A, Wiest P (1973) Atlas zur Wärmebehandlung der Stähle. ZTA-Schaubilder, Max-Planck-Institut für Eisenforschung in Zusammenarbeit mit der Technischen Universität Berlin und dem Werkstoffausschuß des Vereins deutscher Eisenhüttenleute, vol. 3. Orlich J, Pietrzeniuk HJ (1976) vol. 4. Verlag Stahleisen mbH: Düsseldorf
14. Hougardy HP, Sachova E (1986) In: Steel Research 57: 188–198
15. Vöge H, Heat Treatment. In: (1), vol. 2. Wilmes S, Becker HJ, Krumpholz R, Verderber W, Tool Steels. In: (1), vol. 2
16. Wever F, Rose A, Rose A, Peter W, Strassburg W, Rademacher L (1954–1958) Atlas zur Wärmebehandlung der Stähle, Max-Planck-Institut für Eisenforschung in Zusammenarbeit mit dem Werkstoffausschuß des Vereins deutscher Eisenhüttenleute, vol. 1. Rose A, Hougardy HP (1972) vol. 2. Verlag Stahleisen Düsseldorf
17. Pitsch W, Hougardy HP (1984) Gefügeerzeugung in Stählen. In: Stahl und Eisen 6: 259–265
18. Kulmburg A, Korntheuer F, Kaiser E (1987) In: Härterei-Techn. Mitt. 42: 69–74
19. Hougardy HP (1991) In: Residual Stresses. Measurement, Calculation, Evaluation. DGM Informationsgesellschaft. Verlag
20. Yamazaki K, Hougardy HP (1986) In: Steel Research 57: 466–471
21. Hougardy HP, Wildau M (1985) Festausgabe Stahl und Eisen 105: 1289–1296
22. Gergely M, Reti T, Tardy P, Buza G (1984) In: Heat Treatment 84. The Metals Society: London
23. Fischmeister H (1991) In: Gefüge und mechanische Eigenschaften. Symposium 8. and 9.11.1990 in Bad Nauheim. DGM Informationsgesellschaft. Verlag
24. Lübben Th, Bomas H, Hougardy HP, Mayr P (1991) Härterei-Technischen Mitteilungen, part 1: 24–34. Part 2: 155–170

2 Mechanical Properties of Ferrous and Nonferrous Alloys After Quenching

H.-J. SPIES

2.1 Objectives of Quenching

A workpiece or component is an individual part which is defined by its structure, i.e. geometrical and material parameters. The geometrical parameters are the macrogeometry (shapes, dimensions) and the microgeometry (surface contour) of the workpiece. The material parameters are described by the chemical composition of the material and its condition. In this connection the structure of the core and of the surface layer have to be distinguished.

The term of material condition comprises the specific characteristics of the material, such as structure, segregations, and residual stresses. The microstructure is the most important characteristic.

The properties of a material are determined by its constitution parameters, i.e. its composition and condition. The result of heat treatment depends upon the geometry, chemical composition and material condition of the workpiece concerned as input data as well as upon the parameters of heat treatment (Fig. 2.1). In evaluating the results it turned out to be useful to distinguish between desired change of properties and possible side effects, such as quenching stresses, size and shape distortions. These side effects can considerably influence the working behavior of components.

Whether the objective of heat treatment which is to bring about defined properties is fulfilled, it depends on the possibility of definite change of the structure by means of thermal cycles. The number of possible structural reactions and the resulting structural variety determine the property spectrum of a material. Extensive knowledge about the following material—specific relations is necessary to make full use of this potential.

(1) Material properties as a function of structure: Selection of desired structure
(2) Structure as a function of thermal cycles: Determination of necessary thermal cycles
(3) Local thermal cycles in the workpiece in dependence on material geometry and heat treatment process: Prediction of structural and property distribution in the workpiece.

This book deals with the relations mentioned under [3]. The prediction of structural and property distribution in the workpiece from local cooling rates

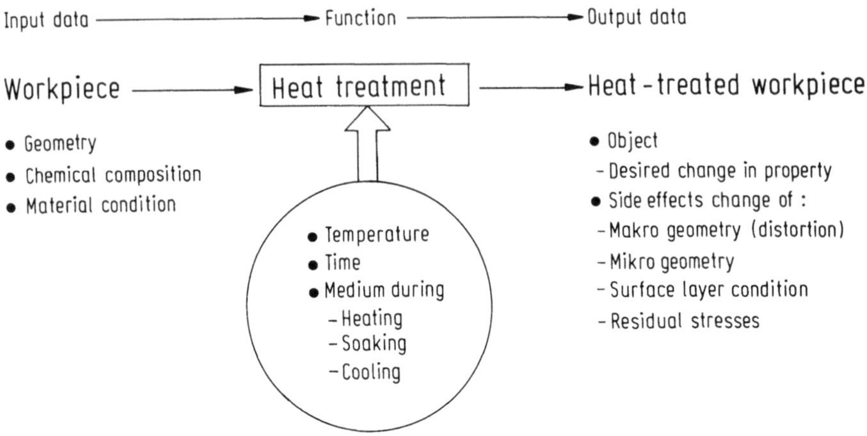

Fig. 2.1. Parameters influencing the result of heat treatment

is based on the preceding steps. These relations are explained in the second and third section of this chapter. First, the goal of quenching is dealt with.

Quenching, i.e. "cooling of an object at a rate faster than in still air" [1], is part of a number of heat treatment processes. One of the most important is transformation hardening of ferrous materials. The cooling rate has to be so fast that the austenite is mainly transformed within the martensite and bainite range. Martensitic and bainitic transformations also occur in a large number of nonferrous alloys [2]. Thus, the importance of transformation treatment far exceeds ferrous materials.

A solution treatment which is followed by rapid cooling to keep the dissolved alloy elements in solid solution is part of precipitation hardening treatment of alloys. This treatment also can be applied to austenitic steels to preserve a homogeneous austenitic structure in order to prevent precipitation from austenitic (quench annealing). Defined cooling rates are relevant for the success of the treatment also in austempering, annealing of ferritic steel, patenting and other processes.

Common to all kinds of quench treatment is that structures are formed which are not in accordance with the thermodynamic equilibrium due to a definite cooling rate.

The change of the internal stress condition is an important side effect of quenching on the material condition and thus material properties which is caused by thermal stresses possibly superimposed by transformation stresses. It leads to changes in size and shape, under unfavorable conditions also to the formation of quenching cracks. Therefore, the management of quenching stresses is an essential part of cooling techniques. The scope to select the cooling rate is determined by the minimum cooling rate necessary for the formation of the desired structure and the necessity to minimize the stresses caused in the workpiece during quenching. Thus, compromises frequently cannot be avoided in practice.

2.2 Influence of Heat Treatment Structures on the Mechanical Properties

Non-equilibrium structures resulting from rapid cooling can roughly be divided into transformation structures and supersaturated solid solutions. In general, transformation and solution treatment are followed by tempering and aging treatment, respectively. With increasing intensity of these treatments the disequilibrium quenching structure comes closer to the equilibrium. The processes are determined by the input data, in particular by the structure after cooling and the tempering and aging conditions. Thus the structure after cooling also influences the material properties after hardening and tempering as well as precipitation hardening.

The relations between mechanical properties and structure depend on material and loading. In the following survey these complicated interactions can only be described in exemplary and simplified form. Starting from the technical importance mainly the properties of the transformation structures of ferrous alloys are dealt with. Thereby, the possibilities of property optimization by means of controlled heat treatment are shown. In addition to that the change of properties by precipitation hardening treatment are explained with aluminium alloys serving as example.

In general, the mechanical behavior can be defined as resistance of a material against plastic deformation and fracture [3]. For its quantitative description the characteristic material parameters of the respective loading case have to be concerned. In the following, the behavior in case of monotonic loading is considered predominantly.

2.2.1 Ferrous Materials

The understanding of the mechanical behavior of ferrous materials is based on the knowledge of the microstructure of the different transformation ranges of austenite (see Chap. 1).

Figure 2.2 shows the mechanical properties of structures formed by isothermal transformation in the pearlite and bainite range of an eutectoid steel.

In the pearlite range the tensile strength and the yield stress grow with increasing undercooling. Simultaneously, the ratio of yield stress to tensile strength and the reduction of area after fracture rise. The strain behavior of pearlitic structures can be explained by the reduction of the interlamellar pearlite spacing due to decreasing transformation temperature. This interlamellar spacing decides the mean free path of dislocation and thus the yield stress [6, 7].

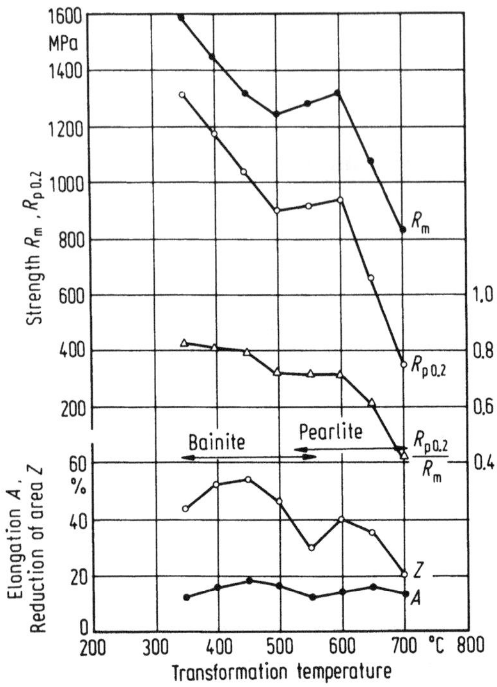

Fig. 2.2. Mechanical properties of an eutectoid steel after isothermal transformation [5]

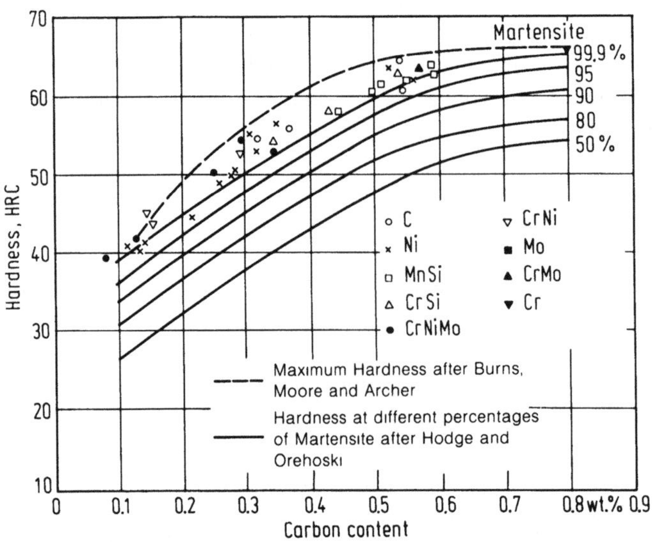

Fig. 2.3. Dependence of hardness on carbon content for cooling structures of different martensite content [9]

The deformability of the cementite lamellas has an influence on the reduction of area. It depends on its thickness which decreases with increasing undercooling [6]. The prior austenitic grain size substantially influences the fracture toughness of pearlite, the upper shelf energy and transition temperature of impact strength [7]. A reduction of austenitic grain size considerably increases the toughness. Thus, it is possible to vary toughness and hardness of pearlitic steels nearly independent of each other by variation of austenite grain size and transformation temperature.

In Fig. 2.2 the first appearance of upper bainite is seen by the decrease of yield strength, tensile strength and reduction of area. After a further decrease of the transformation temperature these values rise again. The reduction of area decreases while approximating to lower bainite with further increasing strength. The scope of properties of bainite results from the variety of its ferrite-carbide morphologies. Its hardness can be explained by increased dislocation density, supersaturation with carbon, precipitated carbides and packet size. Concerning this complex effect of influencing parameters it is only partly possible to weigh up the importance of one of them. Toughness and transition temperature of impact strength are mainly determined by the carbide distribution and packet size. With rising transformation temperature the precipitated carbide becomes coarser and dislocation density and carbon supersaturation decrease. The packet size grows together with transformation temperature and austenitic grain size. Thus, the toughness of bainitic steel is reduced with increasing austenitic grain size and rising transformation temperature [8].

The properties of martensite are mainly determined by the carbon content. In the beginning it rises rapidly with the content of carbon dissolved in austenite before quenching (Fig. 2.3). In case of a carbon content of more than 0.5% it rises only slightly since the temperature for completing the martensite formation in this range falls below room temperature and growing amounts of retained austenite remain. The increase of hardness can be related to the rising dislocation density and carbon supersaturation of martensite [10, 11]. The hardness of martensite can be estimated from the content of dissolved carbon by means of the following empiric relation [12].

$$\mathrm{H}\ (100\%\ \mathrm{Ms}) = 20 + 60\sqrt{C}\ [\mathrm{HRC}] \tag{2.1}$$

Range of validity: 0.1–0.6% of carbon
C: carbon content in wt.-%.

According to the known relations between strength and toughness the martensite toughness rapidly decreases with rising carbon content. Moreover, it has to be considered that with further increase of the strengthening of the supersaturated solid solution high local stresses developed in cooling and transformation are not relieved by a local plastic deformation. These stresses can reach the dimension of cohesive strength and lead to local incipient cracks. The result is an increasing embrittlement with rising carbon content. Therefore, steels of martensitic structure made by conventional hardening are used after tempering only. Only very low-carbon steels have a good toughness in the hardened state so that tempering will not be necessary.

After quenching martensite structure is supersaturated with carbon and has a high density of dislocations. Tempering leads to a reduction of the supersaturation by precipitation, a reduction of the dislocation density by recovery and recrystallization processes as well as to a reduction of quenching stresses. The precipitation and recovery processes influence each other. E.g., the high dislocation density supports the nucleation for precipitates and ensures the dispersion of the nuclei. The resulting pinning of dislocations retards recovery and recrystallization.

Alloying elements delay the reduction of the supersaturation and influence the course and kind of carbide precipitation. Thus, they have an essential influence on the tempering resistance. A higher content of alloy carbide formers leads to distinct secondary hardening. In dependence on steel composition and tempering conditions embrittlement can occur at tempering temperatures of about 300 °C and 500 °C.

The structures of the bainite and pearlite range also change during tempering and come closer to the equilibrium. The degree of structure and

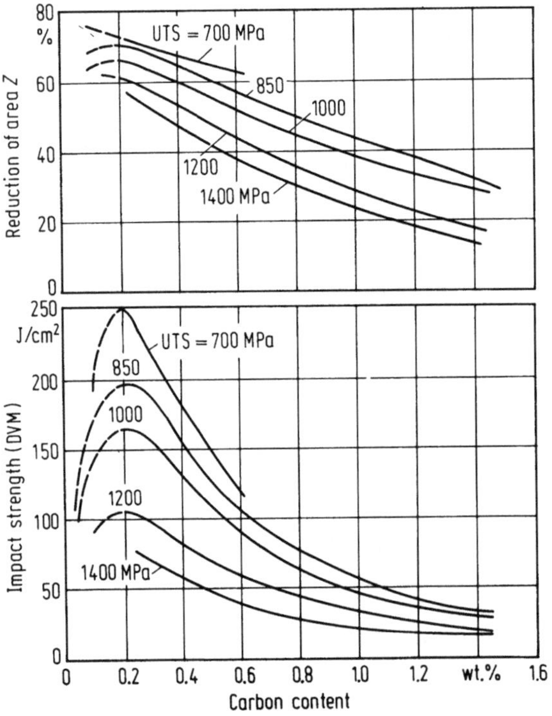

Fig. 2.4. Influence of carbon content on reduction of area and impact strength of martensitic heat-treated steel for strengths of 700 to 1400 MPa [17]

property change depends on the non-equilibrium state after cooling. Therefore, property changes during tempering of pearlitic structures can for the most part be neglected. The high structural stability of bainitic structures is to be pointed out. It results in a higher high-temperature strength which is made technical use of both for creep-resisting steels and hot-work tool steels [13, 14].

With rising tempering temperature the softening and precipitation processes during tempering lead to a reduction of yield stress and an increase of toughness. High demands on resistance to fracture, in particular at very low temperatures, e.g. for low-temperature steels, require recrystallization of martensite. Due to the above mentioned correlation between precipitation and softening the recrystallization temperature is by 100 to 150 K higher than the recrystallization temperature of cold-worked steel and is about 500 °C to 650 °C.

The data listed in Table 1 for 300-M grade of high-strength, vacuum-melted steel give a picture of the change of strength behavior during tempering. Tempering above the recrystallization temperature of martensite leads to a visible increase of fracture toughness. The 0.2% proof stress under monotonic load is relatively low in case of great hardness. At tempering temperatures of above 300 °C the yield stress ratio increases up to the typical range of tempered martensite. Cyclic load results in softening of martensite tempered at higher temperatures. A similar behavior could also be observed by Landgraf for SAE 4142 steel [16]. The fatigue strength decreases parallel to the tensile strength with rising tempering temperature. It is remarkable that even at high strength the ratio σ_D/R_m is about 0.5. This is because of the high grade of purity. With rising tempering temperature the near-threshold fatigue-crack propagation resistance (ΔK_0) increases.

The properties of plain carbon and low-alloy steels with a structure of tempered martensite ($T_T > 400$ °C) are determined by the condition of the ferritic matrix and the amount and distribution of cementite. In case of equal tensile strength reduction of area and impact strength, for example, decrease with growing cementite content, i.e. with growing carbon content of the steel (Fig. 2.4). The other alloying elements influence the strength behavior in particular the tempering behavior.

Table 2.1. Mechanical properties of 300-M high-strength tempered steel (AISI 4340 modified by 1.3% of Si) [15]

Tempering temperature[1] °C	Hardness HRC	0.2% proof stress monotonic MPa	cyclic MPa	U.T.S. MPa	Fracture toughness K_{IC} MPa·m$^{1/2}$	Threshold value K_0 MPa·m$^{1/2}$	Fatigue strength[2] MPa
100	55	1497	2107	2338	36	2.98	1150
300	50	1737	1486	2006	65	3.08	1000
470	45	1497	1198	1683	69	5.10	830
650	37	1074	861	1186	152	8.50	600

[1] tempering time: 1 h
[2] smooth sample bar: $N_L = 10^7$

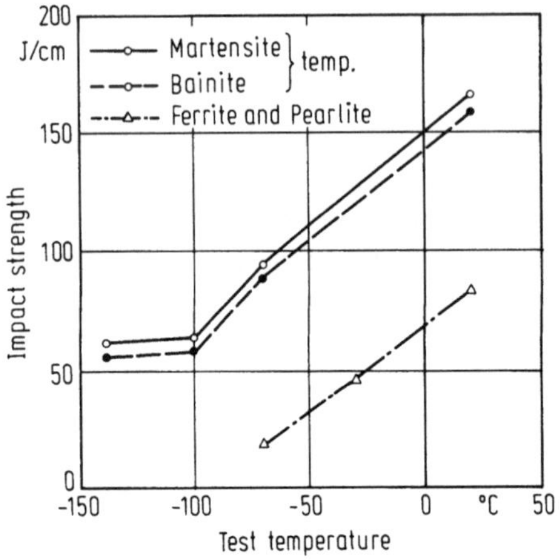

Fig. 2.5. Impact strength-temperature curves for different microstructures of equal tensile strength ($R_m = 950$ MPa) of SAE 4150 steel [19]

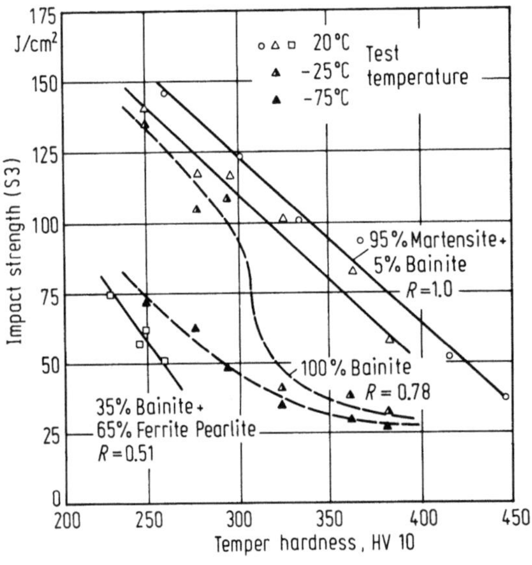

Fig. 2.6. Influence of microstructure and temper hardness on impact strength (S3 test, Charpy-V, notch depth 3 mm); SAE 4142 steel, R = degree of quench hardening [21]

The properties of bainite, e.g. impact strength, depend on the formation temperature [8, 18]. For example, upper bainite has an improved high-temperature strength and a remarkably lower high-temperature toughness [14, 18]. The impact strength of tempered lower bainite can be compared to that of tempered martensite in case of similar tensile strength and carbon content (Fig. 2.5) [19, 20]. Bainitic structures also require an adequately high tempering temperature to reach optimum low-temperature toughness. As shown in Fig. 2.6, only bainite which was tempered for a hardness below 300 HV (tempering temperature > 580 °C) has at a testing temperature of −25 °C the same impact

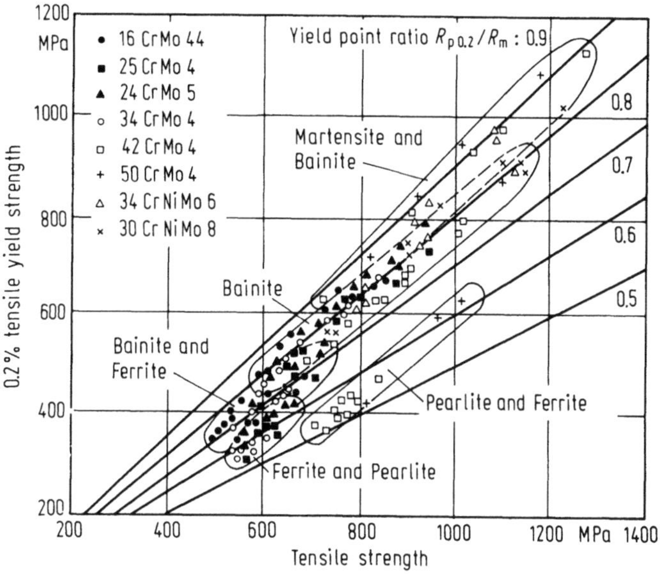

Fig. 2.7. Relationships between tensile yield strength and tensile strength, depending on microstructure [19,26]

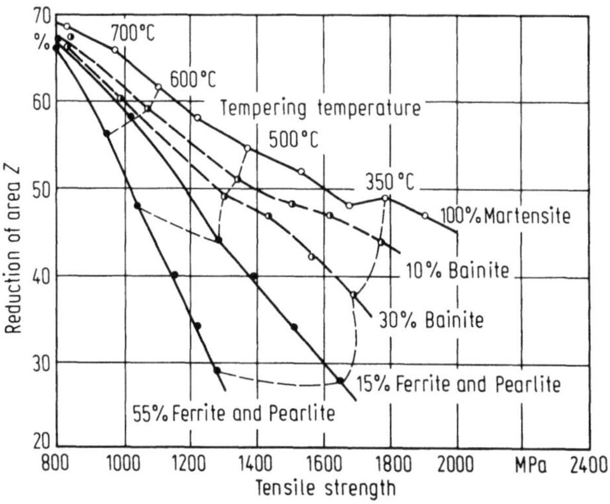

Fig. 2.8. Influence of microstructure on reduction of area; SAE 4142 steel [24]

strength as at room temperature. Under cyclic mechanical load it is superior to tempered martensite because of the distinctly lower notch sensitivity with comparable fatigue strength [22]. With equal strength the cold toughness of pearlitic structures is essentially lower than that of tempered martensite (Fig. 2.5). The yield strength of tempered structures is decisively determined by grain size, shape and distribution of cementite which depend on the cooling structure. In this way the several transition structures have a characteristic yield strength ratio also after tempering at high temperatures (Fig. 2.7) [19, 26].

In continuous cooling which is characteristic of common heat treatment mostly mixed structures are formed in dependence on the input data and cooling conditions (Fig. 2.1). These contain besides martensite structures of the bainite and pearlite range. The properties of these structures strongly depend on the portion and distribution of the accompanying structure of martensite [19, 21, 23–25]. Thus, general prediction of properties is only partly possible. In case of comparable martensite portions the property change strongly depends on the kind of the accompanying structure. As shown in Fig. 2.8 for SAE 4142 steel even small portions of ferrite and pearlite lead to an essential decrease in the reduction of area in contrast to bainite. Only at relatively low hardness, i.e. after tempering at temperatures of more than 600 °C, the differences can be neglected.

It turned out to be useful to mark the structure reached after quenching indirectly by the percentage of the quenched hardness proposed by Wyss [27]. It is explained as the proportion of obtained hardness to maximum hardness possible, i.e. the hardness of the respective steel in case of 100% of martensite, in per cent. Recently this has been indicated as degree of quench hardening R [28]. This is

$$R = \frac{H_{\text{hardened}}}{H_{100\%\,\text{Ms}}} = \frac{H_{\text{hardened}}}{20 + 60\sqrt{C}} \frac{\text{HRC}}{\text{HRC}}. \tag{2.2}$$

The desired degree of quench hardening depends on the requirements of the object properties. It increases with growing strength at guaranteed toughness [23, 27, 28].

When applying the degree of quench hardening, it has to be taken into account that only in case of high quench hardness there is a definite correlation between structure after cooling and degree of quench hardening. For example, the hardness of a structure with 50% of martensite strongly depends on the accompanying structure which can consist of bainite or ferrite-pearlite. The relation between hardness and carbon content shown in Fig. 2.3 for 50% of martensite cannot be generalized as resulted from the assessment of a large number of transformation diagrams.

The prediction of properties on the basis of the degree of quench hardening as possibility of describing the structure after cooling can therefore supply approximate values only. Moreover, it has to be considered that the hardness does not react to small portions of pre-eutectoid precipitates at the grain boundaries. But these precipitates essentially reduce the toughness. Sub-microscopic, pre-eutectoid carbide precipitates at austenite grain boundaries, which can be observed by optical microscope in case of increased etchibility of the grain boundaries only, reduce the reduction of area of H 11 (0.4C-5Cr-1.5Mo-0.4V) heat-treated hot-work steel from about 60% to a value below 20% [14]. After a transformation treatment too, it is necessary, besides the formation of defined transformation structures by sufficiently high cooling rates, to prevent austenite from precipitating alloying elements.

2.2.2 Precipitation-Hardenable Aluminium Alloys

Precipitation hardening is the most important mechanism of increasing the strength of metallic materials, in particular of nonferrous alloys. It is based on the obstruction of the dislocation movement by second-phase precipitates from a supersaturated matrix. Depending on size and hardness (shear modulus) the dislocations can cut the precipitates or are forced to bypass (bow around) them. Another obstruction of the dislocation movement is caused by matrix distortion in the vicinity of precipitates (coherency stress hardening). With given kind of second phase the increase of flow stress resulting from the obstructive effect depends on the volume fraction f and the mean particle radius r_p. The interparticle distance λ_p which is often used to characterize the particle distribution can be calculated from these two values.

With growing particle diameter the resistance against cutting (cutting stress) increases, but the resistance against bowing (bowing stress) decreases. This results in the precipitation hardening curve as function of time and particle size, resp. as shown in Fig. 2.9 for an Al-4% Cu-alloy. The peak strength is reached at the transition point from cutting mechanism to bowing mechanism.

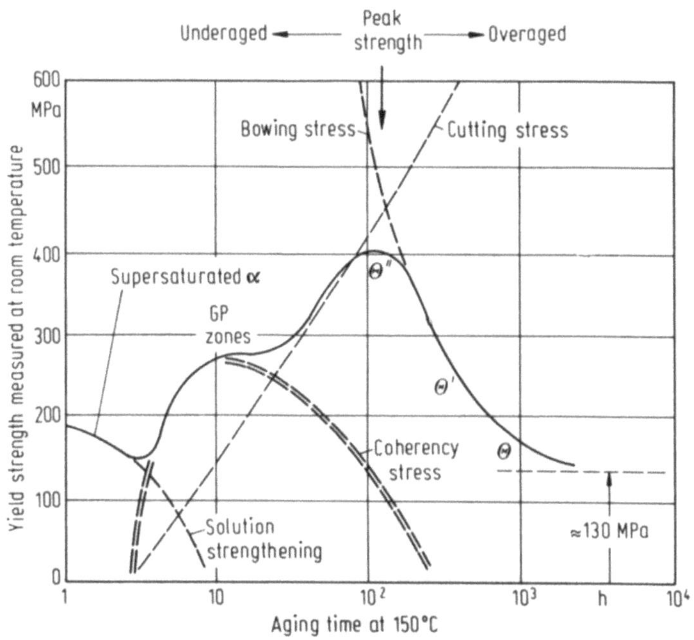

Fig. 2.9. Yield strength of quenched Al-4Cu-alloy during aging at 150 °C; GP zones, disc-shaped 10 nm; θ'' disc-shaped, 100 nm; θ' disc-shaped, 1 μm; θ equilibrium $CuAl_2$ [29]

The microstructure of precipitation-hardenable aluminium alloys becomes visible only after a high degree of magnification of TEM. Figures 2.10a–d show TEM[1] of oxide replicas and thin foils of AlCu-, AlAg- and AlAgCu-alloys after precipitation hardening heat treatment. After solution treatment and aging treatment different states of decomposition occur in the Al-matrix which depend on the alloying elements.

Figures 2.10a–c show the states of precipitation near the peak strength. It is obvious that the superposition of Cu- and Ag-decompositions results in an higher precipitation density than expected for simple addition. The peak strength

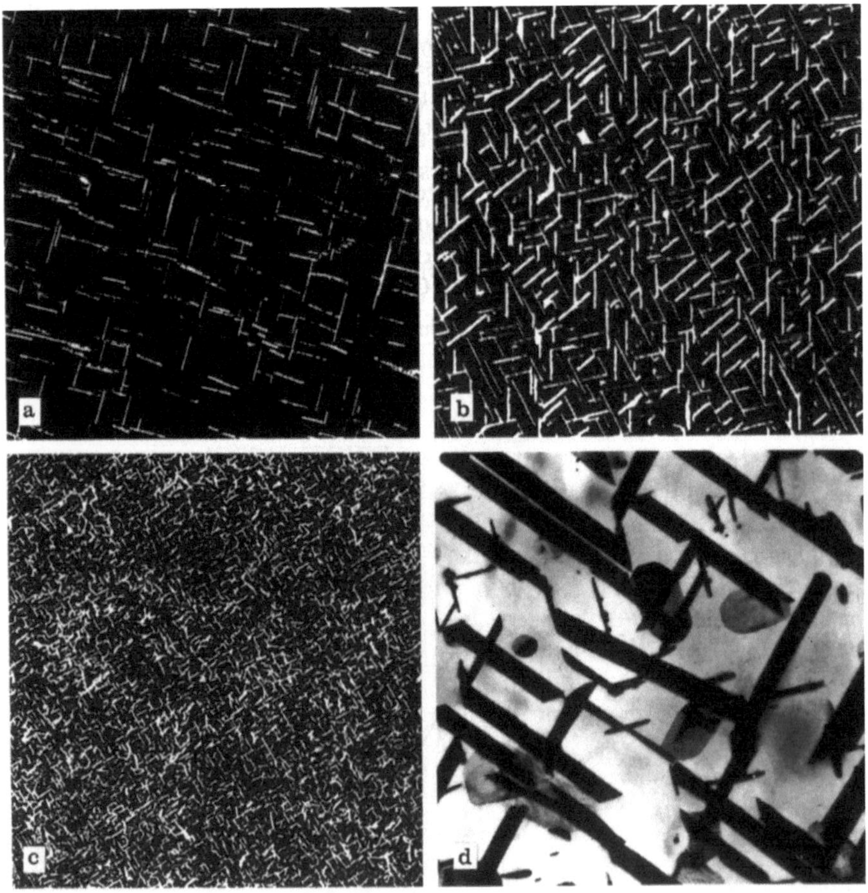

Fig. 2.10. Structure of AlCu-, AlAg- and AlAgCu-alloys after precipitation hardening and overaging treatment. $V = 15.000:1$. Solution treatment: 530 °C, 60 min/water. **a** AlCu3, oxide replica; 200 °C, 1000 min [30]; **b** AlAg10, oxide replica; 200 °C, 100 min [30]; **c** AlAg10Cu3, oxide replica; 200 °C, 1000 min [30]; **d** AlAg10Cu3, transmission foil; 300 °C, 600 min [31]

[1] TEM: transmission electron microscope

of this alloy is therefore essentially higher. In case of overaging the precipitations are strongly coarsened (see Fig. 2.10d) resulting in hardness decrease to the value of quench hardness.

Localized inhomogeneous coarse slip which leads to reduced toughness, low fatigue strength and increased susceptibility to stress corrosion is a characteristic of underaged alloys with shearable precipitates [32]. To avoid these adverse effects in practice overaged alloys are used predominantly. The particle distribution determining the properties of precipitation-hardened alloys is considerably influenced by the solution treatment, too. The cooling from the temperature of solution treatment has to be so fast that decomposition and precipitation of alloying elements are restrained and a homogeneous supersaturated solid solution is formed at room temperature.

The concentration and distribution of vacancies is also influenced by the cooling rate. Place-changes in the closely packed f.c.c. lattice of aluminium take place through the vacancies, only. Therefore, nucleation as well as resulting size and distribution of precipitates depend on concentration and distribution of vacancies as well as on bulk concentration. With falling cooling rate vacancies diffuse to the grain boundaries preferably acting as sinks. Thus, precipitation-free zones are formed. The relatively high rate of heterogeneous nucleation at the grain boundaries favours the formation of coarse grain-boundary precipitates with increasing cooling time. These precipitates predominantly reduce the toughness considerably. Coarser precipitates formed in the midgrain regions as well during slow cooling have only a slight influence on the dislocation movement and have no effect on the increase of strength by precipitation hardening.

The effect of different local cooling rates on the properties depends on the chemical composition of the alloy, the microstructure at the temperature of solution treatment prior to cooling and on the subsequent againg treatment.

In cooling the range between 400 and 290 °C is especially critical for the most aluminium alloys. Figure 2.11 shows the influence of cooling rates on the yield stress of selected high-strength alloys in this temperature range after aging [33]. The 7075 alloy (Al-5.6Zn-2.5Mg-1.6Cu-0.23Cr) has then a very high sensitivity to the quenching rate. Maximum strength after aging is yielded at cooling rates of 330 K/s and more, only. The other alloys have a critical cooling rate of about 100 K/s. The corrosion resistance of high-strength aluminium alloys depends on the cooling rate, too.

Just like the strength under monotonic load the fatigue behavior is influenced by the cooling rate (Tables 2.2 and 2.3) [34]. The microstructure of the AlMgSi0.7 (Al-0.7Mg-0.6Si) alloy was modified by the cooling conditions after solution treatment and soaking time at room temperature. It differs in mean size of the Guinier–Preston zones and the width of precipitation-free zones (PFZ) at the grain boundaries. Grain shape and size as well as size and distribution of primary and high-temperature precipitates were not effected. The extension of the precipitates and the widening of the PFZ lead to reduction of strength under monotonic load and influence crack initiation under cyclic

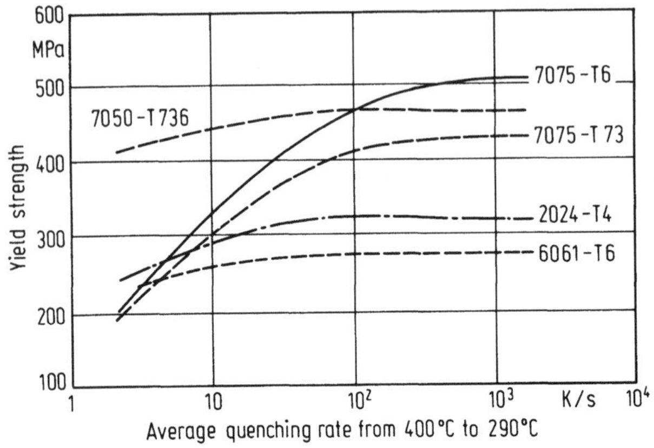

Fig. 2.11. Influence of cooling rate on yield stress of aluminium alloys after precipitation hardening treatment [53]

Table 2.2. Microstructure of 6005 A (AlMgSi0.7) alloy after different kinds of heat treatment [34]

Material condition	Heat treatment[1]	GP zones medium size nm	PFZ medium width nm
I	540 °C/40 min/water RT 24 h/160 °C 48 h	69	72
II	540 °C/40 min/air RT 24 h/160 °C 48 h	73	302
III	540 °C/40 min/water RT 10 s/160 °C 48 h	58	38

[1] quenching intensity (400 – 300 °C) water: 1000 °C/s, air: 2.2 °C/s

Table 2.3. Mechanical properties of 6005 A (AlMgSi0.7) alloy after different kinds of heat treatment [34]

Material condition	U.T.S. MPa	0.2% proof stress MPa	Elongation %	Fatigue strength[1] MPa $K_\sigma{}^2 = 1.0$	$K_\sigma = 1.47$	$K_\sigma = 2.74$
I	308	289	16	115	84	62
II	287	263	15	83	79	61
III	325	310	17	102	93	64

[1] $N_L = 10^7$, fracture probability: 50%
[2] K_σ: elastic stress concentration factor

load by localized plastic deformation. With increasing notch sharpness the plastic deformation processes focus on smaller and smaller parts of material. This leads to a decreasing influence of the PFZ on crack initiation [34].

In dependence on the cooling conditions and the physical properties of the alloy a cooling rate results from the workpiece thickness which decreases in direction of the core. This gradient of the cooling rate can lead to a change of properties across the section in case of alloys with high sensitivity to quenching rate which is known from transformation hardening [35].

2.3 Characterization of Transformation Behavior

The sequence of transformations in thermal cycles is described by transformation diagrams (TTT diagrams). Transformation diagrams are prepared for isothermal transformation as well as for transformation during continuous cooling. The transformation is followed along an isotherm or a given continuous cooling curve. The diagrams are the result of connecting the points of equal transformation state on isotherms or cooling curves. Examples of transformation diagrams of isothermal transformation (I-T diagrams) and continuous cooling (CCT diagrams) are shown in Figs. 2.12 and 2.13. Due to technical limitations the beginning and end of the transformation are generally indicated by structural portions of 1% and 99%, resp. The CCT diagrams of steel additionally contain the structural portions measured at room temperature and the hardness.

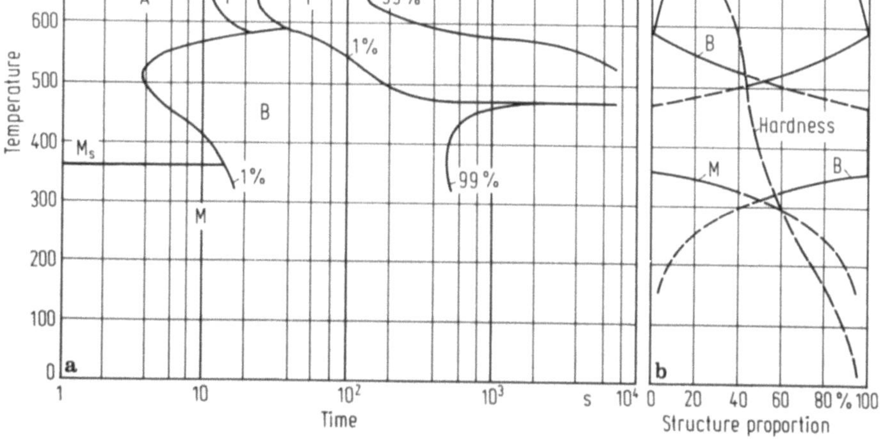

Fig. 2.12 a, b. Isothermal transformation diagram (**a**) and structure-volume fraction diagram (**b**) of SAE 4150 steel [4, 54]

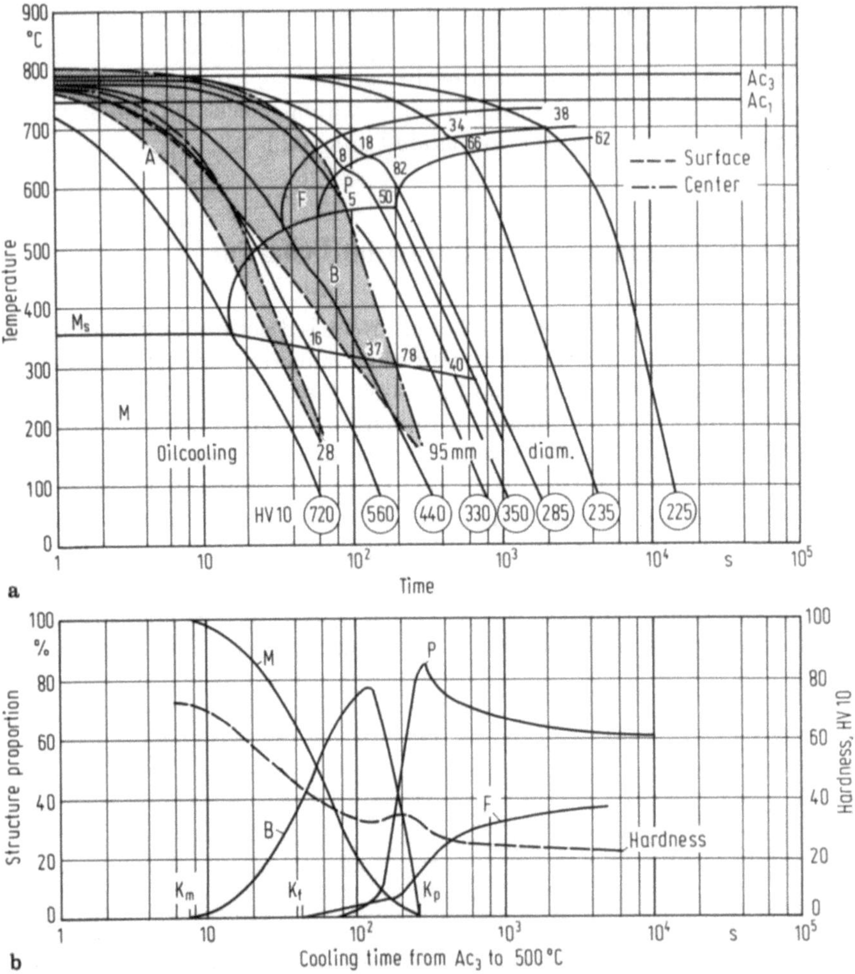

Fig. 2.13 a, b. Continuous cooling transformation diagram of SAE 5140 steel with oil cooling curves of cylinders of 28 and 95 mm diameters, surface distance 0.5 mm (**a**) and structure-volume fraction diagrams (**b**) [4, 54]

The observation of precipitation processes from supersaturated solid solutions requires the application of sophisticated methods of structural and microstructural analysis. Therefore, the sensitive reaction of the mechanical properties to the beginning decomposition of supersaturated solid solutions is used for indirect determination of the precipitation behavior for aluminium alloys, in particular [35–37]. Thereby, the samples are exposed to aging treatment after defined cooling. The resulting strength as percentage of the maximum possible strength is the basis of establishing curves of equal

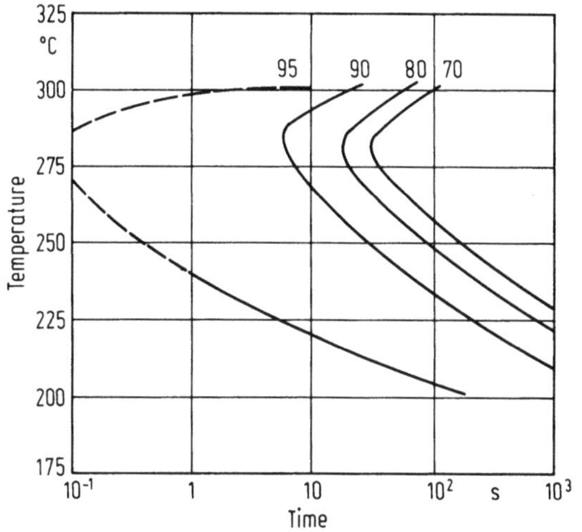

Fig. 2.14. Transformation diagram of the Al-8.65 Mg-alloy for isothermal precipitation, criteria 95, 90, 80 and 70% of maximum strength [36]

transformation state. An example of an isothermal transformation diagram is shown in Fig. 2.14.

In dependence on the their preparation the transformation diagrams have always to be read along isotherms or cooling curves. They often are added by structure-volume fraction diagrams which also contain data of hardness (Figs. 2.12 and 2.13). In the structure-volume fraction diagram of Fig. 2.13 the course of cooling is indicated by the cooling time from the Ac_3 temperature to 500 °C. In this connection the cooling time from 800 °C to 500 °C or the cooling parameter λ[1] is often taken into consideration. Besides the structure-volume fraction diagram the cooling time is also considered as parameter of describing the phase transformation and property change in form of cooling time diagrams [46–48].

From the CCT Diagrams characteristic parameters of the transformation behavior, such as critical quenching rates and cooling times, can be taken. The upper critical quenching rate (cooling time K_m) is the upper limit of the quenching rate at which still plain martensitic structures are formed. The lower critical quenching rate (cooling time K_p) is the lower limit at which structures of the pearlite range are formed only. The cooling time K_f indicates the beginning of pre-eutectoid ferrite precipitations.

The possibility to produce a distinct hardness distribution across the section of the workpiece by transformation treatment is described by hardening capacity and hardenability. The hardening capacity of a steel is its capacity to reach a distinct maximum hardness during quenching under optimum conditions. It is a function of the carbon content of martensite (Eq. (1) and Fig. 2.3). The

[1] λ- cooling time from 800 °C to 500 °C in $s \cdot 10^{-2}$

hardenability is the ability of a steel to be hardened fully, completely martensitic and/or bainitic, to a given depth below the surface. The hardenability is characterized by the upper critical quenching rate. For a large number of loading cases a full martensitic structure after cooling is not necessary. Therefore, the cooling rate for a structure with 50% of martensite is taken into account as criterion of hardenability.

Another possibility of describing the property change during cooling is the quench factor concept. The quench factor Q interrelates quenching variables, the section size of workpieces, and transformation data of nonferrous metals and steels to provide a single number indicating the extent to which a workpiece can be through-hardened or strengthened. For details refer the literature [37, 40, 41].

The knowledge of the relations between the course of cooling and the structure after cooling makes a prediction of structural and property distribution in the workpiece possible based on the structure-property correlations. For this purpose the local course of cooling has to be determined on the basis of the input data workpiece geometry, material and material condition as well as cooling conditions (see Fig. 2.1).

The possibilities and restrictions of property prediction are visible in Fig. 2.13 showing a CCT diagram and the cooling curves of different section sizes. For small dimensions the thermal cycle of natural cooling approximately agrees with the cooling curves used for the preparation of the transformation diagrams. But there are considerable differences in case of large dimensions, in particular. Experimental investigations have shown that the kind of cooling, for example exponential or linear, has an essential influence on the course of phase transformation even if the cooling parameter λ is the same [42, 43]. Thus, the differences in the course of cooling lead to a geometry-dependent effect on the transformation behavior. Additionally, the influence of external stresses on the course of transformation has to be considered. According to Kurdjumov and Sachs [44], e.g. shear stress supports the martensitic transformation. Investigation performed by Kulin, Cohen and Auerbach [45] as well as by Schmidtmann, Grave and Klauke verify the inhibition of other transformations by compressive stresses and their support by tensile stresses. The level and distribution of cooling stresses which depend on the workpiece geometry result in another specific influence. Moreover, restrictions of the prediction of the microstructural formation on the basis of transformation diagrams result from the above mentioned influence of chemical composition and initial state prior to cooling on the transformation behavior. A transformation diagram is therefore only valid for the chemical composition of the samples used for its preparation and the respective initial state. Manufacture dependent deviations of the melt analyses as well as differences in composition caused by macro- and micro-segregations have a strong influence on the transformation behavior. According to investigations of Kulmburg [43], the effectiveness of parameters influencing the transformation behavior of a steel decreases by the following

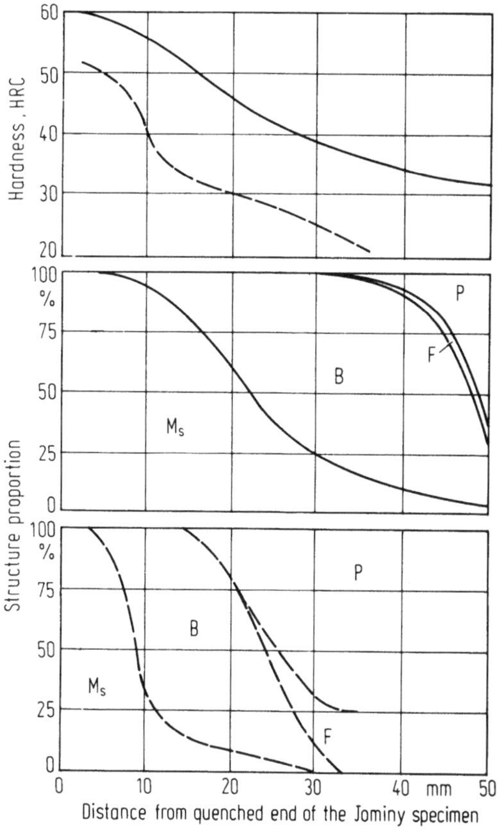

Fig. 2.15. End-quench hardenability band and structure-volume fraction diagram of SAE 5140 steel

sequence

—chemical composition
  average (melt analysis)
  local (macro- and microsegregations)
—austenitizing conditions (temperature and time)
—course of cooling (deviation from natural cooling).

Weighting of the material-specific and technological effects of the transformation behavior is possible by means of hardenability curves derived from end-quench tests (Jominy tests). In the end-quench test the cooling rate continuously decreases with rising distance from the quenched end. The change of hardness as function of the distance from the end face "*J*" indicates the hardenability of a steel. The end-quench test is characterized by uncomplicated experimental procedure. It turned out to be useful in characterizing the effect of deviations in the chemical composition on the hardenability of steel in form of hardenability bands. The lack that hardenability bands do not give

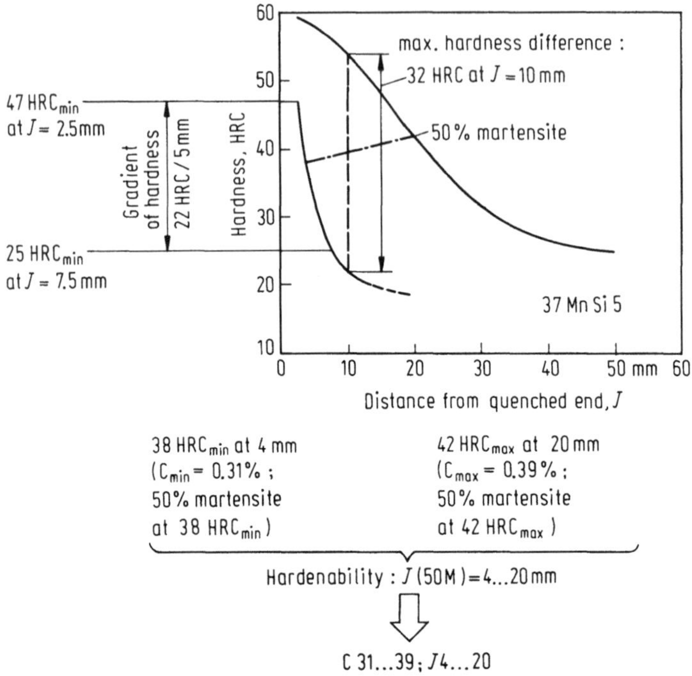

Fig. 2.16. 37MnSi5 (0.37C-1.25Si-1.2Mn) hardenability band and classifying system of end-quench hardenability [47]

information on structural change connected with the cooling rate can be compensated by additional structure-volume fraction diagrams for the upper and lower limit of the band (Fig. 2.15).

A comparison of the hardenability of the melt used for the preparation of the transformation diagram with the limits of the hardenability band allows the assessment of the information given by the diagram. The effective depth of hardening which is defined by a given martensite content can be read from the band. If structure-volume fraction diagrams are not available, the limit values of hardness can be estimated on the basis of the data given in Fig. 2.3 concerning the dependence of hardness on the martensite and carbon content. However, the determination of the effective depth of hardening does not exhaust the information of a hardenability band. An evaluation following the proposal presented in Fig. 2.16 provides besides hardenability information about possible material-dependent spread of hardenability and susceptibility of a steel to technologically caused variations of the cooling rate. The hardness difference in case of equal distance from the end face (equal cooling rate) is the standard of the level of material-dependent deviations. A high hardness gradient points to a high sensitivity of a steel against varying cooling conditions.

The application limits of hardenability bands predominantly result from the relatively short range of cooling rates included in the Jominy test. The lowest cooling rate (the cooling time from 800 °C to 500 °C is about 180 s) corresponds to cooling in the core of a round bar with a diameter of 160 mm in water hardening or 125 mm in oil hardening. The application of the Jominy test is therefore restricted to plain carbon and low-alloy steels. The significance of the Jominy test is limited due to the relatively large spread of the correlation between the distance from the end face and the cooling time [48]. This results from possible change of the temperature curve because of released transformation heat in case of steel in which a phase transformation occurs during the cooling from 800 °C to 500 °C.

The transformation behavior of steels can also be calculated from their chemical composition. Even a few years after the end-quench test had been introduced by Jominy and Boegehold [49] Grossmann [50] presented a first concept of calculating the hardenability from the chemical composition. The great number of investigations performed in reference to this complex can be summarized in the following groups.

—calculation of transformation temperatures
—calculation of hardenability (ideal critical diameter, course of hardness in end-quenched samples, critical quenching rates)
—calculation of transformation diagrams.

The influence of the alloying elements on the transformation behavior is very complex and differs in the several transformation ranges of austenite. Therefore, the transformation behavior has hitherto been described by means of regression analyses only also if the models have a thermodynamic basis. Such systems of equations are valid for the respective data records only. The application outside of this range is connected with remarkable errors. For details of calculating the transformation behavior refer the literature [51–53].

References

1. Tyrkiel E (1986) Multilingual Glossary of Heat Treatment Terminology. The Institute of Metals London
2. Brooks CR (1984) Heat Treatment, Structure and Properties of Nonferrous Alloys. ASM Metals Park
3. Hornbogen E (1977) In: Z. Metallkde 68: 455–469
4. Werkstoffkunde Stahl, Bd. 1: Grundlagen (1984) Verein Deutscher Eisenhüttenleute. Springer Verlag Berlin, Heidelberg, New York; Verlag Stahleisen mbH. Düsseldorf
5. Brick RM, Pense AW, Gordon RB (1977) Structure and Properties of Engineering Materials, 4th ed., p. 163. Mc Graw-Hill Book Comp
6. Flügge J et al. (1978) Arch. Eisenhüttenwes. 47: 635–640
7. Hyzak JM, Bernstein JM (1976) In: Metallurg. Trans. 7A: 1217–1224
8. Nakajima H, Araki T (1973) In: Trans. Nat. Res. Inst. Metals Jap. 15: 47–61
9. Gerber W, Wyss U (1978) In: Von Roll-Mitt. 7: 13–49. Comp. Burns JL, Moore TL, Archer RS (1978) In: Trans. ASM 26: 1–36. Hodge JM, Orehoski MA (1946) In: Trans. AJME 167: 502–512
10. Speich GR, Warlimont H (1968) In: J. Iron Steel Inst. 206: 385–392
11. Macherauch E, Vöhringer O (1986) In: Harterei-Techn. Mitt. 41: 71–91

12. Just E (1968) In: Härterei-Techn. Mitt. 23: 85–100
13. Bennek H, Bandel G (1943) In: Stahl u. Eisen 63: 653–659, 673–684 a. 695–700
14. Spies H-J, Vogt F (1988) In: Neue Hütte 33: 327–331
15. Ritchie RO (1977) In: Metal Sci. 11:368–381; Comp.: Conf. Proc. Fundamentals of tribology, pp. 127–134 (June 1978) Cambridge, Mass
16. Landgraf RW (1970) In: ASTM STP 467: 3–36. Am. Soc. for Testing and Materials
17. Kroneis MR, Gattringer R, Krainer E (1961) In: Berg- u. Hüttenm. Mh 106: 434–446
18. Peter W, Klein A, Finkler H (1967) In: Arch. Eisenhüttenwes. 38: 561–569
19. Rose A, Krisch A, Pentzlin F (1971) In: Stahl u. Eisen 91: 1001–1020
20. Matsuda S et al. (1972) In: Trans. Iron Steel Inst. Jap. 12: 325–333
21. Spies H-J, Wittig S, Münch G (1978) In: Neue Hütte 23:421–423
22. Tauscher H (1968) In: IfL-Mitt. 7: 397–405
23. Baschnin JuA, Rachmonova VJ, Panischin JF (1975) Isvest. vyss. ucebn. zaved. cern. met., pp. 166–169
24. Klärner HF, Hougardy E (1970) In: Arch. Eisenhüttenwes. 6:587–593
25. Legat A, Moser A (1968) In: Härterei-Techn-Mitt. 23: 10–14
26. Hengerer F, Geiger T, Sonderegger B (1979) In: Stahl u. Eisen 90: 1263–1275
27. Wyss U (1953) In: Härterei-Techn. Mitt. 6, No. 2: 9–40
28. Just E (1974) In: VDI-Berichte, No. 214: 75–84
29. Ashby MF, Jones DRH (1986) Engineering Materials 2, p. 100. Pergamon Press. Oxford, New York, Frankfurt
30. Thym G (1968) Dr.-Ing. Diss. Faculty of Machine Engineering. Technical University München
31. Woitscheck A (1968) Dr.-Ing. Diss. Faculty of Machine Engineering. Technical University München
32. Warlimont H (1973) In: Phasenumwandlungen im festen Zustand: 29–57. VEB Deutscher Verlag für Grundstoffindustrie Leipzig
33. Metals Handbook, 9th ed., Vol. 2, p. 34 (1979) Am. Soc. Metals
34. Bomas H, Mayr P (1985) In: Härterei-Techn. Mitt. 40:137–140; the same (1985) In: Z. Werkstofftechn. 16: 88–94
35. Schneider GL, Sacharov ED, Smolenzev WI (1991) In: Metalloved. i therm. obr. met., No. 2: 49–51
36. Tokarski M, Krol S (1985) In: Neue Hütte 30: 137–138
37. Evancho JW, Staley JT (1974) In: Metallurg. Trans. 5A: 43–47
38. Steven W, Mayer G (1953) In: J. Iron Steel Inst. 174: 33–45
39. Kunze E, Brandis H (1965) In: DEW-Techn. Ber. 5: 106–110
40. Staley JT (1987) In: Mater. Sc. and Techn. 3:923–935
41. Bates CE (1988) In: J. Heat Treat. 6: 27–45
42. Thelning K-E (1978) In: Scand. J. Metallurgy 7: 252–263
43. Kulmburg A, Kornteuer F, Kaiser E (1987) In: Härterei-Techn. Mitt. 42: 69–74
44. Kurdjumov G, Sachs G (1930) In: Z. Physik 64:325–343
45. Kulin SA, Cohen M, Averbach BL (1952) In: J. Metals 4: 661–668
46. Schmidtmann E, Grave H, Klauke H (1976) In: Stahl u. Eisen 96: 1168–1176
47. Spies H-J (1978) In: IfL-Mitt. 17: 11–16
48. Brandis H, Preisendanz H (1963) Bänder, Bleche, Rohre, pp. 541–551
49. Jominy WE, Boegehold AL (1938) In: Trans. ASM 26:574–606
50. Grossmann MA (1942) In: Trans. AIME 150: 227–225
51. Tzitzelkov I, Hougardy HP, Rose A (1974) In: Arch. Eisenhüttenwes. 45: 525–532
52. Doane DV, Kirkaldy JS (1978) Hardenability Concepts with Applications to Steel. Conf. Proc., Met. Soc. AJME, New York
53. Gergely M, Reti T (1988) In: J. Heat Treat, 6: 125–140
54. Altas zur Wärmebehandlung der Stähle. (1954–1958) Max Plank Institut für Eisenforschung in Zusammenarbeit mit dem Werkstoffausschuß des Vereins Deutscher Eisenhüttenleute; 1.1 Weber F, 1.2 Rose A, Rose A, Peter W, Strassburg W, Rademacher L. Düsseldorf

3 Thermo- and Fluiddynamic Principles of Heat Transfer During Cooling

F. Mayinger

3.1 Phenomena of Heat Transfer During Immersion Cooling

Heat transfer during immersion cooling is mostly connected with boiling because the temperature of the material to be cooled is usually higher than the boiling temperature of the liquid coolant. With boiling many kinds of complicated fluiddynamic and thermodynamic phenomena are interacting with each other. In spite of early starting research and many experimental and theoretical investigations boiling phenomena cannot be described on a fully theoretical basis until today. First systematic experiments originate from the 30th and here especially the investigations by Jakob [1], Fritz [2], Linke [3], Bosnjakovic [4] and Nukiyama [5] have to be mentioned. Already in these experiments it was observed that vapour bubbles are formed at very distinctive sites at the heated surface increasing in number with higher heat flux. Each bubble grows from a so-called nucleus which is assumed to be present in a small groove or local roughness on the heated surface. The heat is not directly transferred from the surface to the vapour bubble, but it first enters the liquid, being adjacent to the surface, superheats a thin liquid layer and by this creates a thermodynamic metastable situation for a very short period. If the nucleus is big enough or the superheating of the thin liquid layer is high enough, a bubble is formed which gets its heat and mass—vapour—from this superheated liquid layer, called boundary layer.

The vapour in the bubble must be of higher pressure p_v than the surrounding liquid because in addition to the liquid pressure p_l the surface tension σ acts on the phase interface between bubble and liquid as can be shown with a simple force balance. Equation (1) gives this force balance for a spherical bubble being in equilibrium i.e. not growing and not shrinking.

$$p_v - p_l = \frac{2\sigma}{R}. \tag{3.1}$$

If we use the Clausius Clapeyron equation in addition

$$\frac{dp}{dT} = \frac{\Delta h_v}{(v_v - v_l)T} \tag{3.2}$$

well-known from thermodynamics we can derive an expression describing the relationship between the radius of the bubble and the necessary superheating of the vapour in it. By this we get some information about the metastable situation of the liquid in the boundary layer. If we bring Eq. (3.1) and Eq. (3.2) together and integrate with the simple assumption that the vapour can be treated as ideal gas and that the specific volume of the liquid is negligibly small compared with that of the vapour, we can easily get a rough estimation of the superheating temperature $(T_v - T_s)$ being necessary to allow a bubble of the radius R to grow.

$$R = \frac{2\sigma}{\Delta h_v \cdot \rho_v} \cdot \frac{T_s}{T_v - T_s} \tag{3.3}$$

Equation (3.3) shows that with increasing superheating of the liquid beyond the saturation temperature T_s smaller nuclei at the heated surface can become active. On the other side however, the superheating of the surface is a function of the heat flux being transferred through the solid material to it. Usually the surface of a solid material contains roughness sites of different size and, therefore, the number of the activated nuclei—i.e. of the locations, where bubbles are formed at the surface—is increasing with higher heat flux. This increasing of the number of activated nuclei results in a more intensive fluiddynamic mixing of the liquid at the heated surface. Both together, namely the mass transport in the bubble in form of vapour and the microscopic turbulence with its drift flux behind the bubble improve the heat transfer conditions. Therefore, one can expect that the heat transfer coefficient is improved with increasing heat flux during boiling.

A simple experiment which was demonstrated in the literature by Nukiyama [5] proves this deliberation. Nukiyama measured at and above a heated plate the surface temperature T_W and the liquid temperature T_F and correlated both with the heat flux $\dot{q}$. In Fig. 3.1 the heat flux and the heat transfer coefficient α are plotted versus the temperature difference between the heated surface and the boiling liquid in a logarithmic scale as Nukiyama did it. At low heat flux the energy transport is managed at the wall by liquid free convection only, and evaporation occurs only after the superheated liquid reaches the upper surface of the pool where the radius of the phase interface is almost infinite.

The onset of bubble formation at the heated wall, i.e. the bubble boiling causes a sudden change in the slope of the curves, shown in Fig. 3.1, because the heat transfer conditions are now instantaneously improved. Increasing the heat flux more and more, one finally reaches a situation, where the transport phenomenon with bubble boiling becomes hydrodynamically unstable, because due to the dense bubble population and the huge vapour flow, the liquid will be prevented from flowing down to the heated solid surface, and this impairment of liquid flow to the wall changes the boiling situation. Suddenly a thin but coherent vapour film is formed at the heating surface separating the liquid from it and now bubble boiling changes into the so-called film boiling. This sudden change is called "Departure from Nucleate Boiling" (DNB) and the heat flux at which it occurs is referred to as the "Critical Heat Flux" (CHF). The heat transfer coefficient reaches its maximum shortly before DNB is observed.

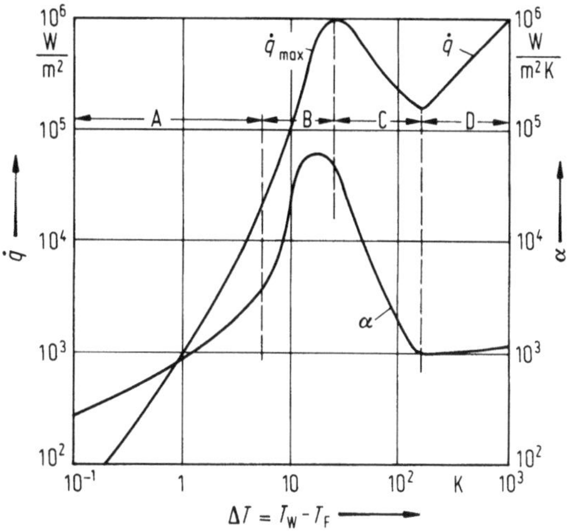

Fig. 3.1. Nukiyama curve for boiling, *A* free convection, *B* nucleate boiling, *C* transition boiling, *D* film boiling

If we increase the heat flux at the DNB-point only a little more we can observe a sudden and large jump in the wall temperature which may result—e.g. with water of ambient pressure—in several 100 K. At higher pressures this temperature-jump is not so large and at very high pressures—near the critical point—the temperature may even continuously change with heat flux after the DNB, but still to a large extent.

After this unstable situation at and immediately above the DNB-point, the temperature of the wall is again continuously rising with increasing heat flux, however, the q-ΔT-curve is now much flatter as it was before departure from nucleate boiling. As seen from Fig. 3.1, the heat transfer coefficient drops down when DNB is exceeded by one or two orders of magnitude.

If we now lower the heat flux, we do not observe the sudden jump at that point, where we arrived at after exceeding the DNB-point when heating, but the temperature is still continuously decreasing until the q-ΔT-curve reaches its minimum. Here the temperature instantaneously falls and the situation returns to the nucleate boiling branch of the curve. With further reduction of the heat flux the temperature is then continuously, but only slightly decreasing along the nucleate boiling line.

Situations in the region *C* on this curve—in the literature called Nukiyama-curve—can only be reached if we change the heating conditions. Instead of imposing a given heat flux, we can imagine that we keep the wall at a given constant temperature by appropriate means, for example by heating with liquid metal. Then we can adjust a temperature in the region *C*, and we shall observe a fluctuating behaviour of the boiling conditions between nucleate and film boiling and with intermediate wetting of the wall.

For situations in the region *D* only film boiling can be observed and the wall remains unwetted by the liquid. The temperature at the minimum is called

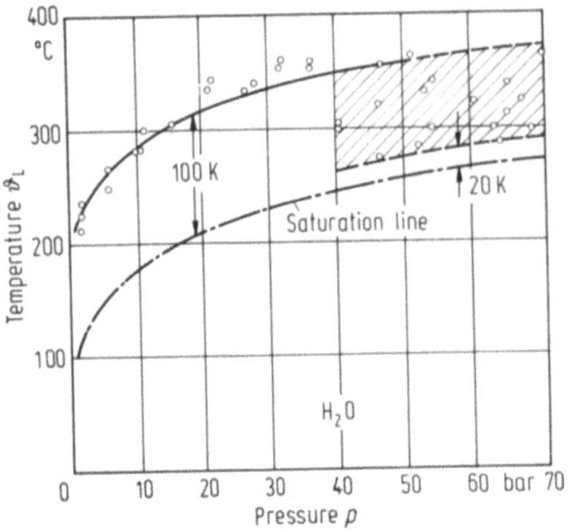

Fig. 3.2. Rewetting temperature (Leidenfrost-temperature) for water according to Hein [7]

Leidenfrost-temperature or rewetting-temperature, because by lowering the heat flux below that value, the wall becomes rewetted again. The temperature at which this rewetting occurs is still under discussion and research in the literature. Sometimes theories predicting the so-called homogeneous nucleus-formation-temperature are used, however, they give much too high rewetting temperatures. Yao [6] confirmed earlier observations that the rewetting temperature is strongly depending on the surface conditions, such as roughness or thin coating layers and also on the thermal conductivity of the material at which rewetting occurs. Yao found that on thermally stable substances—solid materials—rewetting starts much earlier than the homogeneous nucleus-formation-theory predicts. The rewetting- or Leidenfrost-temperature, is also a slight function of the pre-cooling history.

A detailed survey on the rewetting by flooding and on the Leidenfrost-temperature is given by Hein [7]. According to his investigations and measurements the rewetting temperature of water is at low pressure—between 1 and 40 bar—approximately 100 K above the saturation temperature as Fig. 3.2 shows. This means that the vapour layer at the wall breaks down if film boiling or thermal conduction to a near-by rewetted area could cool down the surface of the material to a temperature less than 100 K above the saturation temperature. At higher pressures the difference between saturation temperature and rewetting temperature becomes smaller and it can be reduced down to 20 K. The measurements, however, show a wide range of scattering. This is not so much due to the uncertainty of the experimental readings, but the rewetting temperature is strongly influenced by the surface conditions, roughness and coating layers, e.g. by oxide formation.

With other substances the situation is similar. In Fig. 3.3 rewetting temperatures of the refrigerant R12 are plotted versus the pressure. Similar to water also in this substance the difference between rewetting temperature and

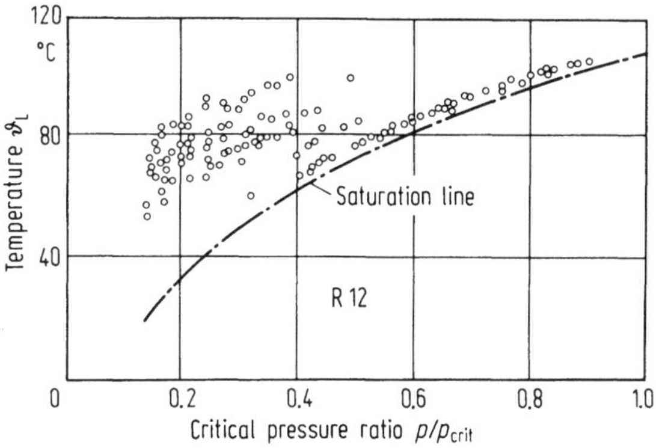

Fig. 3.3. Rewetting temperature (Leidenfrost-temperature) for the refrigerant R12 according to Hein [7]

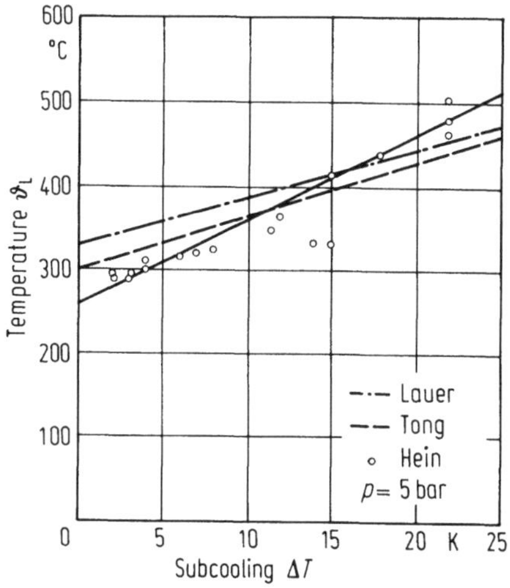

Fig. 3.4. Influence of the subcooling of the liquid onto the rewetting (Leidenfrost-temperature) in water

saturation temperature decreases with increasing pressure. At high pressures this substance even shows a much more homogeneous behaviour than water.

As mentioned before, the heat transfer conditions are influencing the rewetting temperature also. Improved cooling conditions exist when the liquid is subcooled because the condensation of the vapour at the phase interface between vapour layer and liquid produces strong turbulence and acceleration of the liquid mass towards the hot surface. By this, liquid is impinged to the wall and rewetting can start earlier—at higher surface temperatures—compared with film boiling in saturated liquids. This observation was confirmed by Lauer [8], Tung [9] and Hein [7]. As demonstrated in Fig. 3.4 the rewetting-temperature is increasing linearly with the subcooling of the liquid and for example water, 25 K below the saturation temperature and at 5 bar, can wet a hot wall already at 500 °C.

The thermo- and fluiddynamic situation with nucleate boiling is even more complicated than with film boiling. In the literature many theoretical and

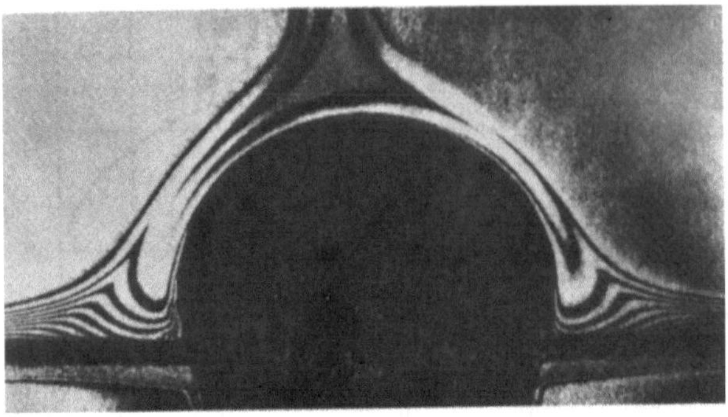

Fig. 3.5. Temperature field around a bubble growing on a wire in water ($p = 0.3$ bar, $q = 30$ W/cm^2, subcooling 2 K)

experimental investigations on bubble formation and bubble growth can be found. One should mention e.g. the papers by Forster and Zuber [10], Hahn and Griffith [11], Plesset and Zwick [12], Beer [13] or Winter [14]. The occurrences with bubble growth are controlled by various parameters and forces. In the very first moment after the necessary superheating of the liquid directly at the wall is reached, the growth of the bubble out of the nucleus is governed by the inertia of the liquid which has to be pushed away by the growing bubble. This first evaporation step causes a temperature drop in the liquid near the phase interface and the pressure in the bubble is lowered equivalently to the cooling of the liquid. For further evaporation, liquid has to be transported to the phase interface from surrounding areas and, therefore, during this second period of bubble growth heat- and masstransport in the liquid is governing the bubble growth. To understand these heat- and masstransport phenomena better, it may be helpful to get information on the temperature conditions near the wall and at the phase interface of the bubble.

Optical measuring techniques are good tools to record the very quickly changing temperature situation near a growing bubble. A convenient method to be handled is the holographic interferometry [15]. An example of a holographic interferogram taken of the temperature field around a bubble which is growing on a heated wire is shown in Fig. 3.5. The light and dark fringes in this interferogram represent—in a first approximation—isotherms in the liquid. The temperature field around the bubble is influenced by a proceeding bubble which left the heated wire a few milliseconds before and which is only "visible" in this picture by its drift flow. The superheated boundary layer around the wire can be clearly seen on the left side of Fig. 3.5. The temperature field around a bubble growing on a heated flat surface and its life-history is shown in Fig. 3.6. When the nucleus becomes active, the bubble starts growing into the superheated boundary layer due to evaporation out of this layer. After 4 ms the bubble starts to shrink and the bubble is disappeared after 7 ms. The reason for this is that the water flowing over the heated surface is subcooled

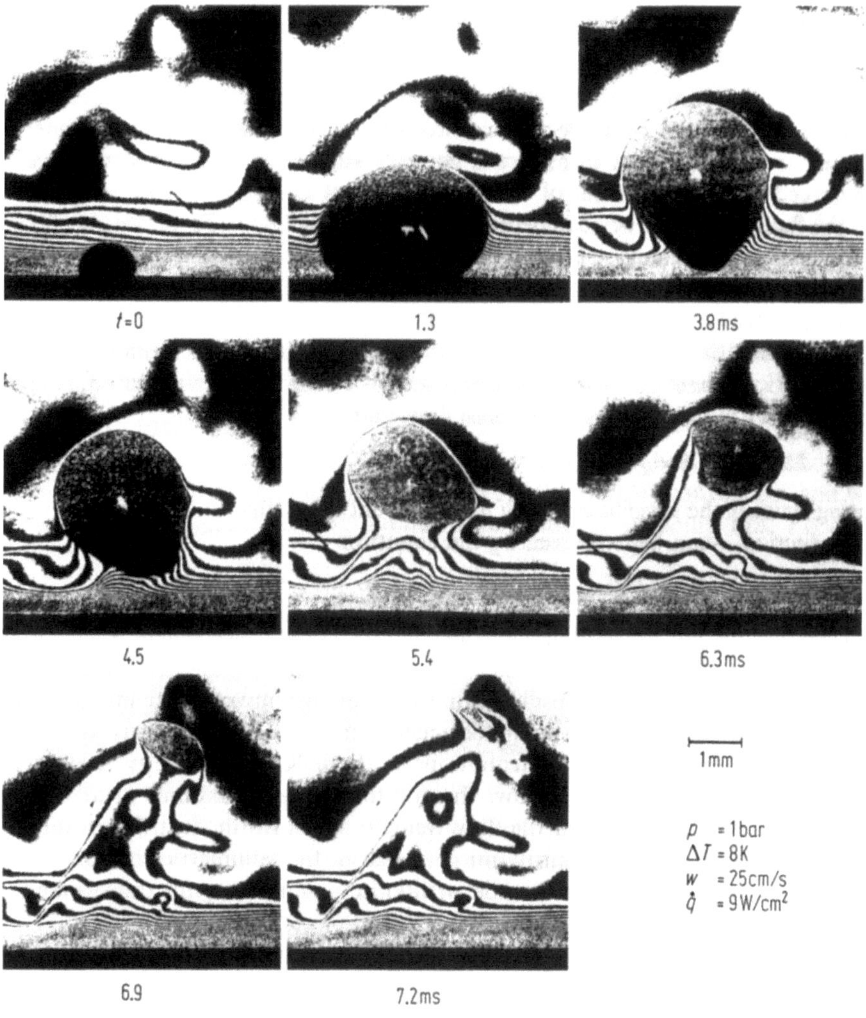

Fig. 3.6. Boundary layer and bubble formation with boiling in water at 1 bar, water-temperature 8 K below saturation temperature, velocity $w = 0.25\,m/s$, heat flux $q = 9\,W/cm^2$

—8 K below saturation temperature—and only an approximately 1 mm thick boundary layer is superheated due to heat addition. In this boundary layer a large temperature gradient exists as the densely situated fringes—representing isotherms—show. Evaluating this interference fringes, one finds that the liquid adjacent to the wall is approximately 10 K superheated above the saturation temperature. The total temperature difference between the wall and the bulk therefore is 18 K.

3.2 Single Phase Convection

Without convection—enforced by natural buoyancy or by pressure differences—heat is transported by molecular conduction only. With convecting fluids—liquid or gas—the heat transport is strongly supported or almost exclusively managed by the movement of the fluid. This heat transport by the movement of the fluid can easily be considered in a laminar fluid flow. To do this, we look at a rectangular element having edge distances dx, dy, and dz, as shown in Fig. 3.7. In z-direction, i.e. vertically to the drawing plain, the temperature is assumed to be constant. Then the heat fluxes $d\dot{Q}_{\lambda,x,\mathrm{in}}$, $d\dot{Q}_{\lambda,y,\mathrm{in}}$, enter the volume element and the heat fluxes $d\dot{Q}_{\lambda,x,\mathrm{out}}$ and $d\dot{Q}_{\lambda,y,\mathrm{out}}$ leave this element. If this volumetric element in addition is penetrated by a fluid flow in x-direction having the velocity w then the entering mass flow-rate

$$d\dot{M} = \varrho w dA = \varrho w dy dz \tag{3.4}$$

brings with it the specific enthalpy h and the enthalpy flux due to this flow—or convenction—can be expressed by the equation

$$d\dot{Q}_{\mathrm{conv,in}} = \varrho w h dA = \varrho w h dy dz \tag{3.5}$$

In steady state flow the enthalpy flux leaving the volumetric element is increased by the temperature rise $d\vartheta$ if heat is added to the volumetric element by conduction. In a detailed consideration we would also have to take into account the kinetic energy and the dissipated energy in this element due to the flow. Here we assume that the contributions of these kinds of energy are small compared with the enthalpy flow. In addition we shall assume steady state conditions which means that the flow velocity and also the heat fluxes due to heat conduction are temporarily not changed in the volumetric element.

The energy balance can then be written in a simple way for this element:

$$(d\dot{Q}_{\lambda,x,\mathrm{in}} - d\dot{Q}_{\lambda,x,\mathrm{out}}) + (d\dot{Q}_{\lambda,y,\mathrm{in}} - d\dot{Q}_{\lambda,y,\mathrm{out}}) + (d\dot{Q}_{\mathrm{conv,in}} - d\dot{Q}_{\mathrm{conv,out}}) = 0. \tag{3.6}$$

With the expressions for the molecular heat conduction and the convective

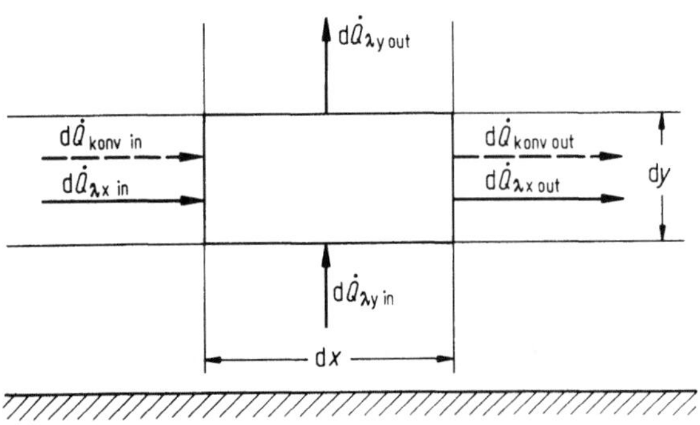

Fig. 3.7. Heat balance with molecular (conductive) and convective transport

enthalpy transport, the energy balance then has the form

$$\frac{\partial}{\partial x}\left(-\lambda dA\frac{\partial \vartheta}{\partial x}\right)+\frac{\partial}{\partial y}\left(-\lambda dA\frac{\partial \vartheta}{\partial y}\right)+\frac{\partial}{\partial x}(\varrho w h dA)=0 \tag{3.7}$$

By rearranging this equation and by substituting

$$h=c d\vartheta \tag{3.8}$$

and with the thermal diffusivity $a=\lambda/\varrho\cdot c$ we get the well-known energy-transport equation in two-dimensional form.

$$a\left(\frac{\partial^2\vartheta}{\partial x^2}+\frac{\partial^2\vartheta}{\partial y^2}\right)=w\frac{\partial\vartheta}{\partial x}. \tag{3.9}$$

With most practical applications and for almost all fluids—with exception of liquid metals—the molecular heat transport, i.e. the heat conduction in x-direction, is small compared with the energy transport by convection, and we therefore can neglect the longitudinal heat conduction. Doing this, Eq. (3.9) reads:

$$a\frac{\partial^2\vartheta}{\partial y^2}=w\frac{\partial\vartheta}{\partial x}. \tag{3.9a}$$

For solving Eqs. (3.9) and (3.9a) we need information on the flow velocity in all 3 directions x, y, and z which we get from the laws of fluiddynamics. A very general fluiddynamic law e.g. is formulated in the Navier–Stokes-equation which is a balance equation for the forces acting on a fluid element

$$\begin{aligned}
w_x\frac{\partial w_x}{\partial x}+w_y\frac{\partial w_x}{\partial y}+w_z\frac{\partial w_x}{\partial z}&=-\frac{1}{\varrho}\frac{\partial p}{\partial x}+\nu\left(\frac{\partial^2 w_x}{\partial x^2}+\frac{\partial^2 w_x}{\partial y^2}+\frac{\partial^2 w_x}{\partial z^2}\right)\\
w_x\frac{\partial w_y}{\partial x}+w_y\frac{\partial w_y}{\partial y}+w_z\frac{\partial w_y}{\partial z}&=-\frac{1}{\varrho}\frac{\partial p}{\partial y}+g+\nu\left(\frac{\partial^2 w_y}{\partial x^2}+\frac{\partial^2 w_y}{\partial y^2}+\frac{\partial^2 w_y}{\partial z^2}\right)\\
w_x\frac{\partial w_z}{\partial x}+w_y\frac{\partial w_z}{\partial y}+w_z\frac{\partial w_z}{\partial z}&=-\frac{1}{\varrho}\frac{\partial p}{\partial z}+\nu\left(\frac{\partial^2 w_z}{\partial x^2}+\frac{\partial^2 w_z}{\partial y^2}+\frac{\partial^2 w_z}{\partial z^2}\right).
\end{aligned} \tag{3.10}$$

From the Navier–Stokes-equation dimensionless numbers as the Reynolds-number

$$Re=\frac{wL}{\nu} \quad \text{or} \quad Re=\frac{wD}{\nu} \tag{3.11}$$

or the Grasshof-number

$$Gr=\frac{L^2 g\beta\Delta\vartheta}{w\nu}Re=\frac{L^3 g\beta\Delta\vartheta}{\nu^2} \tag{3.12}$$

were derived by writing this equation in a dimensionless form [16]. Doing the same with the energy equation (Eq. (9)) we get the Peclet-number.

$$Pe = \frac{wL}{a}. \tag{3.13}$$

Finally the quotient of the Peclet- and Reynolds-number forms the Prandtl-number.

$$Pr = \frac{Pe}{Re} = \frac{\nu \cdot \varrho \cdot c}{\lambda} = \frac{\nu}{a}. \tag{3.14}$$

If we now consider the situation directly at the wall where an infinitely thin stagnant liquid layer exists and introduce the well-known definition for the heat transfer coefficient α

$$\alpha = \frac{\dot{Q}}{A(T_{\text{Wall}} - T_{\text{Bulk}})} = \frac{\dot{Q}}{A(\vartheta_{\text{Wall}} - \vartheta_{\text{Bulk}})} \tag{3.15}$$

we can write a simple energy balance at this position just by taking into account that the heat conducted through this infinite thin stagnant liquid layer must be equal to the total heat transport from the wall to the fluid.

$$\lambda A \frac{\partial T}{\partial y} = \alpha A(T_{\text{Wall}} - T_{\text{Bulk}}). \tag{3.16}$$

In the Eqs. (3.15) and (3.16) T_{Bulk} represents the temperature of the fluid in a position far away from the wall. By rearranging and writing Eq. (3.16) in a dimensionless form we finally can derive the Nusselt-number.

$$Nu = \frac{\partial(T/\Delta T)}{\partial(y/L)} = \frac{\alpha L}{\lambda} = \frac{\alpha D}{\lambda}. \tag{3.17}$$

We shall now consider a very simple example of a heat transfer problem, namely the heat transport to a flat plate from a fluid flowing longitudinally over it, and the solutions of the energy- and Navier–Stokes-equation will be demonstrated in a simple and dimensionless form for these fluiddynamic conditions. We assume that the flat plate as shown in Fig. 8 is positioned in a fluid flow of the

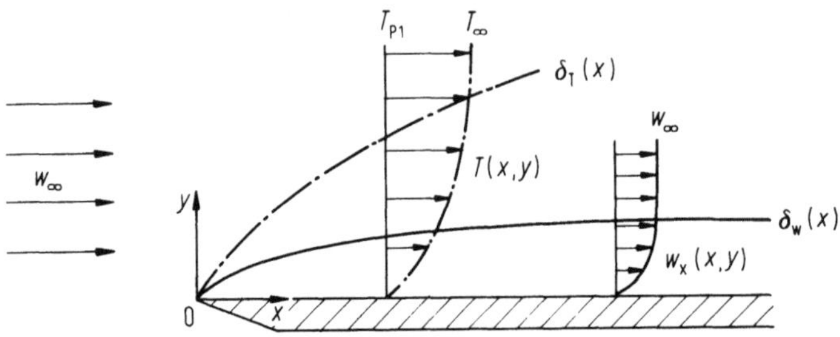

Fig. 3.8. Boundary layers of velocity and temperature on a plate with longitudinal flow

velocity w_∞. In the immediate vicinity of the plate surface this velocity w_∞ is decelerated by the friction at the wall. Only in a distance away far enough from the wall the upstream-velocity w_∞ remains uninfluenced. Between the wall and the uninfluenced environment a velocity profile is formed as sketched in Fig. 3.8. In a similar way the temperature of the fluid above the plate behaves. In a distance far enough away from the plate the temperature of the fluid is not influenced by the heat addition to the plate and is equal to the upstream-temperature. Near the plate surface the fluid temperature approaches the plate temperature.

Starting from the front edge of the plate, a laminar flow zone will be formed, even if the upstream-flow-conditions are turbulent. This laminar zone is increasing in its thickness over the plate length, starting from the front plate edge. This continues until turbulence starts.

It is usual in the literature to define a boundary-layer thickness δ_w as that distance from the plate surface, where the velocity just reaches 99% of the upstream-velocity. In a similar way also a temperature-boundary-layer thickness δ_T is defined as that distance from the plate surface, where the fluid-temperature is 99% of the temperature in the upstream undisturbed region. So we see that the heat transport is mainly restricted to the region where the boundary-layer exists.

If we want to have some information about the boundary-layer thickness we have to solve the Navier–Stokes-equation and the energy equation. In addition we need the continuity equation. In its simplest form we can write the Navier–Stokes-equation by assuming that the pressure gradient vertically to the plate can be neglected and that there is also no friction due to any flow in this perpendicular direction.

$$w_x \frac{\partial w_x}{\partial x} + w_y \frac{\partial w_y}{\partial y} = -\frac{1}{\varrho}\frac{\partial p(x)}{\partial x} + \nu \frac{\partial^2 w_x}{\partial y^2} \tag{3.18}$$

Equation (3.18) is called boundary-layer equation and was formulated by Prandtl in 1904. A first solution for this equation was given by Blasius, who made the additional simplifying assumption that the pressure gradient dp/dx longitudinally to the plate can be neglected also. For the above mentioned boundary layer thickness Blasius found the equation

$$\frac{\delta(x)}{x} \approx \frac{5}{\sqrt{Re}} \quad \text{or} \quad \frac{\delta(x)}{L} \approx \frac{5}{\sqrt{Re_L}}\sqrt{\frac{x}{L}}. \tag{3.19}$$

From this equation for the boundary-layer thickness Blasius finally derived an expression for the friction factor ψ in laminar flow.

$$\psi(x) = \frac{0.664}{\sqrt{Re_x}} \tag{3.20}$$

As well-known, this friction factor is defined in laminar flow as

$$\psi(x) = 2\frac{\tau_{\text{Wall}}(x)}{\varrho w_{\text{Bulk}}^2} \approx 2\frac{\tau_{\text{Wall}}(x)}{\varrho w_\delta^2}. \tag{3.21}$$

For calculating the heat transfer conditions we have to take into account the energy equation in addition, and for simplicity we assume that there is heat conduction only perpendicular to the plate surface.

$$w_x\frac{\partial\vartheta}{\partial x} + w_y\frac{\partial\vartheta}{\partial y} = a\frac{\partial^2\vartheta}{\partial y^2}. \tag{3.22}$$

For $Pr = 1$ and constant wall temperature the exact solution for the heat transfer using the equations by Blasius and the energy equation is:

$$Nu_x = 0.332\sqrt{Re_x}. \tag{3.23}$$

For fluids, where the Prandtl-number is not equal to 1, only approximate solutions exist as for example:

$$Nu_x = \frac{\alpha x}{\lambda} = 0.332\, Re_x^{1/2}\cdot Pr^{1/3}. \tag{3.23a}$$

By comparing Eqs. (3.20) and (3.23), we very easily can see the connection between momentum—and heat transport

$$\frac{Nu}{Re\,Pr} = \frac{\psi}{2}; \tag{3.24}$$

which is called the "Reynolds-analogy".

For turbulent boundary conditions the circumstances are much more complicated. In the literature there are several models showing possibilities for taking into account the turbulent cross flow fluctuations or the eddy diffusivity. For practical use, however mostly empirical correlations are proposed which are similar to Eqs. (3.23) or (3.23a).

$$Nu = C Re^{\mathrm{m}} Pr^{\mathrm{n}}. \tag{3.25}$$

The exponent n of the Prandtl-number is a little different for a heated wall ($n = 1/3$) and for a cooled wall ($n = 0.4$). The flow velocity and by this the Reynolds-number, however, is of stronger influence in turbulent flow and the exponent m, therefore, has values of 0.7–0.8. The Reynolds-analogy in its simplest form is similar to that in laminar flow (Eq. (3.24)), however, the quotient between heat transfer and friction is smaller.

$$\frac{Nu}{Re\,Pr} = \frac{\psi}{8}. \tag{3.26}$$

If we apply this turbulent Reynolds-analogy (Eq. (3.26)) to the well-known friction law by Blasius

$$\psi = \frac{0.3164}{Re^{0.25}} \tag{3.27}$$

we get for fluids with $Pr = 1$

$$Nu = 0.0396 \cdot Re^{0.75} = C Re^{m} \tag{3.28}$$

which corresponds in its form with Eq. (3.25) and which is the basis for many correlations predicting heat transfer in heat exchanging components.

3.2.1 Heat Transfer Equations for Forced Convection

Heat transfer and fluid flow, however, are also depending on the geometrical form of the heated or cooled surface. Therefore, in the literature we find different—mostly empirical—correlations for predicting heat transfer in different configurations. For laminar flow in tubes and channels (up to $Re = 2300$) Schlünder [17] recommends the simple correlation

$$Nu = \frac{\alpha D_i}{\lambda} = \sqrt[3]{3.66^3 + 1.62^3 \, Re \, Pr \frac{D_i}{L}}, \tag{3.29}$$

which is also valid in the entrance region of the channel.

Applying this or other heat transfer-equations, it is important to use the correct temperature—the so-called reference temperature—for selecting the values for the thermodynamic properties in the dimensionless numbers. There is a temperature gradient perpendicular and longitudinal to the heat exchanging wall, and usually the reference temperature—sometimes also called mean bulk-temperature—is calculated as the arithmetic mean value between the entrance and the outlet temperature in the channel or tube.

$$T_{Bulk} = (T_{in} - T_{out})/2. \tag{3.30}$$

Sometimes also the logarithmic mean value is used.

For turbulent flow in tubes or channels Colburn derived from the Reynolds-analogy the simple correlation

$$Nu = \frac{\alpha D_i}{\lambda} = 0.023 \, Re^{0.8} \cdot Pr^{1/3}, \tag{3.31}$$

which gives good results for not too high heat fluxes and in the range $10^4 < Re < 10^5$ and $0.5 < Pr < 100$. It is not applicable in the entrance region, because there the heat transfer coefficiences are higher than Eq. (3.31) predicts. Hausen [18] presented a correlation for a very large range of Reynolds-numbers, namely from laminar flow up to the highly turbulent conditions of $Re = 2.5 \cdot 10^5$, which is also valid in the entrance region.

$$Nu = 0.0235(Re^{0.8} - 230)(1.8\,Pr^{0.3} - 0.8) \cdot \left[1 + \left(\frac{D_i}{L}\right)^{2/3}\right]\left(\frac{\eta_{Bulk}}{\eta_{Wall}}\right)^{0.14}. \tag{3.32}$$

3.2.2 Heat Transfer Equations for Natural Convection

Free convection is driven by buoyancy forces and not by pressure drop. Therefore the heat transfer correlations for convection are formed with the Grashof-number (Eq. 3.12) instead of the Reynolds-number. The heat transfer coefficient is again expressed by the Nusselt-number and the correlations for calculating the heat transfer in free convection have the form

$$Nu = f(Gr, Pr). \tag{3.33}$$

In closed cavities free convection only starts if:

$$Gr\,Pr > 1700. \tag{3.34}$$

The Grashof-number for closed cavities is formed with the distance between the two vertical walls.

For quenching free convection around bodies in a pool is of more interest. For a vertical plate Rohsenow an Choi [20] derived a correlation, using the balance equations, which is valid for laminar flow along the plate

$$\frac{Nu_x}{(Gr_x/4)^{1/4}} = \frac{0.676\,Pr^{1/2}}{(0.861 + Pr)^{1/4}}. \tag{3.35}$$

In (3.35) the Nusselt- and the Grashof-number are functions of the flow path, i.e. of the distance from the lower edge of the plate. In the Grashof-number the difference of the temperatures at the plate surface and in the fluid where it is not yet affected by heat transfer has to be inserted as characteristic temperature difference. If the plate is cooled, the flow goes downward and therefore, the flow path has to be counted from the upper edge of the plate.

One can also rearrange Eq. (3.35) to derive a mean Nusselt-number and Eq. (3.35) then reads

$$\frac{Nu_m}{(Gr/4)^{1/4}} = \frac{0.902\,Pr^{1/2}}{(0.861 + Pr)^{1/4}}. \tag{3.36}$$

Laminar boundary layers are observed at vertical flat plates up to $Gr\,Pr = 10^8$ Closed solutions for turbulent boundary layers are more complicated. For Prandtl-numbers between 1 and 10 one can, however, use the simple correlation

$$Nu_m = 0.13(Gr \cdot Pr)^{1/3} \tag{3.37}$$

with good accuracy.

3.3 Two Phase Heat Transfer

Discussing heat transfer with single-phase forced convection, we realized that the heat transfer coefficient is not a function of the temperature difference between the wall and the fluid. With free convection we found that density differences, caused by temperature differences, are impelling the heat transport and, therefore, the heat transfer coefficient is here depending on the temperature difference between wall and fluid. Boiling and heat transfer in two-phase flow are always connected with phase change which results in very large volume changes due to the density differences between liquid and vapour. Therefore, one can expect a priori that heat transfer is strongly influenced by buoyancy forces, however, also by dynamic forces originating from the bubble growth. These forces are affecting the boundary layer to such an extent that heat transfer with boiling allows high heat fluxes and cannot be improved very much by superimposing forced convection. Therefore, one reaches almost the same values in free and in forced convection. As we have seen when discussing the boiling phenomena, the number of activated nuclei per unit of area increases with rising wall temperature and by this with higher heat flux. Therefore one can expect that the heat transfer coefficient is a function of the heat flux because of the agitating effect of the growing, departing and rising bubbles. From these simple deliberations one can derive the correlation

$$\alpha = Cq^{\mathrm{n}}, \tag{3.38}$$

where, however, the constant C differs from fluid to fluid and is a function of the pressure and the surface roughness. For more general validity, therefore, one would have to extend Eq. (3.38) to the form

$$\alpha = C_{\mathrm{F}} \cdot C_{\mathrm{w}} \cdot F(p) q^{\mathrm{n}} \tag{3.39}$$

where C_{F} represents the properties of the liquid, C_{w} the roughness and the thermal conductivity of the wall, and the influence of the pressure is expressed by $F(p)$.

3.3.1 Free Convection Boiling

Considering the formation and movement of the bubbles in the liquid, it is possible to develop physical models and empirical correlations for the heat transfer coefficient with pool boiling. However, as seen in Fig. 3.1, we always have to observe whether—depending on the heat flux and the surface temperature—the system is in nucleate or in film boiling conditions. Both conditions are separated by the so-called boiling crisis and for knowing the heat flux at which nucleate boiling changes into film boiling, we have to discuss critical heat flux correlations.

Correlations using dimensionless numbers for boiling heat transfer have a more general validity than Eqs. (3.38) and (3.39). Dimensionless numbers for pool boiling are formed with the transport properties of the substance, the heat flux density and the thermodynamic state—i.e. the boiling temperature. Correlations formed with such dimensionless groups are of empirical character, too. However, they have the benefit that they are valid for several substances and for a wide range of pressure and saturation temperature. An additional influence onto the heat transfer with boiling comes from the roughness and the thermal conductivity of the solid surface on which boiling occurs. Most of the correlations in the literature, however, neglect these effects because up to now, they are not well enough understood. On very smooth surfaces, for example on glas, it may happen that nucleate boiling is suppressed and after a high superheating of the liquid suddenly film boiling occurs. Technical surfaces—e.g. of metal, —however, are usually rough enough that a large number of bubble nuclei can form, and there the influence of an additional roughness is small.

For calculating the heat transfer with nucleate boiling on metallic surfaces, here the equation by Stephan and Preusser [21] is presented

$$Nu = \frac{\alpha D_{\mathrm{bub}}}{\lambda_\rho} = 0.1\left[\frac{qD_{\mathrm{bub}}}{\lambda_1 \vartheta_s}\right]^{0.674}\left[\frac{\varrho_v}{\varrho_1}\right]^{0.156}\left[\frac{\Delta h_v D_{\mathrm{bub}}}{a_1^2}\right]^{0.341} \cdot \left[\frac{a_1^2 \varrho_1}{\sigma D_{\mathrm{bub}}}\right]^{0.350}\left[\frac{\eta c}{\lambda}\right]_1^{-0.162} \tag{3.40}$$

Equation (3.40) is written in the power and product form which is familiar from the correlations for single-phase convection. It contains dimensionless groups which are formed with thermodynamic properties, the saturation temperature and the heat flux density. In addition some groups in this correlation contain the diameter of the bubble, when separating from the surface. This bubble diameter can be calculated by considering the equilibrium of the separating force due to buoyancy and of the holding force due to surface tension.

$$D_{\mathrm{bub}} = 0.0146\beta\left(\frac{2\sigma}{g(\varrho_1 - \varrho_v)}\right)^{0.5}. \tag{3.41}$$

For the contact angle β between the bubble and the solid surface, one has to insert into Eq. (3.41) the following values:

For water 45°,
for cryogenic substances 1°,
for hydrocarbons including refrigerants 35°.

In Eq. (3.40) a_1 is the thermal diffusivity of the liquid formed with its thermal conductivity, density and specific heat.

$$a_1 = \frac{\lambda_1}{\varrho_1 c_1}. \tag{3.42}$$

For water and if the claim with respect to accuracy of the predicted values is

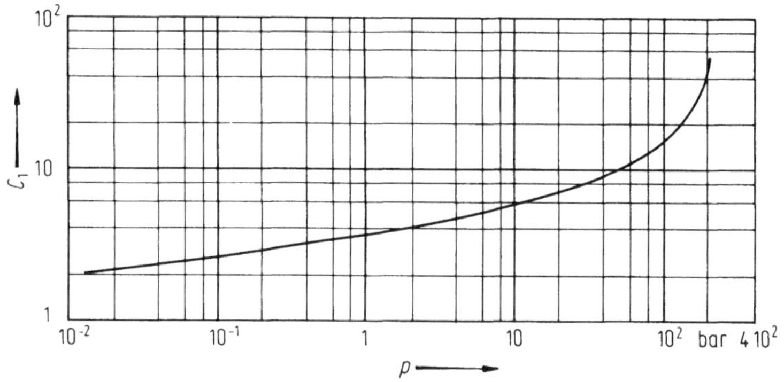

Fig. 3.9. Constant C_1 in Eq. (43) as a function of the pressure (water)

not too high, one can also use the much simpler equation

$$\alpha_{\text{boil.}} = C_1 \dot{q}^{0.673}, \tag{3.43}$$

where C_1 is a factor which depends on the pressure in the system and whose value can be taken from Fig. 3.9.

3.3.2 Forced Convection Boiling

With forced convection boiling the heat transfer situation changes as far as a velocity profile is superimposed onto the microconvection, produced by the departing bubbles near the wall. However, looking more in detail, one realizes that this velocity profile is only weakly influencing the very first phase of the bubble formation, the nucleation. Due to friction forces the velocity at the wall is zero and, in addition, growing and departing bubbles decelerate the velocity in the boundary layer near the wall.

For activating a nucleus only the superheating in the boundary layer in the immediate vicinity of the wall is of influence, also in forced convection flow. Different is the situation for the bubble growth and the bubble departure. If we consider the forces acting on a growing and departing bubble in forced convection, we have to take into account also the resistance and the momentum force due to the flow, in addition to the forces by buoyancy and surface tension. So the heat transfer with nucleate boiling is slightly improved by the forced convection compared with pool boiling. There are attempts in the literature to calculate the heat transfer in forced convection nucleate boiling by superimposing the effect of microconvection, produced by bubble departure and that of forced convection due to the superimposed directed flow. Rohsenow [22] presented a correlation in which the heat flux from the wall is the sum of two parts, one is formulated with the heat transfer coefficient for nucleate pool boiling and the other one with the heat transfer coefficient for single-phase forced convection. Compared with single-phase forced convection, the heat transfer coefficient for nucleate pool boiling, however, is larger by one or two orders of magnitude. So the contribution of the forced convection for improving the heat transfer with boiling is small and can be neglected in most cases. Therefore, Eq. (3.40) is frequently used for forced convection boiling also.

A different situation is given, when dipping a body, i.e. a plate, into a liquid. As long as the surface temperature of the body is not too high—that is below the Leidenfrost-temperature—we can expect nucleate boiling. Here the free convection, due to buoyancy forces, is superimposing the microconvection, generated by bubbles growing and departing in the immediate vicinity of the wall. Based on dimensionless groups derived by Stephan, an empirical correlation was developed by Kaufmann and Vaihinger [23]

$$Nu = \frac{\alpha D_{\mathrm{bub}}}{\lambda_1} = 0.078\, K_a^{0.62} \cdot K_b^{0.133} \cdot K_c^{0.634} \cdot K_d^{0.234} \cdot Pr^{1.032} \cdot Re^{0.078} \tag{3.44}$$

The dimensionless groups in Eq. (3.44) are formulated as follows

$$K_a = \frac{\dot{q} D_{\mathrm{bub}}}{T_s c_1 \eta_1}; \quad K_b = \frac{k}{D_{\mathrm{bub}}}; \quad K_c = \frac{\varrho_1^2 (f D_{\mathrm{bub}})^2 D_{\mathrm{bub}}}{\sigma \rho_v}$$

$$K_d = \frac{\Delta h_v \varrho_1^2 D_{\mathrm{bub}}^2}{\sigma^2 c_1 T_s}; \quad Pr = \frac{\nu_1}{a_1}; \quad Re = \frac{(f D_{\mathrm{bub}}) D_{\mathrm{bub}} \varrho_1}{\eta_1}$$

and the bubble diameter can be calculated with Eq. (3.41). The product of the bubble departing frequency f and the bubble diameter was formulated by Kaufmann and Vaihinger in the equation

$$f D_{\mathrm{bub}} = 0.314 \frac{g(\varrho_1 - \varrho_v)}{\varrho_1}.$$

Completely different, however, is the situation if only a thin liquid film covers the wall which may be the case with cooling by falling film flow or by spraying liquid onto the surface. With this liquid layers at a hot surface which, however, is below the Leidenfrost-temperature, experiments showed that almost no bubbles are formed in this layer, and the evaporation takes place at the free surface of the liquid film. This evaporation mode is called "surface boiling", and the word surface stands here for the interface between the liquid and the ambient gaseous environment and not for the surface of the solid wall to be cooled. The heat is transported from the solid wall to the free surface, mainly by conduction and convection in the liquid film.

Up to now a fully theoretical description of this heat transport is not presented in the literature and therefore, the heat transfer coefficient with this evaporating mode is described by semi-empirical correlations. These semi-empirical correlations are based on the Martinelli-parameter X_{tt} which is formulated for turbulent conditions in the liquid film and in the gaseous environment:

$$X_{tt} = \left(\frac{\varrho_v}{\varrho_1}\right)^{0.5} \left(\frac{\eta_1}{\eta_v}\right)^{0.1} \left(\frac{1-\dot{x}}{\dot{x}}\right)^{0.9} \tag{3.45}$$

and on the boiling number

$$Bo = \frac{q}{\dot{m}\Delta h_v}; \tag{3.46}$$

The equations given in the literature can be separated in two groups of the form

$$\frac{\alpha_{\text{surf. boil.}}}{\alpha_{\text{sing. phase}}} = A\left(\frac{1}{X_{tt}}\right)^b \tag{3.47}$$

or

$$\frac{\alpha_{\text{surf. boil.}}}{\alpha_{\text{sing. phase}}} = M\left[Bo \cdot 10^4 + N\left(\frac{1}{X_{tt}}\right)^n\right]^m. \tag{3.48}$$

Which form should be used depends on the fluiddynamic conditions in the film. If the film itself is purely of single-phase nature Eq. (3.47) can be used, however, if the liquid film is thicker and some bubble nucleation is to be expected, Eq. (3.48) should be preferred. If nothing is known about the single- or two-phase nature of the Film Eq. (3.48) should be preferred. The boiling number *Bo* in Eq. (3.48) takes into account some bubble formation in the liquid film which improves the heat transfer. Values for the constants A, M, N and the exponents b, n, m are given in Tables 3.1 and 3.2 for various substances and different flow directions.

In Eqs. (3.47) and (3.48) the heat transfer coefficient with evaporation is related to the heat transfer coefficient in purely single-phase forced or free convection. This heat transfer coefficient with free or forced single-phase convection can be calculated, using the equations given in Chap. 2.2. In the

Table 3.1. Values for A and b in Eq. (54)

	A	b
water, upflow	2.9	0.66
water, downflow	2.72	0.58
R113 upflow	4.0	0.37
n-butanol	7.5	0.328
org. liquids, nat. conv.	3.4	0.45

Table 3.2. Values for M, N, n and m in Eq. (55)

	M	N	n	m
water, upflow	0.739	1.5	2/3	1
water, downflow	1.48	1.5	2/3	1
R113 upflow	0.9	4.45	0.37	1
R12 horiz.	1.91	1.5	2/3	0.6
n-Butanol	2.45	1.5	2/3	1

Reynolds-number of these convective heat transfer correlations a superficial velocity has to be used which is calculated with the assumption that only liquid would be present in the channel. Using the Colburn-correlation, one can formulate this reference value of the single-phase heat transfer coefficient with the simple equation

$$\alpha_{\text{sing. phase}} = \frac{\lambda_1}{d_{\text{equ.}}} 0.023 \left[\frac{d_{\text{equ.}} \dot{m}(1-x)}{\eta_1} \right]^{0.8} \left[\frac{c_1 \eta_1}{\lambda_1} \right]^{0.4} . \tag{3.49}$$

The Reynolds-number in this equation is formed with the dynamic viscosity of the liquid and by expressing the mass flowrate of the liquid only with the help of the quality $x = m_o/m$ were m is the total mass flowrate density of the liquid and the vapour.

We now need a criterion when to use this thin film or surface boiling correlations with falling film flow or spray-cooling. Under wetting conditions surface boiling always can be assumed if the thickness of the liquid film is smaller than the diameter of bubbles which would be formed with nucleate boiling. With forced convection two-phase flow, usually the Martinelli-parameter X_{tt}, is used as criterion for separating nucleate boiling and surface boiling. Figure 3.10 gives some help to estimate the border-value of X_{tt} between nucleate and surface boiling. From this figure we learn that for

$$\frac{1}{X_{tt}} > 5 \tag{3.50}$$

we always can assume surface boiling, and we can use Eq. (3.48) to calculate

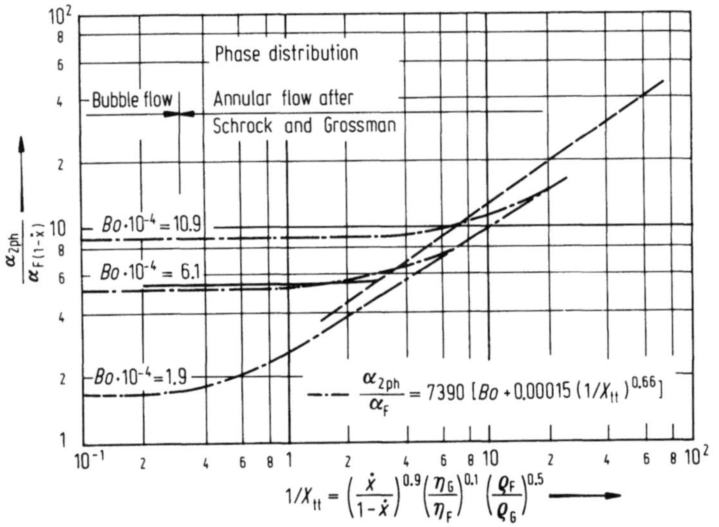

Fig. 3.10. Transition from nucleate boiling to surface boiling in annular flow

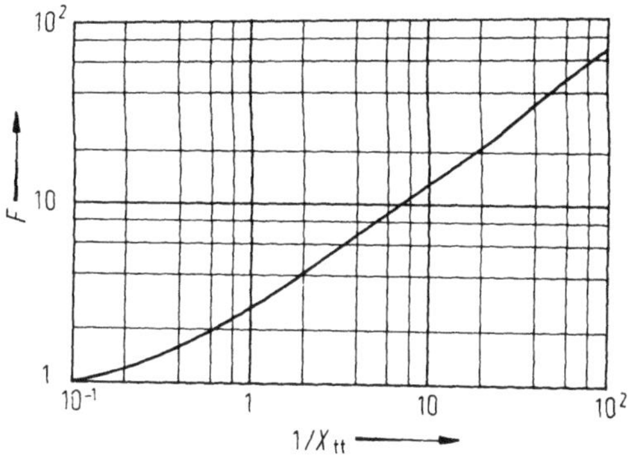

Fig. 3.11. Correction function F in Eq. (3.53) according to Chen [25]

the heat transfer coefficient. Below this value, one has to use pool boiling equations. Figure 3.10 shows also that this border-value of X_{tt} is a function of the boiling number.

For a rough estimation of the heat transfer coefficient with surface boiling in water, we can use Eq. (3.51)

$$\frac{\alpha_{\text{surf. boil.}}}{\alpha_{\text{sing. phase}}} = 0.065\left(\frac{1}{X_{tt}}\right)\left(\frac{T_s}{T_s - T_l}\right)\left(\frac{\sigma_{H_2O}}{\sigma_l}\right)^{0.9} \tag{3.51}$$

which was formulated by Calus [24] and which is much simpler to handle than Eq. (3.48).

In a different way as described above, the heat transfer with forced convection boiling is treated by Chen [25]. Similar as Rohsenow [22] he composes the heat transfer to the gas-liquid-mixture of two components, that of the boiling and that of the forced convection.

$$\alpha_{\text{boil}} = \alpha_{\text{sing. phase}} + \alpha_{\text{pool. boiling}}. \tag{3.52}$$

Unlike to Rohsenow he is not adding the heat flux densities, but the heat transfer coefficients. The reason for this is that he assumes that the driving temperature difference is the same for both mechanisms of heat transport. Chen recommends for calculating the convective contribution to the combined heat transfer coefficient the equation

$$\alpha_{\text{sing. phase}} = \frac{\lambda_l}{d_{\text{equ}}} 0.023\left[\frac{\dot{m}(1-x)d_{\text{equ}}}{\eta_l}\right]^{0.8}\left[\frac{\eta_l c_l}{\lambda_l}\right] F. \tag{3.53}$$

In this equation F is a correction factor which takes into account the different flow conditions in gas-liquid-mixtures compared to that in pure single-phase fluids. The temperature field is strongly influenced by the velocity gradient in the boundary layer near the wall, and this gradient again is depending on the sheer stress situation and the void fraction there. Both fluiddynamic situations can be expressed as a function of the Martinelli-parameter X_{tt}, assuming that both phases are in turbulent conditions. Therefore, it is reasonable to describe this correction factor F as a function of the Martinelli-parameter as shown in

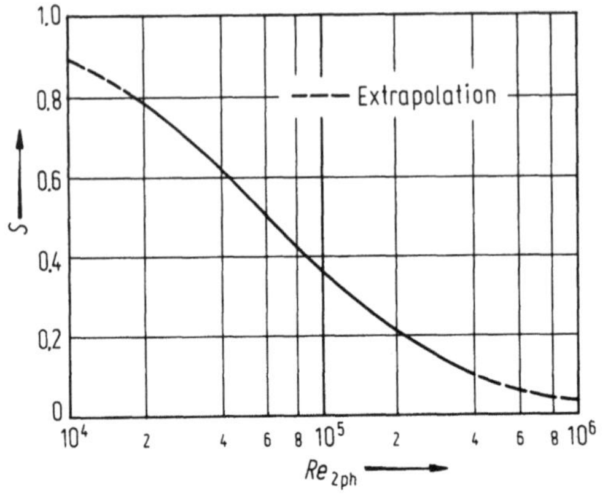

Fig. 3.12. Correction factor S in Eq. (3.54) according to Chen [25]

Fig. 3.11. According to Chen this course of the correction factor is valid over a wide range of void fraction or quality, reaching from nucleate boiling via surface boiling up to the onset of dry-out. With this correction factor and an equation for the heat transfer coefficient with nucleate boiling Chen succeeded in correlating the heat transfer in the whole range of vapour-liquid-mixtures, without splitting it in two regions as described before. The equation, recommended by Chen for the heat transfer coefficient with nucleate boiling is based on a correlation by Forster and Zuber [26]

$$\alpha_{\text{pool, boil}} = 0.00122\left[\frac{\lambda_l^{0.79} c_l^{0.45} \varrho_l^{0.49}}{\sigma^{0.5} \eta_l^{0.29} \Delta h_v^{0.24} \varrho_v^{0.24}}\right](T_{\text{Wall}} - T_s)^{0.24} \Delta p_s^{0.75} S \qquad (3.54)$$

and contains a correction factor S which takes into account the temperature situation in the boundary layer. This correction S depends on the two-phase Reynolds-number

$$Re_{2\text{ph}} = \left[\frac{\dot{m}(1-x)}{\eta_l d_{\text{equ}}}\right] F^{1.25}, \qquad (3.55)$$

as shown in Fig. 3.12. So Chen combines the mechanism of nucleate boiling with that of forced convection and the correction factor F stands for the sheer stress induced velocity field. The temperature difference $\Delta T_s = T_{\text{Wall}} - T_s$ is the superheating of the liquid in the immediate vicinity of the wall and by this is a measure of the driving force for nucleation and bubble growth.

Equation (3.54) is an empirical correlation and was presented by Chen in a non-dimensionless form. Therefore, one has to choose the right dimensions when using it, namely the international system with the mass in kg, the length in m, the force in N and the energy in J. The pressure must be given in N/m^2. The calculation procedure is a little complicated because, depending on the boundary conditions an iterative method has to be used. It is recommended to start with calculating the heat transfer coefficient for forced convection single-phase flow and with evaluating the correction factor F. For boundary conditions with given heat flux then the wall temperature has to be estimated, and the estimation has to be improved in a proper way during the iteration.

3.3.3 Heat Transfer with Film Boiling

Heat transfer with film boiling is of complete different nature compared to that of nucleate boiling. With film boiling the wall is unwetted due to its high temperature and the heat transport has been managed through a thin vapour film from the wall to the saturated liquid. Film boiling occurs at high heat fluxes beyond the so-called critical heat flux and as shown in Fig. 3.13 bubbles separate from the vapour film adjacent to the wall and travel into the saturated liquid. The heat transfer process can be easily described, if one assumes that the vapour in the film flows in a laminar configuration and if one neglects the small sheer stress between the phase interface of vapour and liquid. In addition one can assume that the distance between two bubble columns separating from the film can be expressed by the Taylor- or Helmholtz-instability depending on whether the film is horizontally or vertically orientated. One then ends up with the well-known equations by Bromley [27] or Berenson [28]. These heat transfer correlations are similarly derived as the equations for film condensation which are based on the falling liquid film theory by Nusselt. For a detailed information concerning these theoretical models reference is made to papers by Hsu [29] and Bressler [30].

Assuming turbulent flow in the vapour film, one can also start from the laws for mixed convection for vertically orientated vapour films and one then ends up with a correlation for describing the heat transfer coefficient, which contains a Reynolds-number referred to the thickness of the liquid film, a modified Grashof-number and the Prandtl-number. There are also correlations based on the turbulent boundary layer theory which take in account the effect of interface oscillations between the phases. These oscillations improve the heat transfer.

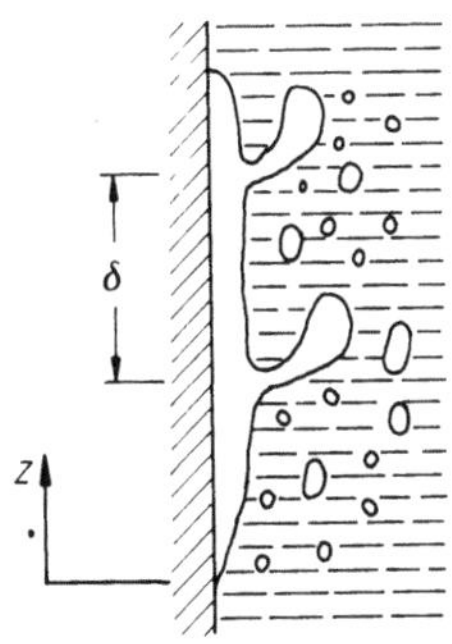

Fig. 3.13. Situation with film boiling at the Ball

Based on the equation by Bromley

$$\alpha_{\text{film, boil}} = 0.62\left[\frac{\lambda_v^3 \varrho_v(\varrho_l - \varrho_v)\Delta h_v g}{\eta_v(T_v - T_s)\delta_{\text{hor}}}\right]^{0.25} \tag{3.56}$$

Hsu [31] developed the equation

$$\alpha_{\text{film, boil, hor.}} = 1.456 \cdot 10^3 \cdot \exp(-3.76 \cdot 10^{-3} \cdot p^{0.1733})\{T_v - T_s\} + 0.62\left[\frac{\lambda_v^3 \varrho_v(\varrho_l - \varrho_v) g \Delta h_v}{\eta_v(T_v - T_s)\delta_{\text{hor}}}\right]^{0.25}. \tag{3.57}$$

This equation is valid for a horizontal orientation of the vapour film and contains the distance δ_{hor} between the vapour columns separating from the film which can be calculated from Eq. (3.58)

$$\delta_{\text{hor}} = 2\pi\left[\frac{\sigma}{g(\varrho_l - \varrho_v)}\right]^{0.5} \tag{3.58}$$

based on the theory for Taylor-instabilities. The second term of Eq. (3.57) represents a pressure correction. For vertical orientation Leonhard [32] uses the Helmholtz-instability for the distance of the separating bubbles representing a characteristic length δ_{vert}

$$\delta_{\text{vert}} = 16.24\left[\frac{\sigma^4 \Delta h_v^3 \eta_v^5}{\varrho_v(\varrho_l - \varrho_v)^5 g^5 \lambda_v^3 (T_{\text{wall}} - T_s)^2}\right]^{0.5} \tag{3.59}$$

which he then implements in Bromley's-equation Eq. (3.56). Sherman and Sabersky [33] made an interesting proposal for correlating heat transfer with film boiling at vertically orientated surfaces. They introduce a dimensionless heat transfer coefficient N_α

$$N_\alpha = \frac{\alpha}{\varrho_v c_v \left(\dfrac{\eta_v g}{\varrho_v}\right)^{1/3}} \tag{3.60}$$

which they describe as a function of the physical properties of the vapour in the film.

$$N_\alpha = 0.22\left(\frac{\varrho_l}{\varrho_v}\right)^{1/3} Pr_v^{-0.65}\left(\frac{c_v(T_{\text{wall}} - T_s)}{\Delta h_v}\right)^{-0.23}. \tag{3.61}$$

Compared with the literature describing nucleate boiling, papers on film boiling are much more rare, and measured values show larger tolerances. The reason for this is the high temperature under which the measurements have to be performed and also the difficulties of the measuring techniques under these high temperatures. The situation with respect to reliable data becomes even worse if one looks for heat transfer measurements in subcooled film boiling. With subcooled liquid. The fluiddynamic phenomena at the phase interface change remarkably because the condensation of the vapour at the subcooled liquid

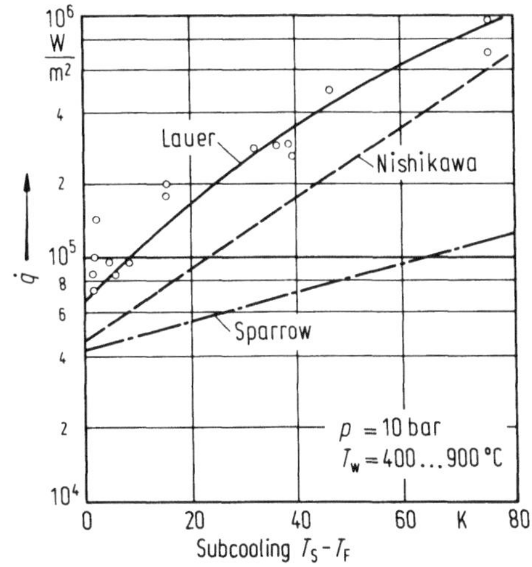

Fig. 3.14. Heat flux with film boiling in subcooled water at 10 bar

surface produces high acceleration forces towards the interface and also towards the wall which can induce local rewetting for a short period. Also the turbulence in the vapour film is strongly increased. Lauer [8], Nishikawa [34] and Sparrow [35] showed within a certain agreement and as demonstrated in Fig. 3.14 that with film boiling in subcooled liquids much higher heat fluxes can be transferred than in saturated liquids.

3.3.4 Transition Boiling

As we saw in Fig. 3.1 when we discussed the Nukijama-curve, there is a region between critical heat flux and fully developed film boiling which is called transition boiling. In this region a heat transfer mechanism can be observed which gives decreasing heat transfer coefficients with increasing temperature difference between wall and fluid. Visual observations of the fluiddynamic situation give the impression as if the hot wall would be temporally wetted so that periods of nucleate boiling change with that of film boiling.

A detailed literature survey on heat transfer under the conditions of transition boiling is given by Groeneveld and Fung [36]. These authors defined the transition boiling as a combination of unstable film boiling and unstable nucleate boiling. The later one plays the more important role for the heat transfer process from the wall to the fluid. This means that correlations used for nucleate boiling could be also of some relevance for describing transition boiling.

Starting from the idea that the heat transfer with transition boiling is strongly depending on the short-time rewetting of the wall and by this on bubble boiling, in the literature frequently correlations can be found of the form

$$q = Ae^{-B(\Delta T)} \tag{3.62}$$

This form implements the difficulty to describe the coefficients A and B in a physically proper way. A simple method is to formulate A and B as a function of the physical properties of the fluid. However, also the temperature difference ΔT between the rewetting temperature—the Leidenfrost-temperature—and the saturation temperature plays a role. As an example of these kinds of equations here the correlation by Tong [9] is presented

$$\dot{q} = \dot{q}_{\text{sing. phase}} \cdot \exp\left[-0.001 \frac{X_{\text{equ}}}{\frac{dx_{\text{equ}}}{dz}} (\Delta T/100) \right]^{1+0.0016\,\Delta T} \tag{3.63}$$

This equation was originally developed to describe the cooling phenomena in nuclear reactors under the conditions of a loss of coolant accident. Applying Eq. (3.63) one has to know the heat flux $q_{\text{sing. phase}}$ which would exist with purely single phase liquid flow, i.e. without boiling. This heat flux can be calculated by using the equations given in the previous chapters. In addition the change of the quality $x_{\text{equ.}}$ with respect to the coordinate in flow direction must be known. This can be calculated by assuming thermal equilibrium.

Another form of correlating heat transfer coefficients with transition boiling was presented by Dhir [37]. Dhir evaluated his equation from his measurements where he immersed cooper- and silver-spheres into water which was subcooled up to 60 K. Dhir used the Jakob-number as describing parameter.

3.3.5 Critical Heat Flux

For applying heat transfer correlations under high heat flux densities one has to know the boiling mode—nucleate, transition or film boiling. There exist numerous correlations in the literature predicting the critical or peak heat flux with pool boiling. Several examples for these correlations could be cited, however, here only the form elaborated by Zuber [38] shall be presented

$$\dot{q}_{\text{crit}} = 0.131\,\Delta h_{\text{v}} \varrho_{\text{v}} \left[\frac{\sigma(\varrho_{\text{l}} - \varrho_{\text{v}})g}{\varrho_{\text{v}}^2} \right]^{1/4} \tag{3.64}$$

The constant in Eq. (3.64) as originally proposed by Zuber was 0.131, however, this value is generally considered to be low and Rohsenow [39] proposed the value 0.18.

While this correlation describes quite well sets of data measured in water it is not accurate for all systems. For example, the predicted critical heat fluxes are widely divergent for boiling liquid oxygen.

Considerable progress has been made in accounting for the effects of heater geometry. Sun and Lienhard [40], Lienhard and Dhir [41] and Lienhard and Riherd [42], re-examined the model by Zuber and found that the vapour-removal configuration varies according to the heater geometry and size. A simple

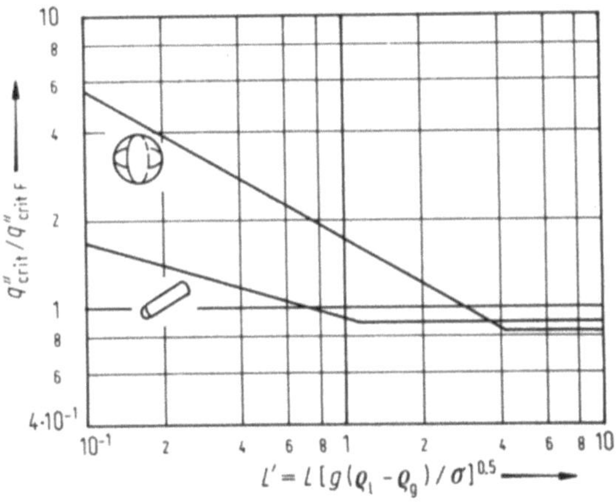

Fig. 3.15. Critical heat flux in water with various heater configurations according to Bergles [48]

correction can be made by using a characteristic length L'

$$L' = L[g(\varrho_l - \varrho_v)/\sigma]^{1/2}. \tag{3.65}$$

The correlation for this characteristic length then serves as a correction factor for the critical heat flux taking into account the heater geometry and its size

$$\dot{q}''_{crit}/\dot{q}''_{crit,F} = f(L'). \tag{3.66}$$

Results of semi-empirical correlations are summarized in Fig. 3.15. They indicate that the correlation curves vary according to geometry for small heaters and are also generally different for large heaters. These correlation curves were based on a large number of data points.

Although the experimental verification is difficult, data for large flat plates (facing up) tend to substantiate the semi-empirical prediction for an infinite flat plate

$$\dot{q}''_{crit}/\dot{q}''_{crit,F} = 1.14. \tag{3.67}$$

Nucleate boiling and the associated hydrodynamic instabilities vanish for very small heaters ($L' < 0.01$ mm) and therefore with increasing heat flux natural convection proceeds directly into film boiling.

For forced convection boiling the prediction of the critical heat flux is much more complicated and strongly depending on the heater geometry as well as on the fluiddynamic conditions. For finding correlations to predict critical heat flux under these conditions reference is made to the book by Collier [43].

3.3.6 Immersion Cooling

With immersion cooling, as it is usual in the quench hardening treatment, several thermo- and fluiddynamic modes—starting from film boiling via transition boiling and nucleate boiling up to single-phase free convection—can exist. The heat transfer process is primarily depending on the surface temperature of the immersed body, however, also on its thermal conductivity near the surface and its surface roughness.

With hardening usually the temperature of the body is known before dipping it into the quenching bath and in most cases this temperature is above the Leidenforst-temperature. So film boiling will start immediately after dipping.

The lower limit of stable film boiling corresponds to the onset of liquid-solid contact. Numerous analyses have been made to predict this condition, generally based on hydrodynamic stability theory similar to that employed in determining the critical heat flux. For a flat horizontal surface the minimum heat flux can be calculated by an equation given by Zuber [44]

$$q_{\min} = C \Delta h_v \varrho_v \left[\frac{\sigma g(\varrho_l - \varrho_v)}{(\varrho_l + \varrho_v)^2} \right]^{1/4} \tag{3.68}$$

where C is variably given as 0.177 [46] or 0.09 [47].

For small bodies—wires—it is necessary to account for curvature effects and especially for the effect of surface tension in the transverse direction upon the Taylor- instability of the interface. Lienhard and Wong [48] have suggested the following semi-empirical equation

$$q_{\min} = 0.057 \frac{\varrho_v \Delta h_v}{R} \left[\frac{2g(\varrho_l - \varrho_v)}{(\varrho_l + \varrho_v)} + \frac{\sigma}{(\varrho_l + \varrho_v)^2} \right]^{1/2} \cdot \left[\frac{g(\varrho_l + \varrho_v)}{\sigma} + \frac{1}{2R^2} \right]^{-3/4}. \tag{3.69}$$

This equation should be used for curvatures of small radii (below 2 nm) only. For larger curvatures Eq. (3.68) may be applied. However, Kovalev [49] notes that this equation overpredicts the data for water on clean surfaces at pressures above atmospheric pressure.

In any case these equations cannot be relied upon for systems where the liquid contains impurities and the surface exhibits some degree of contamination. Oxidation increases wettability.

Also from another point of view it is difficult to calculate the minimum heat flux of film boiling. The energy balance demands that the heat flux transported by conduction in the solid material to its surface must be the same as the heat flux transported by film boiling or nucleate boiling from the surface to the fluid. So the moment when film boiling ends has always to be determined in an iterative way by using Eq. (3.68) and by calculating the conductive heat flux to the surface.

The period of transition boiling—after the minimum heat flux was reached—is usually very short and can be neglected for most practical cases. Therefore, one can assume that when the vapour film has collapsed, nucleate boiling starts. For the following period the heat transfer coefficient can be calculated with the correlations given in the chapter on nucleate boiling. Here again, we have to observe the balance between heat conducted to the wall and heat transported from it by nucleate boiling. When the surface temperature of the wall reaches the saturation temperature nucleation stops and the heat transfer is furtheron managed by liquid free convection only. For calculating this period reference is made to the chapter on free convective heat transfer.

List of Symbols

A	area
A	constant
a	pitch-diameter-ratio perpendicular to flow
a	thermal diffusivity
b	pitch-diameter-ratio longitudinal to flow
Bo	boiling number
c	specific heat capacity
c_F	specific heat capacity of the liquid
C_F	factor representing properties of liquid
C_w	factor representing properties of a heated wall
D	diameter
CHF	critical heat flux
DNB	departure from nucleate boiling
f	bubble departing frequency
F	correction factor
Gr	Grashof number
g	standard acceleration due to gravity
h_l	enthalpy of liquid
h_v	enthalpy of vapour
k	roughness depth
L	characteristic length
M	constant
$\dot{M}$	mass flow
$\dot{m}$	mass flow rate density
N	constant
Nu	Nusselt number
N_α	dimensionless heat transfer coefficient
p_l	pressure of liquid
p_v	pressure of vapour
Pe	Peclet number
Pr	Prandtl number
$\dot{q}$	heat flux density
$\dot{Q}$	heat flow rate
R	bubble Radius
Re	Reynold number

S	correction factor
T	temperature
T_s	saturation temperature
T_v	vapour temperature
$T_v - T_s$	superheating temperature
v_l	specific volume of liquid
v_v	specific volume of vapour
w	velocity
$\dot{x}$	specific vapour flow rate or vapour quality
X	Martinelli parameter
α	heat transfer coefficient
β	contact angle of bubble
β	volume coefficient of expansion
δ_{hor}	distance in horizontal direction
δ_T	thickness of thermal boundary-layer
δ_W	boundary-layer-thickness of velocity field
η	dynamic viscosity
λ	heat conductivity
ν	kinematic viscosity
Φ	two phase flow multiplier
ψ	friction factor
ϱ_l	density of liquid
ϱ_v	density of vapour
σ	surface tension
τ	shear stress
ϑ	temperature

List of Subscripts

bulk	condition in the core
boil	boiling
bub	bubble
conv.	convective
crit	critical
equ	equivalent
hor	horizontal
l	liquid
max	maximum
min	minimum
s	surface
sing.phase	single phase
surf.boil	surface boiling
v	vapour

x x-direction
y y-direction
z z-direction

References

1. Jakob M (1936) In: Mech. Engng. 58: 643–660, 729–739
2. Fritz W (1935) In: Physik. Z. 36: 379–384
3. Jakob W, Linke (1935) In: Physik. Z. 36: 267–280
4. Bosnjakovic F (1930) In: Techn. Mech. Thermo-Dynam. 1: 358–362
5. Nukiyama VJ (1934) In: Jap. Mech. Engr. 37: 53–54, 367–374
6. Yao SC, Henry RE (1978) Experiments of quenching under pressure. Proc. 6th Intern. Heat Transfer Conf., Toronto, Canada
7. Hein D (1980) Modellvorstellungen zum Wiederbenetzen durch Fluten. Diss., Univ. Hannover, Inst. f. Verfahrenstechn.
8. Lauer H (1976) Untersuchung des Wärmeübergangs und der Wiederbenetzung beim Abkühlen heißer Metallkörper. Diss., Univ. Hannover, Inst. f. Verfahrenstechn.
9. Tong LS, Young JD (1974) A phenomenological transition and film boiling heat transfer correlation. Proc. 5th Intern. Heat Transfer Conf., Tokio
10. Forster HK, Zuber N (1954) In: J. Appl. Phys. 25: 474–478
11. Han CY, Griffith P (1965) In: Int. J. Heat and Mass Transfer 8: 887–904, 905–914
12. Plesset MS, Zwick SA (1954) In: J. Appl. Phys. 25: 493–500
13. Beer H (1969) In: Progr. Heat Mass Transfer 2: 311–370
14. Winter ERF, Matekunas FA (1971) An Inteferometric study of nucleate boiling. Int. Symp. on Two-Phase Systems, 29, Aug. – 2, Sept. Haifa, Israel
15. Mayinger F, Panknin W (1974) Holography in heat and mass transfer, Vol. 6, p. 28–43 Proc. 5th Int. Heat Transfer Conf., Tokio
16. Grigull U (1961) Wärmeübertragung, Springer-Verlag, Berlin
17. Schlünder EU (1983) Einführung in die Wärmeübertragung, 4. Aufl., S. 103. Vieweg (Verlag), Braunschweig
18. Hausen H (1959) Neue Gleichungen für die Wärmeübertragung bei freier oder erzwungener Konvektion. In: Allgem. Wärmetechnik 9: 75–79
19. Hausen H (1969) Bemerkung zur Veröffentlichung von A. Hackl und W. Gröll, Zum Wärmeübergangsverhalten zähflüssiger Öle. In: Verfahrenstechn. 3: 355, 480 (Berichtigung)
20. Choi H, Rohsenow WM (1961) Heat, mass und momentum transfer. Englewood Cliffs, Prentice Hall
21. Stephan K, Preußer P (1979) Wärmeübergang und maximale Wärmestromdichte beim Behältersieden binärer und ternärer Flüssigkeitsgemische. In: Chem.-Ing.-Techn. MS 649/79, Synopse Chem.-Ing.-Techn. 51: 37
22. Rohsenow WM (1963) Modern Developments in Heat Transfer, p. 85–158. Academic Press, New York
23. Vaihinger D, Kaufmann WD (1972) In: Chem.-Ing.-Techn.: 921–927
24. Calus WF et al. (1972) In: Chem. Engng. J. 6: 223–250
25. Chen JC (1966) Correlation for boiling heat transfer to saturated liquids in convective flow. Int. Eng. Chem. Process Design and Development 5, p. 322
26. Forster HK, Zuber N (1955) Dynamics of vapour bubbles and boiling heat transfer. In: AIChE Journ. 1(4): 531–535
27. Bromley LA (1950) In: Heat transfer in stable film boiling. Chem. Eng. Progr. 46: 221–227
28. Berenson PI (1962) Experiments on pool boiling heat transfer. In: Int. J. of Heat and Mass Transfer 5: 985–999
29. Hsu YY (1972) In: Adv. Cryogenic Engng. 17, p. 361
30. Bressler RG (1972) Adv. Cryogenic Engng. 17, p. 382
31. Hsu YY (1975) Tentative correlations of reflood heat transfer. LOCA-research Highlights, (Apr. 1 – June 30)

32. Leonhard JE, Sun KH, Dix GE (1977) Solar and nuclear heat transfer. In: AIChE Symp. 73, No. 164: 7
33. Shermann DC, Sabersky RH. Natural convection film boiling on a vertical surface. Persönliche Mitteilung
34. Nishikawa KT, Ito T (1966) In: Int. J. Heat and Mass Transfer 9: 103
35. Sparrow EM, Cess RD (1962) In: J. Heat Transfer 84: 55
36. Groeneveld DC, Fung KK (1976) Forced convective transition boiling. Review of literature and comparison of prediction methods. In: AECL-Report p. 5543
37. Dhir VK (1978) Study of transient transition boiling heat fluxes from spheres subjected to forced vertical flow. Proc. 6th Int. Heat Transfer Conf., Toronto, Canada
38. Zuber N, Tribus M, Westwater JW (1961) The hydrodynamic crisis in pool boiling of saturated and subcooled liquids. Int. Develop. in Heat Transfer, Part II, ASME, p. 230–235
39. Rohsenow WM and Griffith P (1956) Correlation of maximum heat transfer data for boiling of saturated liquids. In: Chem. Eng. Progr., Symp. Series 52, No 18: 47–49
40. Sun KH, Lienhard IH (1970) The peak pool boiling heat flux on horizontal cylinders. In: Int. J. of Heat and Mass Transfer 13: 1425–1439
41. Lienhard IH, Dhir VK (1973) Hydrodynamic prediction of peak pool boiling heat fluxes from finite bodies. In: J. Heat Transfer 95: 152–158
42. Lienhard IH, Dhir VK, Riherd DM (1973) Peak boiling heat flux measurements on finite horizontal flat plates. In: J. Heat Transfer 95: 477–482
43. Collier JG (1981) Convective Boiling and Condensation, 2nd edn. McGraw Hill, New York
44. Zuber N (1958) On stability of boiling heat transfer. In: Trans ASME 80: 711–720
45. Berenson PJ (1961) Transition boiling heat transfer from an horizontal surface. In: J. Heat Transfer 83: 351–358
46. Lienhard IH, Wong PTY (1964) The dominant instable wavelength and minimum heat flux during film boiling on an horizontal cylinder. In: J. Heat transfer 86: 220–226
47. Kovalev SA (1966) An investigation of minimum heat fluxes in pool boiling of water. In: Int. J. of Heat and Mass Transfer 9: 1219–1226
48. Bergles AE (1975) Burnout in Boiling Heat Transfer, Part I Pool Boiling Systems. In: Nuclear Safety 16: 29–42

4 Heat Transfer During Cooling of Heated Metallic Objects with Evaporating Liquids

R. JESCHAR, E. SPECHT, CHR. KÖHLER

4.1 Mechanism of Heat Transfer

Generally, hot metals are quenched by pouring over liquids, by immersing the metals into a liquid bath or by spraying them with a liquid. These three quenching techniques are shown in Fig. 4.1. Independent of the type of quenching a stable vapour film forms over the hot surface. This has an insulating effect so that the cooling during this film boiling is relatively low. The vapour film collapses as soon as the temperature of the surface falls below the Leidenfrost temperature. According to Fig. 2.1 in the subsequent ranges of partial film boiling and nucleate boiling a considerably larger heat flow is transferred, allowing the solid to cool rapidly. The heat loss during film boiling and the possibilities to influence the Leidenfrost temperature are dependent on the type of quenching technique.

Figure 4.2 shows the principle mechanism of heat transfer for the three cooling techniques described above. In all three cases the heat flow $\dot{q}$ emitted from the surface is transported through the vapour film by conduction ($\dot{q}_\lambda$) and by radiation ($\dot{q}_\varepsilon$). At the vapour liquid interface the heat flow $\dot{q}$ is partly given to the liquid by convection ($\dot{q}_\alpha$), the remaining part being combined as enthalpy of vaporization ($\dot{q}_v$). The temperature curves are also given in Fig. 4.2. Within the film the vapour temperature falls from the surface temperature ϑ_w to the boiling temperature ϑ_b of the liquid. From this temperature of the vapour–liquid interface the temperature decreases further to the ambient temperature ϑ_L of the liquid. The thickness δ of the vapour film and, therefore, the transferred heat flow are influenced greatly by the extent of the heat flow $\dot{q}_\alpha$ given into the liquid, as will be explained below. On the other hand, this convective heat transfer is determined from the velocity profile in the liquid. Hence, in the lower part of Fig. 4.2 the velocity profiles are also given qualitatively. It can be seen that these profiles differ for all three quenching techniques. For film quenching, an approximately uniform velocity exists in the draining liquid over the cross section. The vapour flow is impressed by the liquid and as a result, an approximately linear velocity course adjusts. In immersion quenching the flow in the liquid is formed by the vapour flow, insofar as the bath was initially inactive. As a result of the buoyancy the vapour flows in an upward direction, taking the liquid with it, as can be seen from the velocity curve. If required, this liquid

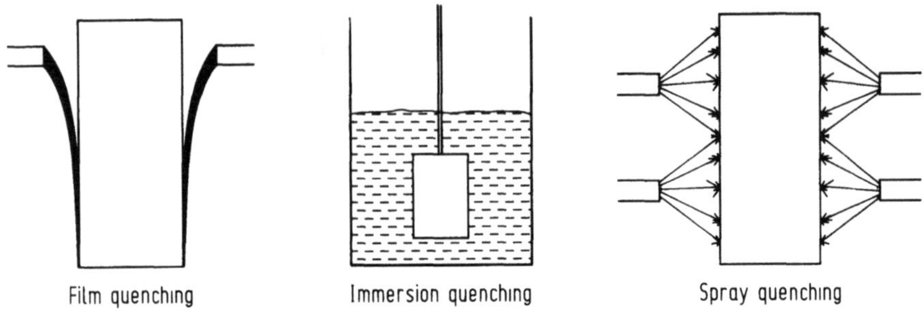

Fig 4.1. Quenching techniques

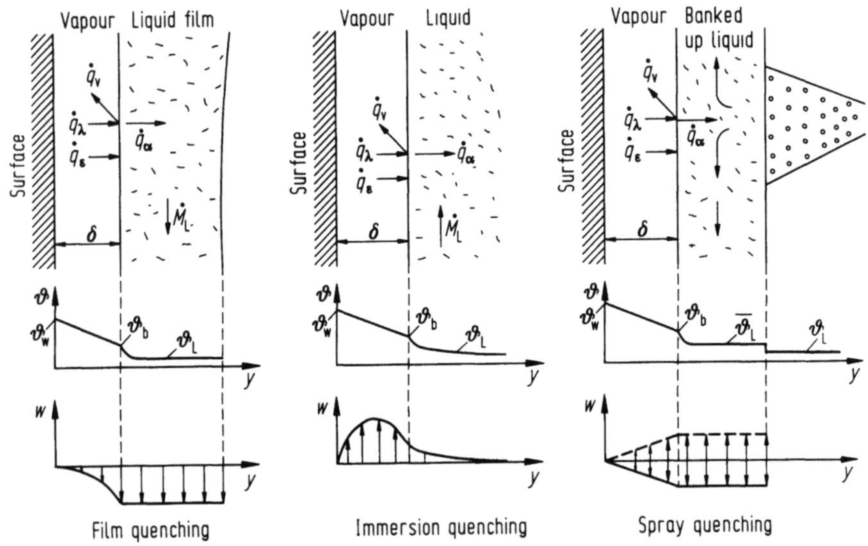

Fig. 4.2. Mechanism of heat transfer during film boiling

flow, which was produced by the vapour, can be superposed by a forced bath flow. In spray quenching the impinging liquid can flow off either in an upward or in a downward direction. For this quenching technique the flow of the vapour is also impressed by that of the liquid.

The hitherto qualitatively explained mechanism of heat transfer is described in the following mathematically for the example of film quenching.

4.2 Film Quenching

For the heat flow conducted through the vapour film the following equation is valid for film quenching and for the other two quenching techniques

$$\dot{q}_\lambda = \frac{\lambda_v}{\delta} \cdot (\vartheta_w - \vartheta_b), \tag{4.1}$$

where λ_v is the mean thermal conduction coefficient of the vapour film, δ its thickness, ϑ_w the wall temperature and ϑ_b the boiling temperature. In most cases the radiation superposing this heat flow can be neglected, as is shown in a following example. At the vapour–liquid interface the heat flow is partly transported further to the liquid by convection and the remaining part is combined as enthalpy of vaporization

$$\dot{q}_\lambda = \dot{q}_\alpha + \dot{m}_v \cdot \Delta h, \tag{4.2}$$

where $\dot{m}_v$ is the evaporating mass flow and Δh the enthalpy of vaporization.

To calculate the heat flow $\dot{q}_\alpha$ it is assumed that the velocity w is uniform over the cross section of the liquid. If w_0 is the initial velocity at which the liquid reaches the workpiece, then the velocity along the workpiece due to gravity is

$$w = \sqrt{w_0^2 + 2 \cdot g \cdot x}, \tag{4.3}$$

where g is the acceleration due to gravity and x the flow route. From the mass balance

$$\frac{\partial w}{\partial x} + \frac{\partial v}{\partial y} = 0 \tag{4.4}$$

the following is obtained for the velocity component v transverse to the direction of flow

$$v = -\frac{g \cdot y}{\sqrt{w_0^2 + 2 \cdot g \cdot x}}. \tag{4.5}$$

From the Fourier differential equation

$$w \cdot \frac{\partial \vartheta}{\partial x} + v \cdot \frac{\partial \vartheta}{\partial y} = a_L \cdot \frac{\partial^2 \vartheta}{\partial y^2} \tag{4.6}$$

(a_L thermal diffusivity of the liquid) results for the temperature distribution in the liquid

$$\frac{\vartheta - \vartheta_b}{\vartheta_L - \vartheta_b} = \operatorname{erf}\left(\sqrt{\frac{3}{4}} \cdot \sqrt{\frac{\frac{g}{a_L} \cdot y^2}{\sqrt{w_0^2 + 2 \cdot g \cdot x} - \frac{w_0^3}{w_0^2 + 2 \cdot g \cdot x}}}\right), \tag{4.7}$$

where ϑ_b is the boiling temperature at $y=\delta$ and ϑ_L the liquid temperature at $y=\infty$ and $x=0$. For the heat-flux density conducted at the phase interface into the liquid (λ conductivity)

$$\dot{q}_\alpha = \lambda_L \cdot \left.\frac{\partial \vartheta}{\partial y}\right|_{y=\delta} \tag{4.8}$$

is valid. Using Eq. (4.7) yields (ρ density, c spec. heat capacity)

$$\dot{q}_\alpha = \sqrt{\frac{3}{\pi}} \cdot (\vartheta_b - \vartheta_L) \cdot \sqrt{\frac{g \cdot \lambda_L \cdot \rho_L \cdot c_L}{\sqrt{w_0^2 + 2 \cdot g \cdot x} - \dfrac{w_0^3}{w_0^2 + 2 \cdot g \cdot x}}}. \tag{4.9}$$

If the acceleration of the liquid due to gravity can be neglected, e.g. if the initial velocity w_0 is high and the flow route x is short, then the above equation results in

$$\dot{q}_\alpha(g \cdot x \ll w_0^2) = \sqrt{\frac{\lambda_L \cdot \rho_L \cdot c_L}{\pi}} \cdot \sqrt{\frac{w_0}{x}} \cdot (\vartheta_b - \vartheta_L). \tag{4.10}$$

This is the equation for the non-steady heat conduction ($t = x/w_0$) in a semi-infinite solid [1].

To calculate the evaporating mass flow $\dot{m}_v$ it is assumed that the velocity distribution in the vapour film is linear

$$w_v = \sqrt{w_0^2 + 2 \cdot g \cdot x} \cdot \frac{y}{\delta}. \tag{4.11}$$

As such, the vapour mass flow in x-direction is

$$\dot{M}_x = \int_0^\delta \rho_v \cdot w_v \cdot b \cdot dy = \tfrac{1}{2} \cdot \rho_v \cdot \sqrt{w_0^2 + 2 \cdot g \cdot x} \cdot b \cdot \delta, \tag{4.12}$$

where ρ_v is the mean vapour density, b the width of the vapour film and y the coordinate vertical to the flow route. The evaporating mass flow is equal to the change in mass of the vapour in the direction of flow

$$\dot{m}_v = \frac{d\dot{M}_x}{b \cdot dx}. \tag{4.13}$$

With this, the differential equation for the local change in the thickness of the vapour film can be determined from (4.1), (4.2) and (4.9)

$$\frac{\delta}{x} \cdot \frac{d\delta}{dx} + \left(\frac{\delta}{x}\right)^2 \cdot \frac{Gn_x}{Pe_x^2 + 2 \cdot Gn_x} + \frac{\delta}{x} \cdot \sqrt{\frac{12}{\pi} \cdot \frac{\rho_L}{\rho_v}} \cdot Ph \cdot \sqrt{\frac{Gn_x}{(Pe_x^2 + 2 \cdot Gn_x)^{3/2} - Pe_x^3}}$$
$$= 2 \cdot \frac{\rho_L}{\rho_v} \cdot \frac{\lambda_v}{\lambda_L} \cdot \frac{Ph}{\theta} \cdot \frac{1}{\sqrt{Pe_x^2 + 2 \cdot Gn_x}}. \tag{4.14}$$

In this equation are:

$$Pe_x = \frac{w_0 \cdot x}{a_L} \quad \text{(Peclet number)}, \tag{4.15}$$

$$Gn_x = \frac{g \cdot x^3}{a_L^2} \quad \text{(Gravity number = Galilei number} \cdot \text{Prandtl number}^2\text{)}, \tag{4.16}$$

$$Ph = \frac{c_L \cdot (\vartheta_b - \vartheta_L)}{\Delta h} \quad \text{(phase change number)}, \tag{4.17}$$

and

$$\theta = \frac{\vartheta_b - \vartheta_L}{\vartheta_w - \vartheta_b} \quad \text{(dimensionless temperature)}. \tag{4.18}$$

The solution of the differential equation is

$$\frac{\delta}{x} = \frac{2}{\sqrt{3 \cdot \pi}} \cdot \frac{\rho_L}{\rho_v} \cdot Ph \cdot \left(\sqrt{1 + \frac{\pi}{Ph \cdot \theta} \cdot \frac{\lambda_v}{\lambda_L} \cdot \frac{\rho_v}{\rho_L}} - 1 \right) \cdot \sqrt{\frac{(Pe_x^2 + 2 \cdot Gn_x)^{3/2} - Pe_x^3}{Gn_x \cdot (Pe_x^2 + 2 \cdot Gn_x)}}. \tag{4.19}$$

The equation for the heat-flux density conducted through the vapour layer can subsequently be derived from (4.1)

$$\dot{q}_\lambda = \frac{\sqrt{3 \cdot \pi}}{2} \cdot \frac{\rho_v}{\rho_L} \cdot \frac{\lambda_v}{x} \cdot \frac{1}{Ph} \cdot \frac{\vartheta_w - \vartheta_b}{\sqrt{1 + \frac{\pi}{Ph \cdot \theta} \cdot \frac{\lambda_v}{\lambda_L} \cdot \frac{\rho_v}{\rho_L}} - 1} \cdot \sqrt{\frac{Gn_x \cdot (Pe_x^2 + 2 \cdot Gn_x)}{(Pe_x^2 + 2 \cdot Gn_x)^{3/2} - Pe_x^3}}. \tag{4.20}$$

The relation

$$\frac{\dot{q}_\alpha}{\dot{q}_\lambda} = \frac{2}{\pi} \cdot \frac{\lambda_L \cdot \rho_L}{\lambda_v \cdot \rho_v} \cdot Ph \cdot \theta \cdot \left(\sqrt{1 + \pi \cdot \frac{\lambda_v}{\lambda_L} \cdot \frac{\rho_v}{\rho_L} \cdot \frac{1}{Ph \cdot \theta}} - 1 \right) \tag{4.21}$$

indicates which part of the heat flow conducted through the vapour layer is transferred further to the liquid by convection. Figure 4.3 shows this relation for water as a function of the temperature of the liquid with the surface temperature of the solid as parameter. Hence, for water temperatures up to ca. 80 °C the heat flow $\dot{q}_\lambda$ is almost entirely transferred to the liquid. It is only at liquid temperatures in the range of the boiling temperature that a notable part of the heat flow is converted to the enthalpy of vaporization.

For technical liquid temperatures far below the boiling temperature it can be assumed that the heat flow conducted through the vapour layer is almost fully transported to the liquid. The fraction in the first root of Eq. (4.19) then accepts values which are considerably smaller than one. For random values z

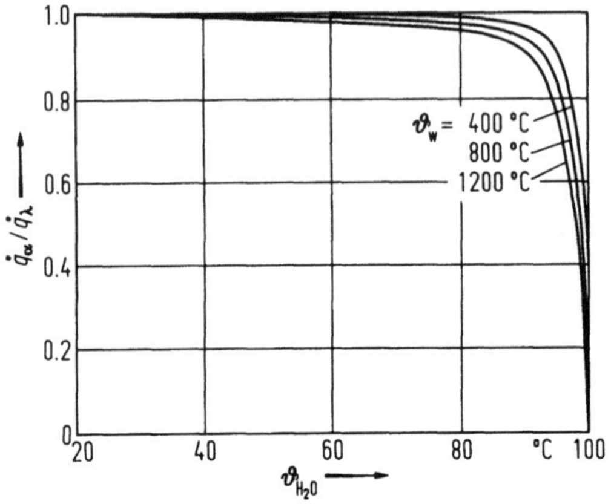

Fig. 4.3. Portion of total heat flux which is transported into the liquid

the approximation

$$\sqrt{1+z}-1 \approx \tfrac{1}{2}\cdot z \tag{4.22}$$

is valid with an inaccuracy of less than 3% for $z < 0.12$. With this approximation Eq. (4.20) is converted into Eq. (4.9) and neglecting the initial velocity w_0 ($Pe_x = 0$) the thickness of the vapour film is derived from Eq. (4.19)

$$\frac{\delta}{x}(\vartheta_L \ll \vartheta_b) = \sqrt[4]{\frac{2\cdot\pi^2}{9}\cdot\frac{\lambda_v}{\lambda_L}\cdot\frac{1}{\sqrt[4]{Gn_x}}\cdot\frac{\vartheta_w-\vartheta_b}{\vartheta_b-\vartheta_L}}. \tag{4.23}$$

From this equation it is evident that the thickness of the film is proportional to the temperature difference $\vartheta_w - \vartheta_b$. Hence, the thickness of the film adjusts such that for a decreasing surface temperature the transferred heat-flux density remains constant according to Eq. (4.1). An influence of the surface temperature is observed only at very high water temperatures, when the heat is no longer transported predominantly to the liquid and as such causes the liquid to vaporize.

Figure 4.4 shows the calculated thickness of the vapour film for water as a function of the surface temperature. It can be seen that the film thicknesses are in the μm-range and are as such extremely thin when compared to the liquid film. The temperature boundary layer in the liquid film itself, i.e. the drop in temperature from the state of boiling to the ambient state, is of the same order of magnitude as in the vapour film.

The radiation emitted from the solid is absorbed in the liquid approximately in the range of the temperature boundary layer. If the liquid is undercooled, the vapour film is not influenced by the radiation. Hence, the total heat flow $\dot{q}$ emitted from the solid is additively composed of the conduction and radiation parts

$$\dot{q} = \dot{q}_\lambda + \dot{q}_\varepsilon. \tag{4.24}$$

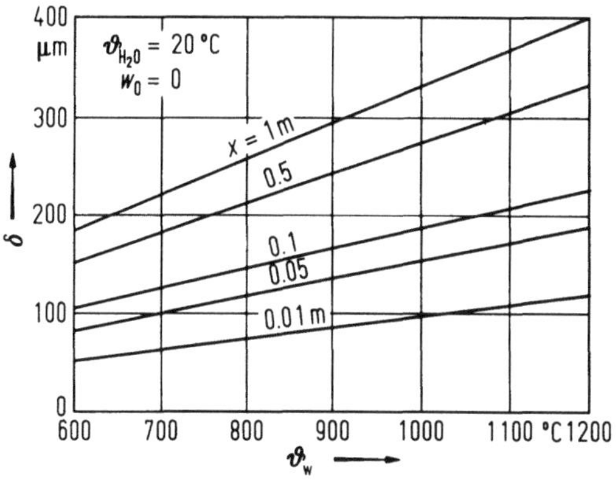

Fig. 4.4. Indications for the magnitude of vapour films in water of 20 °C

In many cases, however, the heat flow emitted by radiation can be neglected, as is shown in the following example. If a film thickness of 70 µm is taken as a basis and assuming that the surface temperature is 800 °C, a heat-flux density of 750 kW/m^2 can be derived from Eq. (4.1) using the mean heat-transfer coefficient of the vapour of 0.075 W/(m·K). Black radiation of the same temperature has, on the other hand, a heat-flux density of only 75 kW/m^2.

The given results are confirmed by further model calculations which are based on the boundary layer equations for the momentum and energy transport and which, furthermore, take the temperature dependency of the material values into account [2].

In many technical cases, it is not so much the local heat-flux density which is of interest rather the mean heat-flux density which is transported along the length of the solid L

$$\dot{q}_\alpha = \frac{1}{L} \cdot \int_0^L \dot{q}_\alpha \cdot dx \tag{4.25}$$

for which the following equation can be derived from Eq. (4.9)

$$\bar{\dot{q}}_\alpha = \frac{2}{\sqrt{3 \cdot \pi}} \cdot \frac{\lambda_L}{x} \cdot (\vartheta_b - \vartheta_L) \cdot \sqrt{\frac{(Pe^2 + 2 \cdot Gn)^{3/2} - Pe^3}{Gn}}. \tag{4.26}$$

The Peclet and Galilei numbers in this equation are to be formed using the length of the solid L, as according to Eq. (4.15) and (4.16), in place of the flow route x. If a heat-transfer coefficient is defined according to

$$\dot{q}_\alpha = \alpha \cdot (\vartheta_b - \vartheta_L) \tag{4.27}$$

then Eq. (4.26) results in the following as Nusselt function for the range of film boiling

$$Nu = \frac{2}{\sqrt{3 \cdot \pi}} \cdot \sqrt{\frac{(Pe^2 + 2 \cdot Gn)^{3/2} - Pe^3}{Gn}}. \tag{4.28}$$

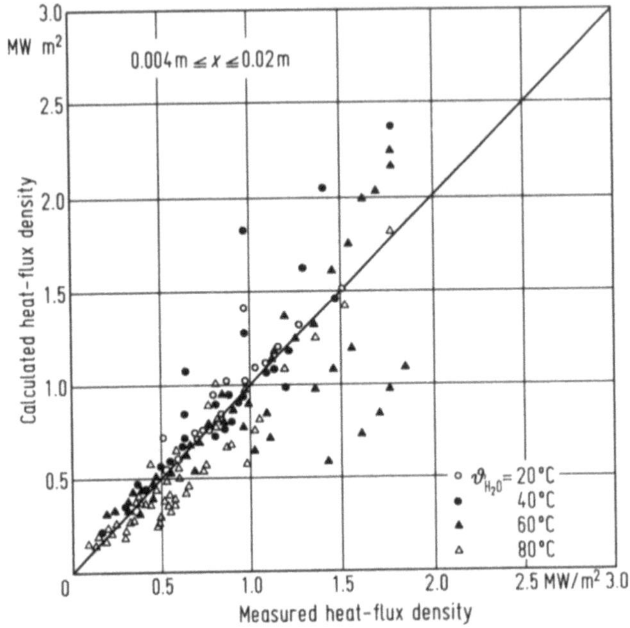

Fig. 4.5. Comparison of measured and calculated heat flux densities in the range of film boiling for film quenching [2]

If the acceleration of the liquid can be neglected, this results in

$$Nu(w_0^2 \gg g \cdot L) = \frac{2}{\sqrt{\pi}} \cdot \sqrt{Pe} \tag{4.29}$$

If, on the other hand, the initial velocity can be neglected,

$$Nu(w_0^2 \ll g \cdot L) = \sqrt[4]{\frac{128}{9 \cdot \pi^2}} \cdot \sqrt[4]{Gn}. \tag{4.30}$$

is valid. In the first case the heat-flux density decreases with the dimensions of the solid proportional to $L^{1/2}$, in the second case only proportional to $L^{1/4}$.

Equation (4.26) is well confirmed by experimental results, as Fig. 4.5 shows. According to [2], however, for flow routes $x < 10\,\text{mm}$ Eq. (4.26) results in too small heat-flux densities, forcing $\dot{q}$ to be multiplied by the correction factor

$$k = 1 + 3.1 \cdot \exp\left(-170 \frac{x}{m}\right). \tag{4.31}$$

The cooling velocity in the range of film boiling can technically only be raised by an increase in the initial velocity w_0, as the flow route x is specified by the dimensions of the workpiece.

Quantitative statements on the Leidenfrost temperature, the DNB-temperature and the maximum heat-flux density in the range of nucleate boiling cannot

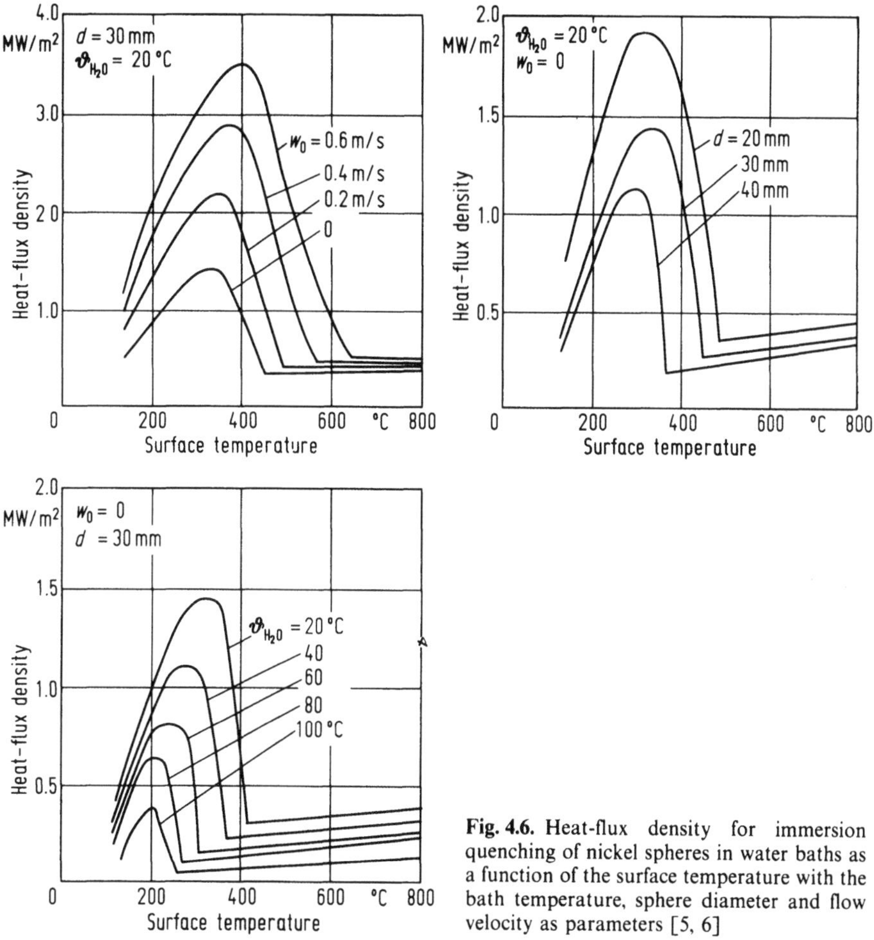

Fig. 4.6. Heat-flux density for immersion quenching of nickel spheres in water baths as a function of the surface temperature with the bath temperature, sphere diameter and flow velocity as parameters [5, 6]

be made at the present time. All three variables, however, have higher values than for immersion and spray quenching which are dealt with below.

4.3 Immersion Quenching

The most important variables in heat transfer for immersion quenching are given in Fig. 4.6 for the example of a nickel sphere dipped into a water bath. In this figure the heat-flux density is plotted as a function of the wall surface temperature. On the left, the parameter is the liquid velocity, in the middle the diameter of the sphere and on the right the bath temperature. In the range of stable film boiling the heat-flux density for static baths ($W_0 = 0$) increases linearly with the surface temperature. As calculations of the boundary layer for immersion quenching have shown [3, 4], this quenching method also involves the transported heat flow up to water temperatures of ca. 70 °C being determined only by the heat flow being transferred to the liquid by convection. The higher the surface temperature is, the stronger is the buoyancy flow in the vapour film and as such its dragging effect on the liquid. An increasing flow velocity of the liquid results in an increase in the convectively transferred heat flow. If the liquid flow caused by the dragging effect is superposed by a forced flow, then, as shown in the left-hand side of the figure, the heat-flux density becomes independent of the surface temperature, as is also the case in film quenching. If a heat-transfer coefficient is defined according to Eq. (4.27) based on the difference between the boiling and liquid temperatures, the result is a Nusselt number which is proportional to the fourth root of the Grashof number [3]. For water is valid:

$$Nu = \frac{4}{3} \cdot \left(0.25 + 0.003 \frac{\vartheta_w}{^0C}\right) \cdot \sqrt[4]{Gr} \cdot \sqrt[4]{\frac{\nu_{L\infty}}{\nu_{Lb}}} \tag{4.32}$$

with

$$Gr = \frac{g \cdot L^3 \cdot (\rho_L - \rho_v)}{\gamma_v^2 \cdot \rho_v}.$$

The influence of the solid dimensions on the quenching rate is also proportional to $L^{-1/4}$. The proportionality factor for Eq. (4.32), which is dependent on the thermophysical material values for the vapour and the liquid and also on the surface temperature, will not be described here in further detail. As can be seen from Fig. 4.6 and Eq. (4.32), the dependence of the heat-flux density during film boiling on the surface temperature and on the solid dimensions is not particularly high. Hence, for practical cases, a constant value of approximately $\dot{q}_F = 0.3\ \mathrm{MW/m^2}$ can be used for the heat-flux density in the range of film boiling.

With regard to the quenching rate Fig. 4.6 shows that with increasing face velocity and decreasing solid size and bath temperature, not only the heat-flux densities become larger but that above all the Leidenfrost temperature shifts to higher values.

Figure 4.7 gives the quenching rates for two examples of immersed spheres. In this figure the surface temperature is plotted as a function of the quenching duration, for which the parameters are the bath temperature and the liquid velocity. In the range of stable film boiling, which is passed through first, the temperature of the sphere falls slowly, almost independently of the bath temperature and the liquid velocity. As soon as the temperature falls below the Leidenfrost temperature and the film collapses, the surface temperature drops rapidly. The higher the Leidenfrost temperature is, i.e. the sooner the film collapses, the shorter is the total quenching time. An increase in the Leidenfrost temperature of a few degrees Kelvin is sufficient to reduce the quenching time of the solid considerably. These correlations show clearly that for a defined quenching of workpieces it is necessary to influence the position of the Leidenfrost point.

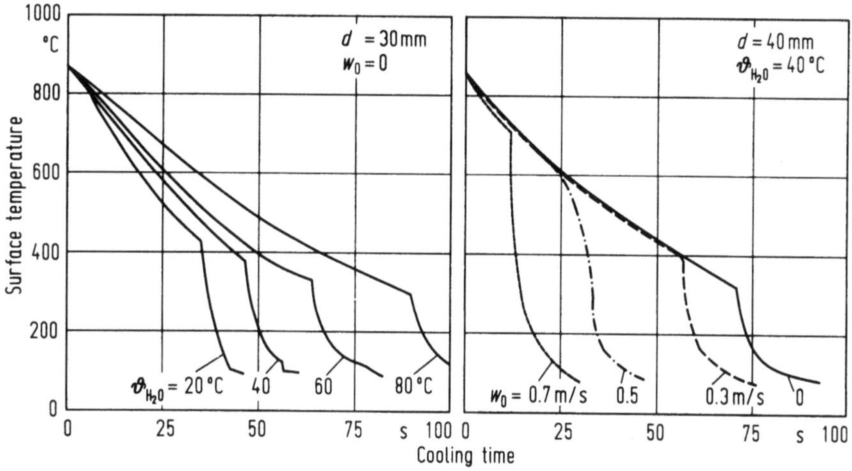

Fig. 4.7. Quenching curves of nickel spheres dipped in water baths at different temperatures and flow velocities [5, 6]

Hence, the Leidenfrost temperature is represented separately in Fig. 4.8 as a function of the liquid velocity with the bath temperature and the diameter of the sphere as parameters. It is evident how for low bath temperatures in particular, the Leidenfrost temperature can be influenced strongly by the liquid velocity.

The above results are valid for spheres. As shown in Fig. 4.9, at high solid temperatures a compact vapour film always forms around the spheres. On reaching the Leidenfrost temperature this film collapses and nucleate boiling begins evenly distributed on the surface. For a cylinder, on the other hand, no compact vapour film forms around the solid, as is portrayed is the right-hand section of Fig. 4.9. The vapour film on the lower edge of the vertically suspended cylinder is destroyed directly after immersion. As a result, the lower part of the cylinder is prematurely wetted which is clearly visible during the boiling phase. The wetting front shifts continuously from the bottom to the top, which results in differing Leidenfrost temperatures along the length of the cylinder [6]. Correspondingly, highly differing quenching rates adjust along the surface of the cylinder.

In the following the conditions at half height are given to allow comparison between sphere and cylinder. On the righthand side of Fig. 4.8 it is evident that the Leidenfrost temperature for the cylinder can also be influenced by the liquid velocity and bath temperature. The diameter of the cylinder has no influence on the Leidenfrost temperature. At higher bath temperatures the Leidenfrost temperatures for sphere and cylinder are approximately equal. At lower bath temperatures, however, considerably higher Leidenfrost temperatures adjust for cylinders.

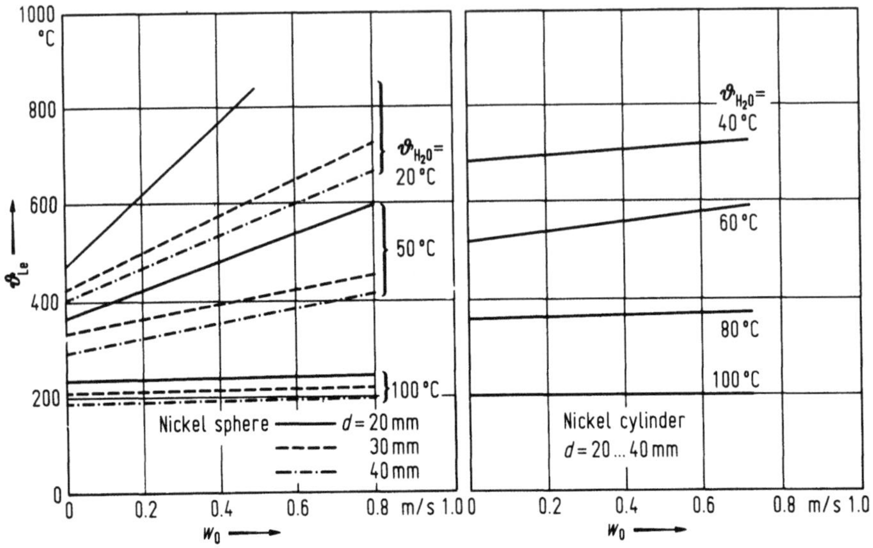

Fig. 4.8. Leidenfrost temperature for immersion quenching of spheres and cylinders in water [6]

Fig. 4.9 a, b. Photographs of film boiling for a sphere **(a)** and a cylinder **(b)** [6]

The variables given so far make evident that an exact prediction of the Leidenfrost temperature for quenching in water is difficult. Although the bath temperature can be adjusted exactly, quantitative consequences of other influences, such as points and edges existing on the surface of the workpiece or oxidation phenomena, are as yet insufficiently known. The results given in the preceding figures render at least an estimation possible of the Leidenfrost temperatures to be expected for the quenching of a workpiece in water.

As mentioned previously, the heat-flux density rises abruptly in the range of partial film boiling from the Leidenfrost point to the DNB-point. It has been shown that the DNB-temperature is determined largely from the position of the Leidenfrost temperature. The same is valid for the maximum heat-flux density. The dependence of both variables is shown in Fig. 4.10 for nickel spheres and cylinders. An influence of the diameter of the solid, of the bath temperature, the liquid velocity and the roughness is not detected.

The preceding results were valid for nickel substances. A noticeable influence of the type of metal on the heat-flux density in the range of stable film boiling and on the Leidenfrost temperature could not be established [6]. Only the maximum heat-flux density is dependent on the type of metal. Aluminium results in somewhat higher values, brass and copper give values of up to 50% of those for nickel [6].

The results given so far are sufficient for a simplified calculation of the heat flow for quenching. For this, the heatflux density curve for the respective operating conditions is approximated by three straight lines, as shown in Fig. 4.11. For line 1 in the range of stable film boiling a constant value of $\dot{q}_F = 0.3\,MW/m^2$ can be assumed for practical calculations. For the

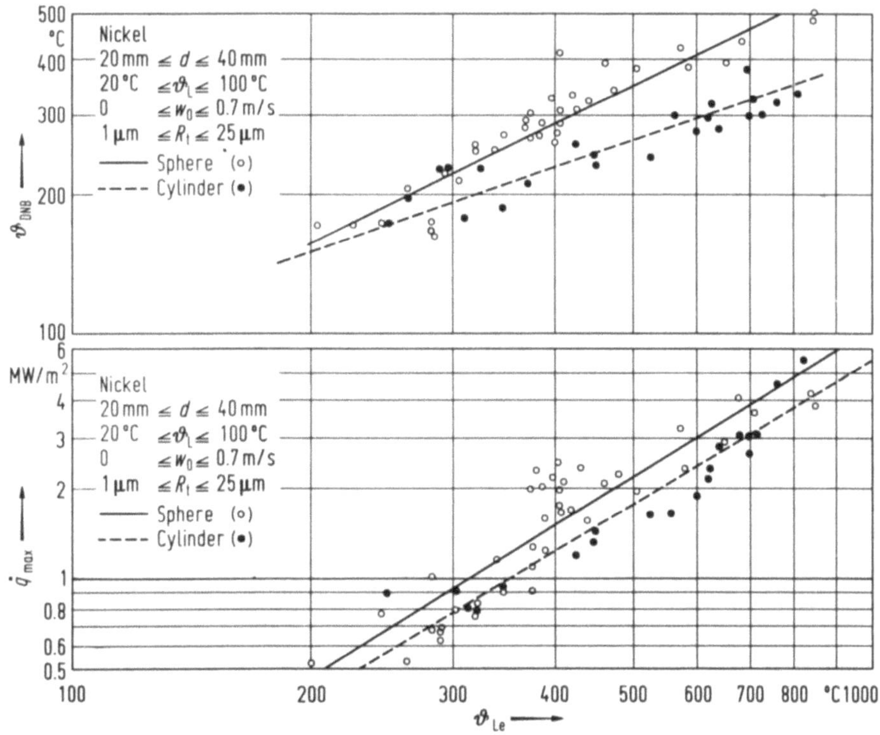

Fig. 4.10. DNB-temperature and maximum heat-flux density for immersion quenching of nickel spheres and cylinders in water [6]

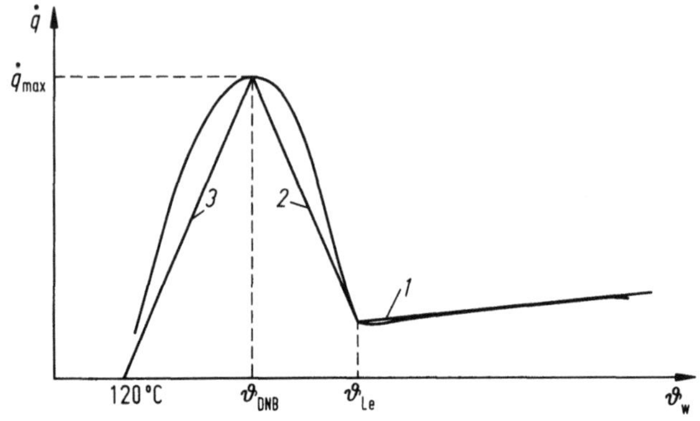

Fig. 4.11. Approximation of curves for heat-flux density using straight lines

determination of line 2, the Leidenfrost temperature, the DNB-temperature and the maximum heat-flux density can be estimated from Figs. 4.6, 4.8 and 4.10. Line 3 is determined by assuming a fixed point of 120 °C for $\dot{q} = 0$. The range of nucleate boiling described by these straight lines is of little importance for quenching in water, as generally the solids have cooled down considerably before reaching this range. If required, the equations in Chap. 2 can deliver more exact calculations for this range. The simplification according to Fig. 4.11 can in principle be carried out for all quenching agents. The influence of the quenching agent on the cooling rate is described below.

In Fig. 4.12 the curves for heat-flux density and the respective cooling rate for a sphere are compared for four different quenching agents [7]. The aquatenside is based on a polyalkalineglycol solution and is assigned to the polymer solutions. The quenching agent polymer 834294 belongs to the polyacrylates. Durixol B4 is a quenching oil which is graded as light-weight and abruptly quenchant. In the range of stable film boiling the quenching agents differ only little in their cooling effect, although the effect of the liquid velocity on the solid is stronger for oils than for water. As in the case of water, the heat-flux density $\dot{q}_F$ for polymer solutions also decreases with the liquid temperature. For viscous quenching oils, however, $\dot{q}_F$ increases with ϑ_L. In this case, because of the decrease in viscosity with the temperature, the dragging effect through the vapour film increases so rapidly that this compensates the reduced heat flow to the liquid, which is caused by the lower temperature difference.

The effect of the different quenching agents on the cooling rate is conditioned mainly by the influence of the Leidenfrost temperature. The higher the Leidenfrost temperature of the quenching agent is, the more intense is its cooling effect. As is the case for water, the Leidenfrost temperature for polymer solutions increases with the liquid velocity and decreases with the liquid temperature. For quenching oils, however, the Leidenfrost temperature is largely independent

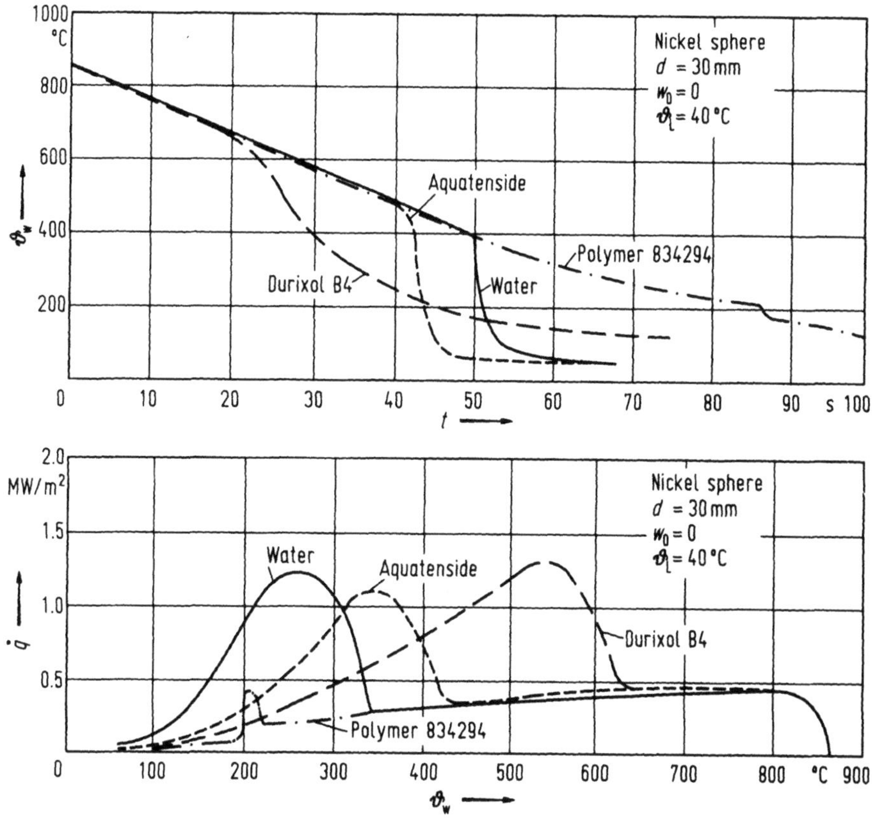

Fig. 4.12. a, b. Temperature vs time and heat-flux density vs surface temperature for different quenching agents [7]

of these two parameters. Data on the DNB-temperature and the maximum heat-flux density as well as on the cooling effect of further quenching agents can be found in the literature, e.g. [2].

Generally, smaller workpieces are quenched not individually but as several pieces together in a wire basket. In such a stock column the top pieces are quenched at a considerably slower rate than those at the bottom (Fig. 4.13a). Figure 4.13 illustrates, however, how the cooling curves of the individual pieces can be homogenized by a specific flow against the stock column. The set of figures show cooling curves for the four spheres marked black in the stock column. As is evident in Fig. 4.13b, the cooling curves of the individual spheres are homogenized by a forced convection in the bath. Approximately the same degree of homogenization is reached by inducing a bubble flow in the bath (Fig. 4.13c). A further improvement is obtained, however, from a direct gas spray, as is shown in Fig. 4.13d. A lateral spray of water also results in a balanced quenching (Fig. 4.13e).

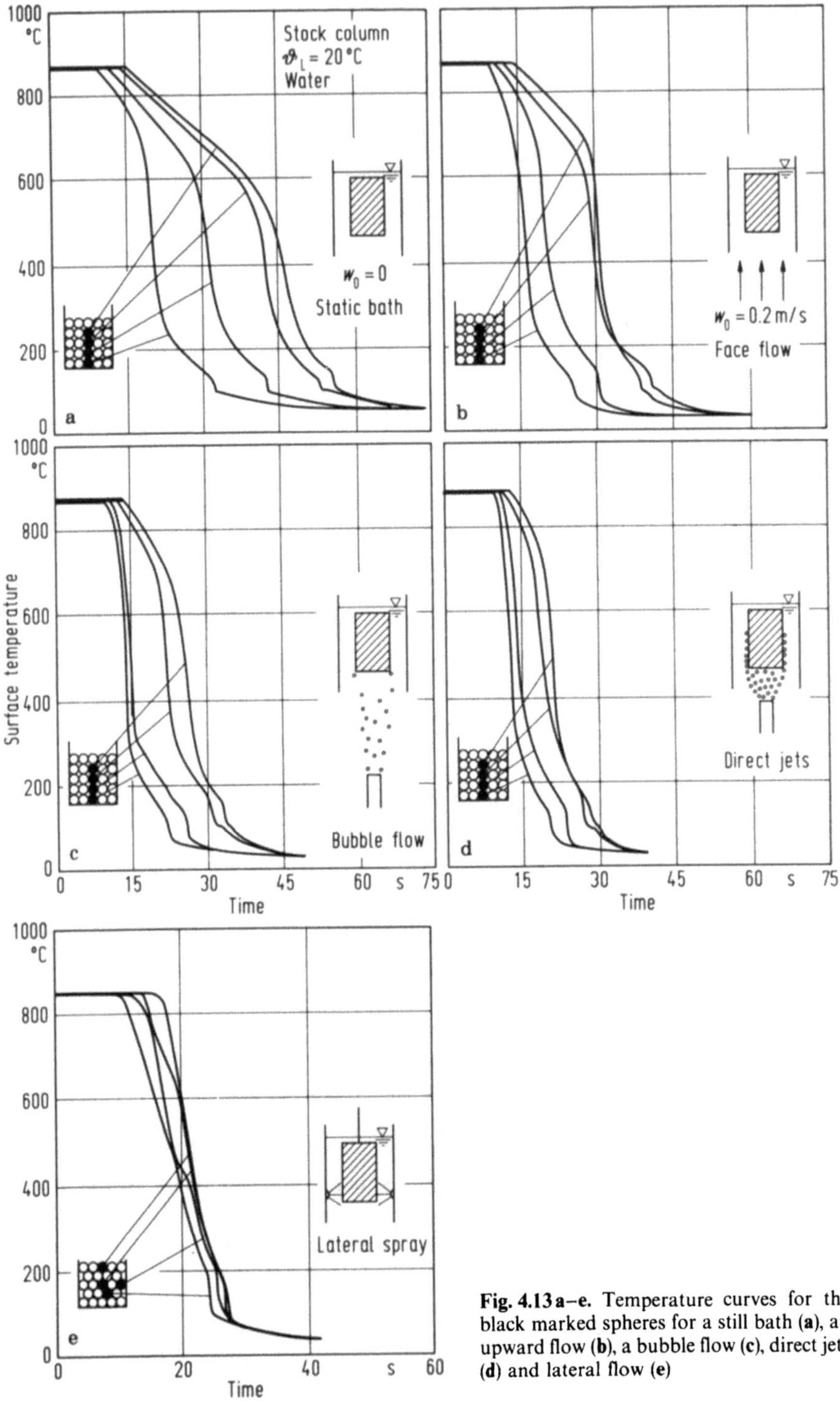

Fig. 4.13 a–e. Temperature curves for the black marked spheres for a still bath (**a**), an upward flow (**b**), a bubble flow (**c**), direct jets (**d**) and lateral flow (**e**)

4.4 Spray Quenching

For spray quenching the eliminated heat-flux density is set proportional to the temperature difference between surface and liquid

$$\dot{q} = \alpha \cdot (\vartheta_{\mathrm{w}} - \vartheta_{\mathrm{L}}). \tag{4.33}$$

The heat-transfer coefficient is given in Fig. 4.14 for water as a function of the surface temperature. In Fig. 4.14a the parameter is the impingement density which indicates how much liquid reaches the hot solid per unit of time and area. An increase in the impingement density causes the heat transfer to improve and the Leidenfrost temperature to shift to higher values. Figure 4.14b indicates that the heat transfer is also dependent on the type of metal, albeit only below the Leidenfrost temperature.

In the range of stable film boiling the heat-transfer coefficient is independent of the surface temperature. In this range it can only be influenced by the impingement density. Other parameters, such as drop velocity, distance nozzle to surface and nozzle type, do not require to be considered separately, as they are contained in the impingement density [13, 14]. Figure 4.15 shows the dependence of the heat-transfer coefficient on the impingement density for water. This can be reproduced by the equation

$$\frac{\alpha_{\mathrm{F}}}{W/(\mathrm{m}^2 \cdot \mathrm{k})} = 200 + 1.82 \frac{\dot{m}_{\mathrm{L}}}{\mathrm{kg}/(\mathrm{m}^2 \cdot \mathrm{min})} \tag{4.34}$$

which is valid for the range 100 to 2000 kg/(m^2·min) [9]. The above equation only holds for water temperatures under 20 °C. For higher water temperatures

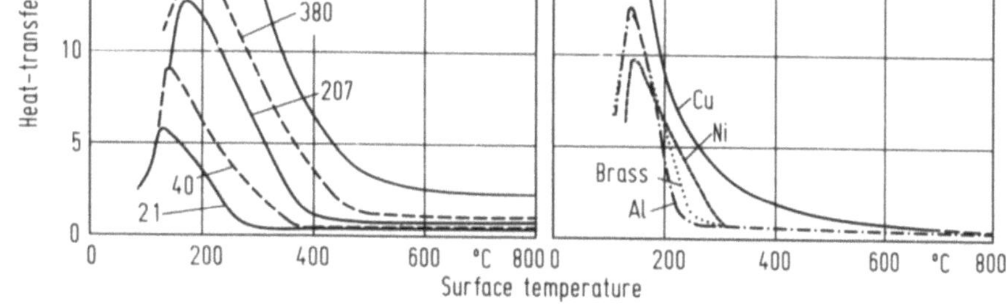

Fig. 4.14 a, b. Heat-transfer coefficient for spray quenching as a function of surface temperature. **a** Influence of impingement density; **b** influence of type of metal [9]

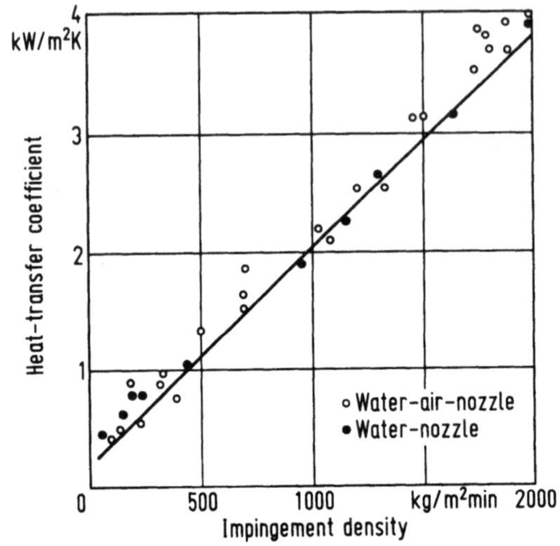

Fig. 4.15. Heat-transfer coefficient in the range of film boiling for spray quenching with water as a function of the impingement density [10, 11, 13]

the heat-transfer coefficient decreases more, the higher the impingement density is [14].

As the impingement density is the only variable controlling the heat flow in the range of stable film boiling, the local distribution of the impingement density is of particular importance. In industrial practice, two-component nozzles, i.e. water-air nozzles, are being increasingly employed. The addition of air as second quenching medium serves mainly to influence the nozzle characteristics. The cooling effect of these nozzles is, however, only determined by the magnitude of the impingement density [13, 14].

According to Fig. 4.14 the heat flow in the range of partial film boiling depends not only on the impingement density but also on the type of metal. This influence can be described by the coefficient of heat penetration $\sqrt{\lambda \cdot \rho \cdot c}$. Figure 4.16 shows for four metals the dependence of the Leidenfrost temperature, the DNB-temperature and the maximum heat-transfer coefficient on the impingement density. According to [9] all three variables increase approximately exponentially with the impingement density. Hence, the maximum heat-transfer coefficient, the Leidenfrost temperature and the DNB-temperature can be described by the relation

$$\alpha_{\max}, \vartheta = B \cdot \left[\frac{\dot{m}_L}{\text{kg/(m}^2 \cdot \text{min)}} \right]^n. \tag{4.35}$$

The variables B and n are compiled in Table 4.1 for the four given metals.

To calculate the coefficient of heat penetration it must be taken into account that the material values at the surface are decisive. For copper, in particular,

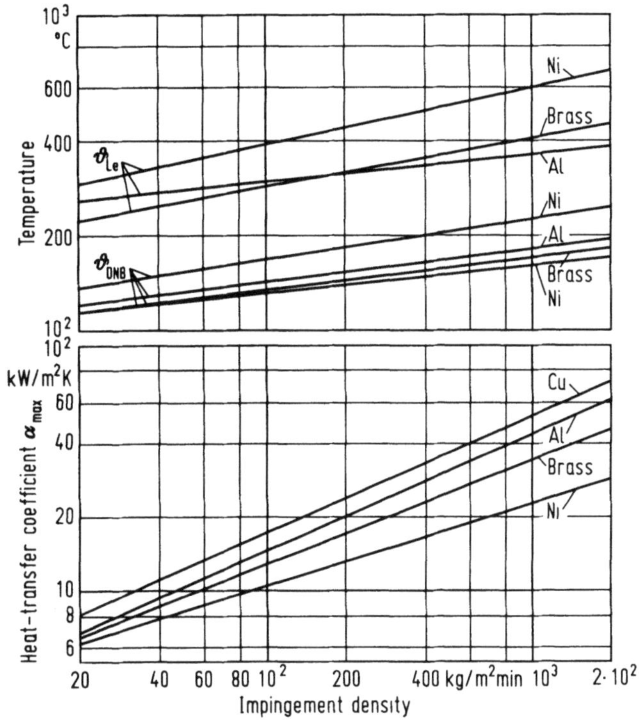

Fig. 4.16. Leidenfrost temperature, DNB-temperature and maximum heat-flux density as a function of the impingement density for different metals according to [9]

Table 4.1

Metal	$\sqrt{\lambda\cdot\rho\cdot c}$	ϑ_{Le}		ϑ_{DNB}		α_{max}	
		B	n	B	n	B	n
Metal	$kW\cdot s^{0.5}/(m\cdot K)$	°C	—	°C	—	$W/(m^2\cdot k)$	—
Copper	35.5	—	—	97	0.08	1840	0.48
Aluminium	23.6	190	0.09	83	0.11	1530	0.48
Brass	21.3	126	0.17	80	0.11	1890	0.41
Nickel	14.3	156	0.19	72	0.16	2360	0.32

an oxide layer always forms at the surface in which $\sqrt{\lambda\cdot\rho\cdot c}$ contains considerably smaller values than in the metal itself.

The data listed in Table 4.1 are valid for smooth surfaces. A roughness of more than ca. 2 µm (according to DIN 4768) causes the Leidenfrost temperature and the DNB-temperature as well as the maximum heat-transfer coefficient to increase [14]. A roughness of e.g. 30 µm results in an increase in the Leidenfrost temperature of ca. 100 K, in the DNB-temperature of ca. 20 k and in the maximum heat-transfer coefficient of ca. 20%.

References

1. Carslaw HS, Jaeger JC (1959) Conduction of heat in solids. Oxford at the Clarendon Press
2. Köhler Chr (1990) Wärmeübertragung von heißen Oberflächen durch Wasserfilmkühlung im Bereich der stabilen Filmverdampfung. Dissertation Technische Universität Clausthal
3. Jeschar R, Specht E, Heidt V (1991) Mechanismen der Wärmeübertragung beim Kühlen von Metallen mit verdampfenden Flüssigkeiten. Abhandlung der Braunschweigischen Wissenschaftlichen Gesellschaft, Jahrbuch
4. Jeschar R, Köhler Chr, Specht E, Heidt V (1989) Methoden zur definierten Abkühlung metallischer Werkstoffe. In: Die Industriefeuerung 47: 34–39; Gaswärme int. 38: 223–229
5. Jeschar R, Maaß R (1986) Wärmeübertragung beim Abschrecken kugelförmiger Körper in Wasser. In: Gaswärme int. 35: 370–375
6. Maaß R, Jeschar R (1989) Einflußgrößen des Wärmeüberganges beim Abschrecken von Metallen in Wasser. In: Gaswärme int. 38: 142–150
7. Maaß R, Jeschar R (1989) Bestimmung des Wärmeüberganges beim Abschrecken von Metallen in Härtemitteln. In: Gaswärme int. 38: 427–434
8. Maaß R, Jeschar R (1988) Wärmeübertragung beim Abschrecken von Haufwerken. In: Gaswärme int. 37: 385–389
9. Müller HR, Jeschar R (1983) Wärmeübergang bei der Spritzwasserkühlung von Nichteisenmetallen. Metallkunde 74: 257–264
10. Jeschar R, Scholz R, Reiners U (1984) Heat transfer during one and two-phase water-spray cooling in hot metals. In: The Sixth Japan–Germany Seminar. Proceedings: 146–159
11. Jeschar R, Reiners U, Scholz R (1984) Wärmeübergang bei der zweiphasigen Spritzwasserkühlung. In: Gaswärme int. 33: 299–308
12. Reiners U, Jeschar R, Scholz R, Zebrowski D, Reichelt W (1985) A measuring method for quick determination of local heat transfer coefficients in spray water cooling within the range of stable film boiling. In: Steel research 56: 239–246
13. Jeschar R, Reiners U, Scholz R (1986) Heat Transfer During Water and Water–Air Spray Cooling in the Secondary Cooling Zone of Continuous Casting Plants. In: Proceedings of the 69th Steelmaking Conference in Washington, Vol. 69: 511–521
14. Reiners U, Jeschar R, Scholz R (1989) Wärmeübertragung bei der Stranggußkühlung durch Spritzwasser. In: Steel Research 60: 442–450
15. Reiners U (1987) Wärmeübertragung durch Spritzwasserkühlung heißer Oberflächen im Bereich der stabilen Filmverdampfung. Thesis. Technical University of Clausthal (Germany)

5 Wetting Kinematics

HANS M. TENSI

5.1 Introduction

The wetting processes taking place on the surface of metallic samples during immersion cooling in fluids with boiling temperatures lower than the sample temperature determine the cooling of the sample. The same is true for film cooling during continuous casting or induction hardening.

Hardening by Martensitic transformation of steels usually requires a temperature of $>800\,^{\circ}C$, from which the sample must be quenched in a defined way. Age hardening (Wilms hardening) also requires sufficiently rapid quenching from high temperatures so that the supersaturated mixed crystal forms, which makes this type of hardening possible. The quenching temperatures for Al alloys, the most strongly age hardened materials, are in the range of 500 °C.

The quenchants used today have boiling temperatures far below the quenching temperatures described above. Salt melts and metal melts are rarely used as quenchants nowadays to avoid exposure at the work place and because it is difficult to dispose of them. For the same reason, the so-called hardening oils (with flash points in the range of 160 °C to 250 °C) are increasingly being replaced by fluids less harmful to the environment, such as aqueous polymer solutions.

Most quenchants have boiling temperatures between 100 °C and 300 °C at normal pressure. When work pieces are quenched in these fluids, the wetting of the surface is usually time dependent. This influences the cooling process and thus the achievable hardness. J.G. Leidenfrost [1] described the process of wetting more than 200 years ago. The Leidenfrost temperature named after him is the temperature on the surface of a body at which the vapour film is broken through and the surface wetted by the liquid.

The process of wetting, referred to as wetting kinematics, can be seen as the control variable for the cooling of a work piece and, at the same time, as the characteristic feature of the quenchant.

5.2 Definition of the Wetting Process

Having non-steady state conditions experimental ascertainment of the Leidenfrost temperature (or wetting temperature) and its analytical models involve considerable problems. The literature describes values between 150 °C and 300 °C merely for water at normal pressure. Calculated Leidenfrost temperatures for saturated water range between 160 °C [Berenson, 2] and 300 °C [Henry, 3]. The Leidenfrost temperatures used in different rewetting model calculations vary between 150 °C [Ymanouchi, 4], 200 °C [Thomson, 5], 190 °C to 250 °C [Duffly and Porthouse, 6] and 214 °C [Künzel, 7]. Directly measured values for water are between 210 °C and 230 °C, as per Hein [8] (see also Chap. 2, Fig. 2.2).

It is thus apparent that the Leidenfrost temperature is influenced by a variety of factors, part of which cannot precisely be quantified even today.

As per Ladisch [9], immersed body and fluid come into contact due to the break-down of the vapour blanket, which takes place when the minimum temperature of the vapour blanket phase is reached. This temperature is not equal to the Leidenfrost temperature for surfaces cooling in a non-steady manner, since during intermittent wetting the temperature of the wetting site is additionally influenced by lateral heat conduction (relative to the surface).

The reason for this is the simultaneous presence of the various heat transfer periods "vapour blanket", "heat conduction" and "evaporation" on the surface (Fig. 5.1).

The simultaneous presence of the different phases

- vapour blanket, or film boiling,
- nucleate boiling, and
- convective heat transfer

with significantly varying heat transfer coefficients α_{FB}, α_{NB} and α_{conv} (or differing heat flow densities) must therefore be taken into account both during immersion cooling and film cooling, during continuous casting or induction hardening.

Wall temperatures higher than the Leidenfrost temperature (in water) cause film boiling with heat transfer coefficients α_{FB} in the range of 100 to 250 W/(m^{2}K) in water. The variations in wall temperature, resulting from sample geometry, surface depositions, oxide layers, pressure deviations and locally differing immersion periods, lead to locally confined wettings characterized by greatly differing heat transfer coefficients. Because of its short duration, this partial vapour blanket can be considered part of the subsequent nucleate boiling phase. The evaporation of the fluid, in connection with the gravity-driven upward movement of the vapour bubbles, causes extremely strong convection in this area, which results in maximum heat delivery to the fluid. The heat transfer coefficients measured in this area range between 10 and 20 kW/(m^{2}K) in water

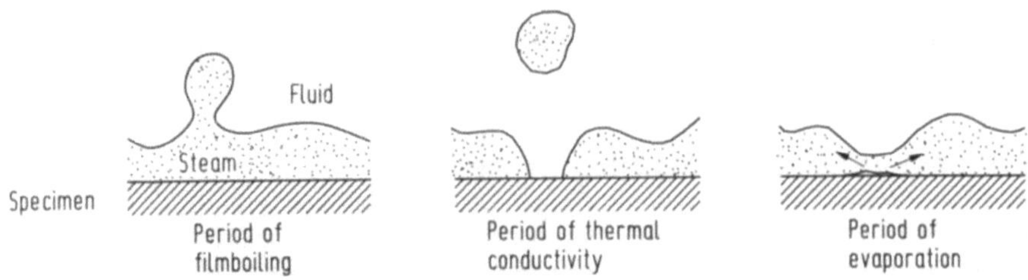

Fig. 5.1. Heat transition having phases of wetting during film boiling [3]

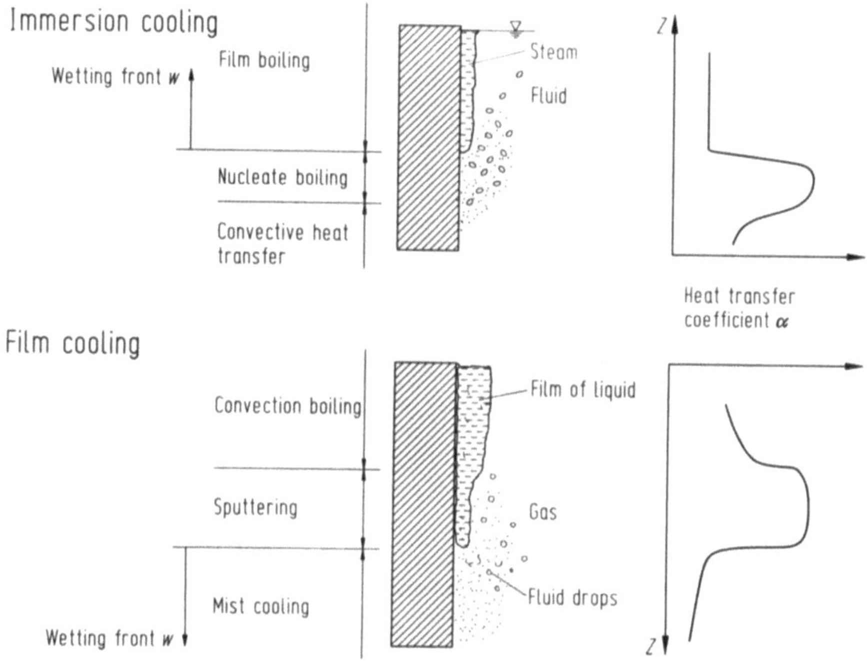

Fig. 5.2. Situation of heat transition during immersion cooling and film cooling (i.e. cooling by running down liquid film) with assigned heat transfer coefficient [7]

(see Chap. 3, Fig. 3.14). When the surface temperature drops below the fluid's boiling temperature, the surface is permanently wetted by the fluid. Thermomigration takes place through heat conduction and convection with considerably reduced heat flow density (α_{conv} approx. 700 W/(m^{2}K)).

The sequence of phases occurring on sample surfaces during immersion cooling or film cooling and the correspondingly changing heat transfer coefficients are schematically shown in Fig. 5.2 [7]. The speed of the spreading wetting front is determined by the physical properties of the body and the fluid. These are,

- the temperature distribution on the sample surface,
- the thermal diffusivity *a* of the body,
- the surface roughness of the body,
- the layers present on the surface, such as oxides or organic substances,
- the sample geometry and its effect on the heat flow towards the surface and on the gravity-driven convection in the fluid;

in addition,

- the boiling temperature of the quenchant and its Leidenfrost temperature,
- its dynamic viscosity, thermal capacity, and surface tension,
- the temperature of the bath, and forced convection.

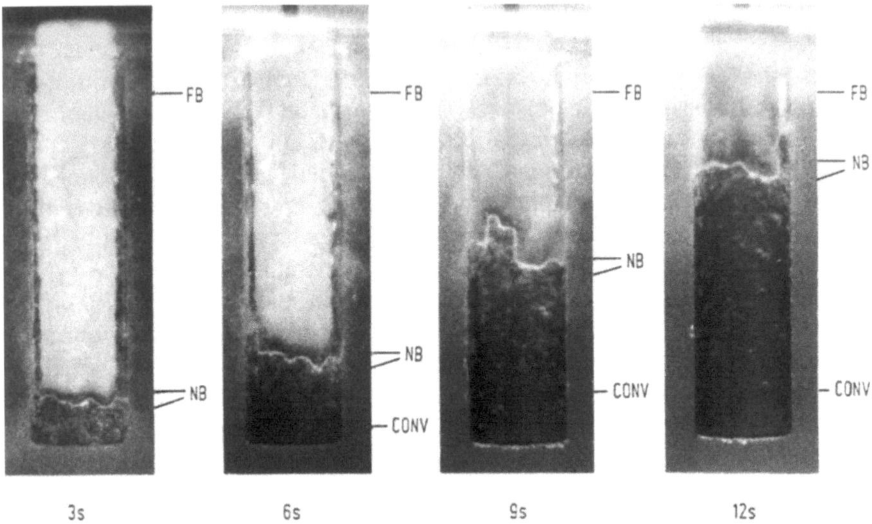

Fig. 5.3. Process of transition between the three phases film boiling (FB), nucleate boiling (NB) and convective heat transfer (CONV) during immersion cooling of a CrNi-steel sample, $\phi 25 \times 100$ mm; immersion temperature: 860 °C; quenchant: distilled water; temperature of quenchant: 35 °C [11, 12]

The calculation of the wetting front's velocity *w* during immersion cooling can in principle be based on a formulation made by Yamanouchi [4], but fails to yield an explicitly usable solution; it is useful to identify those parameters, which have an influence on the wetting and cooling behaviour.

Thus, the velocity is experimentally determined, as required for the given situation (see also the parameters listed above). A typical example of the wetting sequences on a steel sample quenched in water, oil, and aqueous polymer solutions[1] is depicted in Fig. 5.3. Specimens which are completely submerged in the quenchants show an additional wetting front, running from the upwards.

The cylindrical sample is made of austenitic CrNi steel, the quenchant is water at 35 °C. The migration of the wetting front (as the interface between vapour blanket and nucleate boiling) from the lower edge of the sample to its top is clearly visible. The time required for completely wetting the sample is approx. 20 s in this case. The three phases of cooling are thus simultaneously present on the suface for a significant period of time. The cooling of the sample is therefore subject to great local variations.

The time interval of wetting (or the velocity of the wetting front) can be strongly influenced by putting chemical additives in the water. Figure 5.4 shows two wetting processes which were observed in water with different admixtures. In relation to the picture frequence of filming, the wetting time is "infinitely

[1] Some polymer solutions show an explosive wetting behaviour

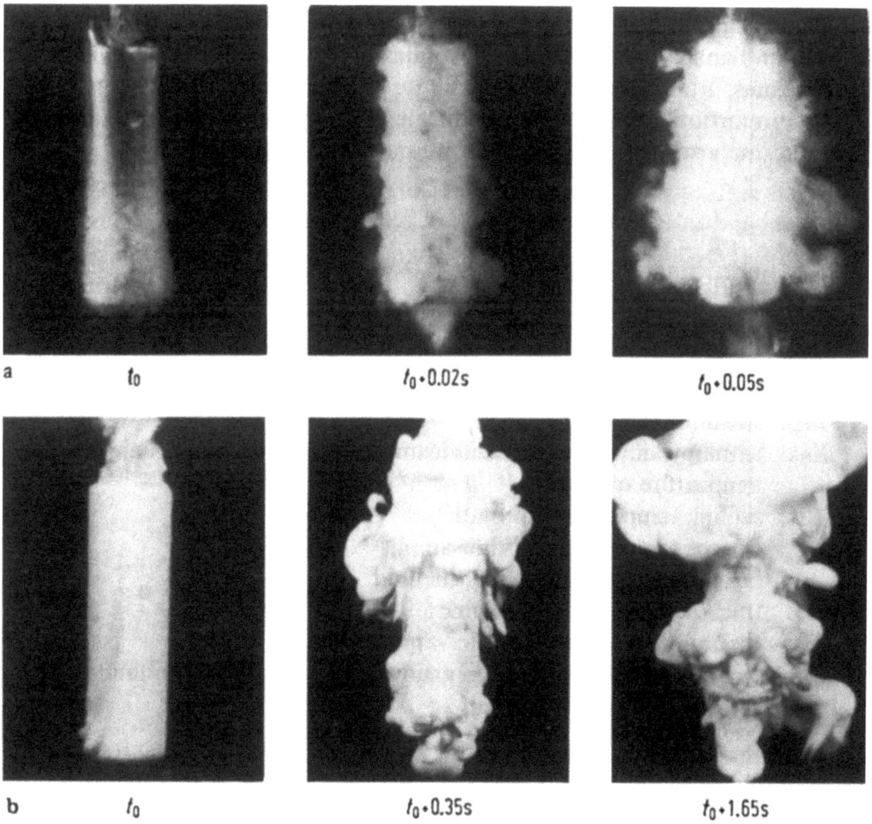

Fig. 5.4 a, b. Transition from film boiling to nucleate boiling during immersion cooling of a silver-sample ($\phi 15 \times 45$ mm); [22]. Movie pictures with the speed of 6o pict./s; t_0: Starting time of wetting. **a** quenchant: Water solution of a polymer with $c = 10\%$; temperature of quenchant: 25 °C; temperature and time of transition: $T_1 \approx 360$ °C, $t_t \approx 20$ s; **b** quenchant: water solution of a polymer with $c = 10\%$ with a chemical additive; temperature and time of transition: $T_t \approx 620$ °C, $t_t \approx 6$ s

short" (see the time interval between the pictures: 0.02 s and 0.35 s). No lateral heat flow beneath the sample surface occurs in this case.

5.3 Model of Vapour Blanket Breakdown During Immersion Cooling of Metallic Bodies

The fact that wetting does not take place upon reaching of the boiling temperature on the sample surface, but already upon reaching of the considerably higher Leidenfrost temperature is explained by the regularities of the heat flow from the sample. During the vapour blanket phase, heat is given off via evaporation, thermal conduction through the vapour blanket into the bath and thermal radiation.

Balancing the heat flows in conjunction with the variable thickness δ of the vapour blanket yields a term which indicates direct proportionality between the thickness δ and the heat conductivity of the vapour blanket phase and inverse proportionality between the thickness δ and the heat absorption in the fluid and the heat absorbed by evaporation (Eq. 5.1) [10, 11]:

$$\delta = \frac{\lambda_d}{\alpha_F \cdot \dfrac{(T_S - T_b)}{(T_W - T_S)} + \dfrac{m_d \cdot \Delta h_F}{A \cdot (T_W - T_S)} - \varepsilon_{W,F} \cdot \sigma \cdot T_S^3 \cdot f\left(\dfrac{T_W}{T_S}\right)}; \qquad (5.1)$$

with

A: surface
m_d: steam flux
Δh_F: enthalpy difference between steam and fluid with boiling temperature
T_b: tempeature of bath
T_S: boiling temperature of bath
T_W: surface temperature of the sample
α_F: heat-transfer coefficient of the fluid
δ: thickness of film boiling phase
λ_d: thermal conductivity of the vapour phase
$\varepsilon_{W,F}$: emission coefficient between submerged specimen and fluid
σ: Stefan Boltzmann constant

This relation also indicates that the boundary layer δ increases as the supercooled condition of the bath $(T_S - T_b)$ decreases and that it only takes on a value of zero (at which wetting takes place) when the surface temperature T_W of the immersed sample arrives at the boiling tempeature T_S. The thickness δ of the vapour blanket thus adjusts itself in such a way that there is an equilibrium between its insulating effect an the heat flow from the sample.

Having a sudden vapour blanket break-down with a surface temperature $T_W > T_S$ (see Chap. 2, Fig. 2.2), there is to suppose a minimum value δ_C of the blanket thickness, at which the validity of Eq. 5.1 is abolished (Fig. 5.4a). This critical thickness δ_C is determined by the hydrodynamic stability of the vapour blanket which itself is influenced by the physical and chemical attributes of the bath [12].

Contrary to this having a slow wetting of the sample surface (according to Fig. 5.3) we can assume that on narrow local areas (for example edges and tips of the submerged specimen) the surface temperature T_W falls off in such a way, that at once at these areas wetting occurs. From these points the front of wetting runs over the whole specimen's surface.

According to this hypothesis all wetting processes during submerge cooling can be described [13–15].

5.4 Effect of Wetting Process on Cooling Behaviour

The simultaneous presence of the different kinds of boiling phases on a cylindrical sample causes not only a radial heat flow, but also a heat flow lateral to the mantle line. If the heat transfer coefficients of the different phases are known and the velocity of the wetting edge on the sample surface has been measured (e.g., optically), the cooling of the sample can be calculated from the change in heat transfer along the mantle line over time (see Fig. 5.5). The mean heat transfer coefficient α_{FS} for film boiling may be determined on the basis of the thermal balance, as per [9, 16–20], or merely on the basis of the temperature change over time in the sample centre, since the temperature difference between the centre and the lower edge of the sample can be neglected in the case of small Biot numbers ($Bi < 0.2$) and great Fourier numbers ($Fo > 0.25$). Cooling is in this case merely ascertained by the function $T_C(t)$ [7, 14]. For the cylinder, cooling is then describable by the exponential function:

$$\theta = \exp(-2\,Bi \cdot Fo), \tag{5.2}$$

with

$$\theta = \frac{T - T_b}{T_0 - T_b}$$

and

Bi: Biot-number
Fo: Fourier-number
T: actual temperature in sample
T_b: temperature of bath
T_0: starting temperature in sample

This involves an error of $< 1\%$. The value of α_{FS} can be determined via the Biot number by applying the Eq. (5.2) to the measured thermal profile. The change of the temperature field over time up to the first contact between sample and fluid

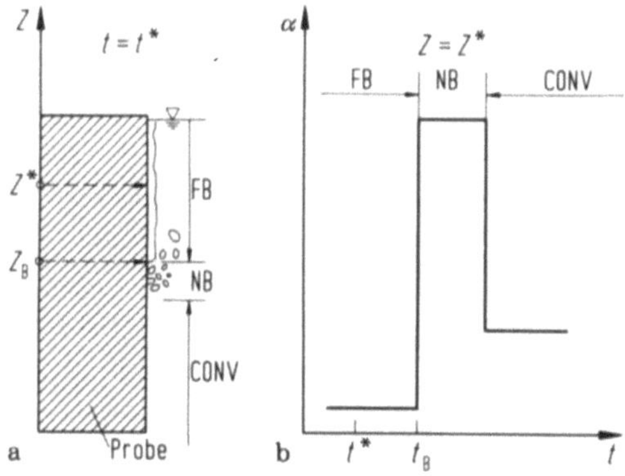

Fig. 5.5 a, b. The wetting state of sample surface at time t^* and the change of heat transfer coefficient α at the place Z^* with the time t (schematically) [11]. Z_B place of transition from film boiling to nucleate boiling (i.e. point of wetting front); Z^* random surface point; t_B time when the wetting front has reached the point Z^*

(wetting) can be calculated on the basis of this value in conjunction with the original temperature distribution. The heat transfer coefficient α_{BS} of nucleate boiling can also be ascertained on the basis of the heat balance for a high wetting edge velocity (see Fig. 5.4a), or taken from literature [8] (e.g., $\alpha_{BS} \approx 15\,\mathrm{kW/(m^2K)}$).

The heat transfer coefficient α_{conv} of free convection is ascertained by means of the following approximation equation, as per [7]:

$$Nu = Nu_0 + 0.668k \cdot f(\mathrm{Pr}) \times Ra^{0.25} \tag{5.3}$$

with

Nu: Nusselt-number
Pr: Prandtl-number
Ra: Rayleigh-number

having specimens with cylindric geometry and water: $Nu_0 = 0.68$; $k \cdot f = 1$; $Pr = 7.03$; $f(Pr) = 0.913$

The values $Nu_0 = 0.68$, $k \cdot f = 1$, $Pr = 7.03$, and $f(Pr) = 0.913$ apply for water and cylindrical geometry. Although this equation has been formulated for the cooling of a plate, it is also sufficiently precise when used for cylindrical samples with a diameter of, for example, 15 mm and a length of 45 mm [7]. The transition from nucleate boiling to convective heat transfer in the case of water takes place when a surface temperature of 150 °C is attained, as per Andreoni [21].

As per Künzel [7], in addition to the heat transfer coefficients of the three kinds of boiling, only the currently measured velocity of the wetting front is required for calculating the temperature distribution during the entire cooling process. Temperature measurement in the sample centre only serves as a check in this case. The measured progression of temperature in the specimen and the progression calculated on the basis of the wetting process match well, as can be seen in Fig. 5.6 (when measuring the wetting speed by means of temperature

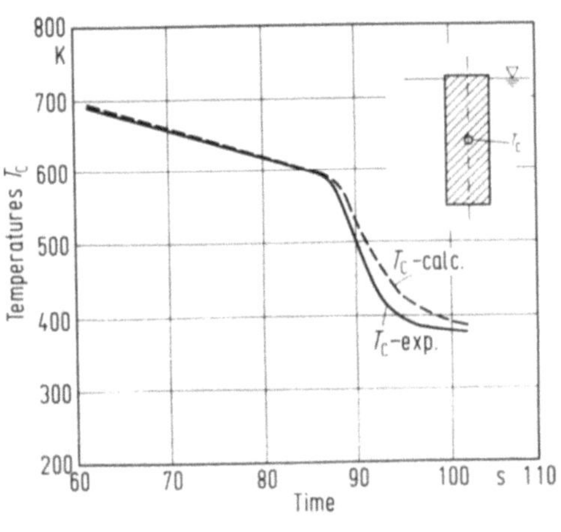

Fig. 5.6. Drop in temperature $T_c(t)$ in the centre of a CrNi-steel specimen ($\phi 15 \times 45$ mm) during immersion cooling in water [14]; quenchant: water with $T_b = 97\,°C$; broken line: measurements; unbroken line: calculated with parameters; $\alpha_{FB} = 140\,\mathrm{Wm^{-2}K^{-1}}$; $\alpha_{NB} = 15\,\mathrm{kWm^{-2}K^{-1}}$; $\alpha_{conv} = 730\,\mathrm{Wm^{-2}K^{-1}}$; $w = 3.2\,\mathrm{mm\,s^{-1}}$; $t_S = 81.3$s (the two last parameters were measured relevantly)

a
b
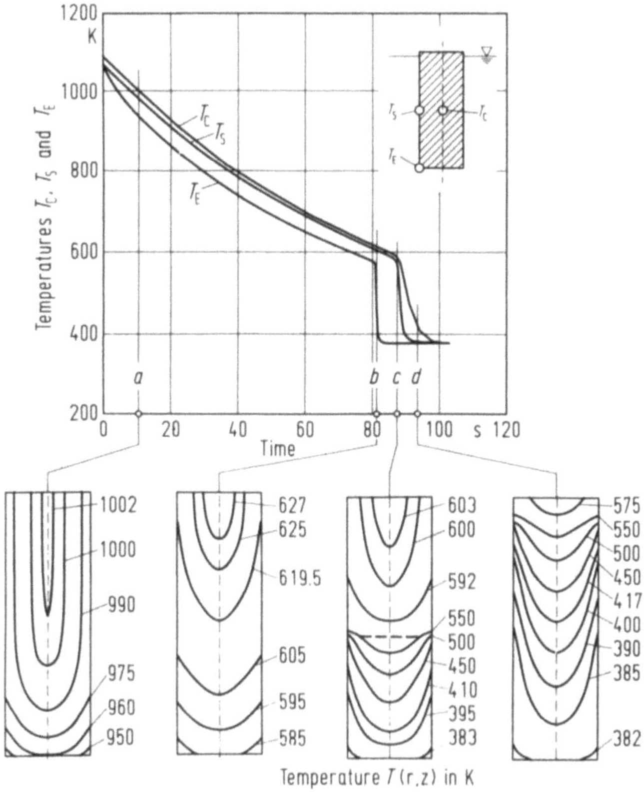

Fig. 5.7 a, b. Calculated drops in temperature in a CrNi-steel-specimen ($\phi 15 \times 45$ mm) during immersion cooling in water with $T_b = 97\,°C$ (7.14). **a** Drops in temperature at the point centre (C), surface (S) and edge (E); **b** Isotherms in the specimen at the moments *a*, *b*, *c* and *d* in Fig. **a**

probes, the sub-surface temperature would have to be measured in at least three locations along a given mantle line of a sample of the same size and a loss of accuracy would have to be expected, cf. also Chap. 8.1.2, cf. Fig. 8.1.11). The calculated temperature distribution during cooling, referring to four points in time, is shown in Fig. 5.7. It becomes apparent that the axial temperature difference during cooling is up to 220 K even with the small-sized sample used in this case (this has a very great impact on hardening, as will be shown in Chap. 11.1.3, cf. Fig. 11.1.4). Allocation of the respective (measured) position of the wetting edge to the calculated surface temperature results in a wetting temperature value of about 210 °C (dispersion of results < 8 K). This value is in good agreement with the values of the Leidenfrost temperature given by literature.

5.5 Impact of Quenchant Properties on Wetting Process

Influence of Bath Temperature

During the stable vapour blanket phase, particularly having a low supercooling of the liquid, vapour bubbles are escaping from the interface into the liquid and finally to atmosphere. The heat given off by the sample causes additional evaporation of the fluid at the interface between the vapour phase and the fluid. The fluid evaporates into this bordering layer, which is thus sustained. As the surface temperature drops, however, the thickness δ of the film boiling is reduced. The upward flowing vapour causes a recognizable disturbance of the interface, taking the form of upwardly moving waves. This disturbance is intensified with further reduction of the surface temperature and spreads downwards (Fig. 5.8). Energy requirements for evaporation lessen as the temperature of the bath increases. The heat transfer coefficient α_{FS} decreases in this case. Thus, the stable film boiling zone achieves a higher thickness (see Eq. 5.1) and can occur at lower temperatures in the specimen's center. The cooling curves $T_c(t)$ measured during immersion cooling of an Ag cylinder in water with $25\,°C < T_b < 95\,°C$ show that the range of film boiling can be expanded by more than 300 K when the temperature of the bath is increased (Fig. 5.9; the high thermal diffusivity a of Ag makes it possible to ascertain, in rough approximation, the start of wetting on the basis of the temperature drop measured in the sample centre).

The high heat flow density during nucleate boiling is due to the two following factors: The fluid is evaporated upon contact with the sample surface; and the gravity driven rising of the vapour bubbles close to the sample surface causes intense convection, which increases thermomigration within the fluid by several magnitudes. With increased bath temperature, the energy required for

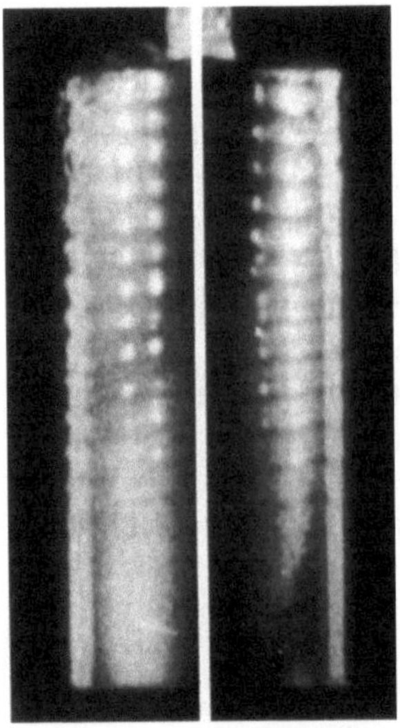

Fig. 5.8. Waves in the interface between liquid and vapour during film boiling at the surface of a CrNi-steel specimen (ϕ40 × 80 mm), immersion cooled in water solution of polymer with $c = 10\%$; temperature of quenchant 25 °C [22]. Left part: 10 ms before wetting; right part: 5 ms before wetting

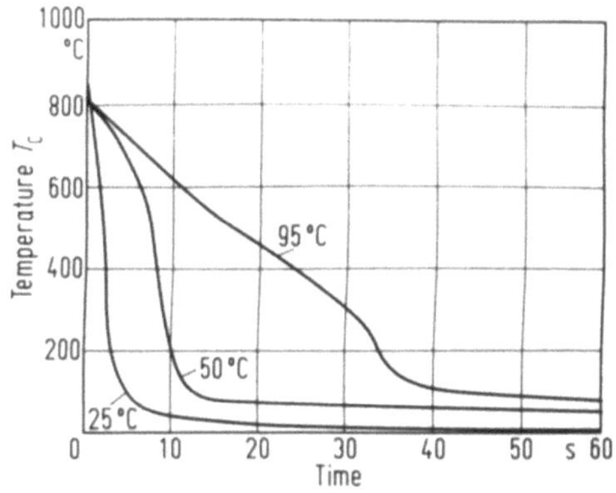

Fig. 5.9. Influence of quenchant temperature T_b on the drops of temperature T_c in the centre of a silver specimen ($\phi 15 \times 45$ mm), immersion cooled in water [22]

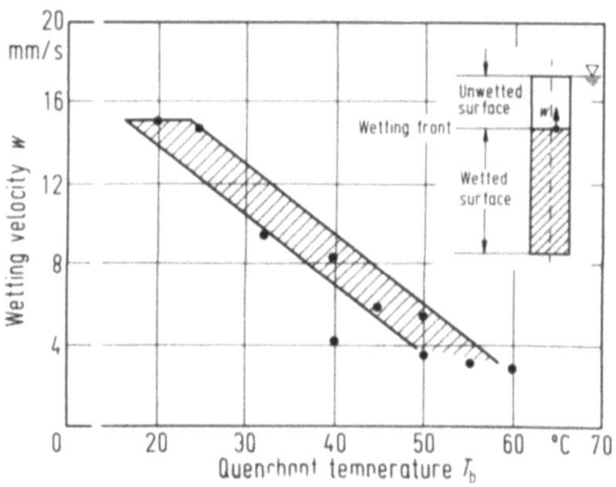

Fig. 5.10. Influence of quenchant temperature T_b on the wetting velocity during immersion cooling of a Ni-base alloyed specimen ($\phi 15 \times 45$ mm with sharp lower edge) [7, 11]. Quenchant: water; immersion temperature: 920 °C

evaporating the fluid is reduced and the thickness δ of film boiling is increased. At the same time, the gravity-driven rising of the bubbles is reduced because of the diminished density difference and the sharply reduced surface tension. Convection, and thus thermomigration, is reduced.

Increasing the bath temperature therefore results in a reduction of the velocity of the wetting front, as becomes apparent in immersion cooling of a Ni base alloy sample (Fig. 5.10).

Different measurements concerning the impact of the bath temperature on the transition temperature T_{trans} (value of the cooling curve, usually measured in the centre of the sample at the transition point between flat and sharp decline) are compared in Fig. 5.11: The transition temperature generally drops sharply with increasing bath temperature under all test conditions, some of which greatly differ concerning the material and shape of the sample as well as the temperature measurement locations [22].

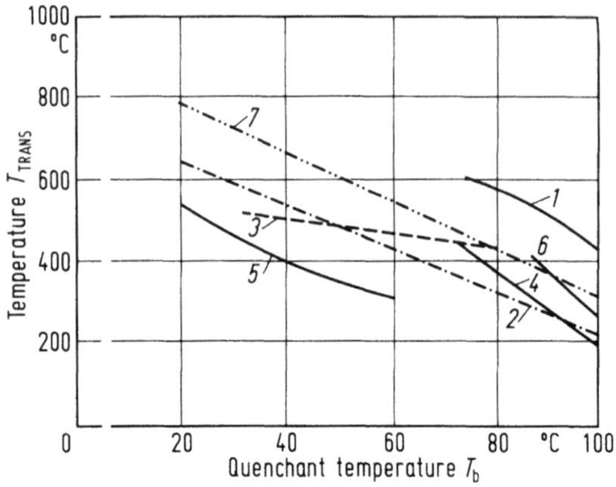

Fig. 5.11. Influence of quenchant temperature T_b on the temperature T_{trans} of the transition from lower to higher cooling velocity in the centre of different specimens, immersion cooled in water [7, 11]. *1* CrNi-steel cylinder, thermocouple in the specimens' centre [16]; *2* Molybdenum cylinder, thermocouple near the specimens' surface [11]; *3* CrNi-steel ball, thermocouple near the specimens' surface [17]; *4* CrNi-steel ball, thermocouple near the specimens' surface [18]; *5* Nickel ball, several thermocouples near the specimens' surface [19]; *6* Inconel ball, plated with chromium, thermocouple in the centre [20]

Influence of Organic Admixtures to Water

Admixtures to water, such as aqueous polymer solutions, greatly influence wetting behaviour. Although there are some model notions concerning the wetting process in aqueous polymer solutions (see also Eq. 5.1), it has not been possible to ascertain the influence of surface tension, viscosity, and/or macromolecular structure on the wetting process with universal validity, even when analytically pure tenside solutions were used [7, 11, 23].

When 5% NaDs (sodium dodecyl sulfate solution) is added to water, the heat transfer during nucleate boiling and convection is below 450 °C practically unchanged in comparison to water, while the stable vapour blanket is maintained down to 650 °C. The breakdown of the vapour blanket takes place at an immeasurably fast wetting speed (in an "explosive" manner), which is not the case in water (Fig. 5.12a). The transition temperature T_{trans} however, which is measured in the centre of the sample, remains almost unaffected by the concentration and shows the imperfection of temperature measuring in the specimens center (see also Chap. 8.1) (Fig. 5.12a, b).

An increase of the concentration of commercial aqueous polymer solutions usually reduces the transition temperature and accelerates the wetting process (cf. Fig. 5.4).

The following schematic temperature/time and conductance/time curves are intended to illustrate fundamental differences in the wetting behaviour of commercial polymer fluids (Fig. 5.13a–d). When the change in conductance $G(t)$ between the immersed sample and a given counterelectrode is measured in

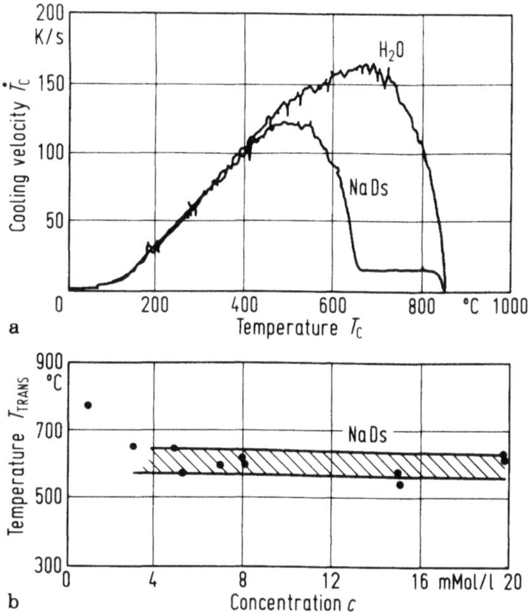

Fig. 5.12 a, b. Influence of water solution of NaDs (Natriumdodecylsulfat) on the velocity of cooling $\dot{T}_c$ and the temperature T_{trans} for transition from lower to higher cooling velocity, measured in the centre of a CrNi-steel specimen ($\phi 15 \times 45$ mm; lower edge broken according to D2 DIN 6784) [7, 11]. **a** Comparison of the cooling velocity during immersion in water and water solution of NaDs with $c = 5$ mMol/l; **b** Influence of the concentration c on the temperature for transition T_{trans}

addition to the temperature change over time in the sample centre (cf. also Chap. 8.1), the wetting processes can also be integrally recorded, assuming that the conductance observed is approximately proportional to the wetted surface. In Fig. 5.13a, wetting starts at temperature T_s (measured in sample centre) and is only completed at temperature T_f (slowly proceeding wetting). In Fig. 5.13b, wetting takes place within a short time interval (quickly proceeding wetting). In Fig. 5.13c, there also occurs sudden wetting, but bubbles remain attached to the surface, which reduce conductance between sample and counter-electrode and also the heat flux; thus, cooling also decelerates (usually in a locally confined area). In Fig. 5.13d, the transition to nucleate boiling is also "explosive", but after brief wetting a new vapour blanket builds up over the entire surface, followed by a renewed rapid wetting. This sequence may be repeated several times. Figure 5.14 and 5.15 depicts the measured wetting profile of a sample immersed in a commercial quenchant with a concentration of 30% and a temperature of 25 °C (corresponding to Fig. 5.13d).

The differences in the wetting process cannot be discerned from the temperature measurements in the sample centre with high enough precission. Since the wetting behaviour controls the temperature distribution in the cooling sample, it directly influences the hardness distribution. Thus it is particularly important to consider the wetting of the sample as a fluidspecific function.

Influence of Bath Convection

Two different convection states in quenching baths are usually distinguished: free convection (buoyancy-driven convection) and forced convection. It is especially

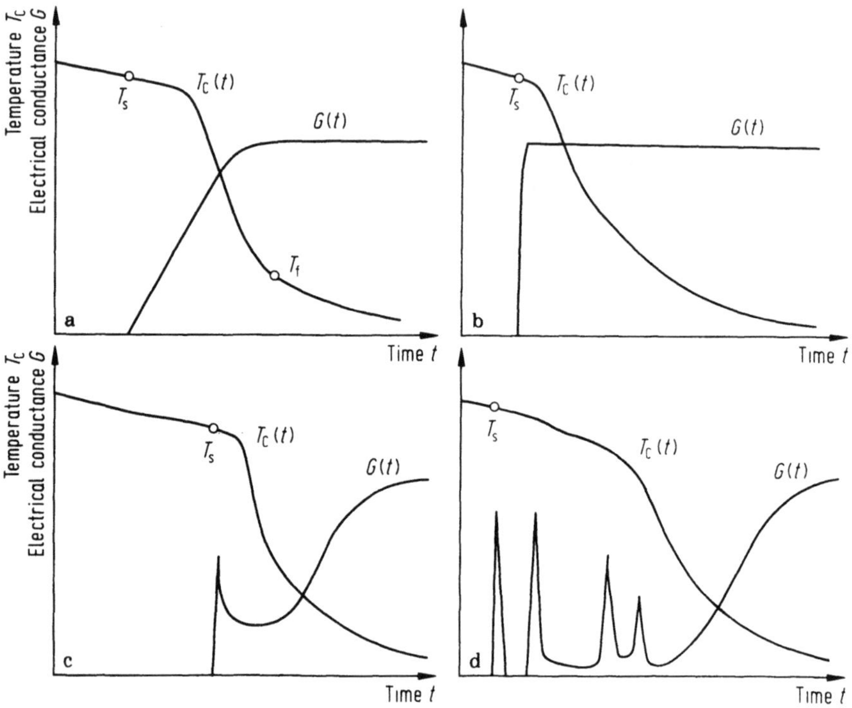

Fig. 5.13 a–d. Drop of temperature T_c (measured in the centre of the specimen) and increasing of electrical conductance G (measured between the submerged specimen and counter electrode; see also Chap. 8.2) during quenching in water solution of different polymers (schematically). The four diagrams show the different characteristic behaviour of water solution polymers [11]. **a** Slowly going wetting; **b** Quickly going wetting ("explosive wetting"); **c** Quickly going wetting with following isolation by bubbles adhering at the sample surface; **d** Quickly going wetting with repeated formation of film boiling, after this process (as in Fig. **c**)

forced convection that increases heat transfer during the three kinds of boiling phases. In aqueous polymer solutions and oils, forced convection has an additional effect on wetting kinematics: the temperature and duration of each of the different phases change considerably. In particular, the time intervals of simultaneous presence of the different phases are reduced. These changes in wetting kinematics have a significant effect on the cooling behaviour of immersed samples. Consequently, it is important to determine the influence of forced convection of a given intensity on the cooling performance of quenching media in the laboratory [24].

The wetting behaviour of another commercial aqueous polymer solution in the case of free and forced convection has been compared in Fig. 5.16. The quenchant exhibits rapidly proceeding wetting in Fig. 5.16a (sharp ascent of the conductance/time curves). In three repeated experiments (in which the experimental parameters were precisely maintained) wetting took place in a temperature area ranging between 720 °C and 650 °C (measured in the sample

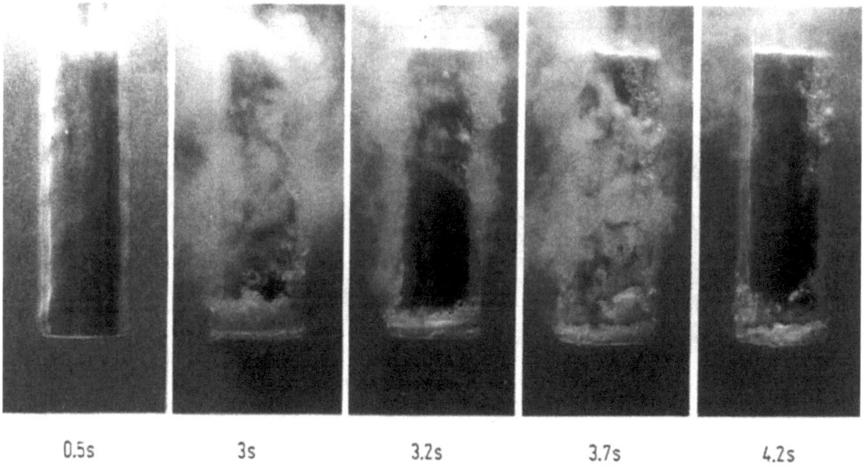

Fig. 5.14. Wetting kinetic with repeated collapse and new formation of film boiling on a CrNi-steel specimen ($\phi 15 \times 45$ mm) during quenching in water solution of a polymer with $c = 10\%$; quenchant temperature 25 °C without forced convection [11]; wetting behaviour according to electrical conductance in Fig. 5.13d

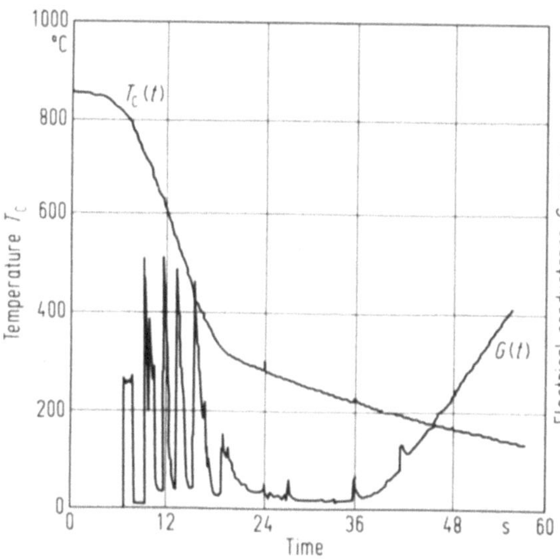

Fig. 5.15. Drop of temperature T_c (measured in the centre of a CrNi-steel specimen ($\phi 15 \times 45$ mm) and increasing of electrical conductance G (measured between the specimen and the counter electrode) during quenching in a water solution of a polymer with $c = 30\%$. Quenchant temperature 25 °C without forced convection [11, 12] (cf. Fig. 5.13d)

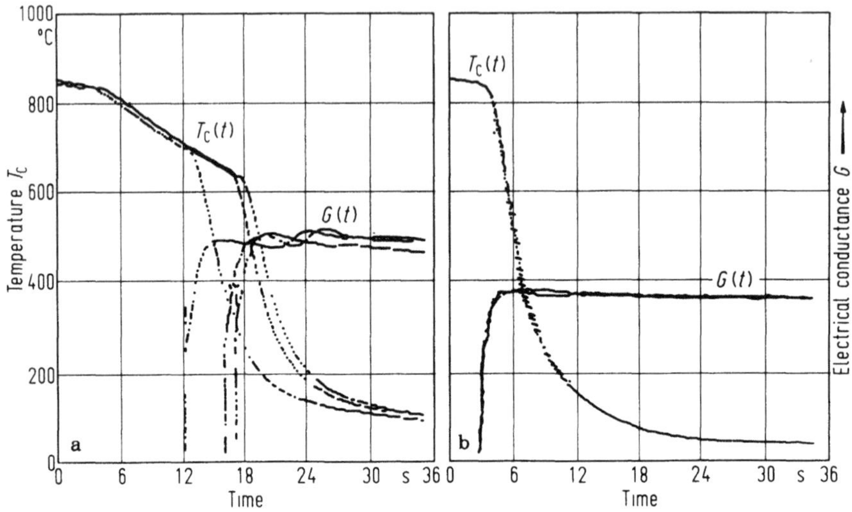

Fig. 5.16 a, b. Influence of bath convection on cooling behaviour: Drop of temperature T_c (measured in the centre of a CrNi-steel specimen ($\phi = 15 \times 45$ mm) and increasing of electrical conductance G (measured between the specimen and the counter electrode) during quenching in water solution of a polymer with $c = 5\%$ (every diagram shows three measurements) [11]. **a** Quenchant without forced convection. **b** Quenchant with a forced convection of $w = 0.3$ m/s

centre). Wetting began at a time interval of 12 to 17 s after the start of cooling. The dispersion areas of the temperature and conductance curves are largely diminished because of forced convection at 0.3 m/s (Fig. 5.16b). Wetting speed has been increased. The time interval of wetting has been shortened to approx. 3 s after immersion without measurable dispersion[1].

When quenching a prismatic cylinder (15 × 15 × 45 mm), the wetting front is parabolically shaped (Fig. 5.17). The wetting along the cylinder edges preceeds the wetting front of the flankes. As can be seen in Fig. 5.18, the starting time of wetting is not influenced by convection, while the finishing time drops from approx. 18 s to approx. 6 s, when forced convection is increased from 0 to only 20 cm/s. The time interval ($t_f - t_s$) of simultaneous presence of vapour blanket and nucleate boiling drops from 15 to 3 s!

Figure 5.19a, b shows the influence of convection on the wetting profile of a cylindrical CrNi sample quenched in commercial hardening oils. An increase of the flow velocity from 0 to 40 cm/s in a not accelerated quenching oil causes the starting and finishing times of wetting to drop almost synchronously by 6 s respectively, so that the wetting duration ($t_f - t_s$) remain unchanged at 5 s. The accellerated quenching oil exhibits an almost instantaneous start of wetting

[1] The different levels of the final value of G in the diagrams a) and b) are caused by adjustments of the measurement apparatus. The respective final values correspond to 100% wetting

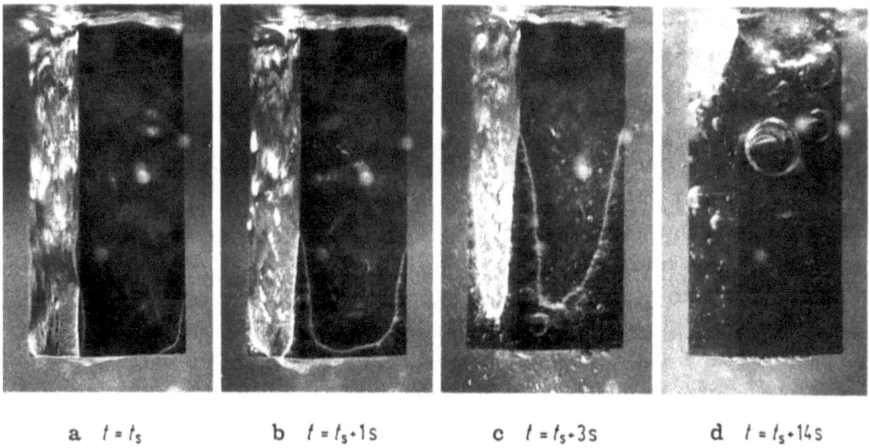

Fig. 5.17 a–d. Process of transition between the phases film boiling (FB), nucleate boiling (NB) and convective heat transfer (CONV) during immersion cooling of a prismatic CrNi-steel sample (15 × 15 × 45 mm); immersion temperature 860 °C; quenchant: destilled water with $T_b = 35\,°C$ [11, 12]

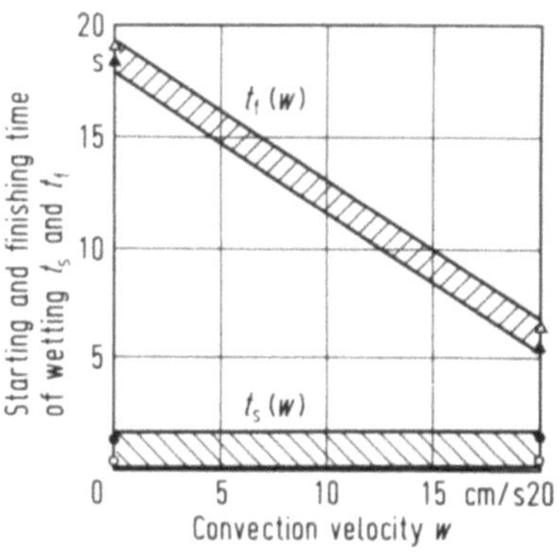

Fig. 5.18. Influence of convection velocity w on the starting and finishing time (t_s and t_f) of wetting of a prismatic CrNi-steel specimen (15 × 15 × 45 mm), during quenching in water with $T_b = 60\,°C$ [11]. ○; △ measuring points obtained from electrical conductance; ●; ▲ measuring points obtained from temperature measurements close to the specimen's surface (see also Chap. 8.1, Figs. 8.8 and 8.11)

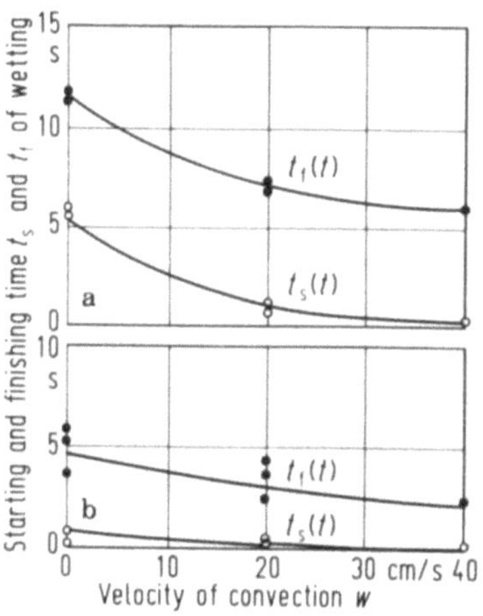

Fig. 5.19 a, b. Influence of forced convection on the starting and finishing time (t_s and t_f) of wetting a CrNi-steel specimen (ϕ15 × 45 mm) during immersion cooling in different kinds of oil quenchants; bath temperature $T_b = 40\,°C$ [15]. **a** normal quenching oil; **b** accellerated quenching oil

($t_s \approx 1$ s) and a significantly more moderate reduction of the finishing time under the same conditions. Wetting time is reduced from 5 s to 3 s.

5.6 Impact of Sample Properties on Wetting Process

The most important properties of the sample which influence wetting behaviour have been enumerated under Chap. 5.2.

Figure 5.20 illustrates to which degree wetting behaviour is influenced by the radius of the lower surface of a cylindrical steel sample. Figure 5.20a shows how the starting and finishing temperature of wetting are influenced by the quenchant temperature in the case of a cylindrical sample which has an extremely small lower surface radius (Cl according to DIN 6784). Wetting starts immediately upon immersion and remains unaffected up to a quenchant temperature of 60 °C. From 60 °C up to the boiling temperature, the dispersion range of the temperature in the centre of the sample expands at the start of wetting. The starting temperature drops by approx. 300 K at $T_b > 85$ °C. The temperature

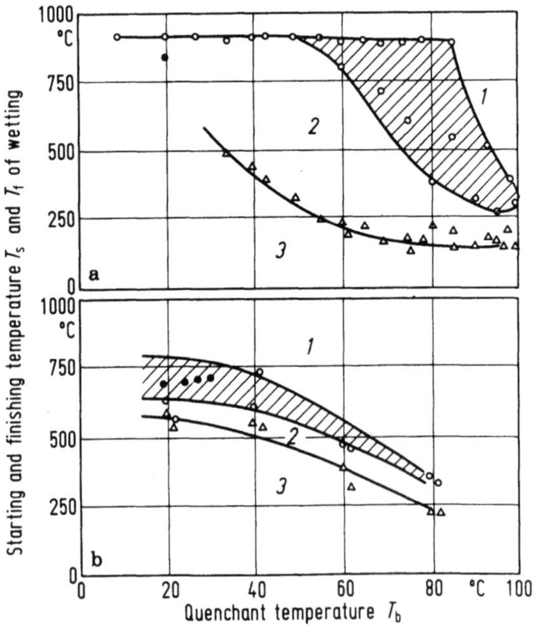

Fig. 5.20 a, b. Influence of quenchant temperature T_b on the starting and finishing temperature (T_s and T_f) of wetting a CrNi-steel specimen ($\phi 15 \times 45$ mm) with different kind of lower edge; [13]. *1* Film boiling (FB); *2* nucleate boiling with FB and CONV; *3* convective heat transfer. **a** Sharp lower edge; **b** lower edge with $r = 7.5$, (i.e. semispheric). Immersion temperature: 920 °C; ○ starting temperatures; △ finishing temperatures; ● Starting temperature of the wetting of a Mo-cylinder with different kind of lower edge

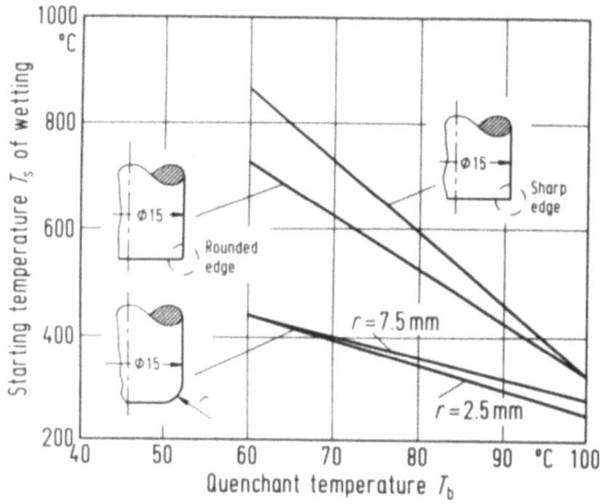

Fig. 5.21. Influence of quenchant temperature T_b on the starting temperature T_s of wetting of CrNi-steel specimens ($\phi 15 \times 45$ mm) with different kinds of lower edge geometry during immersion cooling in water [7, 11]. Sharp edge according to "Cl DIN 6784", rounded edge according to "D2 DIN 6784"

interval of wetting ($T_f - T_s$) is thus considerably reduced. Wetting behaviour is fundamentally different in the case of a lower surface radius of 7.5 mm (semispherical sample end Fig. 5.20b). An increase in bath temperature causes an equivalent reduction of the temperatures at the beginning and the end of wetting. The temperature interval describing the wetting process is in this case very small and, in addition, it remains unaffected by the temperature of the quenchant. Comparison measurements concerning the start of wetting in the case of a Mo cylinder with the same geometry are also shown in the diagrams [as per 9]. The good agreement between measured values is due to the largely identical thermal diffusivities of the materials Mo and CrNi steel.

Figure 5.21 summarizes the influence exerted by the radius of the lower surface on the starting temperature during the wetting of CrNi alloyed steel samples. The influence of the bath temperature on the starting temperature of wetting is diminished with increased lower surface radii, because of increasing thickness δ of the film phases.

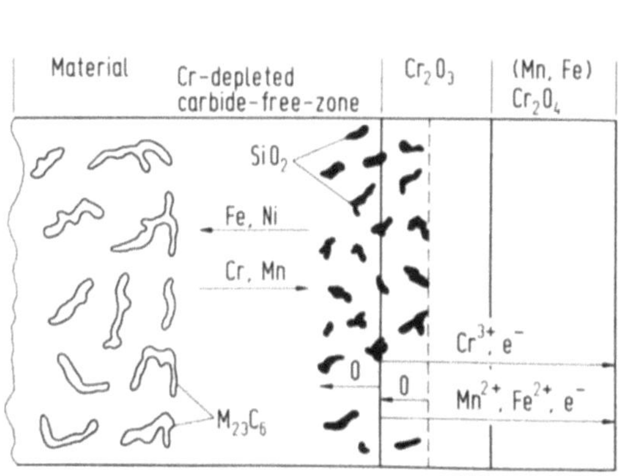

Fig. 5.22. Oxide coatings (passive film) on CrNi-steel specimen, produced by annealing in oxiding atmosphere [25]

It is a well-known fact that during oxidation chromium alloy bodies are passivated by the concentration of chromium oxide on the surface. An oxide surface layer is formed, as shown in Fig. 5.22 [as per 15, 25]. Figure 5.23 shows to what degree the temperatures of the beginning and the end of wetting, the starting time of wetting, and the time interval of wetting are influenced by increasing oxidation of the surface of a cylindrical sample. The thickness of the passive layer is determined by the amount of time the sample is annealed at 920 °C in an unprotected atmosphere. It becomes apparent from the impact on the temperature values and wetting times that additional oxidizing annealing treatment does not considerably reduce the starting time of wetting (or substantially increase the temperature at the beginning of wetting) after approx. 10 hours.

Wetting behaviour is also greatly influenced by the surface roughness. Clear influences on the wetting profile can be observed during immersion cooling (quenchant: water, temperature >60 °C) of cylindrical samples with surface structures ranging between smooth (surface roughness approx. 20 μm) and extremely rough (corresponding to fine-pitch thread, $R_t = 0.9$ mm). There is no observable influence on the starting temperature or wetting time at 20 °C, but beginning with $T_b = 60$ °C starting temperatures drop and the time interval of wetting becomes longer with increasing surface roughness (Fig. 5.24).

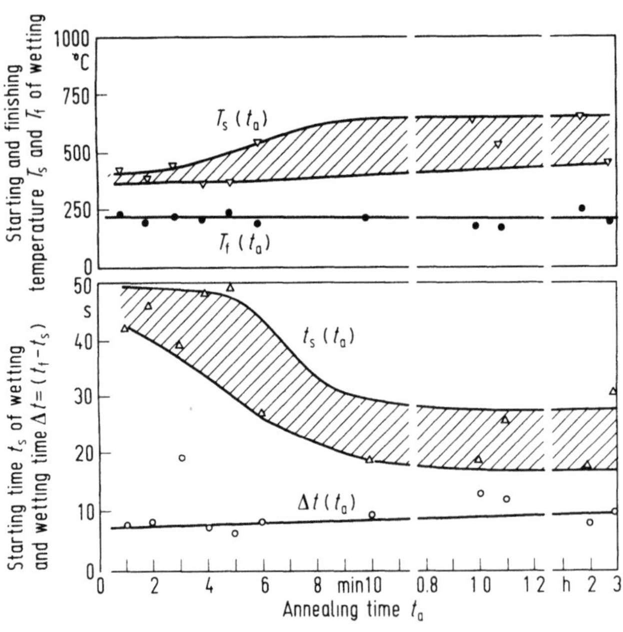

Fig. 5.23. Influence of oxide coatings on CrNi-steel specimens on the wetting during immersion cooling in water [11, 12]. Specimens: $\phi 15 \times 45$ mm, lower edge radius 5 mm; annealing condition: 920 °C, 60 s $\leq t \leq$ 3 h in unprotected atmosphere; quenchant: water with $T_b = 60$ °C, without forced convection

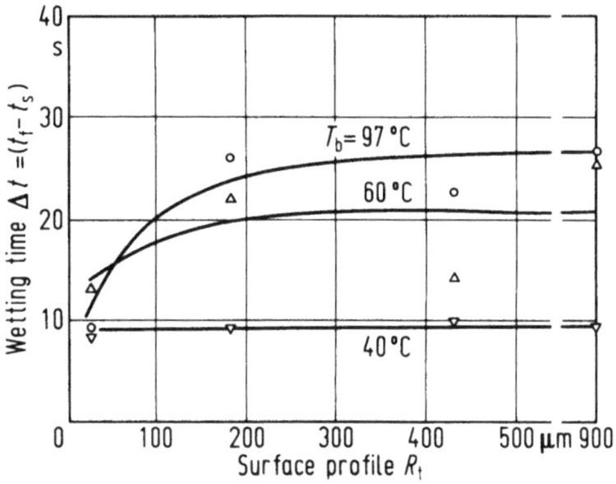

Fig. 5.24. Influence of the surface roughness R_t on the wetting time $(t_f - t_s)$ during immersion cooling in water with different temperatures T_b [11]. Specimens: CrNi-steel, $\phi 15 \times 45$ mm; quenchant: water without forced convection

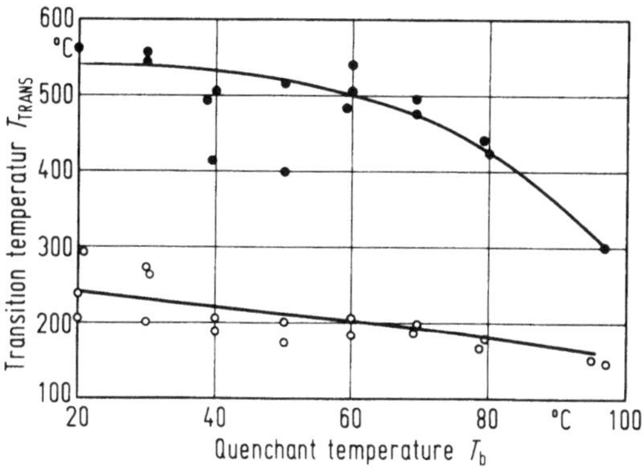

Fig. 5.25. Influence of quenchant temperature T_b on the transition temperature T_{trans} during immersion cooling of specimens with different oxide coatings [7, 11]. Specimens: AlMg5 alloy, $\phi 15 \times 45$ mm, lower edge softened; ● anodized surface with thickness of coating 5 µm; ○ bright surface (polished before annealing); immersion temperature 560 °C; quenchant: water with forced convection

The situation is similar with materials characterized by extreme high thermal diffusivities, such as Ag and Al alloys.

In conclusion, the influence of the surface quality on the wetting of a cylindrical AlMg5 sample cooled in baths with different temperatures is described (Fig. 5.25). The transition temperature measured in the sample centre indicates that when bath temperatures are low, wetting of the anodized sample starts immediately upon immersion and, at a bath temperature of aprox. 95 °C, drops to 300 °C. The transition temperature of the sample covered only by a thin oxide layer[1] is approx. 350 K lower, and can further be reduced to approx. 150 °C with increasing bath temperature.

The stability of the oxide layers varies extremely during annealing, which should particularly be considered when comparing the Al and Ag samples with their very similar heat diffusivities. Annealing a silver sample above 600 °C causes the destruction of the oxide layer. A sample treated in this way will be

[1] This oxide layer was created in the furnace after polishing the specimen before annealing

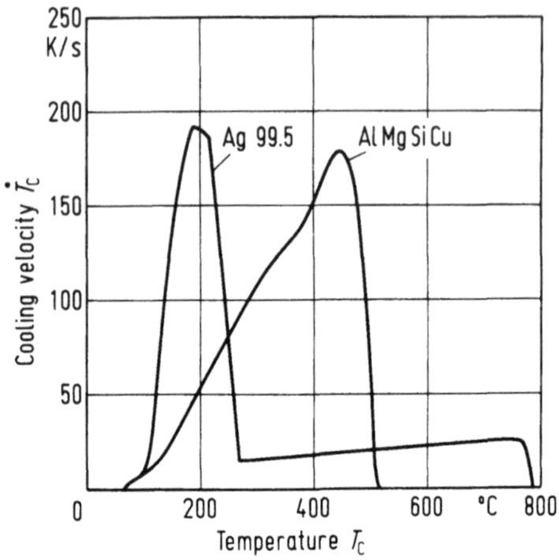

Fig. 5.26. Comparison of the velocity w of specimens cooling made of silver and aluminium [26]. Specimens: Ag99, 5 and AlMgSi $\phi 15 \times 45$ mm, lower edge softened, immersion temperature 800 °C for Ag, 520 °C for AlMgSi; quenchant: water solution of a polymer ($c = 10\%$); $T_b = 25$ °C, without forced convection

quenched in an "extremely bare" condition, which has the already mentioned effects on wetting behaviour. This is made clear by comparing the cooling rates plotted against the sample temperature T_c, as they occurred during the cooling of an Ag sample (temperature at immersion 800 °C). While the oxide layer of the aluminum sample is even strengthened by annealing and the sample is thus instantaneously wetted, the Ag sample becomes extremely bare after annealing and there is a stable vapour blanket down to approx. 250 °C. The maximum values of the cooling rates of both samples are at virtually the same level because of the materials' similar thermal diffusivities (Fig. 5.26).

5.7 Summary

During the immersion cooling of hot metals in fluids whose boiling temperatures are below the immersion temperatures, the contact between the fluid and the surface of the immersed body is initially interrupted by a vapour blanket. The transfer of heat into the fluid is largely restricted during this phase. As soon as the surface of the metal is wetted, the direct contact between surface and fluid causes the heat transfer to increase by at least two magnitudes compared to the vapour blanket phase. The events taking place during wetting and their connection with the Leidenfrost temperature are described.

The wetting process (wetting kinematics) influences the temperature field in the body during its cooling and, in the case of quench hardening, the heat distribution in the body.

The way to influence the wetting kinematics through the properties of the bath (such as chemical composition, temperature, convection) and of the metallic bodies (thermal and geometrical characteristics, surface condition) are described with examples.

5.8 List of Symbols

A	surface, m^2
a	thermal diffusity $a = \lambda/c_p \cdot \varrho$, $m^2 s^{-1}$
Bi	Biot-number $= \alpha \cdot l/\lambda_s$
c	concentration of polymer in fluid, vol. $-\%$
D	diffusion coefficient, $m^2 s^{-1}$
c_p	specific heat capacity, $J kg^{-1} K^{-1}$
Fo	Fourier-number $= a \cdot t \cdot 1^{-2}$
g	gravity, ms^{-2}
G	electrical conductance, S
Gr	Grashof-number $= g \cdot \beta \cdot \Delta T \cdot l^3/v^2$
l	typical length, m
Δh_F	enthalpy difference between steam and fluid with boiling temperature
m_d	steam flux, $kg \cdot s^{-1}$
Nu	Nußelt-number $= \alpha \cdot l \cdot \lambda_L^{-1}$
Pr	Prandtl-number $= v \cdot a^{-1}$
Ra	Rayleigh-number (Gr·Pr)
R_t	depth of profile, µm
r	radius in specimen, m
t_f	time at which wetting is concluded (finishing time of wetting), s
t_s	time at which wetting starts (starting time of wetting), s
T_b	temperature of bath, °C
T_c	temperature in centre of sample, °C
T_f	temperature in sample centre at which wetting is concluded (finishing temperature of wetting), °C
T_0	starting temperature in sample, °C
$T_{c,s}$	temperature in sample centre at which wetting starts (starting temperature) °C
T_{trans}	temperature in sample centre at the point from lower to higher cooling velocity (transition temperature), °C
$\dot{T}$	variation of temperature with time, Ks^{-1}
w	velocity of flow of cooling fluid, ms^{-1}
z	distance from the lower edge of specimen, m
α_{conv}	heat transfer coefficient during convection, $Wm^{-2} K^{-1}$
α_{FB}	heat transfer coefficient during film boiling, $Wm^{-2} K^{-1}$
α_{NB}	heat transfer coefficient during nuclear boiling, $Wm^{-2} K^{-1}$
β	elongation coefficient, K^{-1}
λ_d	thermal conductivity in vapour, $Wm^{-1} K^{-1}$
λ_{fl}	thermal conductivity in liquid, $Wm^{-1} K^{-1}$
λ_s	thermal conductivity in specimen, $Wm^{-1} K^{-1}$
δ	thickness of film boiling phase, m
v	kinematic viscosity, $m^2 s^{-1}$
ϱ	mass density, $kg m^{-3}$
$\varepsilon_{W,F}$	emission coefficient between submerged specimen and fluid
σ	Stefan Boltzmann constant, $Wm^{-2} K^{-4}$

References

1. Leidenfrost GJ (1966) De Aqua Communis Nonnullis Qualitatibus Tractatus 1756. In: Int. J. Heat Mass Transfer 9: 1153–1166, translated by C. Waves
2. Berenson PJ (1961) Film Boiling Heat Transfer from a Horizontal Surface. In: J. Heat Transfer 83c: 351–358
3. Henry RE (1973) A Generalized Correlation for the Minimum Point in Film Boiling. 14th National Heat Transfer Conference, AICHE-ASME Atlanta, Georgia
4. Yamanouchi A (1968) Effect of Core Spray Cooling in Transient State After Loss of Cooling Accident. In: J. Nuclear Science and Technology 5: 547–558
5. Thompson TS (1974) On the Process of Rewetting a Hot Surface by a Falling Liquid Film. In: Nuclear Engeneering and Desigue 31: 234–245
6. Duffly RB, Porthouse DTC (1973): The Physics of Rewetting in Water Reactor Engeneering Core Cooling. Nucleation Engineering and Design 25: 379–394
7. Künzel Th (1986) Einfluß der Wiederbenetzung auf die allotrope Modifikationsänderung tauchgekühlter Metallkörper. Doctoral Thesis. Technical University of Munich, Faculty of Mech. Engeneering
8. Hein D (1980) Modellvorstellung zur Wiederbenetzung durch Fluten. Doctoral Thesis. Technical University Hannover
9. Ladisch R (1980) Untersuchung der minimalen Filmsiedetemperatur auf keramischen und metallischen Leitern. Kernforschungszentrum Karlsruhe KfK 2970
10. Kommnos A (1981) Ein thermo-hydrodynamisches Modell zur Wiederbenetzung. Doctoral Thesis. Technical University of Munich, Faculty of Engeneering
11. Tensi HM, Stitzelberger-Jakob P, Künzel Th, Stich A (1989) Wetting Kinematic and Influence on the Metallurgical Structures. Final DFG-Report (Contract Number Te 65/27-1, 2). Deutsche Forschungsgemeinschaft (DFG), FRG-Bonn
12. Stitzelberger-Jakob P (1990) Härtevorherbestimmung mit Hilfe des Benetzungsablaufes beim Tauchkühlen von Stählen. Doctoral Thesis. Technical University Munich, Faculty of Mech. Engeneering
13. Tensi HM, Welzel G, Künzel Th (1986) Problems of Getting Characteristicas of the Quenching Experiments Useful to Control the Heat Treatment of Metallic Materials. In: 8th Int. Heat Transfer Conf.: 3031-35 San Francisco
14. Tensi HM, Künzel Th (1987) Bedeutung der verschiedenen Siedephasen für das Abkühlen tauchgekühlter Körper—numerische Simulation und experimentelle Überprüfung. In: Neue Hütte 34: 354–359
15. Künzel Th, Tensi HM, Welzel G (1986) Rewetting Rate—the Decisive Characteristic of a Quenchant; Proceedings of the 5th Int. Congress on Heat Treatment of Materials: 1806–1813 Bůdapest
16. Tensi HM, Steffen E (1985) Neue Methode zur quantitativen Bestimmung der Abschreckung flüssiger Härtemedien, hier speziell wässrige Kunststofflösüngen. In: Wärme- und Stoffübertragung 19: 279–286
17. Jeschar R, Maaß R (1985) Bestimmung des Wärmeüberganges beim Abschrechkverhalten von Metallen in Wasser. In: Gas Wärme International 34/9: 348–354
18. Farahat MM, Armstrong DR (1977) Transient heat transfer between hot metal Spheres and Subcooled Water. In: Atomenergie (ATKG) 29/1: 17–22
19. Dhin VK, Puronit GP (1977) Subcooled Film Boiling and the Onset of Transition Boiling. In: ASMG Paper 77-HT-78, 1.22
20. Bradfield WS (1967) On the effect of supercooling on Wall superheat in Pool Boiling. In: Trans. ASMG J. Heat Transfer 89: 269
21. Andreoni D (1975) Echanges thermiques lors renoyage d'un cour de reacteur a eau. Doctoral Thesis. University of Grenoble
22. Tensi HM, Schwalm M (1980) Wirkung von Abschreckflüssigkeiten unter Berücksichtigung spezieller wäßriger Polyäthylenoxide. In: HTM 35: 122–129
23. Pilling NM, Lynch TD (1920) Cooling Properties of Technical Quenching Liquids. In: Trans. A. Inst. Min. Met. Eng. 62: 665–688
24. Tensi HM, Stitzelberger-Jakob P (1989) Evaluation of Apparatus for Assesing Effect of Forced Convection on Quenching Characteristics. In: Materials Science and Technology 5: 718–724
25. Rahmel A (1983) Aufbau von Oxidschichten auf Hochtemperaturwerkstoffen und ihre technische Bedeutung. DGM-Berichtsband
26. Tensi HM, Stitztelberger-Jakob (1988) Bedeutung der Wiederbenetzung für das Abkühlen tauchgekühlter Aluminiumproben. HTM 43: 148–154

6 Residual Stresses After Quenching

E. MACHERAUCH and O. VÖHRINGER

6.1 Introduction

In the course of special heat treatments of metallic components and during the fabrication of intermediate products, quenchings from high to low temperatures are frequently applied. In such cases, quenchants must be brought into contact with the surfaces of the quenched materials, allowing a sufficiently rapid heat transfer. The mostly employed quenching methods involve either immersion cooling, splash cooling or film cooling. Immersion cooling for which a workpiece is submerged in an unmoved or agitated liquid, is, for example, applied in the field of martensitic hardening of steels and after annealing prior to precipitation hardening of age-hardenable aluminium and nickel alloys. Splash cooling which involves the spraying of the coolant through suitable nozzles onto hot workpieces is commonly used in connection with continuous casting and with flame hardening. Film cooling, which involves heat transfer by the aid of liquid films, is utilized for particular casting processes of non-ferrous metals.

In practice, quenchings in connection with special heat treatments of steels are of particular engineering importance, since in that way these materials can easily be brought into a wide variety of conditions with characteristic mechanical properties. In the following, only such quenchings are considered. The most important process in this respect is martensitic hardening, i.e. the formation of martensite from the high temperature austenitic phase during rapid cooling to room temperature. Also quenching and tempering treatments as well as case hardening treatments of steels are basically determined by martensitic transformations. But transformation-free quenchings of steel parts are of practical interest, too. All these treatments inevitably yield residual stresses and distortions in the afflicted parts, which result either from coupled thermo-mechanical interactions or from the complex superposition of thermo-mechanical with transformation effects during the usually rapid cooling from sufficiently high temperatures down to room temperature (see e.g. 1–11). In the following, however, distortions will not be further taken into consideration.

In order to understand the occurrence of residual stresses at the end of a quench, one must first of all be acquainted with the principles of how temporal and local differences in cooling and phase transformation produce shrinking (thermal) strains and transformation strains and change the structural state. Thereby, transformation-free and transformation-influenced quenchings have

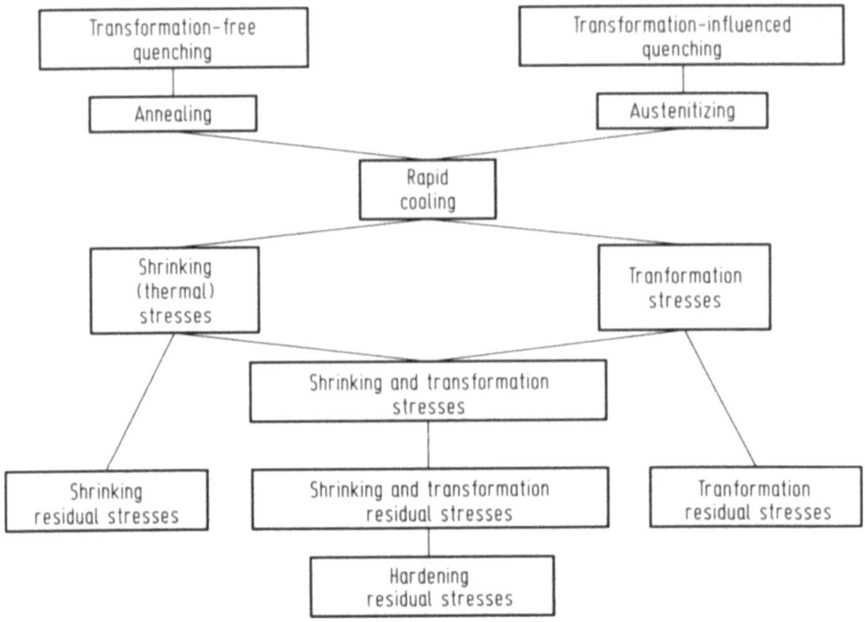

Fig. 6.1. Transformation-free and transformation-influenced quenching

to be distinguished as schematically outlined in Fig. 6.1. During cooling in a transformation-free quench for example, at any time local differences of temperature occur within a component creating triaxial shrinking (thermal) stresses. These stresses locally cause plastic deformations if the corresponding equivalent stresses exceed the local yield strength of the material state. If on the other hand non-equilibrium transformations of the austenitic phase occur during quenching, then additionally the kinetic of these phase transformations and their interactions with the local stress states have to be considered. As a consequence, transformation induced volume changes and more or less pronounced transformation plasticity effects have to be taken into account. Obviously, the field of quenching of steels is of great complexity. Therefore, a reasonable approach is first to consider separately the development of pure shrinking (thermal) stresses and of pure transformation stresses during quenchings under distinct boundary conditions and second to discuss the generation of shrinking and transformation residual stresses. Then, the next consequent step should be to deal with the combined effects of shrinking and transformation stresses and with their consequences on the development of hardening residual stresses. Finally, a few comments on characteristic types of surface residual stresses due to special surface hardening techniques for steels seem to be suitable. Case-hardening of carburized (carbonitrided) steel parts with thermo-chemically produced changes in the carbon (carbon and nitrogen) content near the surface requires rapid quenching to receive a high-carbon hard case and a low-carbon tough core. In

Table 6.1. Material abbreviations of the steels investigated, if available, to AISI according to DIN and their nominal chemical compositions

DIN	AISI	Nominal composition in wt.-%
C 15	C 1015	0.15 C
C 22	C 1022	0.22 C
Ck 35	C 1135	0.35 C
Ck 45	—	0.45 C
C 150	—	1.50 C
15 NiCr 3	—	0.15 C; 0.75 Ni; 0.5 Cr
16 NiCr 5	—	0.16 C; 1.25 Mn; 1.0 Cr
18 CrNi 8	—	0.18 C; 2.0 Cr; 2.0 Ni
17 CrNiMo 6	—	0.17 C; 1.5 Cr; 1.5 Ni; 0.2 Mo
20 MoCr 4	—	0.20 C; 0.4 Mo; 0.4 Cr
20 MnMoNi 55	(A 508 class 3)	0.20 C; 1.25 Mn; 0.5 Mo; 0.6 Ni
20 NiMoCr 3	8620	0.20 C; 0.8 Ni; 0.2 Mo; 0.5 Cr
22 CrMo 4	4122	0.22 C; 1.0 Cr; 0.2 Mo
26 MnCr 4	—	0.26 C; 1.0 Mn
28 CrMo 10	—	0.28 C; 2.5 Cr; 0.4 Mo
34 CrMo 4	4134	0.34 C; 1.0 Cr; 0.4 Mo
41 Cr 4	5140	0.41 C; 1.0 Cr
42 CrMo 4	4142	0.42 C; 1.0 Cr; 0.2 Mo
55 NiCrMoV 6	—	0.55 C; 1.5 Ni; 0.7 Cr; 0.3 Mo; 0.1 V
90 MnV 8	O2	0.90 C; 2.0 Mn; 0.1 V
X5 CrNi 189	304	0.05 C; 18.0 Cr; 9.0 Ni
X20 Cr 12	420	0.20 C; 12.0 Cr
X30 NiCrMo 51	(835 M 30)	0.30 C; 5.0 Ni; 1.0 Cr

comparison with this, flame hardening, induction hardening and laser hardening utilize martensitic transformations due to rapid heating and quenching of only the surface layers of steels having no gradients in the chemical composition. The stresses remaining after such treatments are called case hardening residual stresses, flame hardening residual stresses, induction hardening residual stresses and laser hardening residual stresses.

The present work will be guided by the above mentioned aspects. Initially, however, some introductory points will be discussed concerning the definition of residual stresses and some basic relationships between temperature and structure and fundamental properties of plain carbon and low alloy steels. All steels mentioned in this work are indicated with the German materials designations according to DIN. Their nominal compositions and existing equivalent designations according to AISI are summarized in Table 6.1.

6.2 Some Fundamentals

6.2.1 Definitions of Residual Stresses

Residual stresses are stresses existing in a body which is not externally loaded by forces and/or moments. Consequently, the resultant force and the resultant

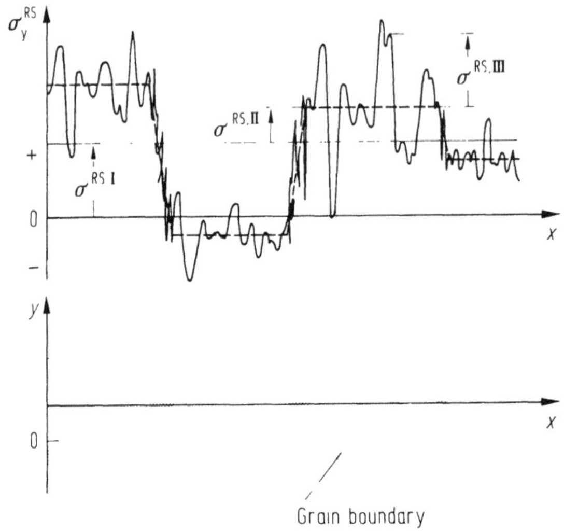

Fig. 6.2. Scheme for the definition of 1st, 2nd and 3rd kind residual stresses

moment produced by residual stresses must be zero. In order to obtain an appropriate and standardized system of designations, the classification of three different types of residual stresses, i.e. residual stresses of the 1st, 2nd and 3rd kind, is expedient. Using the term "homogeneous" for "constant in magnitude and direction", these stresses are defined as follows [12]:

— Residual stresses of the 1st kind are approximately homogeneous across large areas, e.g. several grains of a material, and are equilibrated with respect to the whole body. Any intervention in the equilibrium of forces and moments of a volume element containing such residual stresses will change its macroscopic dimensions.
— Residual stresses of the 2nd kind are approximately homogeneous across microscopically small areas (one grain, or sub-granular regions) and are in equilibrium across a sufficient number of grains. Macroscopic changes in the dimensions of a volume element possessing these stresses may only become apparent if distinct disturbances of this equilibrium occur.
— Residual stresses of the 3rd kind are inhomogeneous across submicroscopically small areas of a body (i.e. regions with dimensions of several atomic distances within single grains) and are in equilibrium across sub-granular sections. No macroscopic changes of the dimensions of the stressed material will result at all, upon disturbing such equilibria.

1st kind residual stresses are also called macro residual stresses, 2nd and 3rd kind residual stresses are often referred to as micro residual stresses.

It can be shown that these definitions suffice to describe all residual stress states occurring in components in the absence of mechanical effects due to non-mechanical influences. In Fig. 6.2 a possible local superposition of residual stresses of the three different types is schematically illustrated. The *y*-component of the total residual stress is drawn as a function of x at $y = \text{const}$. Using RS as an abbreviation for residual stresses, the following relationships exist:

$$\sigma^{RS} = \sigma^{RS,I} + \sigma^{RS,II} + \sigma^{RS,III} \tag{6.1}$$

with

$$\sigma^{RS,I} = \left.\frac{\int \sigma^{RS}\, dA}{\int dA}\right|_{A}, \tag{6.2}$$

where $A = nA_K$ with A_K the mean grain area and n an arbitrary large number,

$$\sigma^{RS,II} = \left.\frac{\int \sigma^{RS}\, dA_K}{\int dA_K}\right|_{\text{one grain } A_K} - \sigma^{RS,I} \tag{6.3}$$

and

$$\sigma^{RS,III} = \left[\sigma^{RS} - (\sigma^{RS,I} + \sigma^{RS,II})\right]\Big|_{\substack{\text{at one point } x \\ \text{at a grain}}}. \tag{6.4}$$

The macroscopic equilibrium of the forces with respect to 1st kind residual stresses for any plane section area A of the whole body requires that

$$\int_A \sigma^{RS,I}\, dA = 0 \tag{6.5}$$

and the equilibrium of the moments for any arbitrary reference line in the section area A demands that

$$\int_A \sigma^{RS,I} r\, dA = 0. \tag{6.6}$$

Thereby r represents the reference line distance for an element dA of the area A. Analogically, for the equilibrium of 2nd kind residual stresses, the last two equations are valid with $A = nA_K$. In the case of 3rd kind residual stresses $A = A_K$ holds.

These definitions sufficiently describe all residual stress states occurring in practical cases. Usually, a superposition of residual stresses of the 1st, 2nd and 3rd kind determines the total residual stress state acting locally at a given material state. A typical example is sketched in Fig. 6.3. A cylindrical section of a two-phase material is considered, in which stable phases A and B possess different coefficients of thermal expansion. A stress free cylinder state is assumed at a sufficiently high temperature. When quenching from this temperature to room temperature, a 1st kind or macro residual stress state develops as a consequence of the shrinkage stresses induced over the cross-section of the cylinder. The longitudinal stresses which exist along the cut $x - x'$ of a section plane perpendicular to the axis of the cylinder are schematically shown on the left in the middle of Fig. 6.3. The picture in the centre of the figure illustrates the 2nd kind residual stresses induced by the different magnitudes of thermal strains in the grains of both phases after temperature equalization. The amounts of these stresses increase proportional to the temperature interval passed through by

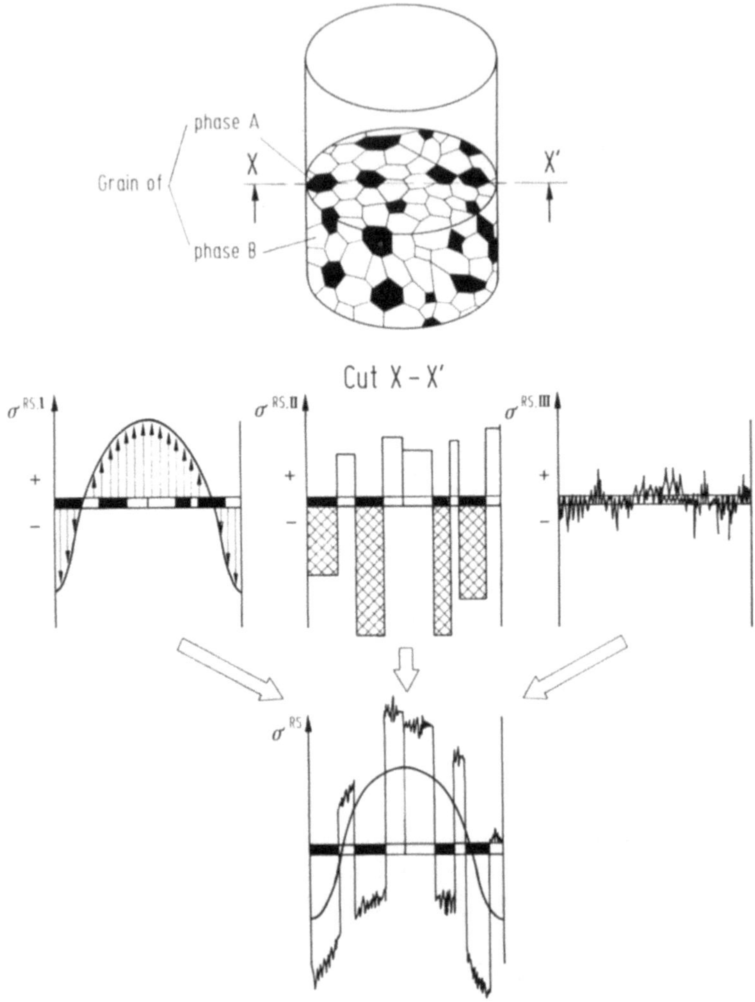

Fig. 6.3. Different kinds of residual stresses in a two-phase material after quenching (schematically)

quenching. Furthermore all types of lattice imperfections within the grains of the two phases, e.g. dislocations or dissolved foreign atoms, cause 3rd kind residual stresses as illustrated on the far right in the middle of the figure. Finally, the superposition of all three kinds of residual stresses leads to the situation illustrated at the bottom of Fig. 6.3.

From an engineering point of view, 1st kind (or macro) residual stresses are nearly always those concerned when discussing residual stress problems. Hence, all further considerations will be restricted to 1st kind or macro residual stresses, respectively.

6.2.2 Quenching of Steel Cylinders

Before components can be quenched, they have to be heated to an adequate temperature. This may either be accomplished by means of a direct energy transfer (e.g. by inductive heating) or an indirect one (e.g. heating in a furnace). Afterwards, a sufficiently rapid extraction of heat is of particular interest. Both processes require knowledge of the temperature distribution within the heat treating appliance, the heat transfer to the component, the heat conductivity which subsequently occurs in the component and the heat transition behaviour during quenching in suitable media. If no interior heat sources or sinks exist, the temperature distributions in a body underlying heating or cooling can be obtained from solutions of the Fourier equation

$$\frac{\partial T}{\partial t} = \frac{\lambda}{\rho c_p}\left[\frac{\partial^2 T}{\partial x^2} + \frac{\partial^2 T}{\partial y^2} + \frac{\partial^2 T}{\partial z^2}\right] \tag{6.7}$$

for appropriate boundary and initial conditions. λ is the coefficient of thermal conductivity in J/smK, ρ the density in kg/m^3, c_p the specific heat under constant pressure in J/kgK, T the temperature in K. t is the time in s and x, y and z are local coordinates. The parameter $a = \lambda/\rho c_p$ in m^2/s is known as the thermal diffusivity. ρ, λ and c_p, and therefore also a, dependend on temperature.

Only a few complete solutions are available for Eq. 6.7 [13, 14]. Of course, problems, which are only dependent on a single coordinate, can more easily be handled than three-dimensional ones. Considering e.g. a quenching process for a cylinder with radius r_0, at $t = 0$ the existence of a homogeneous temperature distribution with $T = T_0$ across all radii $r \leqq r_0$ can be stated. Then, in the case of an adequately long cylinder, the temperatures at constant distances r from the cylinder axis are pure functions of r and the radial heat flux across the cylinder surface with the surface temperature $T_{surface}$ to a medium of the ambient temperature $T_{ambient}$ is given by the Newton heat transition law

$$\dot{q} = -\lambda\left[\frac{\partial T}{\partial r}\right]_{surface} = h(T_{surface} - T_{ambient}). \tag{6.8}$$

Here, $\dot{q}$ represents the specific heat flux and h the heat transfer coefficient in J/sm^2K.

The local temperature-time relations and the temperature distributions along the diameter of a steel cylinder in course of cooling after a quench from 840 °C in water and in oil of 20 °C are exemplarily presented in Fig. 6.4 [15]. The local cooling curves 1–5 differ considerably for water quenching (Fig. 6.4a) and oil quenching (Fig. 6.4b). In the lower parts of Fig. 6.4 the appertaining radial temperature distributions existing at various instants across the diameter in the middle sections ($z = 0$) of the cylinders are shown. It can be seen that due to the faster temperature balance, quenching in water creates higher temperature gradients across the cylinder diameter than quenching in oil.

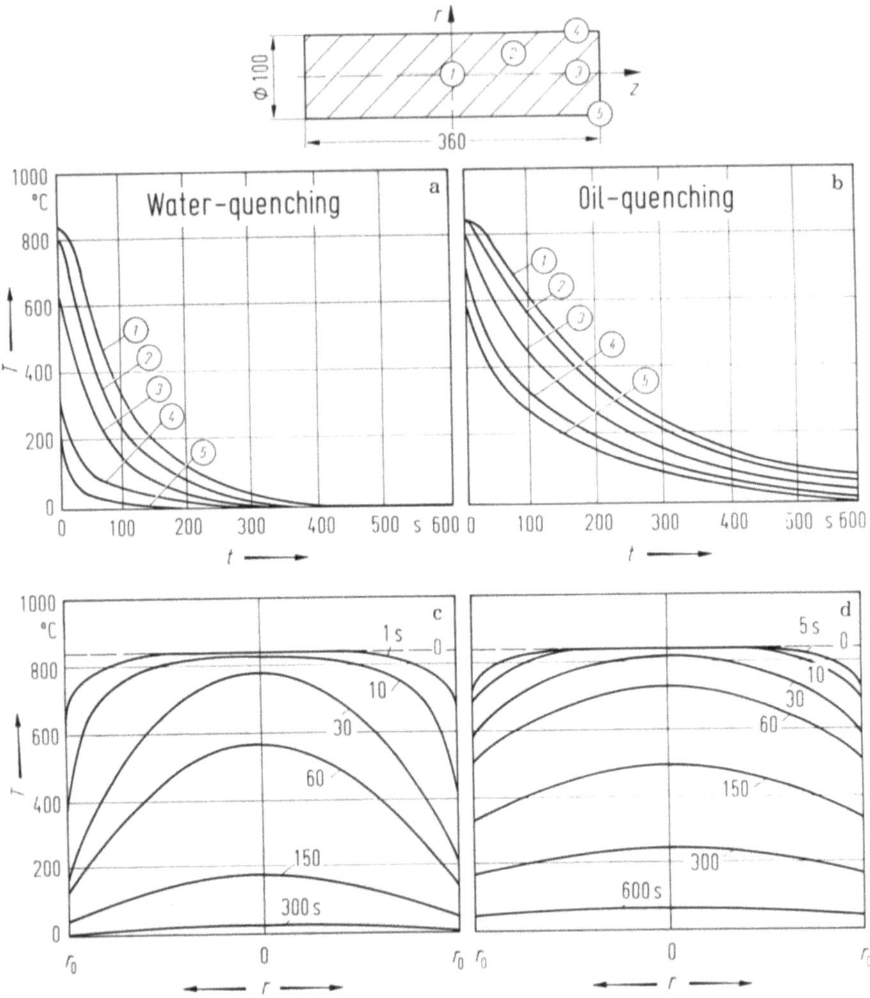

Fig. 6.4 a–d. Temperature-time relationships for a water- and an oil-quenched cylinder of 34 CrMo 4

Of course such temperature distributions are changed if the cylinder diameter is varied. Figure 6.5 shows the development of the temperature distributions for steel cylinders of 30, 50 and 100 mm diameter when quenched from 800 °C in water of 20 °C [16]. The temperature gradients near the surface of the specimens and the absolute temperature differences between centre and surface are enhanced with increasing diameter *D*. The corresponding temperature differences ΔT during water quenching are presented in Fig. 6.6 as a function of time.

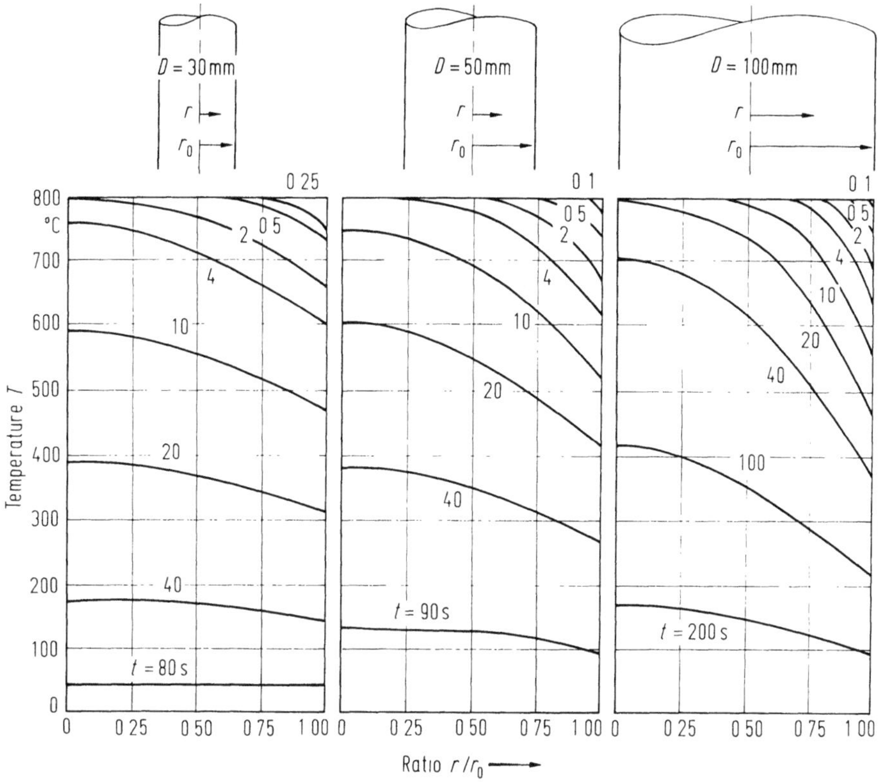

Fig. 6.5. Temperature distributions vs. radius ratio at different times after start of quench for cylinders with diameters of 30, 50 and 100 mm (steel 34 Cr 4, $h = 1500\,J/sm^2\,K$)

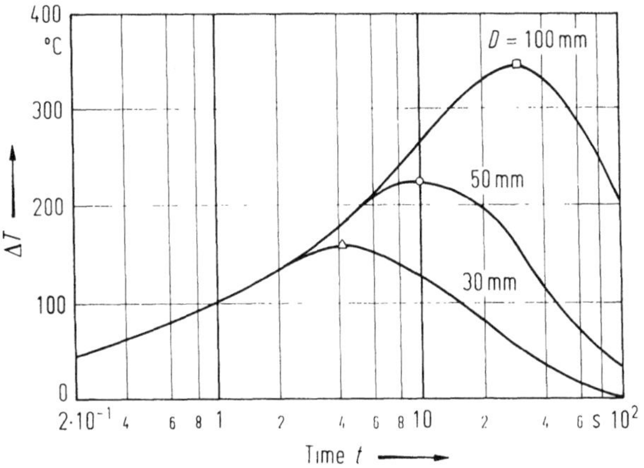

Fig. 6.6. ΔT vs. t for the cylinders considered in Fig. 6.5

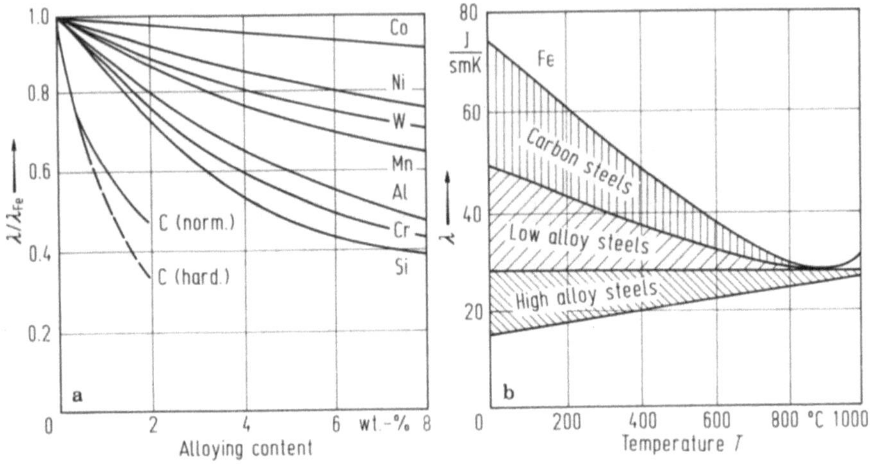

Fig. 6.7 a, b. Influence of alloying elements and temperature on the thermal conductivity of iron-based alloys

The thermodynamic data necessary for the correct assessment of the heating and cooling behaviour of steels are only partially known. Indeed, accurate information concerning the influence of temperature, chemical composition and structural changes on these quantities are lacking altogether. The most serious problems concern the properties of quenchants and their influence on the heat transfer coefficients.

The coefficient of thermal conductivity of pure iron λ_{Fe} is, at room temperature, approximately 70 J/smK. As Fig. 6.7a shows for binary iron-based alloys, this value is significantly reduced by carbon and to a lesser extent by other alloying elements [15]. Carbon steels possess smaller λ values in hardened than in normalized states. Data about the temperature dependence of the coefficients of thermal conductivity of various ferrous-based materials are summarized in Fig. 6.7b. Whilst λ decreases with increasing temperature for pure iron, plain carbon steels and low-alloy steels, an inverse temperature depencence is observed for austenitic steels and austenitic steel states [15].

The heat transfer coefficient h strongly depends on the heat flux between the cooling medium in contact with the surface of the material and therefore

Table 6.2. Approximate values of heat transfer coefficients of different quenchants

Cooling medium (quenchant)	Heat transfer coefficient [J/sm^2K]
furnace atmosphere	15
still air	30
moving air	60
quenching oil	600
water	3600

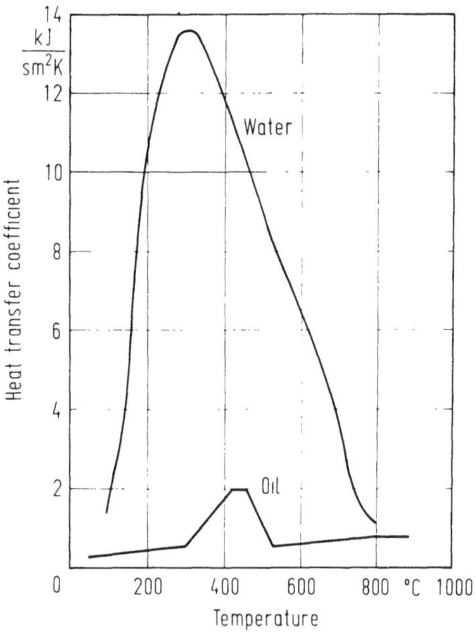

Fig. 6.8. Temperature dependence of the heat transfer coefficient of an austenitic steel for quenching in water and oil

also on the surface condition of the material. Depending on the type of quenchant applied, h incorporates the actual shares of heat transfer due to conduction, convection and radiation. Usually, the heat transfer by conduction is small in comparison to that by convection and radiation. The last-mentioned quantities are strongly temperature dependent and may vary by as much as one order of magnitude between room temperature and the austenitizing of steels. Reliable h-values for quenching processes involving large heat fluxes under non steady-state conditions are not available. Agitated quenchants always possess larger heat transfer coefficients than unagitated ones. For orientation, Table 6.2 summarizes some reference data about heat transfer coefficients for steels, however ignoring their dependence on temperature [15]. Actually, h-values show a pronounced temperature dependence. Often h-values are modelled by empirical relationships of the type $h = a_0 + a_1 T + a_2 T^2 + \cdots + a_n T^n$, whereby a_0, a_1, a_2 and a_n represent system-dependent coefficients [17, 18]. The results of a recent determination of the temperature dependence of h during quenching of an austenitic steel in water and oil are shown in Fig. 6.8 [19].

6.2.3 Transformation Processes of Austenitized Steels During Quenching

a) Steels with Homogeneously Distributed Alloying Elements

The transformation behaviour of steels cooling down from a distinct austenitizing temperature T_A with given temperature-time relations can be assessed using continuous time-temperature-transformation (TTT) diagrams. Figure 6.9

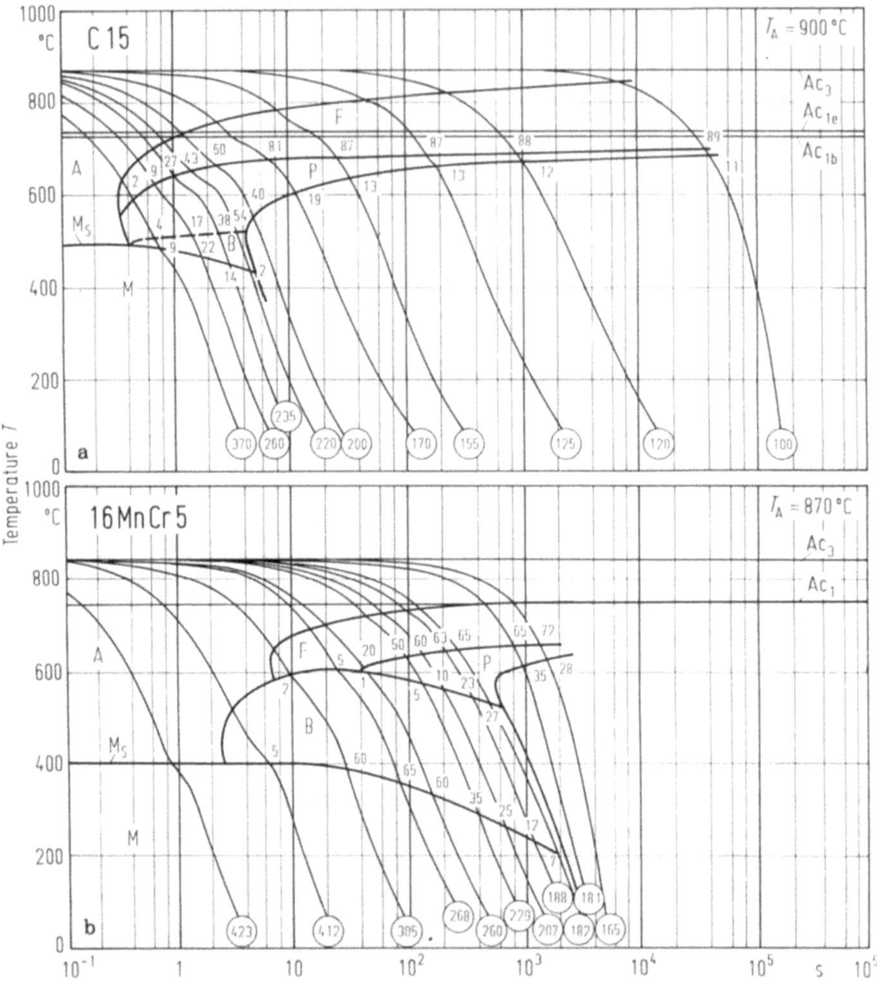

Fig. 6.9 a, b. Continuous time-temperature-transformation diagrams (TTT-diagrams) for (a) C 15, (b) 16 MnCr 5

shows two typical examples [20]. The upper diagram holds for the plain carbon steel C 15 ($T_A = 900$ °C), whilst the lower depicts that of the low-alloy steel 16 MnCr 5 ($T_A = 870$ °C). In the circles at the end of any cooling curve Vickers hardness numbers produced by the respective treatments are indicated. The martensite start temperature M_s of 16 MnCr 5 is about 100 °C smaller than that of C 15. The first diffusional transformations of 16 MnCr 5 begin nearly a power of ten later than those of C 15. Provided that the cooling down of various points of a cylinder from a given temperature agree with the curves available in a

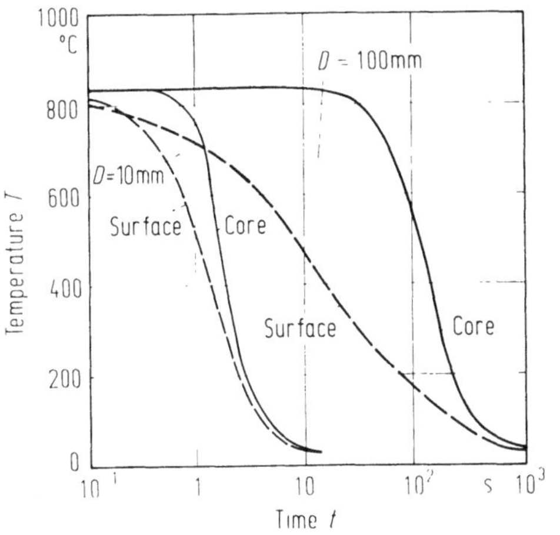

Fig. 6.10. Temperature of surface and core during cooling of water quenched cylinders (diameters 10 mm and 100 mm, 820 °C/20 °C)

corresponding TTT-diagram, the local transformation processes can be assessed. If, for example, one conveys the cooling curves of surface and core regions of cylinders of 16 MnCr 5 with 10 and 100 mm diameter shown in Fig. 6.10 [21] into the TTT-diagram of Fig. 6.9b, it becomes apparent that surface and core of the cylinder with 10 mm diameter will be totally converted to martensite. In the cylinder with 100 mm diameter, however, the surface is transformed by approximately 50% to bainite and martensite, respectively, whilst the core transformation is ferritic, pearlitic and bainitic, with just around 5% martensite.

b) Steels with Inhomogeneously Distributed Alloying Elements

In steels, concentration gradients of alloying elements are frequently produced in the very surface regions either through special thermo-chemical treatments or through unintended surface damages. In such cases, the knowledge of the TTT-diagram at each respective surface distance is necessary to assess the transformation behaviour. An example of this is given by Fig. 6.11 [20]. It is assumed that after carburizing 16 MnCr 5 in a salt bath, the local carbon content near the surface is given by the diagram shown on the left of the figure. Whilst at distances of 0.15 mm and 0.50 mm from the surface, carbon contents of 1 wt.-% and ~ 0.5 wt.-% exist, respectively, the original carbon content of the untreated material of ~ 0.16 wt.-% still exists at depths above 1.6 mm. Hence, the transformation at 0.15 mm depth is described by another TTT-diagram than that at 0.5 mm. The corresponding TTT-diagrams with broken and dotted lines, respectively, are compared with that of the non-carburized 16 MnCr 5 (full lines) on the right of Fig. 6.11. After transferring similar cooling curves as shown in Fig. 6.10 in this diagram, again an assessment of the locally occurring transformation processes is possible, whereby the cooling curve of the specimen surface has to be compared with the dotted lines of the TTT-diagram and that of its core with the full lines. It can be concluded that the core of a 10 mm

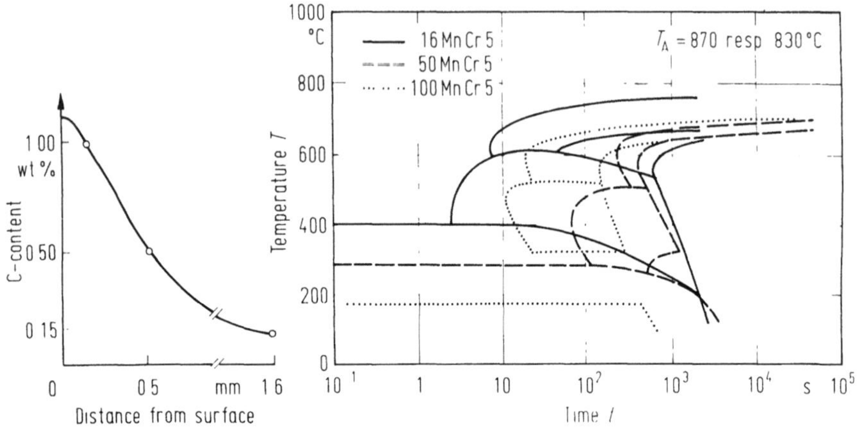

Fig. 6.11. Assumed near surface carbon content and TTT-diagram of 16MnCr5 with shifts of the transition curves due to carburization to 0.50 wt.-% carbon (50 MnCr 5) and 1.00 wt.-% carbon (100 MnCr 5)

diameter cylinder begins to transform to martensite before its surface and also before the areas at 0.5 mm depth. On the other hand, the surface and sub-surface regions of the 100 mm diameter cylinder are initially transformed to martensite before the core starts with ferritic-pearlitic transformations and is later on subjected to a small amount of martensitic transformations. When the transformation is completed, at a depth of 0.2 mm approximately 70 vol.-% martensite and 30 vol.-% retained austenite exist, whilst in the core only 5 vol.-% martensite are present.

c) Volume Changes Due to the Transformation of Austenite

Depending on the cooling rate and the composition, the high-temperature fcc austenite of steels is either transformed to ferrite-pearlite, pearlite, cementite-pearlite, bainite or martensite. As schematically shown in Fig. 6.12, all these phase transformations are accompanied by volume expansions. The occurring volume changes $\Delta V/V$ at room temperature due to total martensitic transformation of plain carbon steels with various carbon contents follow the relationship

$$\Delta V/V|_{20^\circ\mathrm{C}} = 3.216 + 0.859\,\text{wt.-\%}\,\mathrm{C} - 0.343\,(\text{wt.-\%}\,\mathrm{C})^2 \quad [\text{vol.-\%}] \qquad (6.9)$$

up to carbon contents of approximately 1.0 wt.-% [22]. Lower relative expansions occur if complete transformations take place in the lower or upper bainite levels.

An important phenomenon during the transition of austenite to martensite is the so-called transformation plasticity. Usually it is only considered that after cooling down austenite with an adequate rate below the martensite start temperature M_s, martensite forms with a specific volume increase. However, if the transformations occur within a specimen subjected to tensile stresses, larger

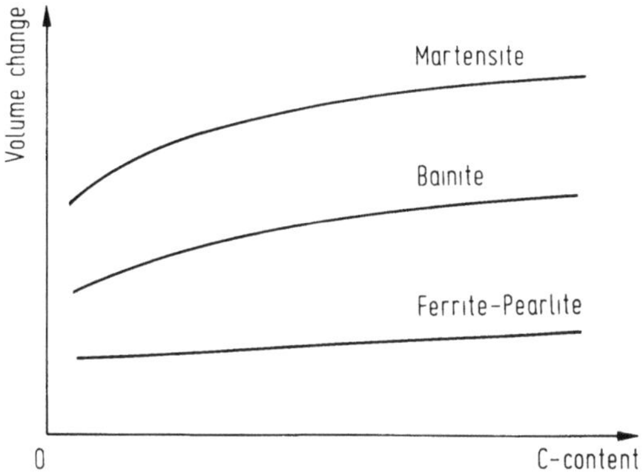

Fig. 6.12. Volume changes vs. carbon content of plain carbon steels due to martensitic, bainitic and ferrite-pearlitic transformation (schematically)

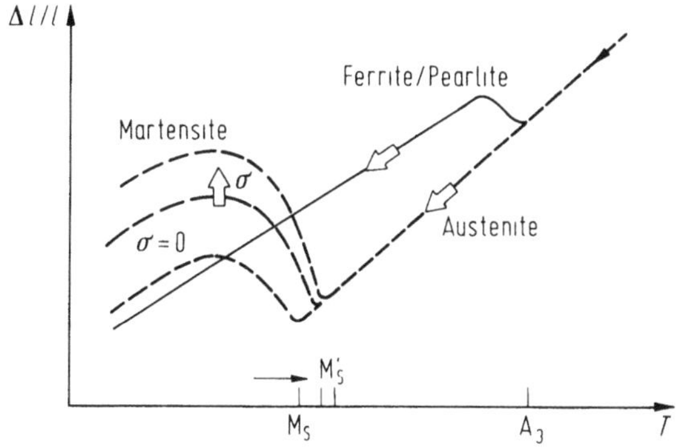

Fig. 6.13. Consequences of tensile stresses on the relative changes of length during rapid cooling (---) of austenitized steel specimens compared with slow cooling (—) behaviour

changes of dimensions are observed than in the case of an unstressed specimen. Simultaneously, the martensite start temperature is raised from M_s to M_s'. The opposite is true if compressive stresses become effective. The differences between the behaviour of rapidly cooled unstressed and tensile stressed steel specimens and that of slowly cooled ones are schematically sketched in Fig. 6.13.

The consequences of different magnitudes of uniaxial tensile and compressive stresses on the transformation strains of two different steels are shown in Fig. 6.14 [23, 24]. The above described effects are well established. Theoretically, the magnitude of the deformation strains should increase with increasing stress. The tendencies of the experimental observations are correctly modelled by the Greenwood–Johnson-concept which predicts the relationship [25]

$$\varepsilon_{pt} = \frac{5}{6} \cdot \frac{\sigma}{R_y} \cdot \frac{\Delta V}{V}. \tag{6.10}$$

ε_{pt} is the plastic transformation strain, σ the effective stress, R_y the yield strength of the austenite and $\Delta V/V$ the specific volume change coupled with the transformation.

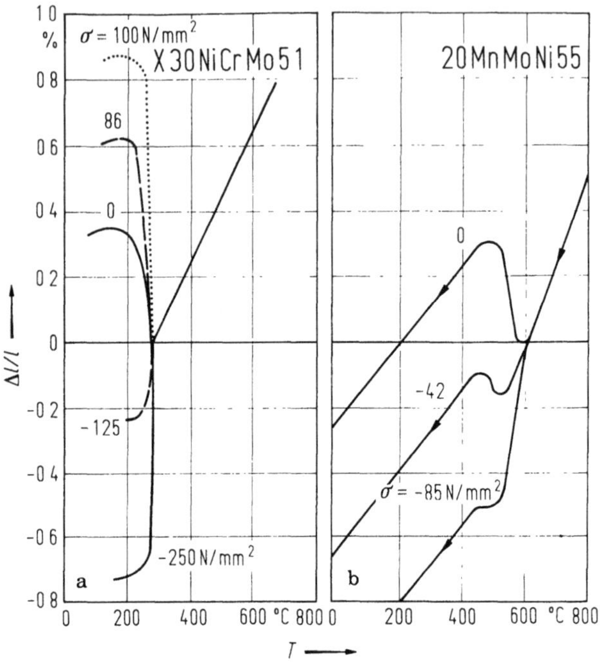

Fig. 6.14. Development of transformation strains in X30 NiCrMo 51 and 20 MnMoNi 55 during formation of martensite under tensile and compressive loadings

During rapid cooling of externally stress-free specimens, transformation induced plastic deformations [see e.g. 26] also occur when the equivalent thermal stress acting at the transition temperature is lower than that of the yield strength of the austenitic phase. The deformations again result from the specific volume changes associated with the transformations. Furthermore, regarding the elastic energy of the really acting stress states, the phases which develop during the martensitic transformation are orientated in such a manner that the free enthalpy is minimized.

d) Retained Austenite

Plain carbon steels with carbon contents higher than approximately 0.5 wt.-% are not completely converted to martensite when rapidly cooled from arbitrary austenitizing temperatures to room temperature [see e.g. 27, 28]. A certain quantity of retained austenite remains, which increases with the soluted carbon content. In plain carbon steels, for example, C 150 shows about 40 vol.-% retained austenite. In low-alloy steels, alloying elements differently influence the amount of retained austenite. Mo, Ni, Cr and Mn, for example, enlarge the amount of retained austenite in the given sequence. Hence, 18 CrNi 8 will tend to possess more retained austenite than 20 MoCr 4 if an identical hardening treatment is applied.

6.3 Stresses During Quenching of Cylinders with Ideal Linear-Elastic Deformation Behaviour

When dealing with stress states occurring in quenched cylinders, it seems expedient to consider first idealized borderline cases. In all quenched materials initially shrinking (thermal) stresses arise. If during quenching phase transformations additionally occur, then transformation stresses superimpose those caused by pure cooling. To simplify matters, at the beginning of this chapter, ideal linear-elastic cylinders are treated in which no plastic deformations can arise during the development of shrinking or transformation stresses or of both of them.

6.3.1 Shrinking Stresses Due to Local and Temporal Differences in Thermal Shrinking

Transformation-free cooling of cylinders is accomplished by the development of a sequence of inhomogeneous temperature distributions which as a consequence of the thermal shrinking behaviour cause in turn locally and temporally different thermal strains and hence shrinking stresses. Linear-elastic cylinders as assumed can elastically accomodate these stresses for all temperatures. At the beginning of quenching, the surface of such a cylinder contracts more rapidly than its core. As a result, the surface zones of the cylinder are subjected to tensile stresses in longitudinal and tangential directions whilst radially compressive stresses are created [16]. In order to establish equilibrium, these stresses are counterbalanced by longitudinal, tangential and radial compressive stresses within the core of the cylinder. Figure 6.15 schematically illustrates the whole situation. The temperatures at the very surface and at the core of the cylinder

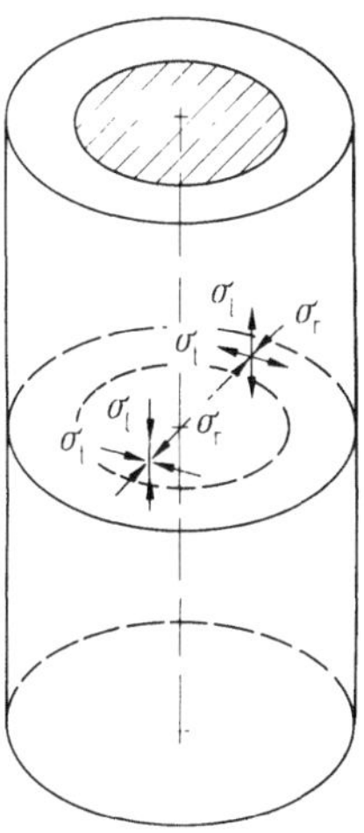

Fig. 6.15. Shrinking stress states near surface and core of a cylinder during rapid cooling

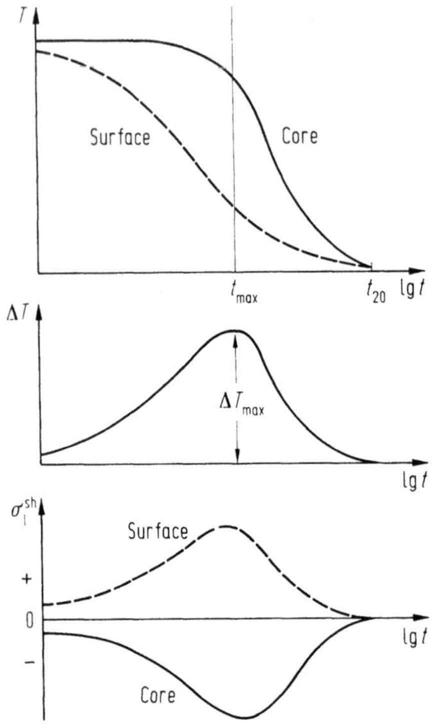

Fig. 6.16. Development of temperature and longitudinal stresses during transformation-free cooling of an ideal linear-elastic cylinder (schematically)

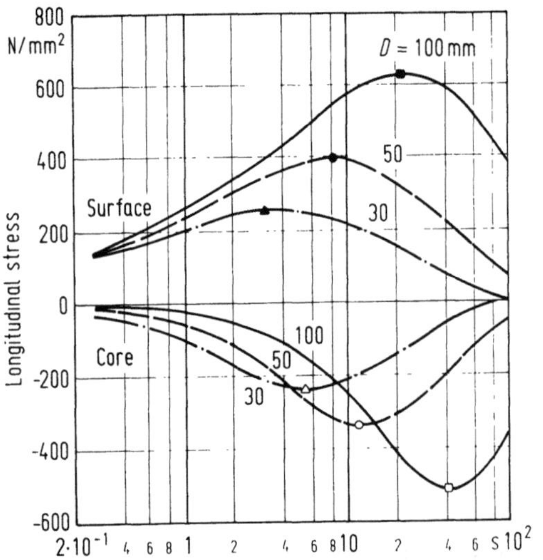

Fig. 6.17. Generation of longitudinal stresses at surface and core of different-sized ideal linear-elastic cylinders water-quenched from 800 °C to 20 °C. The physical data used for the calculations correspond to those of plain carbon steels with medium carbon content

as well as the temperature differences between both sites and the corresponding longitudinal stress components are schematically sketched in Fig. 6.16 as a function of the logarithm of cooling time. The largest temperature difference ΔT_{max} is attained at $t = t_{max}$, where the slopes of the T-lg t-curves are identical for both core and surface. Obviously, the surface reaches its maximum thermal stress before $t = t_{max}$, the core, however, later than $t = t_{max}$.

Figure 6.17 shows the magnitudes of the longitudinal stresses which develop in linear-elastic steel cylinders with diameters of 30, 50 and 100 mm during cooling in the course of quenchings from 800 °C in water of 20 °C [16]. Since according to Fig. 6.6, the maximum temperature differences between surface and core occur later for larger than for smaller cylinder diameters, it is to be expected that the maximum stresses and the time lags in reaching these values at surface and core depend systematically on the cylinder diameters. The longitudinal surface stress maxima always occur at $t < t_{max}$, whereas those of the cylinder core develop afterwards. This results from the fact that for a given cylinder the existence of the same magnitudes of temperature differences at $t \lessgtr t_{max}$ is always coupled with different temperature distributions across the cylinder diameter. At $t < t_{max}$, steep temperature gradients are present near the cylinder surface which cause high tensile stresses. In contrast, when $t > t_{max}$, relatively small surface temperature gradients are established which cause large compressive core stresses. Upon completing the temperature balance (20 °C) at $t = t_{20}$, the cylinders are, in accordance with the assumptions made, free of residual stresses.

6.3.2 Transformation Stresses Due to Local and Temporal Differences in Phase Transformations

In order to describe the development of pure transformation stresses, a hypothetical material has to be considered whose coefficient of thermal expansion is zero. Furthermore, it is assumed that if in course of quenching the martensite start temperature M_s is passed, complete martensitic transformation occurs with the corresponding volume increase. The temperature-time curves for surface and core of a cylinder of such a material are illustrated in the upper part of Fig. 6.18. At the surface of the cylinder, increasing compressive transformation stresses are produced after passing $t = t_1$ as a consequence of the transformation induced volume changes. These surface stresses must be compensated by tensile transformation stresses within the core of the cylinder. The magnitudes of both stresses increase in the course of further surface cooling.

If $t = t_2$ is reached transformation induced volume increases occur in the core, which lead to a reduction of the tensile stresses acting there. Correspondingly, also the surface compressive stresses are reduced. After reaching temperature equalization at $t = t_{20}$, the same amounts of martensite are present across the whole cylinder so that finally a residual stress-free state will exist.

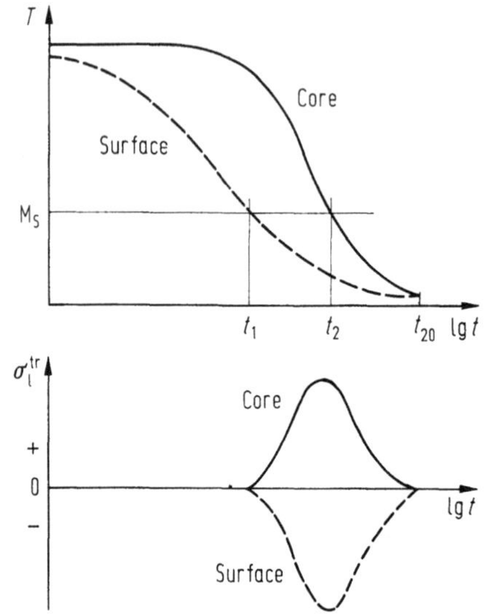

Fig. 6.18. Development of temperature and longitudinal stresses at surface and core of an ideal linear-elastic cylinder which only transforms martensitically during rapid cooling

In addition to the longitudinal stresses, also tangential and radial residual stresses are caused by the transformation processes. At the surface area, tangential compressive and radial tensile stresses are additionally to be expected, whilst in the core, all components should be tensile stresses. Obviously, upon completing temperature balance, no residual stresses appear when the cylinder is totally converted to martensite, or when everywhere equal amounts of retained austenite remain. If, however, different amounts of martensite form within distinct areas of the cylinder, also under the idealized assumptions made here, transformation residual stresses will remain.

6.3.3 Superposition of Shrinking and Transformation Stresses

If shrinking and transformation stresses are simultaneously effective during rapid cooling, then for an ideal linear-elastic material, the situation sketched in Fig. 6.19 should emerge. The upper graph depicts the time dependence of the longitudinal components of shrinking and transformation stresses at surface and core of a cylinder. The formal superposition of both stresses leads to the time dependent total stress course plotted below. The initiation of martensitic transformations immediately reduces the absolute stress values within both core and surface. Later on, increasing martensitic transformations cause a stress inversion in both regions. Provided that the transformations occur uniformly across the whole cylinder, then at $t = t_{20}$, the tensile core stresses and the compressive surface stresses approach zero. Hence, when temperature equilization is reached, a stress-free condition is achieved and no residual stresses remain.

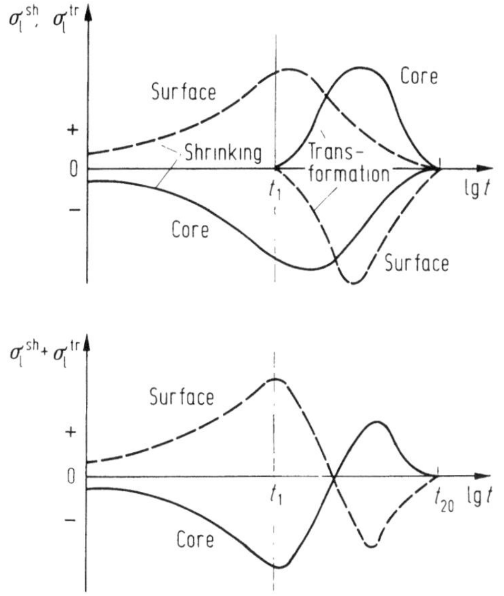

Fig. 6.19. Combined consideration of shrinking and transformation stresses during rapid cooling of an ideal linear-elastic material accompanied by phase-transformations

6.4 Residual Stresses After Quenching of Cylinders with Real Elastic–Plastic Deformation Behaviour

6.4.1 Plastic Deformations Due to Shrinking and Phase Transformations

Ideal linear-elastic deformation behaviour as assumed in Chap. 6.3 is an abstraction. Metallic materials always exhibit under respective high external loadings elastic–plastic deformations. The yield strength which limits the elastic deformation range is strongly temperature dependent and decreases with increasing temperature. At any temperature, elastic–plastic deformations develop after surpassing the yield strength according to the individual workhardening behaviour of the material considered. The temperature dependent ultimate tensile strength always limits the uniaxial loading capacity of the material. According to these facts, steels are relatively soft at high temperatures and cannot withstand without plastic deformations shrinking and transformation stresses occurring in the course of quenching. Exemplarily, the temperature dependencies of yield strength R_y and tensile strength R_m for 16 MnCr 5 and 17 CrNiMo 6 are shown in Fig. 6.20.

As stated already, rapid cooling of a cylinder produces at its surface biaxial longitudinal and tangential stresses whereas in the cylinder core triaxial longitudinal, tangential and radial stresses develop. Under the action of these stresses plastic deformations can only occur if the local equivalent stresses equal or exceed the yield strengths of the material at the temperatures of interest.

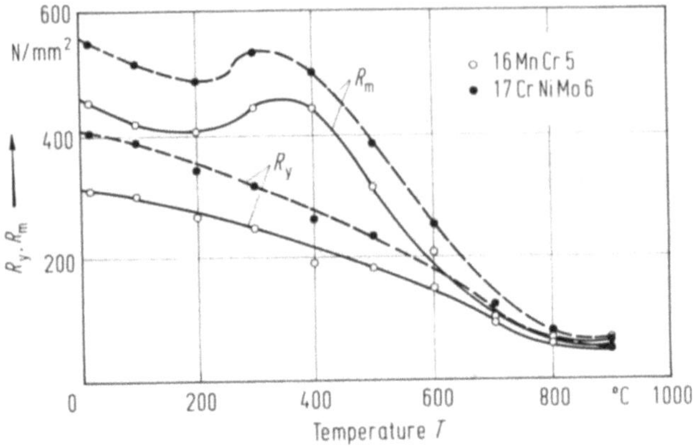

Fig. 6.20. Yield strength R_y and tensile strength R_m of 16MnCr5 and 17CrNiMo6 as a function of temperature

Equivalent stresses can be calculated according to various hypotheses. Assuming the validity of the van Mises' criterion, the equivalent stress of a triaxial stress state given by the principal stresses σ_1, σ_2 and σ_3 is

$$\sigma_{eq} = \frac{1}{\sqrt{2}} \sqrt{(\sigma_1 - \sigma_2)^2 + (\sigma_2 - \sigma_3)^2 + (\sigma_3 - \sigma_1)^2}. \qquad (6.11)$$

Then, the condition for the onset of plastic deformations can be written as

$$\sigma_{eq} = R_y \qquad (6.12)$$

with the yield strength R_y. At the surface of a rapidly cooled cylinder is $\sigma_1 = \sigma_l$ and $\sigma_2 = \sigma_t$ whilst at the core, $\sigma_1 = \sigma_l, \sigma_2 = \sigma_t$ and $\sigma_3 = \sigma_r$. Because the yield strength R_y and the local shrinking and transformation stress components are temperature and time dependent as well as influenced by the material state and its transformation behaviour, the condition for the local occurrence of plastic deformations may generally be given as

$$\sigma_{eq} = f\{\text{temperature, cooling conditions, geometry, mechanical and thermal properties}\}$$
$$= R_y(\text{temperature, structure})$$

Consequently, the temperature dependence of the yield strength is of particular importance for the stresses which result upon quenching. Another important point in this connection is the extent of transformation induced plastic deformations (see Chap. 6.2.3). However, to simplify matters, this aspect will be neglected here. Then, in dependence on time, temperature and yield strength in the core and at the surface of a steel cylinder during cooling can schematically be plotted as in Fig. 6.21. In the left part of the figure, the yield strength of surface and core in the case of a transformation-free cooling is

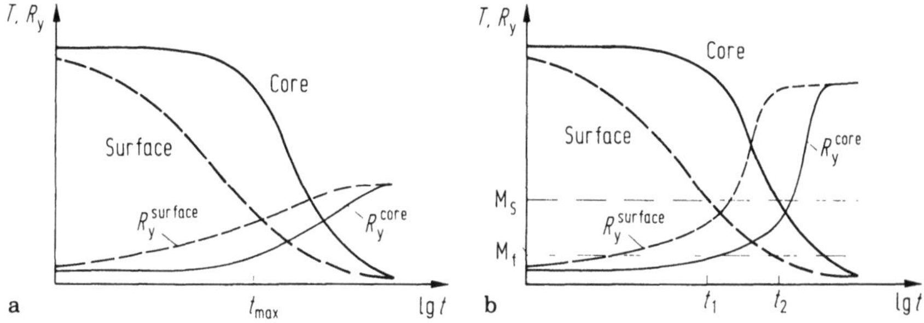

Fig. 6.21 a, b. T and R_y vs. lg t of surface and core for rapidly quenched cylinders without (**a**) and with (**b**) martensitic transformation

considered. More significant changes of the yield strengths during cooling are expected as a result of the generation of martensitic transformations passing M_s as demonstrated on the right. In order to determine the occurrence of plastic deformations, at any instant the local yield strengths must be compared with the local equivalent stresses. Obviously, plastic deformations can be caused either by shrinking stresses or transformation stresses or of course by a combination of both. Because these plastic deformations never occur homogeneously over the whole cross-section of the cylinder, always after reaching the complete temperature compensation between surface and core at the end of the quenching process, residual stresses remain [29, 30]. These will suitably be designated either as shrinking residual stresses or as hardening residual stresses, depending on whether transformation-free cooling was realized or not (see Fig. 6.1). A martensitically transformed cylinder, which was not subjected to plastic deformations during the transformation processes may nevertheless not be assumed to be free of transformation residual stresses. If, for example, different quantities of martensite are generated in surface and core regions, residual stresses must obviously result from the differences in the local volume changes.

6.4.2 Generation of Shrinking, Transformation and Hardening Residual Stresses

In the following steel cylinders are considered whose alloying elements are homogeneously distributed across the whole cylinder volume and whose tensile and compressive yield strengths are equal. At first, the generation of pure shrinking residual stresses and pure transformation residual stresses will be discussed. Afterwards, the combined occurrence of both stress types, which leads to hardening residual stresses, is commented on. For reasons of simplicity, the discussion will again be focused on the longitudinal stress components only [31, 32].

a) Shrinking Residual Stresses

At the top of Fig. 6.22, the cooling curves for surface and core of a transformation-free quenched cylinder are schematically shown. Sketched are also for both zones the time dependent yield strengths, which at the same temperature are assumed to be identical for tensile and compressive loading. When the cylinder starts cooling, the surface temperature decreases faster than the core temperature. As a result, longitudinal tensile and compressive stresses form at surface and core, respectively. If these stresses are elastically accomodated, they develop as drawn in the middle part of Fig. 6.22. However, under the assumed temperature dependence of the yield strengths, $R_y^{surface}$ and R_y^{core}, neither the surface nor the core can withstand their stresses without plastic deformations. As far as only longitudinal stresses are considered, plastic deformations begin when the yield strength is reached. Then, the surface of the cylinder is plastically extended and the core is plastically compressed. Neglecting any workhardening processes, each of these stresses is therefore limited by the respective yield strength.

After $t = t_{max}$, the temperature of the core decreases faster than that of the surface. This leads to a reduction of the quantities of the shrinking stresses in both regions. The stress values of core and surface reach, however, the value zero at different instants, since they can no longer co-exist at the same time in

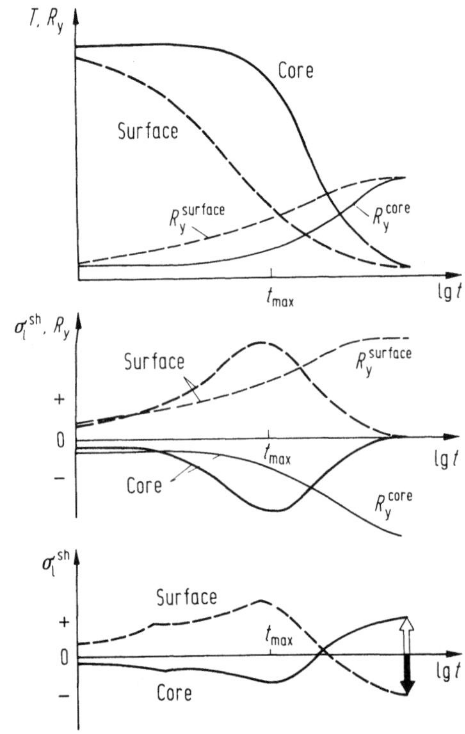

Fig. 6.22. Generation of longitudinal shrinking residual stresses due to rapid cooling of a cylinder

a stress-free state as a result of plastic extension at the surface and plastic compression in the core. Upon further cooling, the plastic extension of the surface and the plastic compression of the core cause compressive and tensile stresses, respectively, which are opposed by those due to the temperature differences still existing between core and surface. The last-mentioned stresses ultimately vanish with progressive cooling. Hence, after reaching the temperature balance at the end of the cooling process, shrinking residual stresses remain which are compressive at the surface and tensile at the core.

In summarizing, the following facts are important for shrinking residual stresses:

1. The magnitudes of the residual stresses are determined by the occurring plastic deformations.
2. In general, plastic deformations occur both at the surface and at the core of a quenched cylinder.
3. For a given yield strength, plastic deformations will be the stronger the higher the magnitudes of the stresses during cooling are. The amounts of these stresses increase both with increasing coefficient of thermal expansion and increasing temperature difference between surface and core.
4. The temperature difference between surface and core increases with the temperature from which quenching is started, with the heat transfer coefficient and with the diameter of the cylinder.
5. Plastic deformations during a quench increase with decreasing high temperature yield strength of the material. Consequently, residual stresses are expected to be the larger the lower the yield strength is. Since, however, the yield strength at low temperatures has to withstand these stresses, fundamentally, the highest shrinkage residual stress values occur at intermediate tensile yield strengths.

b) Transformation Residual Stresses

The upper part of Fig. 6.23 shows cooling curves for surface and core of a steel which upon quenching below the martensite start temperature M_s is completely converted into martensite. The figure also includes the yield strengths for surface and core. Both quantities are time dependent and increase strongly with the onset of martensitic transformations. For reasons of simplicity it is assumed that no shrinking stresses occur and that transformation plasticity effects can be neglected. The surface of the cylinder starts to transform to martensite at $t = t_1$. The volume expansions which arising are impeded by the core being not yet transformed. As a result, compressive transformation stresses are established at the surface which are compensated by tensile stresses at the core. The assumed yield strength-lg t-relations in the middle part of Fig. 6.23, show that both areas plastically deform. In the course of further cooling, the tensile stressed core reaches M_s at $t = t_2$. The volume increase following immediately reduces both the tensile stresses of the core and the compressive stresses of the surface. Due to the differently sized and opposing plastic

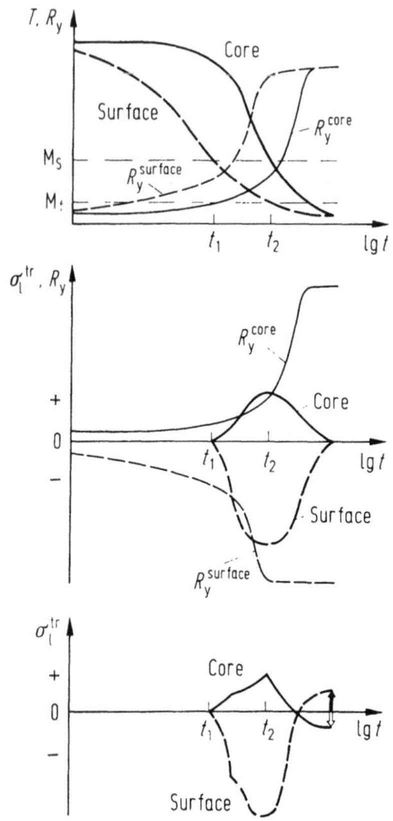

Fig. 6.23. Generation of longitudinal transformation residual stresses due to rapid cooling of a cylinder

deformations generated, the appertaining stresses at surface and core pass zero values at different times. Upon further cooling, the still existing volume incompatibilities between surface and core create transformation stresses with reversed signs than those which are produced by the last-mentioned plastic deformations. After reaching temperature equalization (visualized at the bottom of Fig. 6.23 by arrows) compressive residual stresses remain at the core and tensile residual stresses at the surface of the cylinder.

In summarizing, the following facts are important for transformation residual stresses:

1. The magnitude of the residual stresses is determined by the plastic deformations which occur due to the volume changes during the transformation processes. They increase with decreasing local yield strength.
2. In the case of full-hardening, the volume increases due to transformation do not produce residual stresses, provided the transformations occur to equal extents in all regions of the quenched part, and the yield strength is sufficiently high. Transformation residual stresses are generated when volume increases

develop which are locally and timely displaced and accompanied by plastic deformations of different magnitudes.

3. Since both the volume extension caused by martensitic transformations and the yield strength grow with increasing carbon content, plastic deformations are only then favoured when the effect of the volume increase is greater than that of the yield strength.
4. When transformation-induced plastic deformations occur under local tensile or compressive stresses, the local strains are enhanced with respective consequences for the final residual stress state.

c) Hardening Residual Stresses

Rapid quenching of austenitized steel cylinders to room temperature results in a hardening residual stress state which can not be described by simply superimposing the stresses of the individual processes discussed in the preceeding two chapters. Statements such as e.g. "in a full hardened workpiece, the opposing thermal and transformation stresses are superimposed" are, in actual cases, completely unfounded. Of fundamental importance, however, is the fact that any local martensitic transformation is coupled with volume increase and always shifts the existing stress (irrespective of its sign) to more negative magnitudes. Then, for reasons of equilibrium, the unaffected material zones react with positive stress changes. Transformations which occur in tensile-stressed material regions therefore inevitably reduce the stresses whilst transformations which take place in compressive-stressed zones always enhance the values of the stresses. Consequently, since the shrinking stresses of core and surface change their signs in course of cooling during the time interval $t_{c,o} - t_{s,o}$ (see Fig. 6.24), the relative positions of the initiation time of transformation in the surface $t_{s,i}$ and in the core $t_{c,i}$ to this time interval are of key importance for the hardening residual stresses remaining at the end of the cooling process. The average time which lapses before the quenching stresses invert can be written as

$$t_o = 1/2(t_{s,o} + t_{c,o}). \tag{6.13}$$

Obviously, cooling processes with surface transformations starting long before, near by or long after t_o must be viewed separately. Since for full-hardening steel cylinders of a homogeneous chemical composition, $t_{s,i}$ always occurs earlier than $t_{c,i}$, it seems reasonable to differentiate between

$$t_o < t_{s,i} < t_{c,i},$$
$$t_o \approx t_{s,i} < t_{c,i},$$
$$t_{s,i} < t_{c,i} \approx t_o$$

and

$$t_{s,i} < t_{c,i} < t_o.$$

According to this, Fig. 6.24 schematically shows the development of the longitudinal stresses as a function of the logarithm of time and the remaining longitudinal residual stress distribution across the diameter of cylinder specimens

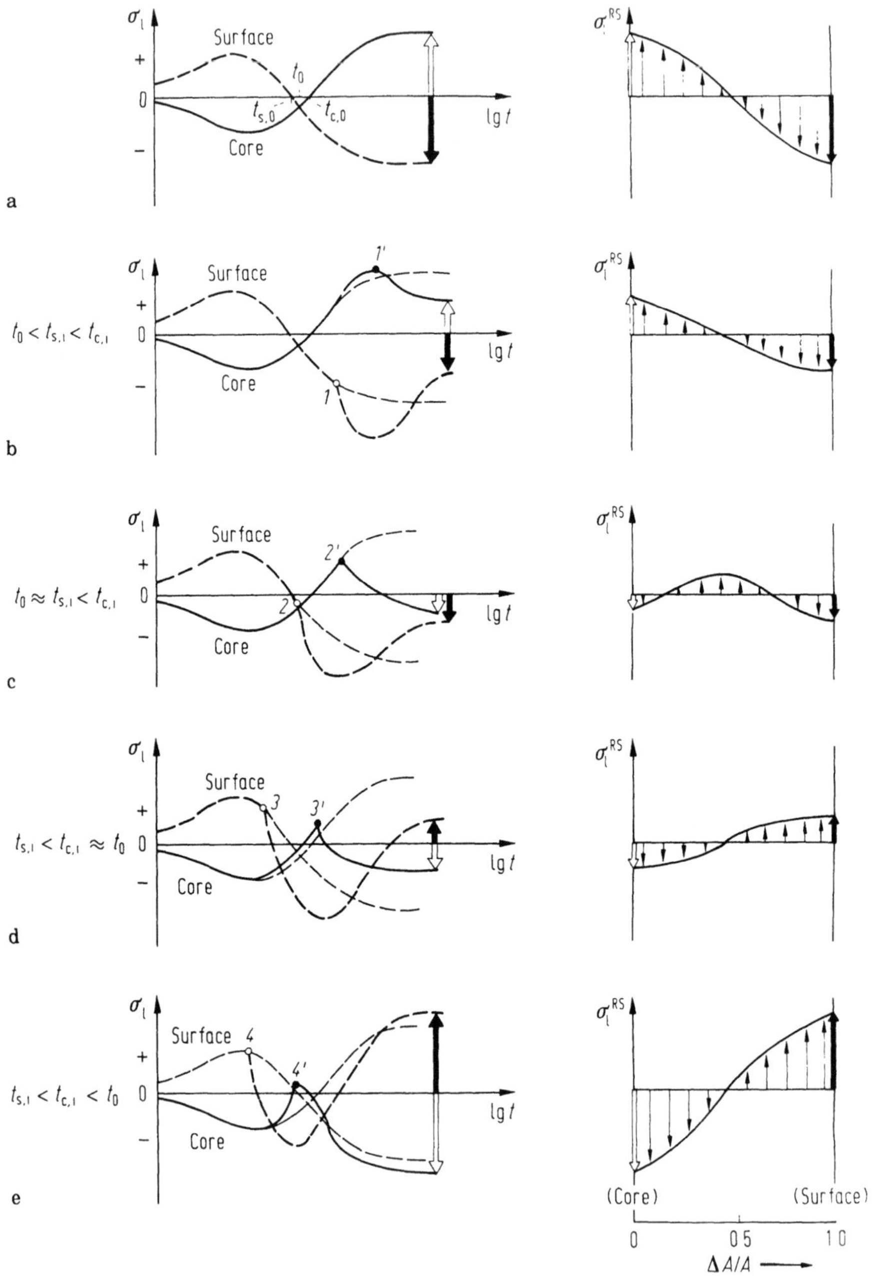

Fig. 6.24 a–e. Various possibilities of the generation of hardening residual stresses (**b–e**) in comparison with the generation of shrinking residual stresses (**a**)

after complete balance of temperature. Figure 6.24a describes a rapid transformation-free cooling. The following Figs. 6.24b–e demonstrate the combined effects of shrinking and transformation processes. The numbers 1, 2, 3 and 4 depict the transformation initiations at the surface, whilst 1′, 2′, 3′ and 4′ represent those of the core of the cylinders. Figure 6.24b illustrates the case where both core and surface transform after t_0. Figure 6.24c demonstrates the stress development when the surface transforms slightly before t_0 and the core thereafter. At the end of this cooling process, both core and surface remain under compressive residual stresses. The cylinder regions in-between are subjected to tensile residual stresses. The conditions under which the surface transforms before t_0 and the core after t_0 are described by Fig. 6.24d. Here, tensile surface residual stresses are equilibrated by compressive core residual stresses. Finally, Fig. 6.24e shows the instance where both core and surface transform before t_0. The start of transformations at the surface leads in this case to a rapid reduction of the tensile stresses followed by the formation of compressive stresses. Therefore, for reasons of equilibrium, the longitudinal stresses at the core must also change during further cooling. The core transforms martensitically when tensile stresses are acting there. This compels again stress inversions in the surface and core areas. The final residual stress state after cooling incorporates tensile stresses at the surface and compressive stresses at the core. As the residual stress states depicted in Figs. 6.24d and e correspond best with those which would be expected in the case of pure transformation processes (see Chap. 6.3.2), they are usually considered as being typical of full-hardening steel cylinders. However, this viewpoint has to be modified by the above mentioned facts.

Experience has shown that full-hardening of equal-sized cylinders which have different martensite start temperatures M_s can produce different residual stress distributions with both positive or negative surface residual stresses. This apparently disagrees with the conclusions drawn from Fig. 6.24. Cylinders with low M_s-temperatures show tensile surface residual stresses whilst cylinders with higher M_s-temperatures take compressive surface residual stresses. This behaviour can be explained with the aid of Fig. 6.25 illustrating the consequences of three different values of M_s (determined by the content of carbon and other alloying elements) on the longitudinal surface stresses and the final surface residual stresses, schematically. Since the high temperature yield strength usually increases with decreasing martensite start temperature, the largest tensile shrinking stresses develop at the surface of steel 3 and the smallest at the surface of steel 1. The martensitic transformations, however, begin earliest for the steel with the highest and latest for the steel with the lowest M_s-temperature. If high shrinking stresses and high M_s-values work together, no secondary stress inversion occurs during further cooling and ultimately compressive residual stresses remain within the surface zones.

Obviously, the statements at the beginning of this section concerning the effects of transformation processes on the shrinking stresses are equally valid for all non-martensitic transformations, which are accompanied by volume changes. Hence, similar reflections as presented in connection with Fig. 6.24 are possible taking account of just those volume changes which occur during each specific transformation process. The principle that the local stresses are shifted to more negative values due to transformation-induced volume increases still holds. In the individual case, the effects of volume changes on the final residual stress state again depend on the time at which the transformations start at core and surface with respect to t_0. However, in general, non-martensitic transformations are accompanied by smaller volume changes (see Fig. 6.12) so that smaller transformation stresses are created than in martensitically transformed parts.

The situations described by Fig. 6.24 mirror the essential aspects of the complex relationships which exist when quenching steel cylinders. As a matter of fact, σ_l-lg t-diagrams cannot be empirically derived. However, considering the relevant facts and material parameters, the processes during rapid cooling can be modelled and σ_l-lg t-relations can be calculated using a FE-method. Examples of such an approach are discussed later.

The preceeding discussion suggests to subdivide the whole range of expected residual stress distributions in quenched steel cylinders into three main groups, as demonstrated in Fig. 6.26. Obviously, the pure shrinking residual stress type and the pure transformation residual stress type are the limiting cases. Furthermore, a transition residual stress type exists with a tensile residual stress maximum and a compressive residual stress minimum, respectively, between surface and core. In each case, the longitudinal stress distributions are actually controlled both by the plastic deformations due to shrinking stresses and the transformation-induced volume increases. The arrows shall indicate how local transformations under existing stress states will affect the residual stress distributions.

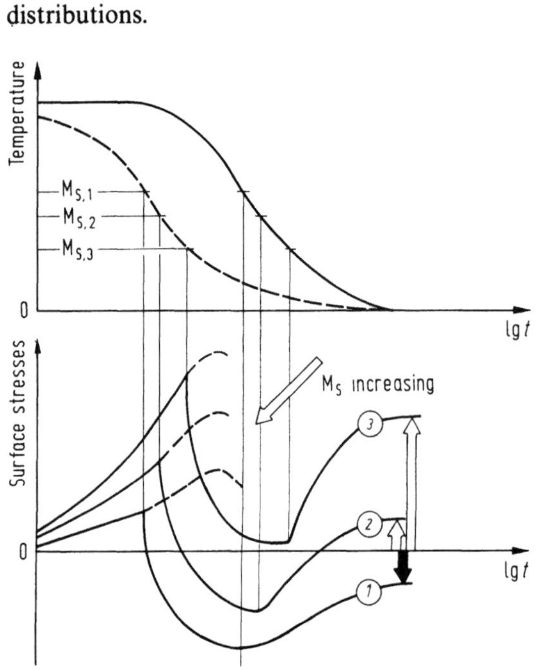

Fig. 6.25. Influence of different M_s-temperatures on the development of surface residual stresses (schematically)

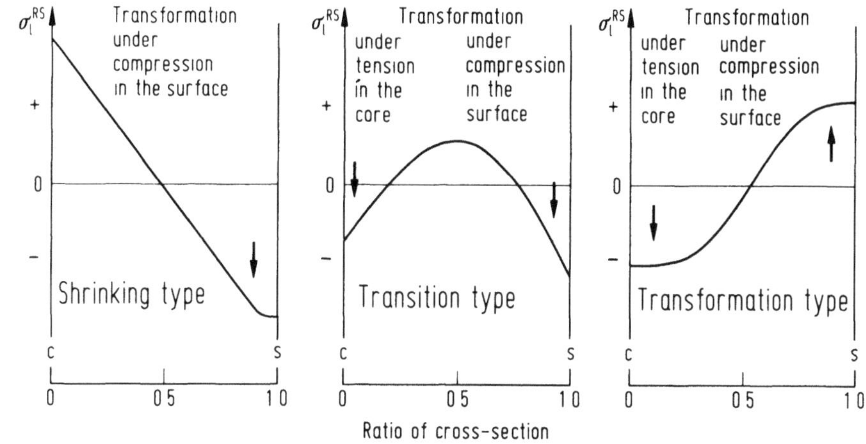

Steel	Quenching process	Cylinder diameters in mm for residual stresses of		
		shrinking type	transition type	transformation type
Ck 45	850 °C → 20 °C, H_2O	100...30	15	5
	850 °C → 60 °C, oil	50	30	10

Fig. 6.26. Basic types of hardening stresses

It was mentioned already that the volume changes accompanying incomplete martensitic transformations even in the absence of plastic deformations influence the formation of residual stresses of steels cylinders. Furthermore, it is important that the residual stress distributions which are created during rapid quenching of differently sized cylinders of the same steel can shift from the transformation type to the shrinking type with increasing cylinder diameter. Accelerating the cooling rate yields the same tendency. A change from oil quenching to water quenching for a steel cylinder shifts the produced residual stress distribution in small-sized cylinders in direction of the shrinking type. This results from the larger temperature differences between the T-lg t relations of surface and core of water-quenched cylinders and from the corresponding more inhomogeneous cooling. At the bottom of Fig. 6.26, some cylinder diameters and cooling treatments are specified for Ck 45, where the above-mentioned basic residual stress types occur. Further examples are discussed in the next section.

6.4.3 Characteristic Examples of Stresses and Residual Stresses in Differently Quenched Plain Carbon and Low Alloy Steels

a) Development of Stresses During Transformation-Free Cooling of Ck 45

When Ck 45 steel cylinders of 20 mm diameter are quenched from 680 °C in oil at 20 °C, no transformation processes occur. The developing temporal

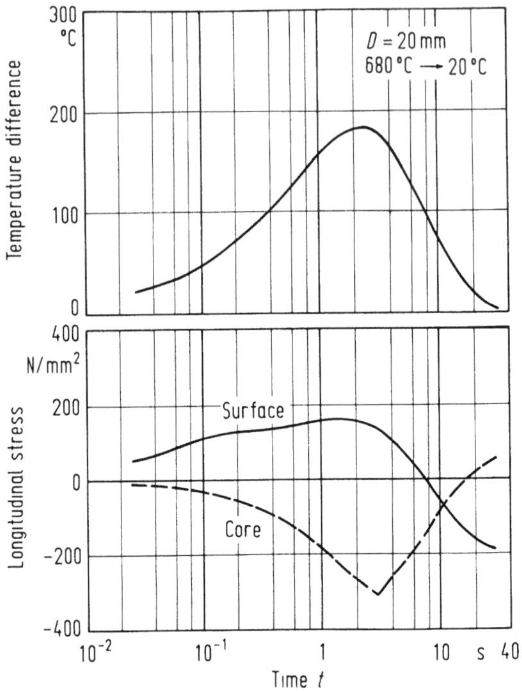

Fig. 6.27. Temperature and longitudinal stress difference between surface and core vs. cooling time of a transformation-free quenched cylinder (Ck 45, diameter 20 mm, 680 °C/ oil 20 °C)

temperature differences ΔT between core and surface are illustrated in the upper part of Fig. 6.27 [16]. A maximum value $\Delta T = 180\,°C$ is calculated 2.5 s after starting the quenching process. The associated longitudinal stresses for surface and core are plotted in the lower graph. It can be seen that the curves comply with the principles discussed in Chap. 6.4.2. Tensile stresses initially induced at the surface are held in equilibrium by compressive stresses in the core. Also, the surface and the core maximum stresses develop at different times during cooling. The first one occurs before the greatest temperature difference is reached, the second one comes out thereafter. When the temperature compensation is complete, the core remains under tensile residual stresses whilst compressive residual stresses prevail at the surface. In this particular case calculated and X-ray determined surface residual stresses agreed quite well.

The occurrence of plastic deformations in the discussed example can be assessed with the data of Fig. 6.28 [16]. The equivalent stresses σ_{eq} are calculated according to the van Mises' hypothesis for surface (upper plot) and core (lower plot) and are compared with the respective values of the yield strength R_y. According to Eqs. 6.11 and 6.12, plastic deformations are expected when $\sigma_{eq} = R_y$. Although in course of cooling the surface of the specimen always possesses higher yield strengths as a consequence of its lower temperatures, plastic deformations start approximately 0.15 s after beginning the quench. Since the calculations were performed assuming ideal elastic–plastic material behaviour,

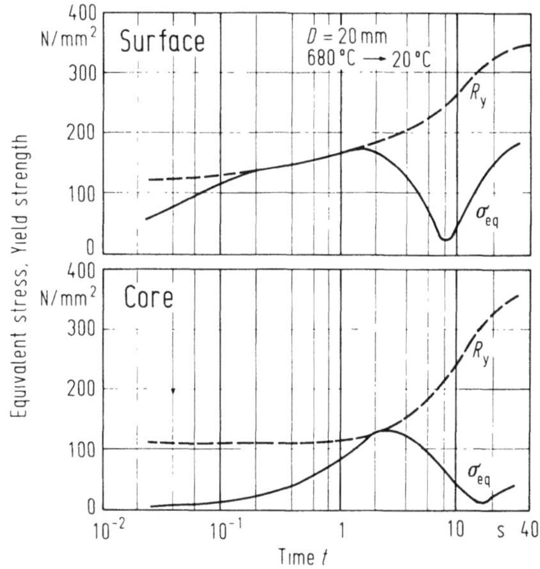

Fig. 6.28. Equivalent stress and yield strength vs. cooling time of a transformation-free quenched cylinder (Ck 45, diameter 20 mm, 680 °C/oil 20 °C)

the equivalent stresses can never take higher values than the yield strengths. During the time interval $0.15 < t < 1.55$ s, the condition for plastic deformations at the surface remains fulfilled. Above $t = 1.55$ s, however, the yield strengths exceed the equivalent stresses, hence ceasing plastic deformations. During the cooling process, the reduced stress states of the core approach the yield strengths relatively late and only for a brief period. Altogether, in the example sketched, plastic deformations are induced both in the core and at the surface of the quenched specimen.

b) Residual Stress Distributions in Cylinders of 22CrMo4 with Various Diameters After Water Quenching from the Same Austenitizing Temperature

Figure 6.29 exemplarily illustrates for cylinders of 10, 30 and 100 mm diameter the consequences of core and surface cooling curves in continuous TTT-diagrams (left-hand part) on the residual stress distributions established across the cross-section ratio of the cylinder (right-hand part) after temperature equalization [20]. When quenching a cylinder of 100 mm diameter from 850 °C to water of 20 °C, the created residual stress distribution is of the shrinking type. The surface is subjected to large longitudinal compressive residual stresses of approximately $-500\ \mathrm{N/mm^2}$ whereas in the core tensile residual stresses of roughly $+250\ \mathrm{N/mm^2}$ are measured. After the same quenching procedure, the cylinder with a diameter of 10 mm is fully hardened. As expected, a residual stress distribution of the transformation type is established with longitudinal stresses of about $+80\ \mathrm{N/mm^2}$ at the surface and of about $-170\ \mathrm{N/mm^2}$ in the core. The cylinder with 30 mm diameter transforms bainitic-martensitically so that at the surface and in the core compressive residual stresses were measured

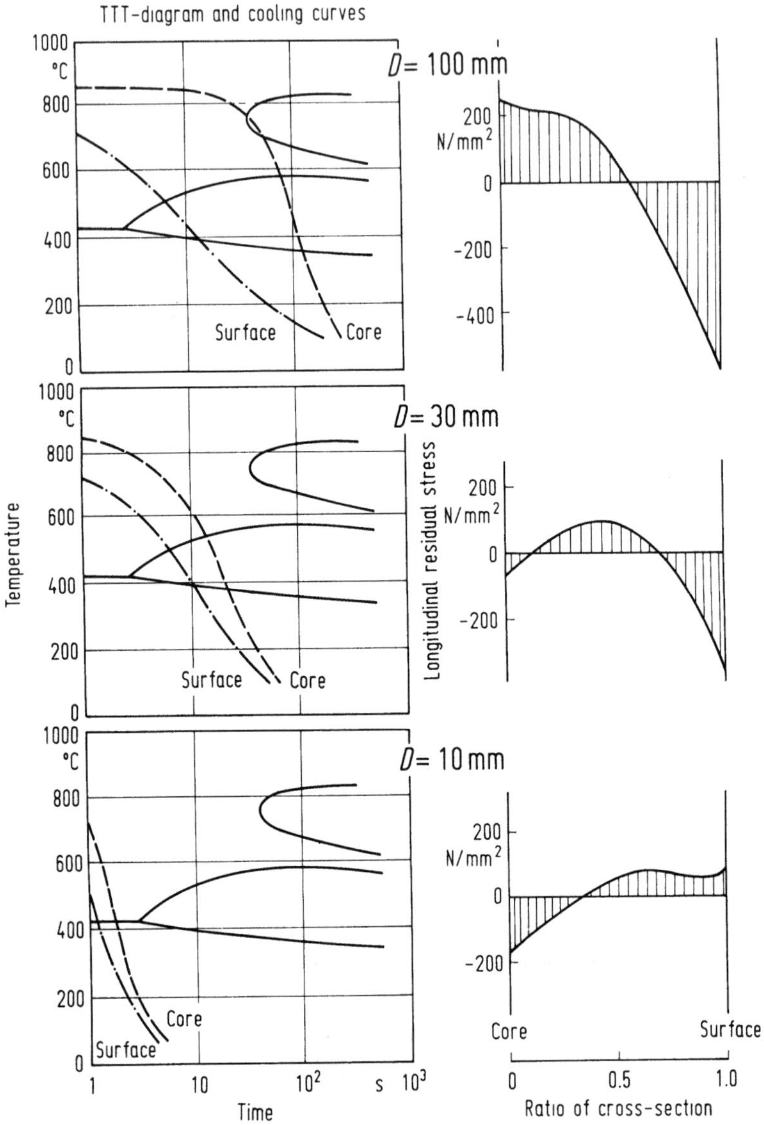

Fig. 6.29. Surface and core cooling curves of differently sized cylinders and longitudinal residual stress distributions after equalization of temperature

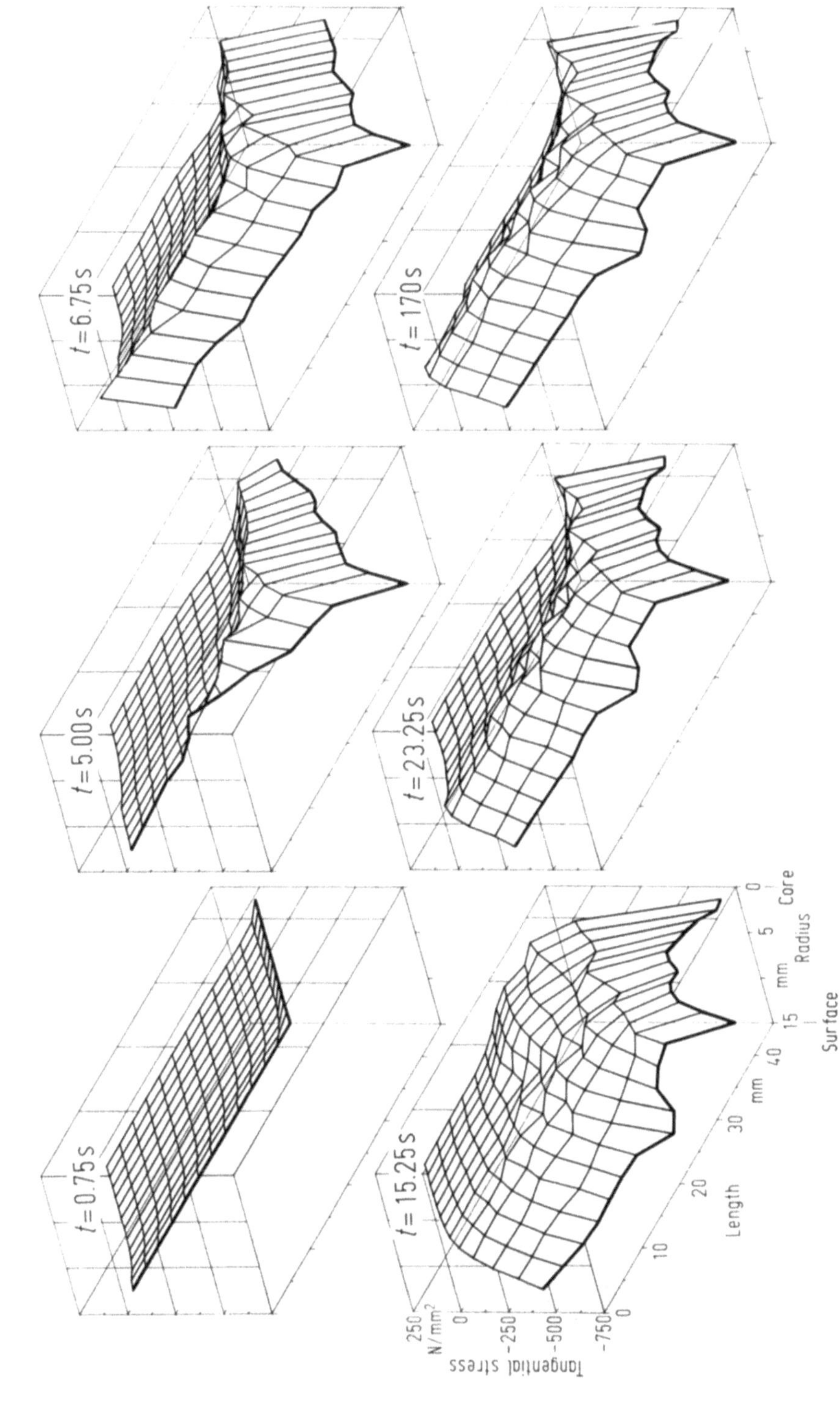

Fig. 6.30. Tangential stress distributions at a quarter-section of an axial cut through a cylinder for various times after starting a quench from 860 to 20 °C (42 CrMo 4, diameter 30 mm, length 90 mm)

after cooling to 20 °C. Since tensile residual stresses exist in the intermediate cylinder sections, a residual stress distribution of the transition type has developed.

c) Generation of Tangential Stresses During Transformation-Influenced Quenching of 42 CrMo 4

The temporal development of the tangential stresses within a 42 CrMo 4 cylinder with a diameter of 30 mm and a length of 90 mm quenched from 860 °C in water of 20 °C is plotted in Fig. 6.30 [33]. The tangential stress distributions of a quarter-section cut through the cylinder axis are shown for various times after starting the quench. As expected, the shrinking processes immediately induce tensile stresses at and near the surface of the cylinder so that compressive stresses are generated in the core. After 5.00 s parts of the surface areas already have expanded due to martensitic transformations leading to stress inversions. Further cooling incites a "pressure front" which shifts from the surface of the cylinder towards its core. Between 6.75 s and 23.25 s two stress reversals occur in the core area so that ultimately, the surface and the core areas of the cylinder remain under residual compressive stresses, whilst tensile stresses prevail within the intermediate regions. The residual stress distribution existing in the middle cross section of the cylinder after temperature equalization is of the transition type. At and near the end face of the cylinder, however, only tangential compressive residual stresses are found.

d) Generation of Longitudinal Core and Surface Stresses During Transformation-Influenced Cooling of Ck 45

In the upper part of Fig. 6.31, the temperature difference between core and surface of a 20 mm diameter Ck 45 cylinder quenched from 830 °C in oil of 20 °C is plotted as a function of time [16]. The lower part of Fig. 6.31 depicts the corresponding longitudinal stresses of surface and core of the cylinder. It can be seen that the signs of these stresses invert four times before the cylinder is fully cooled. Then the surface and near surface regions finally remain under residual compressive stresses, whilst the core takes a state of tensile residual stresses. Thus, the residual stress distribution is of the shrinking type. The plastic deformations occuring in the core and at the surface may be assessed with the aid of Fig. 6.32 where the temporal changes of the local stresses and the yield strengths are plotted for the surface (upper graph) and the core (lower graph), respectively. In the surface areas, plastic deformations appear immediately upon quenching. With a time delay of approximately 0.5 s plastic deformations also occur in the core.

e) Residual Stress Distributions in Equal-Sized Steel Cylinders of Various Hardenability After Quenching from 830 °C in Oil of 20 °C

The continuous T T T-diagrams with the cooling curves for core and surface of cylinders of 28 CrMo 10 and Ck 45 with 50 mm diameter are illustrated in

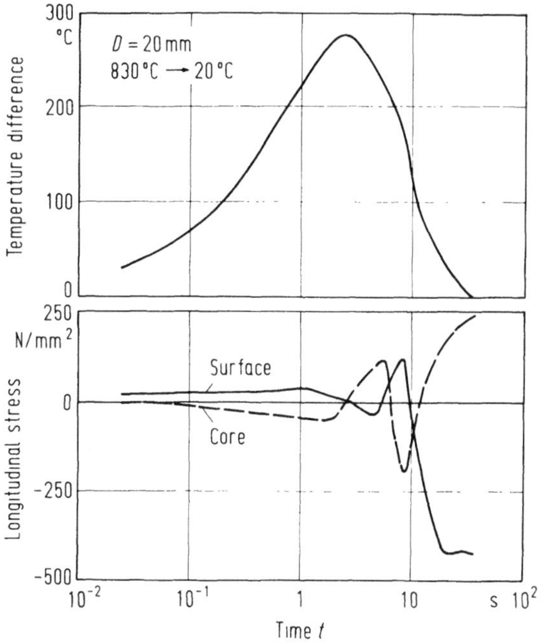

Fig. 6.31. Temperature difference between core and surface and longitudinal stress vs. cooling time for a transformation-influenced quench of a cylinder (Ck 45, diameter 20 mm, 830 °C/oil 20 °C)

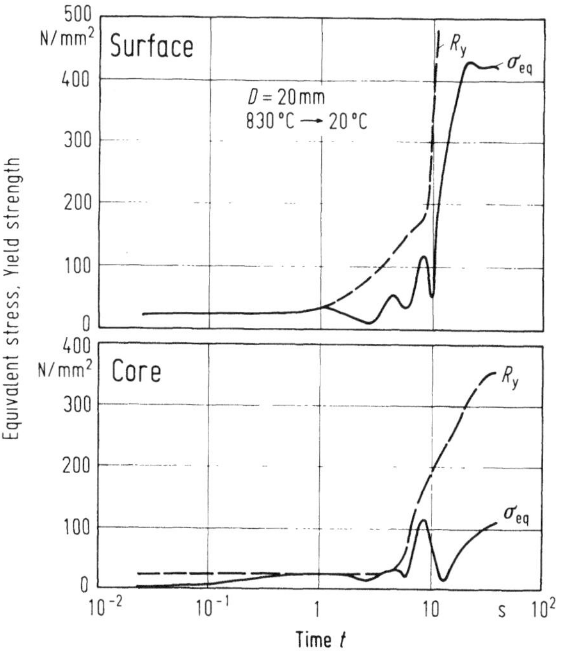

Fig. 6.32. Equivalent stress and yield strength vs. cooling time of a transformation-influenced quenched cylinder (Ck 45, diameter 20 mm, 830 °C/oil 20 °C)

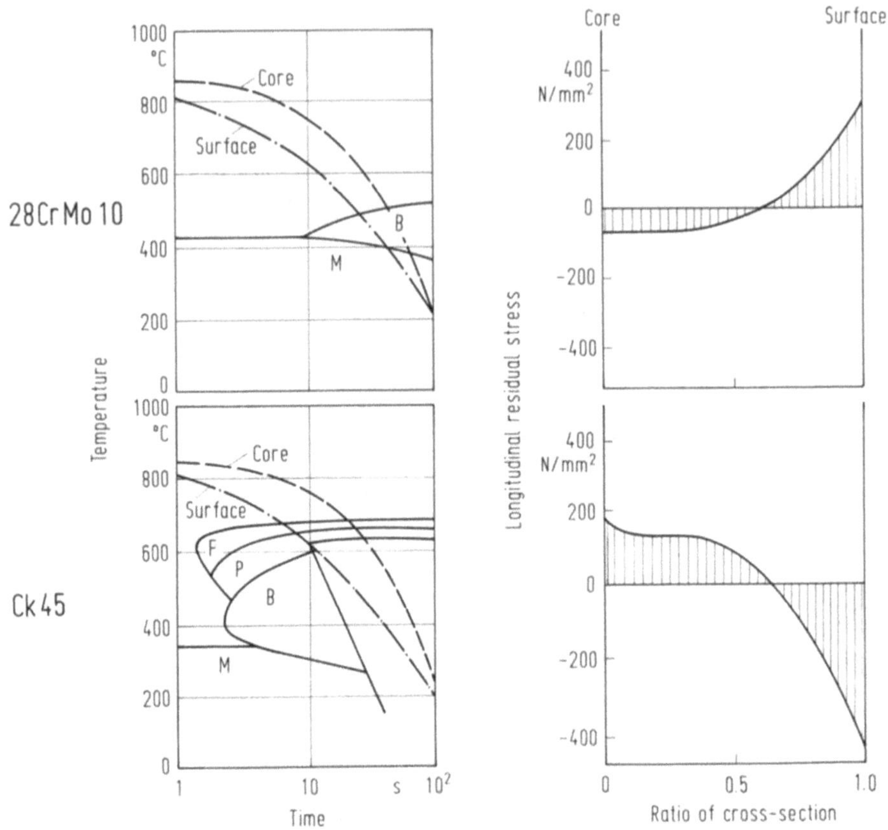

Fig. 6.33. TTT-diagrams of 28 CrMo 10 (top Ck 45 (bottom) with surface and core cooling curves for quenching of 50 mm diameter cylinders and distributions of longitudinal residual stresses after temperature balance at 20 °C

Fig. 6.33 together with the residual stress distributions which are built up across the middle planes of the cylinders when fully cooled [20]. Whilst core and surface of the Ck 45 cylinder are completely converted to ferrite/pearlite, both areas of the 28 CrMo 10 cylinder are subjected to a mixed bainitic/martensitic transformation. As to be expected, the residual stress distribution of Ck 45 is of the shrinking type whilst that of 28 CrMo 10 is of the transformation type.

f) Residual Stress Distributions in Equal-Sized Cylinders of 15 CrMoV 4 8 After Quenching from 830 °C in Water and Oil of 20 °C

In the left-hand part of Fig. 6.34 the *T*-lg *t*-curves for surface and core of 30 mm diameter cylinders quenched from 830 °C in water and oil at 20 °C, respectively, are plotted in continuous TTT-diagrams [20]. In contrast to oil-quenching (lower graph), water-quenching (upper graph) produces larger temperature differences between the core and the surface. Hence, upon quenching in oil, the

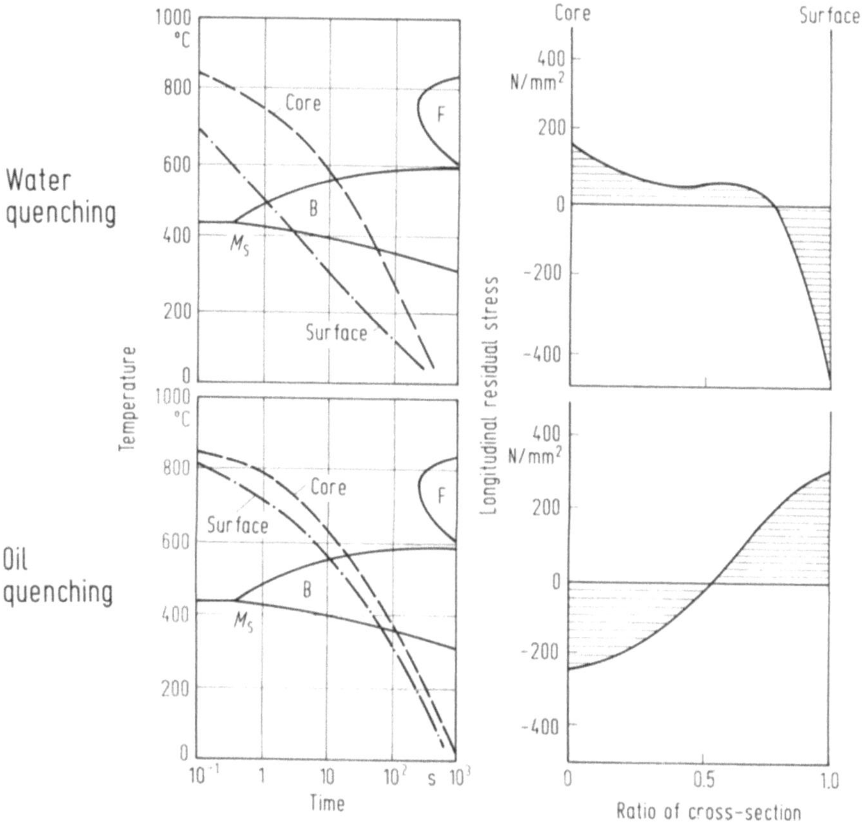

Fig. 6.34. TTT-diagram of 15CrMo48 with surface and core cooling curves for water and oil quenching of a cylinder with a diameter of 30 mm and distributions of longitudinal residual stresses after temperature balance at 20 °C

shrinking stresses remain relatively small. As a result, it is the transformation process which specifies the final residual stress state of the transformation type (bottom right). When quenching in water, however, relatively high compressive residual stresses are generated in the surface zones, which must be compensated by tensile residual stresses in the core. Consequently, the residual stress distribution which is established is of the shrinking type (top right).

6.5 Residual Stresses After Quenching of Carburized Steels

6.5.1 Some Fundamentals

Figure 6.35 schematically illustrates the mostly applied processing if suitable steels are subjected to a case-hardening treatment. After heating and subsequent diffusion treating, which produces an inhomogeneous distribution of carbon within the surface areas with a distinct carburization depth, the specimens are

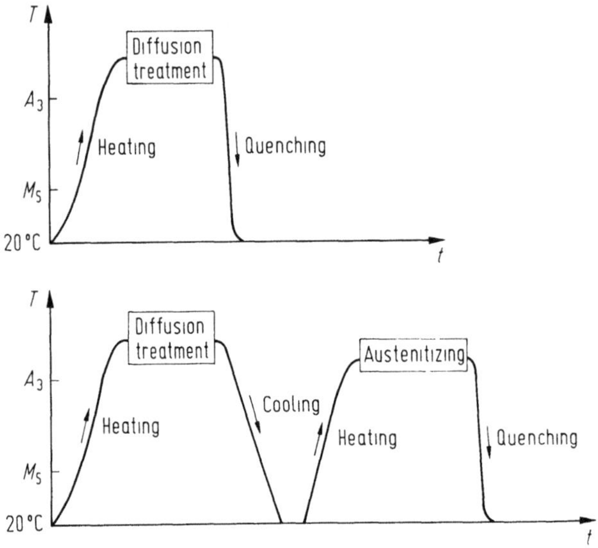

Fig. 6.35. Processing in direct hardening (top) and single hardening (bottom) of case-hardening steels

either directly quenched from the same temperature (direct hardening) or after cooling and again austenitizing from a rather lower temperature (single hardening) to room temperature. If quenching proceeds rapidly enough, case-hardening residual stresses develop. These stresses differ from the hardening residual stresses which would develop in the same steel without a near surface carbon gradient. Of course, the composition of the material influences its physical characteristics which determine the magnitudes of the final residual stresses. Again shrinking stress effects are superimposed on transformation effects which occur temporally and locally displaced and are accompanied by concentration-dependent volume changes. Always, regions of high carbon content show low martensite start temperatures and large volume changes. Conversely, in regions of reduced carbon content, M_s is increased and the volume changes become smaller. Consequently, the respective time-lags in the martensitic transformations occurring at and near the surface and in the core of case-hardening steel cylinders are of key significance for the residual stress formation. Obviously, it is important whether the core of the cylinder starts to transform before or after the surface. The consequences of core transformations prior to that of the surface on the stresses developing during cooling is illustrated by the diagrams in Fig. 6.36. It can be seen from the assumed cooling curves for surface and core of a cylinder plotted in the continuous TTT-diagram that the ferritic-pearlitic transformation of the core begins at $t_{c,i}$, when already compressive shrinking stresses are acting at the core, equilibrated by corresponding tensile surface stresses. For $t_{c,i} < t < t_{s,i}$, the stress state illustrated at the bottom left of Fig. 6.36 exists. Not only the tensile surface stresses but also the compressive core stresses may plastically

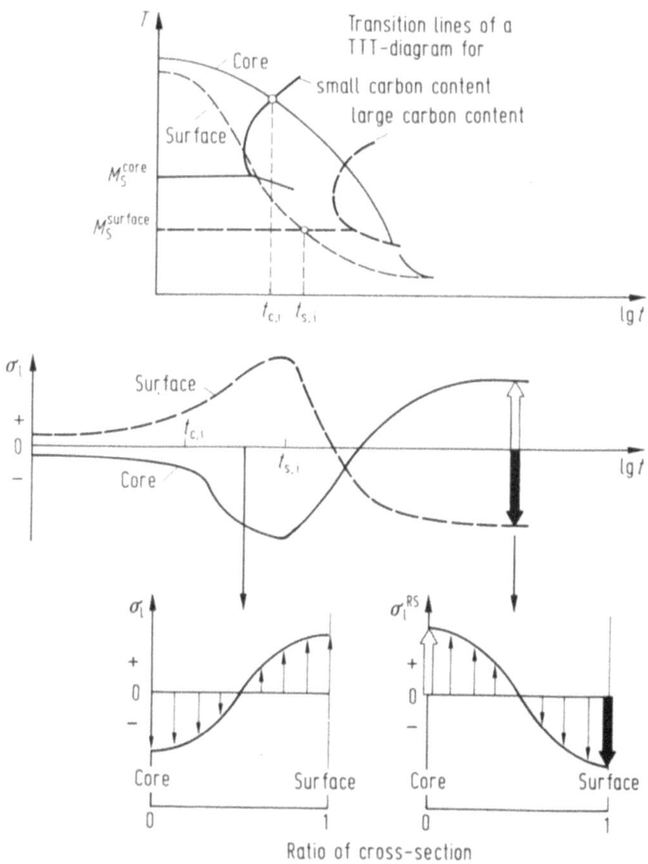

Fig. 6.36. Temperatures and stresses of surface and core during quenching of a carburized steel as a function of lgt when ferritic-pearlitic (bainitic) transformations start in the core and are followed by martensitic transformations in the surface

deform both core and surface due to the relatively high temperatures still existing. Then, in the further course of cooling the near-surface cylinder regions with lower carbon contents starts to transform martensitically followed by the surface of the cylinder at $t_{s,i}$. However, the surface and near-surface regions, which are now relatively cool and rigid, restrict the expansions due to the local martensite formations. Consequently, the stresses are shifted to compression whilst tensile stresses generate in the core. Therefore, when fully cooled, the surface is subjected to compressive residual stresses whilst the core bears tensile residual stresses. Hence, as illustrated at the bottom right in Fig. 6.36, a shrinking type residual stress distribution remains.

The second case, when the surface of the cylinder transforms before the core, is schematically illustrated in Fig. 6.37. In addition to the cooling curves of surface and core, also curves valid for intermediate locations within the

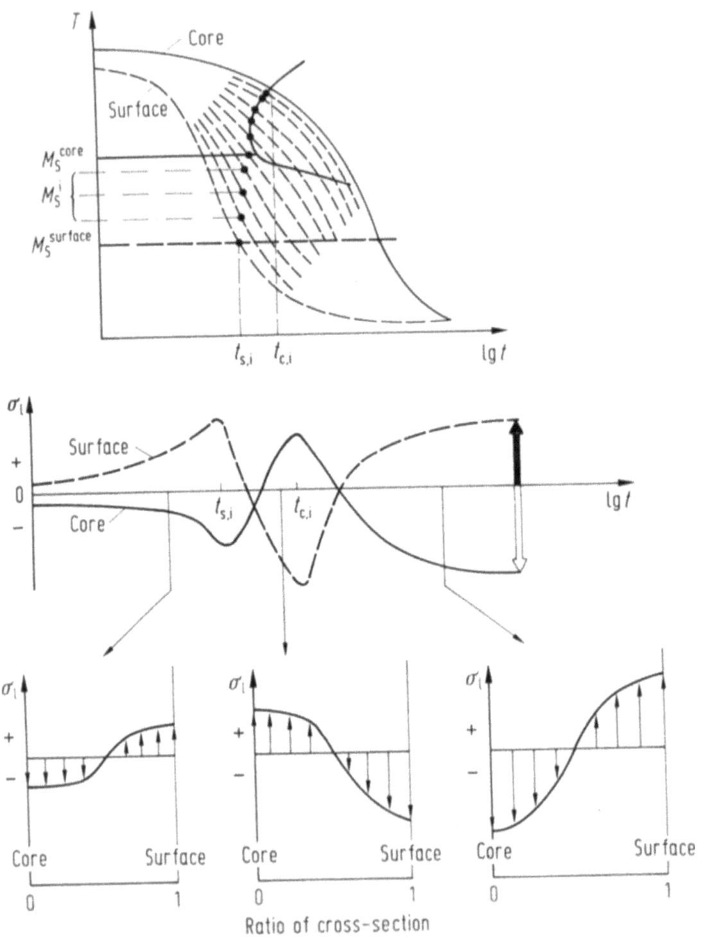

Fig. 6.37. Temperatures and stresses of surface and core during quenching of a carburized steel when martensitic transformations start in the surface and are followed by ferritic-pearlitic (bainitic) transformations in the core

cylinder are drawn in the TTT-diagram at the top of the figure. The points of intersection (dots) with the transformation lines of the TTT-diagrams represent in each case the start of the respective local transformations. The impeded shrinkage of the surface initially leads to tensile stresses which are compensated in the core by compressive stresses. When the surface transformations commence at $t_{s,i}$, the stresses acting there are reduced and inverted to compressive stresses. The sub-surface stresses follow with a similar but weaker trend according to the decrease of the carbon content with increasing depth. A small time interval after passing $t_{s,i}$, tensile core stresses are found to be compensated by compressive surface stresses. At this time, the core is still austenitized at a relatively high temperature, and can therefore easily be plastically strained. Finally, at $t_{c,i}$, also the core starts to transform after the outer core regions have already begun. The transformation-induced expansion of the core is, however, impeded by the already transformed surface regions so that compressive core stresses will form which are held in equilibrium by tensile surface stresses. Thus, altogether, during the cooling process, the signs of the stresses at surface and core of the cylinder invert twice. When the temperature compensation is complete, the core areas remain under compressive residual stresses whilst the surface areas are subjected to tensile residual stresses. The result is a residual stress distribution of the transformation type.

Extremely large transformation stresses are expected when the transformations within the surface zones occur simultaneously. This is possible for specific alloy compositions and specimen sizes if the conditions demonstrated in Fig. 6.38 are fulfilled. Then, tremendous incompatibilities generate between core and surface during the cooling process which mostly lead to internal cracking.

Of course, there exist several intermediate residual stress distributions between the above-mentioned extremes. Their character depends on the extent to which the prevailing conditions deviate from those assumed above. The really existing residual stress distributions are fundamentally influenced by the carburization depth, the case-hardening temperature, the material composition and hence the hardenability, the quenching conditions and lastly by the specimen size [34]. Therefore, in each specific case, the influence of the individual stress generating processes has to be assessed with the aid of the principles outlined above. In this regard, a more distinctive examination of the start of local transformations with respect to the inversion point of the pure shrinking stresses may be helpful.

Up to now, the discussion has not taken into account the fact that after case hardening, the surface and the subsurface zones of both carburized plain carbon and low-alloy steels with carbon contents larger than about 0.5 wt.-% will still contain a certain amount of retained austenite. As a result, the surface zones will never be completely transformed. The retained austenite areas will

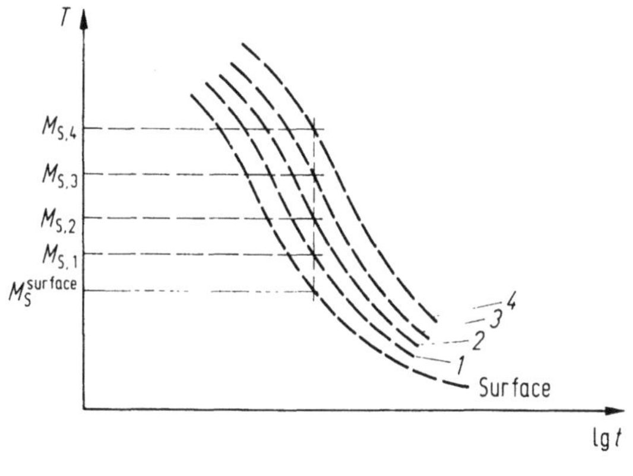

Fig. 6.38. Cooling curves of near-surface areas with different carbon contents and M_s-temperatures producing maximum transformation stresses

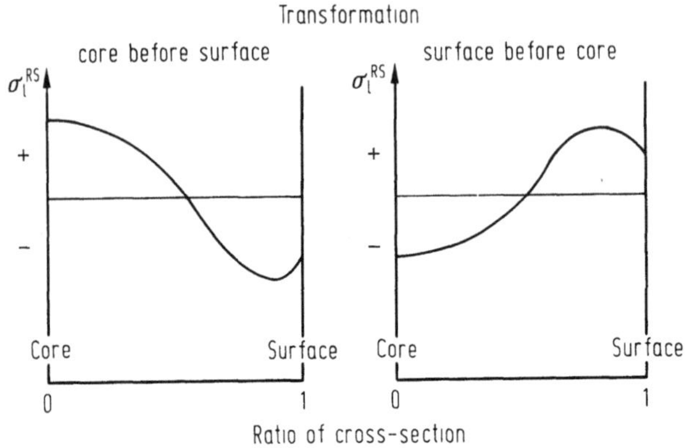

Fig. 6.39. Distributions of longitudinal residual stresses vs. ratio of cross-section of case-hardened cylinders influenced by retained austenite effects in near-surface and surface regions

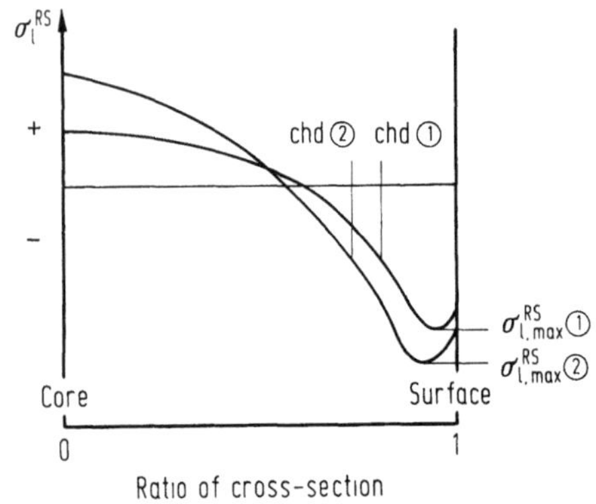

Fig. 6.40. Influence of case-hardening depth (chd) on the distribution of longitudinal residual stresses vs. ratio of cross-section of case-hardened cylinders

always tend towards smaller amounts of residual stresses compared to those where martensitic transformations completely took place. The residual stress distributions shown in Fig. 6.39 are therefore to be expected when additional retained austenite effects occur [35, 36].

The thickness of the near-surface layer possessing an enhanced carbon content determines the case-hardening depth (chd) which is defined as that distance from surface, where a hardness of 550 HV exists. This value often corresponds to a carbon content of about 0.35 wt.-%. With increasing chd, the part of the specimens in which martensite is generated with correspondingly large volume expansions increases. Therefore, including retained austenite effects, one has to expect larger longitudinal residual stresses in the surface zones of a case-hardened specimen if chd increases [35]. This is shown in Fig. 6.40. With increasing chd the locations with the residual stress maxima are shifted towards

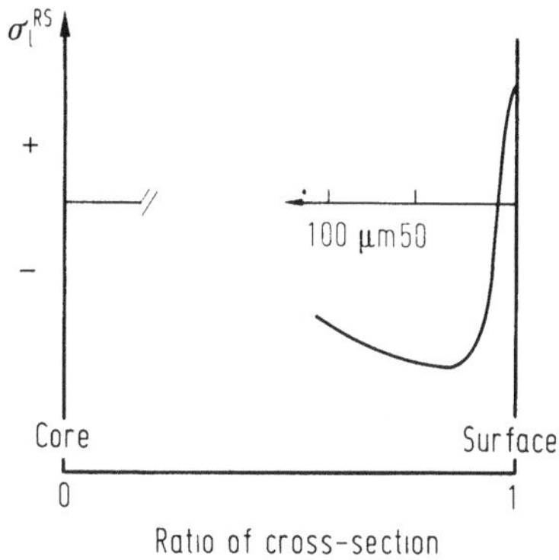

Fig. 6.41. Distribution of longitudinal residual stresses in the near-surface region of a case-hardened cylinder influenced by decarburization and oxidation

the core of the specimen. The residual stress maxima are located considerably nearer to the surface than the case-hardening depths. Frequently, they are found at depths where carbon contents of approximately 0.5 wt.-% exist. Always, the location where the longitudinal residual stresses change their signs are in-between the carburized zone and the core.

Very often, there exists also a strong influence of decarburization and oxidation on the residual stress distribution near the surface of case-hardened parts. In such cases, characteristic surface effects will emerge in zones of approximately 100 μm thickness [36]. A typical residual stress distribution is schematically illustrated in Fig. 6.41. At the very surface, tensile residual stresses exist, accompanied by a steep residual stress gradient just below the surface.

6.5.2 Characteristic Examples

Concerning the formation of case-hardening residual stresses, several investigations have been performed, mostly, however, with simple-shaped specimens. In the following, some characteristic examples are discussed. In transferring these results to components of a more complex nature, always careful considerations of the existing type of material and the geometric and process-specific parameters are necessary.

a) Residual Stresses in Hardened and Case-Hardened Plates of 26 MnCr 4

Plates of the shape and size illustrated in Fig. 6.42 were hardened and case-hardened, respectively [35]. Carburizing was performed in a gas-atmosphere at 930 °C. Both plates were quenched from a hardening temperature of 840 °C down to 20 °C. After a subsequent annealing at 180 °C for 2 hours, the longitudinal residual stress distributions illustrated in the figure were measured. The measurements were carried out in the middle of the plates. It can be seen that the hardened specimen displays a residual stress distribution of the transformation type, whilst that of the case-hardened sample is of the shrinking type.

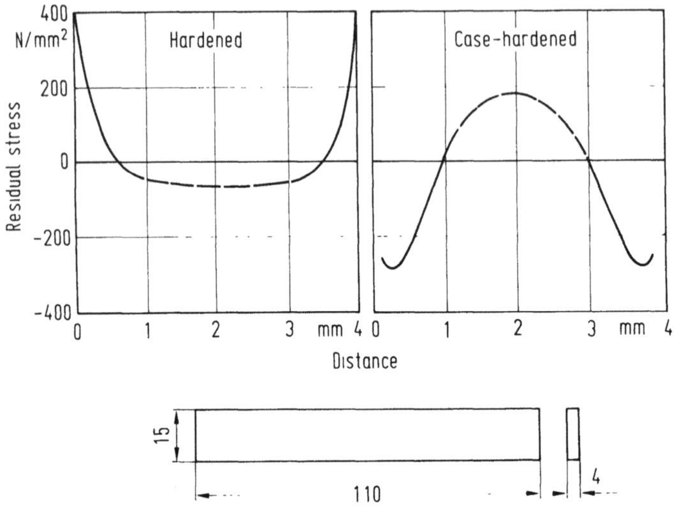

Fig. 6.42. Distribution of residual stresses across the thickness of a hardened and a case-hardened plate of 26 MnCr 4

b) Residual Stresses in Near-Surface Regions of Case-Hardened Bolts of 15 NiCr 3 with Different Diameters

Cylindrical specimens of 17 and 30 mm diameter were carburized to a surface carbon content of 0.8 wt.-%. At a distance of 1 mm from the surface, the carbon content was 0.2 wt.-%. After carburizing, the specimens were quenched from 820 °C in oil at 60 °C. The occurring cooling and transformation processes were simulated with the aid of a FE-program. In the calculations all relevant material parameters were included. The calculated longitudinal, tangential and radial residual stresses are plotted in Fig. 6.43 [37, 38]. In the surface regions of both bolts, nearly equal longitudinal and tangential stresses occur, which change similarly with increasing distance from the surface. The radial residual stresses increase progressively from zero at the surface to higher values towards the core. The maximum compressive residual stresses are located in distances of about 1 mm from the surface. The larger diameter bolt possesses larger maximum compressive residual stresses.

c) Temporal Development of Tangential Stresses in a Case-Hardened Cylinder of 20 NiCrMo 3

A cylindrical specimen of 12.7 mm diameter was carburized to a surface carbon content of 1.2 wt.-% and quenched from 927 °C in oil of 20 °C. The stresses which were created during the quenching process were determined with the aid of a specially developed FE-program. Figure 6.44 shows a three-dimensional plot of the temporal tangential stress generation along the radius of the cylinder [39]. As expected, the inversion of the signs of stresses during cooling occurs

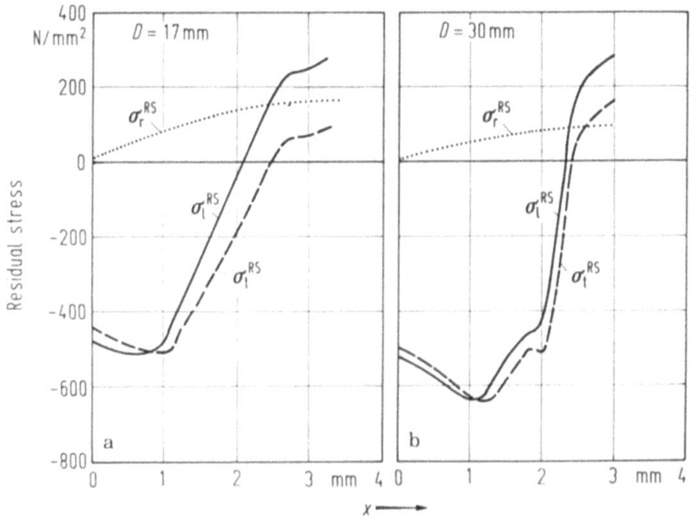

Fig. 6.43 a, b. Residual stress states in near-surface regions of case-hardened bolts of 15 NiCr 3 with diameters of 17 mm (**a**) and 30 mm (**b**), respectively

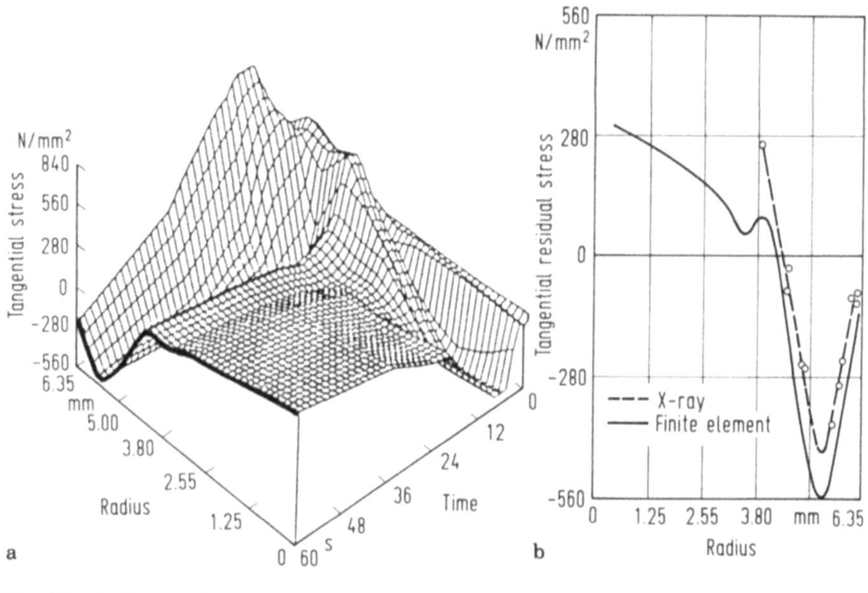

Fig. 6.44 a, b. Tangential stresses and tangential residual stresses of a carburized cylinder quenched from 927 °C in oil at 20 °C. **a** Temporal stress generation; **b** Comparison between experimentally determined and calculated tangential residual stresses

in distinct regions at different times. The distribution of the tangential residual stresses which results 60 sec after starting the quench is plotted in the right-hand graph as a function of the radius. It is of the shrinkage type with a strong reduction of compressive stresses in the near-surface regions caused by retained austenite. Figure 6.44 also shows an X-ray determined residual stress distribution in the near surface regions of the cylinder. As can be seen, the calculated and the experimentally derived values agree quite well.

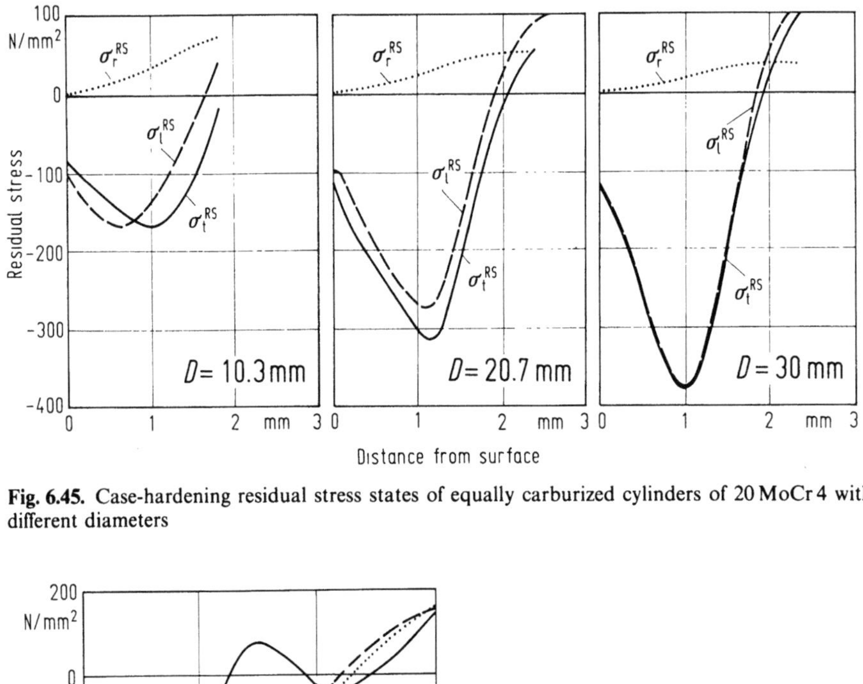

Fig. 6.45. Case-hardening residual stress states of equally carburized cylinders of 20 MoCr 4 with different diameters

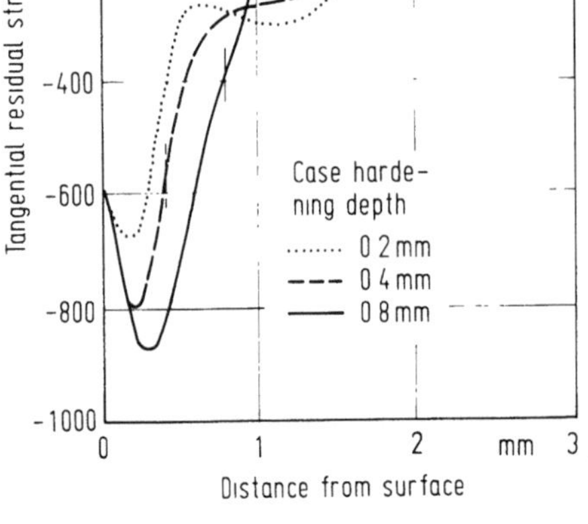

Fig. 6.46. Tangential residual stress vs. distance from the surface of cylinders of C22 with different case-hardening depths

d) Residual Stress Distribution in Case-Hardened Cylinders of 20 MoCr 4 with Different Diameters

Cylinders of 10.3, 20.7 and 30 mm diameter were gas carburized for 3 hours at 930 °C, followed by single hardening (860 °C/60 °C oil). After a two-hours annealing at 180 °C, a complete X-ray residual stress analysis revealed the stress distributions depicted in Fig. 6.45 [40]. Whilst the surface residual stresses remain approximately constant for all three cylinders, at a distance of 1 mm from the surface the maxima of the compressive residual stresses increase with increasing diameter. As can be seen, the maximum compressive residual stress of the 30 mm specimen is approximately twice that of the sample with a diameter of 10.3 mm. The smaller diameter specimens show differences in the magnitudes of the longitudinal and tangential residual stresses. In the largest specimen, such differences cannot be discerned up to a distance of approximately 1.7 mm from the surface.

e) Tangential Residual Stresses in Equal-Sized Cylinders of C 22 with Various Case-Hardening Depths

Figure 6.46 shows the tangential residual stress distributions measured at and below the surface of cylindrical specimens with 11.3 mm diameter and 80 mm length carburized to depths of 0.2, 0.4 and 0.8 mm and quenched from 880 °C in water [40]. As expected, the compressive residual stress maxima increase and shift towards the core with increasing case-hardening depth. At the surface, in all cases, the same compressive residual stresses of $-600\ \mathrm{N/mm^2}$ emerge.

f) Residual Stress Distributions in Different-Sized Case-Hardened Rings of 16 MnCr 5

16 MnCr 5 rings with different external diameters and thicknesses and several surface carbon contents and carburization depths were investigated after quenching from distinct hardening temperatures. For a ring with an external diameter of 400 mm, an internal diameter of 320 mm and a length of 80 mm the longitudinal, tangential and radial residual stresses were calculated in the

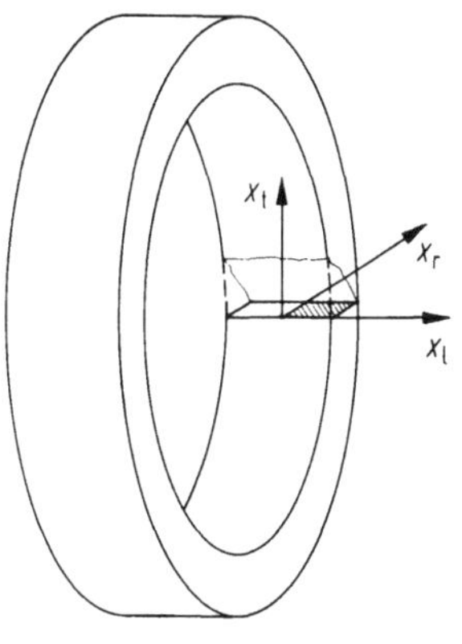

Fig. 6.47. Case-hardened ring with section of interest (shaded) for FE-calculations

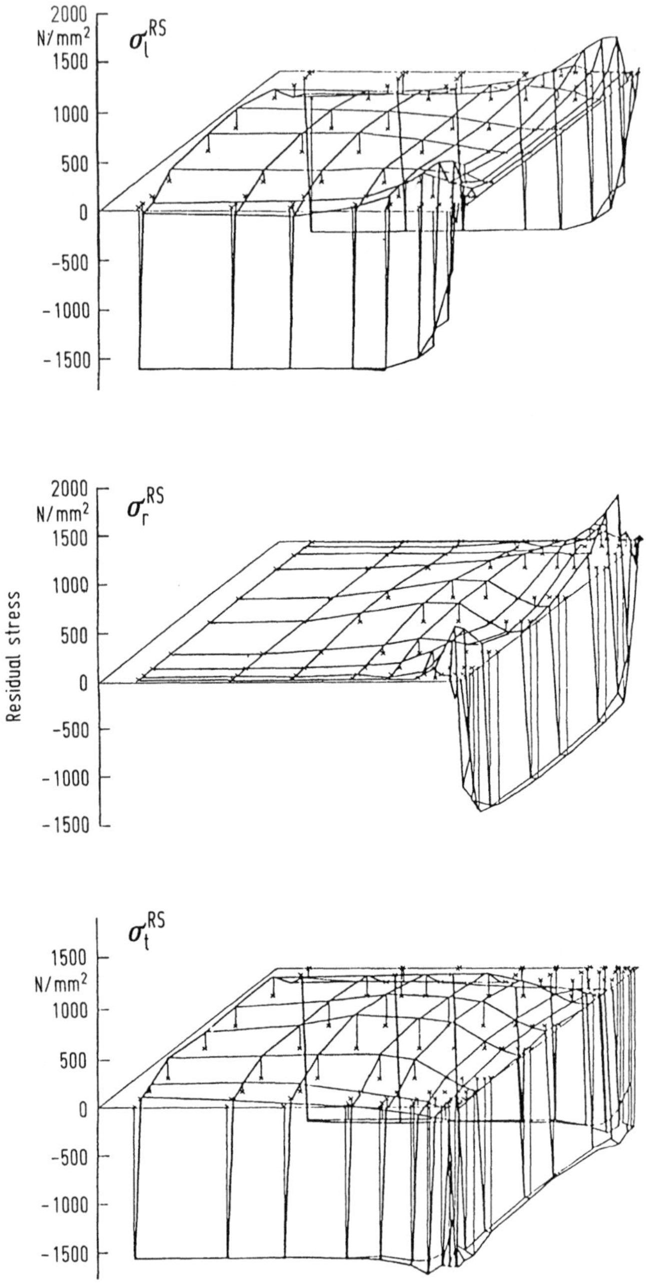

Fig. 6.48. Distribution of longitudinal, radial and tangential case-hardening residual stresses in the hatched wall section of the ring shown in Fig. 6.47

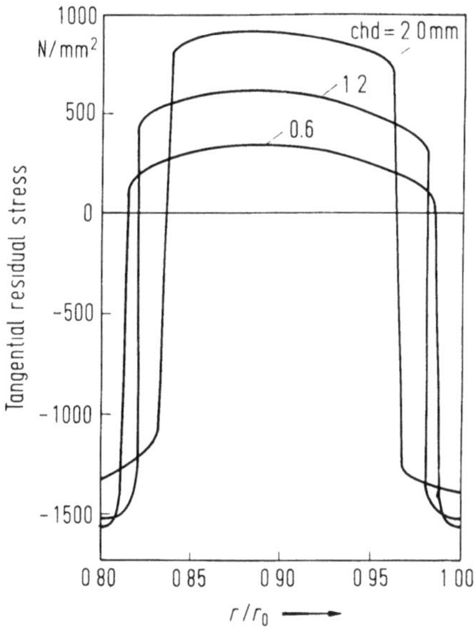

Fig. 6.49. Tangential residual stress across the wall thickness of equal-sized rings with different case-hardening depths

hatched section area in Fig. 6.47 by a FE-method [41, 42]. The distributions of the residual stress components for a ring with a surface carbon content of 0.8 wt.-% and a carburization depth of 1.2 mm after quenching from 880 °C in oil of 20 °C are presented in Fig. 6.48. The local residual stresses are plotted perpendicularly to the ring section considered. Tensile stresses are directed upwards and compressive stresses downwards. The stress distributions at the front side of the diagrams ($x_r = 0$) always correspond to those existing at the inner ring surface whilst those at the back side represent the stresses at the outer ring surface. The stresses at the right-hand side of the graphs correspond to those existing at the face of the ring. At the centre end of the analysed section high surface compressive longitudinal and tangential residual stresses exist which are in equilibrium with tensile residual stresses in the interior. At the face, the longitudinal stresses vanish in agreement with the surface conditions. Radial residual stresses only occur at and near the face of the ring, assuming extremely large values at the very surface where also the compressive tangential residual stresses take large magnitudes.

Figure 6.49 shows distributions of the tangential residual stresses in the mid-section of the ring. The data hold for rings with an external diameter of 120 mm, an internal diameter of 96 mm and a length of 24 mm, possessing the same surface carbon content of 0.8 wt.-% but differing carburizing depths [41,42]. With increasing case hardening depth the magnitude of the stresses at the outer and inner surface decreases and the thickness of the ring-zone with compressive residual stresses increases.

6.6 Residual Stresses After Quenching of Steels with Induction Heated Surface Layers

Surface heat treatments are of great practical importance. Thereby, the mostly applied processes are flame heating, induction heating and laser heating. In this chapter, the behaviour of surface areas which are quenched after induction heating will be discussed.

6.6.1 Quenching Without Transformation

Since induction heating is inevitably linked with large temperature gradients between surface and core of a part, the residual stresses emerging from trans-

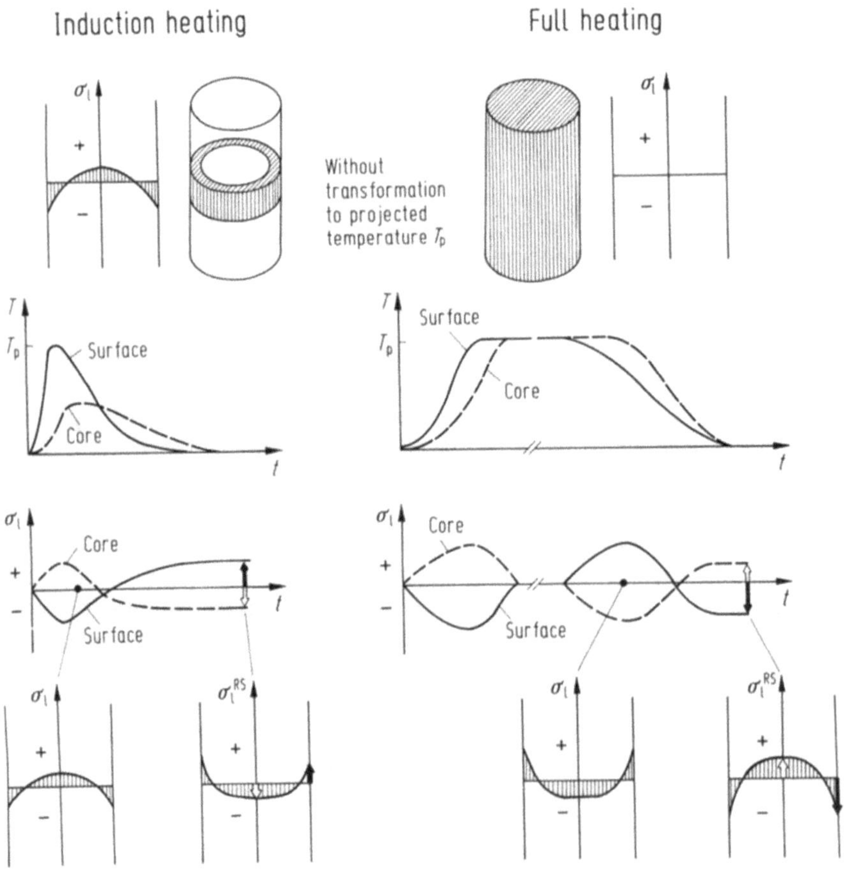

Fig. 6.50. Comparison of the stress-states of a cylinder during induction heating and full heating followed by transformation-free quenching

formation-free quenching differ fundamentally from those of a fully heated cylinder of identical shape and size. This is sketched in Fig. 6.50 schematically [43]. In the middle of the figure, the temperature-time relations and the longitudinal stress-time relations for surface and core in course of heating up to a projected temperature T_p with a following direct cooling (left) or a retarded cooling (right) down to room temperature are shown. Furthermore, the distributions of the longitudinal stresses across the cross section are sketched at the end of the heating process (graphs at the top). Also the stress distributions at distinct instants during cooling and the residual stress distributions after the final temperature balance are presented (graphs at the bottom). From the sequence of the graphs at the right-hand side it is seen that during full-heating the surface takes initially higher temperatures than the core so that compressive surface stresses are created which are compensated by tensile stresses in the core. During holding at the projected temperature, these stresses completely relax. Then, subsequent quenching to room temperature will cause the same processes as described in Chap. 6.4.2a. Accordingly, compressive residual stresses remain at the surface whilst tensile residual stresses occur in the core.

On the other hand, when the cylinder surface is rapidly induction heated to the same projected temperature (left-hand side of Fig. 6.50), thermal compressive stresses are generated within the plastically deformed surface area, whilst tensile stresses act in the core. Hence, when quenching is started, the cylinder does not only possess an extremely inhomogeneous temperature distribution but additionally an inhomogeneous state of stress. Furthermore, the temperature-time relations illustrate that the core temperature is still rising whilst that of the surface is already falling. As a result, the respective temperature-time curves intersect and the core of the specimen becomes hotter than the surface. Hence, the stress conditions just existing at the beginning of quenching are quite different to those of a fully heated cylinder. Finally after temperature equalization tensile residual stresses remain at the surface, which are compensated by compressive residual stresses in the core. Only if in course of quenching a cylinder, large plastic extensions of the surface areas occur, then also residual compressive surface stresses may remain after equalization of temperature.

6.6.2 Quenching Combined with Transformation

Rapid quenching of an induction heated steel cylinder with austenitized surface areas will induce martensitic transformations within the surface and in the near-surface zones. The resulting local stress-time-relations can be interpreted with the aid of the principles previously discussed in Chap. 6.4.2c. During cooling, the thermal compression of surface and near-surface areas will be superimposed by the volume expansion caused by the formation of martensite. Since only the surface and near-surface regions are subjected to transformation processes, the surface stresses will inevitably be shifted to higher negative values. As a result, the core stresses react with a corresponding shift towards more positive values. When fully cooled, the surface zones are finally subjected to longitudinal compressive residual stresses which are held in equilibrium by tensile residual stresses in the core. Additionally tangential and radial residual stresses are also produced during induction hardening of steel cylinders.

Figure 6.51 originates from one of the first publications which quantitatively analysed the residual stresses within inductively hardened components [44]. Hollow cylinders of Ck 45 with an outer diameter of 45 mm and an inner diameter of 16 mm, hardened to a depth of 6 mm were investigated. The residual stress measurements were performed with the classic boring-out method. The figure illustrates the distributions of the longitudinal, tangential and radial residual stresses of such a cylinder. At the outer surface compressive longitudinal and tangential stresses arise which invert to positive values at the inner surface. At both locations, the radial stresses are zero, assuming positive values in-between. At the outer surface, the magnitude of the longitudinal residual stresses is smaller than that of the tangential residual stresses.

The first systematic investigations of the influence of material type and hardening depth on the residual stress state of inductively hardened cylinders are given by the experimental results shown in Fig. 6.52 [45]. The left-hand graph demonstrates for cylinders of 41 Cr 4 with a diameter of 25 mm how the magnitudes of the compressive surface residual stresses decrease and how the stress distributions change with increasing hardening depth. The right-hand graph illustrates the differences in the surface and near-surface longitudinal

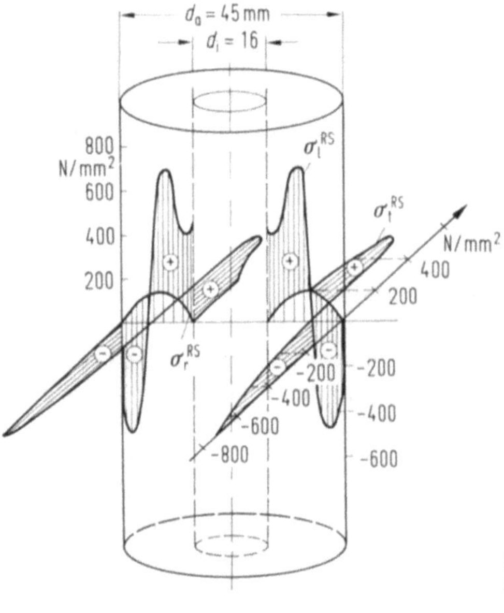

Fig. 6.51. Distribution of residual stresses in an induction hardened hollow cylinder of Ck 45

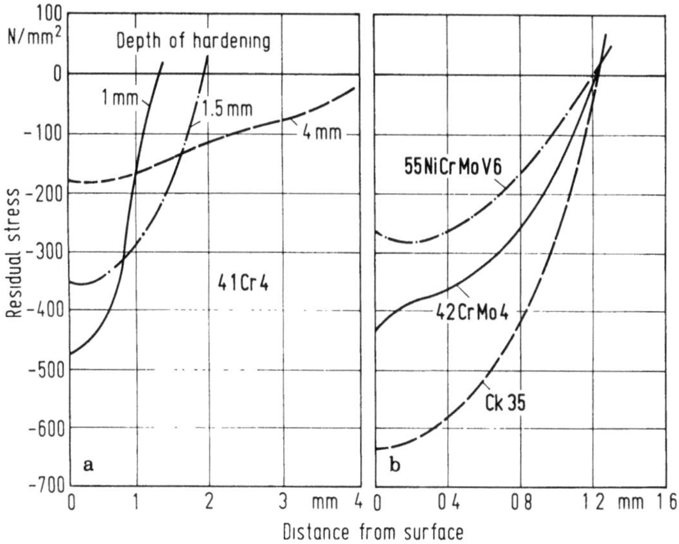

Fig. 6.52 a, b. Near-surface distributions of longitudinal residual stresses in induction hardened steel cylinders. **a** Different hardening depths. **b** Hardening depth 1.0 mm

residual stresses of equal-sized cylinders (diameter 25 mm) of Ck 35, 42 CrMo 4 and 55 NiCrMoV 6 with approximately the same hardening depth of 1.0 mm. The magnitudes of the compressive residual stresses increase in the sequence 55 NiCrMo 6, 42 CrMo 4 and Ck 35. This may be explained by the fact that the high temperature yield strengths of these materials decrease in the mentioned sequence, whilst the martensite start temperatures increase.

The previously discussed older experimental results have been supplemented with more recent studies modelling the induction hardening process. For example, the non steady-state temperature distributions and the stress distributions as a function of time across the radius of a cylindrical disc during induction heating and subsequent quenching were studied. Figure 6.53 shows the results of such a simulation for a cylindrical disc of Ck 45 with a diameter of 120 mm [46], which was inductively heated by a source creating an energy density of 2.4 J/sm³ at 1 mm depth. The temperature increase at and beneath the surface in dependence on time after starting induction heating is shown at the left of the figure. With increasing heating time as can be seen from the right-hand part of the figure, decreasing compressive stresses appear in the near surface regions, whose maxima shift towards the interior of the specimen, simultaneously reducing their magnitudes. Synchronously, tensile core stresses develop which increase with increasing surface temperature. After a 14 sec heating at the surface a temperature of 890 °C and 4 mm below the surface the A_{c3}-temperature is reached. Then, the centre of the disc is subjected to a triaxial stress state with $\sigma_l = 420\ \text{N/mm}^2$, $\sigma_t = 180\ \text{N/mm}^2$ and $\sigma_r = 180\ \text{N/mm}^2$. The subsequent quench-

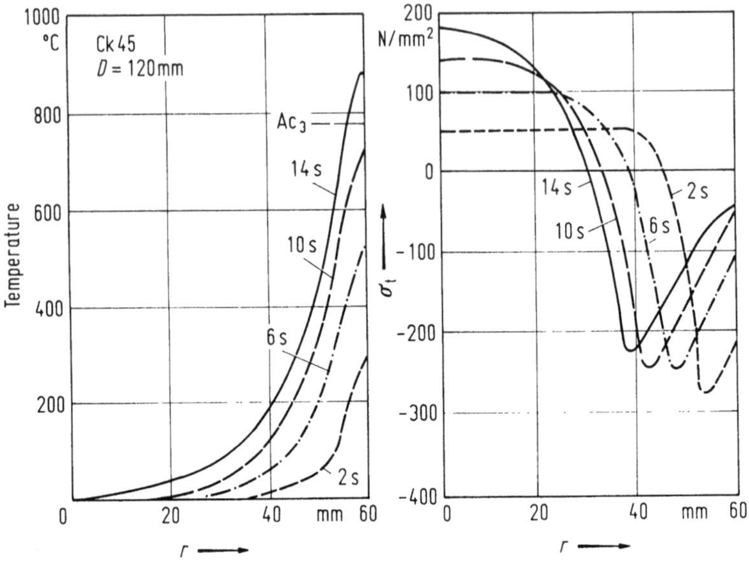

Fig. 6.53. Development of temperatures and tangential stresses during induction heating of a cylindrical disc of Ck 45 with a diameter of 120 mm

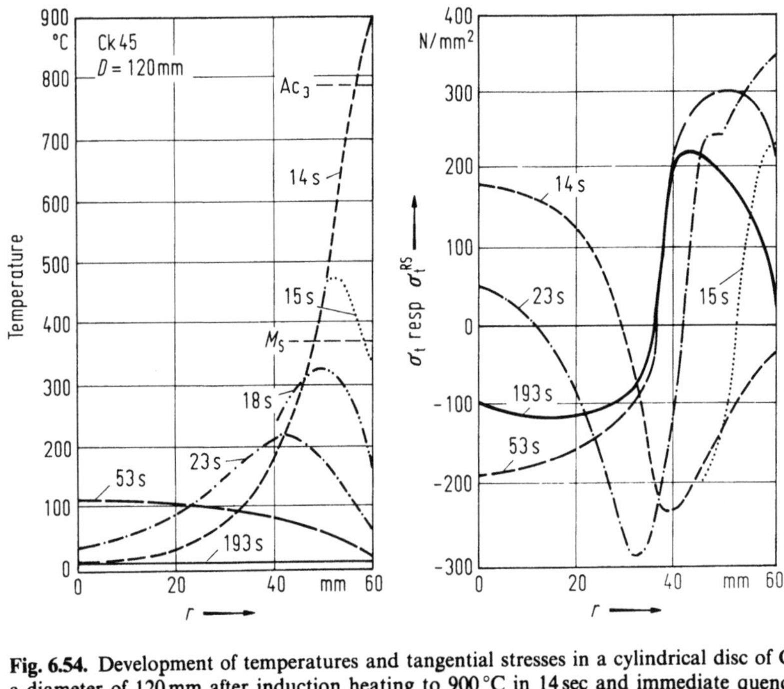

Fig. 6.54. Development of temperatures and tangential stresses in a cylindrical disc of Ck 45 with a diameter of 120 mm after induction heating to 900 °C in 14 sec and immediate quenching. The bold curve in the right-hand figure represents the final residual stress distribution after equalization of temperature

ing was simulated using a temperature dependent heat transfer coefficient. Figure 6.54 visualizes the rapid changes of temperature and tangential stresses along the radius of the disc. As the left-hand graph demonstrates, the external disc regions are already cooled down to temperatures below M_s just 1 s after starting the quench. Thus, very quickly, the shrinking of the surface areas is superimposed by the effects of martensitic transformations. From the right-hand graph it can be seen that the surface stresses at first rapidly change from compressive into tensile stresses. With increasing time the intermediate regions of the cylindrical disc initially subjected to compressive stresses remain compressed. Later on the whole inner part of the disc with $r \lesssim 35$ mm takes up compressive stresses.

Results of a more recent theoretical investigation of induction heating and hardening processes are demonstrated in Fig. 6.55 [47]. A 40 mm diameter cylinder of the steel 42 CrMo 4 was heated in 33 s at 300 kHz to 890 °C, quenched in 4.5 s to 780 °C and afterwards sprayed with water during a period of 43.5 s for temperature equalization. The temperature-time relations of the surface and

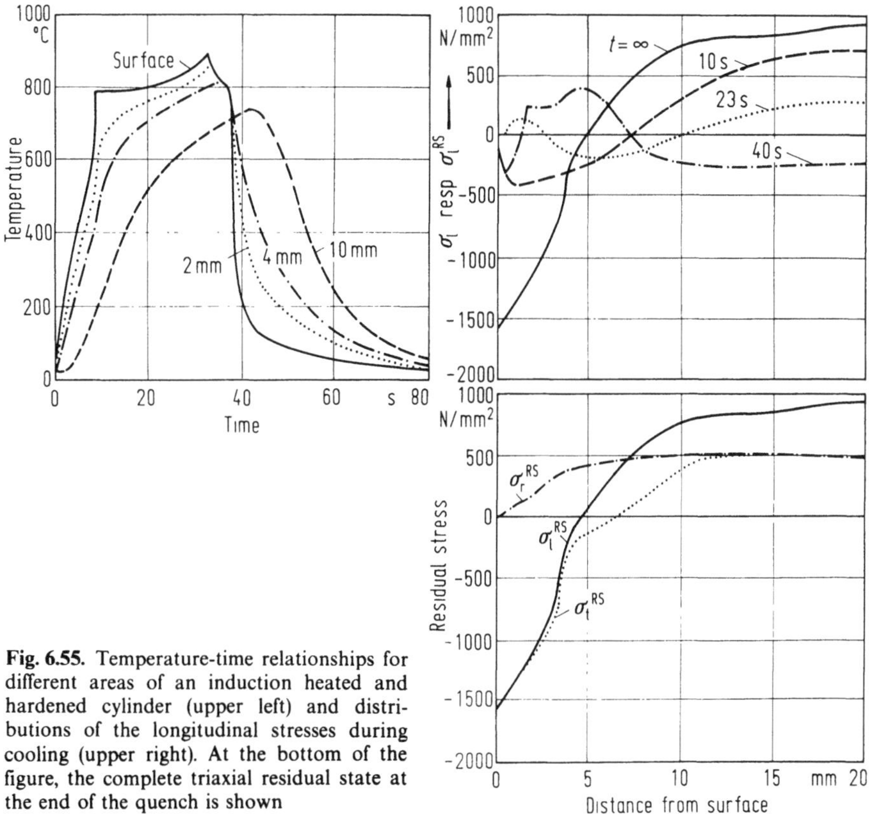

Fig. 6.55. Temperature-time relationships for different areas of an induction heated and hardened cylinder (upper left) and distributions of the longitudinal stresses during cooling (upper right). At the bottom of the figure, the complete triaxial residual state at the end of the quench is shown

three areas with distances of 2, 4 and 10 mm from the surface are shown in the upper graph at the left-hand side. It clearly appears how the temperature peaks of the different areas develop at different instants. The upper graph at the right-hand side reveals the generation of the longitudinal stresses across the cylinder radius at various times after quenching whereby the rather complex temporal stress development at and near the surface becomes apparent. When fully cooled ($t > 80$ s), large longitudinal and tangential compressive residual stresses exist at the surface, whilst the core is subjected to tensile stresses which are approximately 30% and 60% smaller, respectively. This can be seen from the bottom graph which depicts the longitudinal, tangential and radial residual stress distributions. The longitudinal and tangential residual stresses are practically equal in a 4 mm thick zone beneath the surface.

6.7 Residual Stresses After Self-Quenching of Steels with Laser-Heated Surface Layers

Surface heating of metallic materials by light has found particular interest in the course of recent developments in laser technology. In the following, only two aspects of laser treated steels will be considered, that is quenching of laser austenitized surfaces and quenching of laser melted surfaces.

6.7.1 Quenching After Austenitizing

The energy of incident laser beams on metallic surfaces will partly be absorbed in form of heat, leading to a rapid temperature rise in a thin surface layer with a thickness of a few tenth of a millimeter. Power densities up to 10^4 J/smm^2 across laser beam breadths of 1 – 100 mm produce heating rates up to 10^6 K/s with appertaining temperature gradients of several 10^3 K/mm. Hence, by means of laser irradiations, localized short-time increases in the temperature of surface and subsurface regions of metallic materials can be realized. Such short-time local heatings are always accompanied by a severe local heat removal conduction, leading to cooling rates up to 10^8 K/s. In steels, depending on the maximum temperatures reached, these self-quenching processes will occur with or without martensitic transformations [48]. In the course of a transformation-free cooling, at the surface of a laser-heated steel specimen, tensile residual stresses develop which are compensated by compressive residual stresses in adjacent areas. The residual stress producing processes are similar to those discussed in Chap. 6.6.1 for quenched induction heated specimens. However, if laser heatings produce surface austenitizations, then the ensuing quenchs induce martensitic transformations in the laser-heated zones, which, as in the case of induction-hardening, generate compressive residual stresses in the very surface.

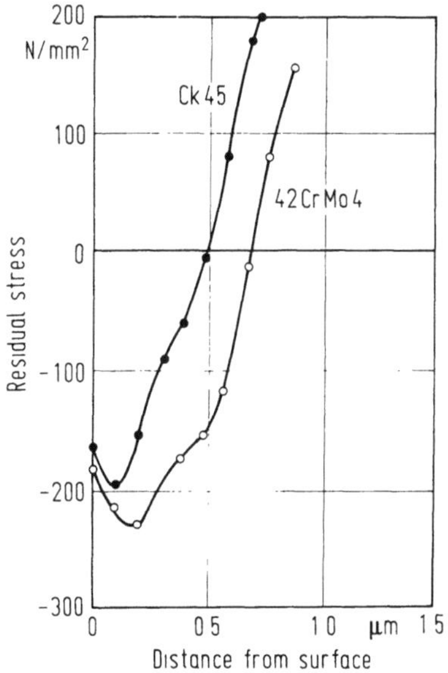

Fig. 6.56. Residual stresses vs. distance from surface in laser hardened Ck 45 and 42 CrMo 4

Typical residual stress distributions near the surface of laser-hardened plates of Ck 45 and 42 CrMo 4 are shown in Fig. 6.56 [49]. The data hold for the same operative parameters (laser beam 14 × 14 mm, 27.5 W/mm^2 for 0.56 sec) and an moderate additional liquid jet cooling at the side opposite to the treated one. In both cases, nearly the same compressive residual stresses exist at the very surface. Below the surface peak values of -200 N/mm^2 for Ck 45 and -230 N/mm^2 for 42 CrMo 4 are observed. Then, with increasing distance from the surface, the magnitudes of the residual stresses rapidly decrease. In the case of Ck 45, the transition to tensile residual stresses takes place at a distance of about 0.5 mm from the surface. The corresponding transition point for 42 CrMo 4 has a larger surface distance.

The surface residual stress state across a single laser beam trace of 8 mm width in annealed Ck 45 is demonstrated in Fig. 6.57 [50]. The residual stresses parallel to the trace are plotted as function of the distance from the trace axis. Near the axis, compressive residual stresses are measured. Near the boundary of the trace, a tensile residual stress maximum appears. In the neighbouring heat influenced zones, again compressive residual stresses prevail. It is interesting to note that modelling the butt-welding process for Ck 45 led to a residual stress distribution which agreed quite well with the discussed residual stress observations after laser hardening [48, 51].

The results of a more recent modelling of the generation of stresses due to laser heating of a 42 CrMo 4 steel are shown in Fig. 6.58 [52]. With a total energy consumption of 9.7 kJ/s, laser traces with a width of 8.1 mm were produced with a speed of 150 mm/min along a test plate. The graphs on the left illustrate the temperature-time relations and the longitudinal stress-time

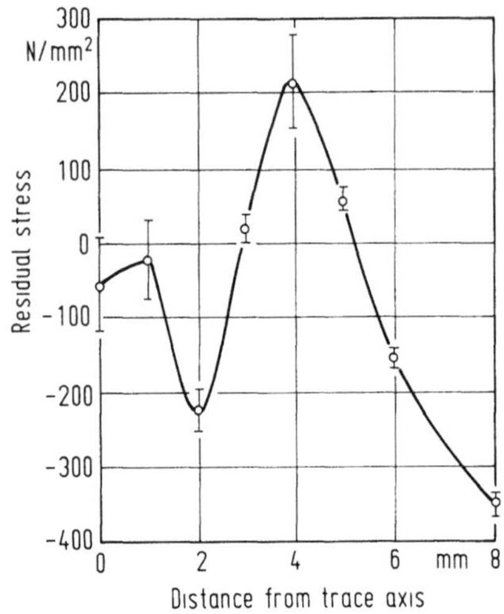

Fig. 6.57. Longitudinal surface residual stresses vs. distance from the trace center of a laser hardened Ck 45

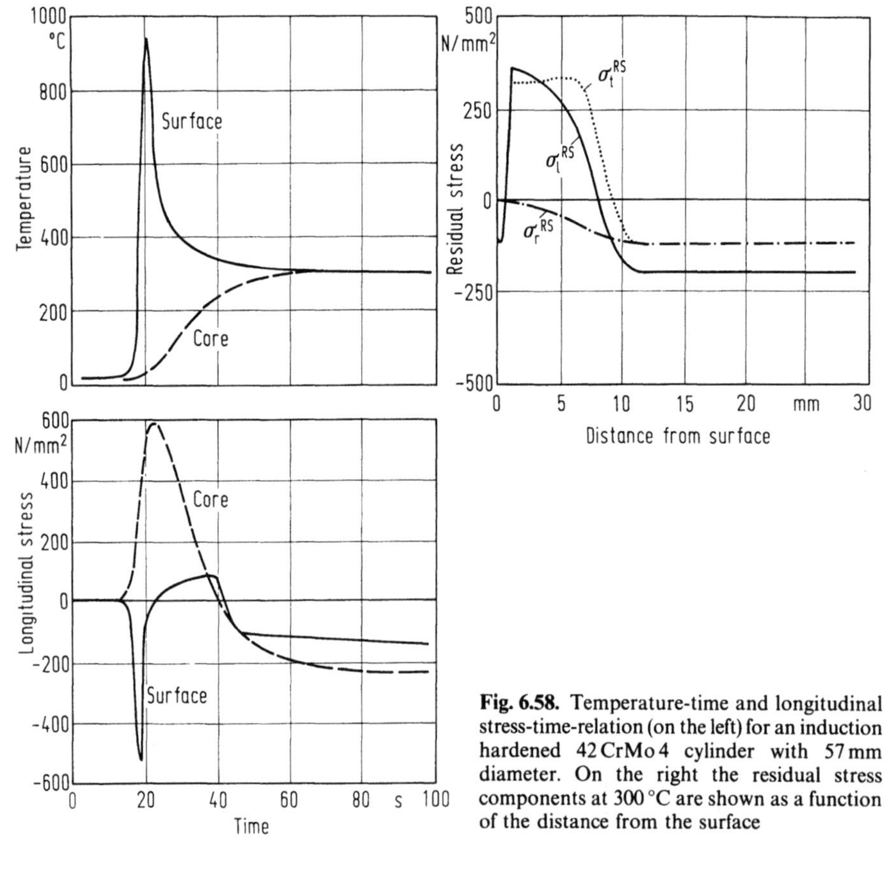

Fig. 6.58. Temperature-time and longitudinal stress-time-relation (on the left) for an induction hardened 42 CrMo 4 cylinder with 57 mm diameter. On the right the residual stress components at 300 °C are shown as a function of the distance from the surface

relations for surface and core of the material. The surface reaches a maximum temperature of 950 °C after approximately 20 s, then cools and approaches together with the core during 40 sec a projected temperature of 300 °C. When heating the surface zones, the restricted thermal expansion immediately creates compressive stresses which are held in equilibrium by tensile stresses in the core. This can be seen in the graph at the bottom on the left. When austenite begins to form at the surface, the appertaining volume contractions completely reduce the surface compressive stresses. Even small tensile stresses are created. The martensite start temperature in the surface is reached after about 40 s. These martensitic transformations are accompanied by a shift of the surface stresses to negative values. The distributions of the longitudinal, tangential and radial stresses, which remain when the metal has cooled to 300 °C after about 100 s are shown on the right of Fig. 6.58. These distributions roughly correspond with the expected residual stress distributions after cooling to room temperature. Surprisingly, at the very surface relatively small compressive stress components occur in longitudinal and tangential direction. But both stress components show at larger distances from the surface distinct tensile stress maxima. At larger distances from the surface, all three stress components possess relatively small negative values.

More complex surface residual stress distributions result when laser traces overlap. In such cases, the material in previously hardened areas will be re-heated by subsequent laser traces. If the maximum temperatures of the traces reach values near A_1, pre-existing laser-hardening residual stresses will be extensively reduced. Due to the subsequent self-quenching, however, new tensile shrinkage residual stresses will be formed in the surface zones. An example is presented in

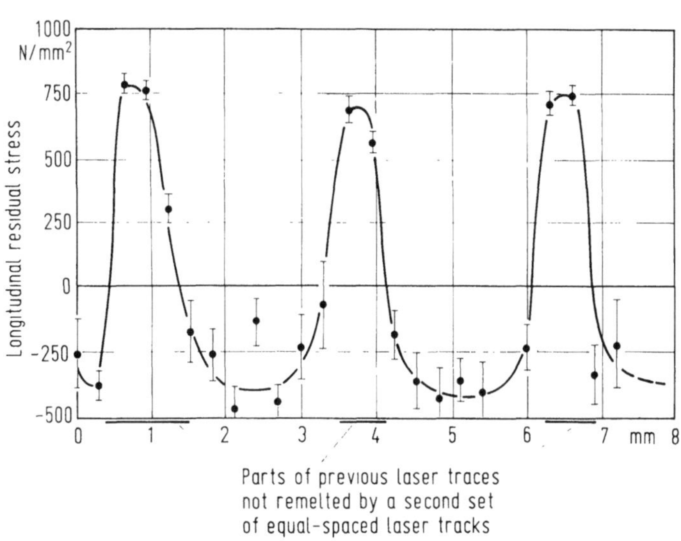

Fig. 6.59. Longitudinal residual stresses parallel to the traces at a two-stage laser treated surface of Ck 45

Fig. 6.59 [53]. The data belong to a 10 mm thick plate of Ck 45 which was subjected to a two-stage laser treatment. After producing a sequence of parallel laser tracks with a beam diameter of 2.8 mm, in-between these tracks new traces were positioned. The laser beam was moved with a velocity of 100 mm/min. The figure shows that the initially treated regions, which are not reaustenitized in the course of the second treatment, display large tensile residual stresses of up to 800 N/mm^2. Within the secondary traces, however, compressive residual stresses up to $-400\ N/mm^2$ are measured.

6.7.2 Quenching After Melting

If the energy conveyed to a steel component by an incident laser beam is sufficiently high, the surface layer will be melted. Then self-quenching will generally create more complex residual stress distributions than in the course of simple heating or austenitizing of the surface areas. In the case of well-separated laser-melted single tracks the residual stresses develop in a similar manner as in the case of quenching of laser austenitized surfaces. For example the residual stress of a single track with a width of 800 μm at the surface of a steel X20 Cr 12 takes a value of $-150\ N/mm^2$ [54]. However, if laser traces were produced parallel to pre-existing ones with decreasing unmelted spacings in-between, the surface residual stresses of the tracks change their sign and increase continuously with decreasing distance between the tracks as shown in Fig. 6.60. A reduction of the distance between the traces causes the neighbouring trace to be more and more heated so that its hardening residual stresses can relax to a larger extent. The subsequent self-quenching, however, induces new quenching stresses. The appertaining plastic deformations increase with increasing heat transfer from the neighbouring traces, as a consequence of which larger tensile residual stresses emerge.

The mutual interactions between parallel laser-melted traces also completely change the distributions of residual stresses beneath the surface of steels which

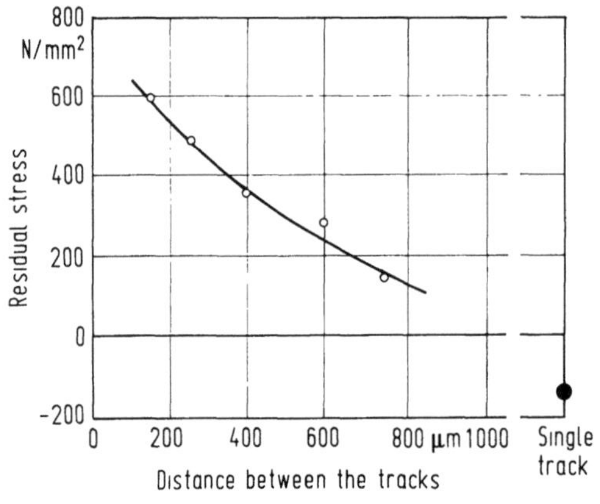

Fig. 6.60. Track residual stresses vs. distance between laser beam tracks of X 20 Cr 12 after laser-melting

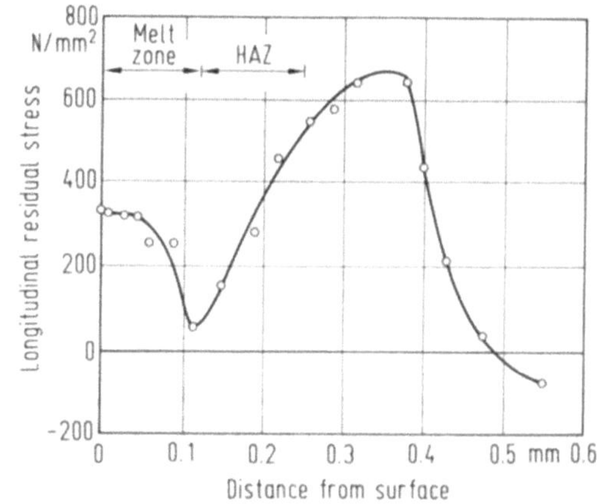

Fig. 6.61. Longitudinal residual stresses vs. distance from surface of a multistage laser melted X 20 Cr 12

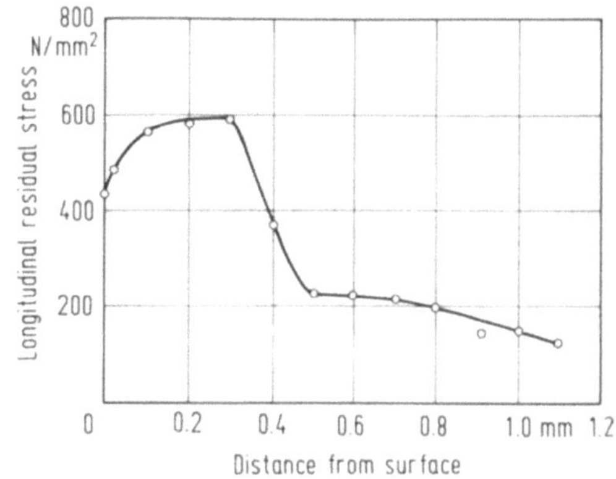

Fig. 6.62. Longitudinal residual stresses vs. distance from surface of Ck 45 after laser-melting

after a single-stage laser treatment are subjected to martensitic transformations in the course of cooling from austenitizing temperatures. This can be seen by comparing the experimental findings in Fig. 6.56 with those in Fig. 6.61. Figure 6.61 is valid for X 20 Cr 12 with several parallel laser traces separated by unmelted spacings of 0.4 mm [54]. The extension of the melt zone perpendicular to the surface is 0.12 mm and that of the heat affected zone (HAZ) 0.13 mm. At all distances from the surface smaller than 0.5 mm, tensile residual stresses are observed. A minimum value of 50 N/mm^2 exists at the interface between the melt zone and the HAZ. A maximum value of 650 N/mm^2 appears at a depth of 0.35 mm. Beyond this maximum, the residual stresses decrease continuously and change their sign at a depth of about 0.47 mm. Figure 6.62 shows the depth distribution of the longitudinal residual stresses for overlapping laser-melted traces in a plain carbon steel Ck 45 [55]. The overlapping was 33% for 0.3 mm wide traces. Tensile residual stresses with values upto 600 N/mm^2 are created within the martensitically hardened zone, which extend to approximately 0.4 mm below the surface.

To summarize, theoretical and experimental investigations have shown that laser treatments can create both surface tensile and compressive residual stresses of various magnitudes, depending on the materials and the process parameters selected. However, further systematic investigations are necessary which allow a more correct prediction and quantification of the residual stresses emerging from such treatments.

References

1. Macherauch E, Hauk V (Eds.) (1987) Residual Stresses in Science and Technology. ICRS 1, Vol. 1 and 2. DGM Informationsgesellschaft mbH, Oberursel
2. Beck G, Denis S, Simon A (Eds.) (1989) Proceedings of the Second International Conference on Residual Stresses. ICRS 2. Elsevier Appl. Sci., London
3. Fletcher AJ (1989) Thermal Stress and Strain Generation in Heat Treatment. Elsevier Appl. Sci., London
4. Ericsson T, Bergström J (Eds.) (1987) Swedish Symposium on Residual Stresses. Linköping Institute of Technology, Sunne/Sweden
5. Niku-Lari A (Ed.) (1987) Advances in Surface Treatments. Vol. 4, Residual Stresses. Pergamon Press, Oxford
6. Attebo E, Ericsson T (Eds.) (1984) Calculation of Internal Stresses in Heat Treatment of Metallic Materials, Vol. 1 and 2. Linköping, Sweden
7. Macherauch E, Hauk V (Eds.) (1983) Eigenspannungen, Entstehung – Messung – Bewertung, Band 1 und 2. Deutsche Ges. f. Materialkunde e. V., Oberursel
8. Kula E, Weiss V (Eds.) (1982) Residual Stress and Stress Relaxation. Proc. 28th Sagamore Army Materials Research Conference. Plenum Press, New York, London
9. van de Walle LJ (Ed.) (1981) Residual Stress for Designers and Metallurgists. ASM, Metals Park/Ohio
10. Hauk V, Macherauch E (Eds.) (1982) Eigenspannungen und Lastspannungen. Beiheft HTM. Carl Hanser Verlag, München
11. Tietz HD (1983) Grundlagen der Eigenspannungen. VEB Deutscher Verlag Grundstoffindustrie, Leipzig
12. Macherauch E, Wohlfahrt H, Wolfstieg U (1973) In: Härterei-Techn. Mitt. 28: 201–211
13. Chapman AJ (1974) Heat Transfer. Maxmillan Publ. Co
14. Uhlitzsch H (1959) In: Neue Hütte 4: 277–287
15. Eckstein H-J (Ed.) (1980) Technologie der Wärmebehandlung von Stahl. VEB Deutscher Verlag für Grundstoffindustrie, Leipzig
16. Schröder R (1985) Dr.-Ing.-Dissertation. Universität Karlsruhe
17. Hildenwall B, Ericsson T (1978) In: Doahe, Kirkaldy. Hardenability Concepts with Applications to Steel, p. 579–605, Met. Soc. AIME, Warrendale, P.A.
18. Graja P (1987) Dr.-Ing.-Dissertation. Universität Karlsruhe
19. Price RF, Fletcher AJ (1980) Metals Technology 7: 203
20. Rose A, Hougardy H (1972) Atlas der Wärmebehandlung der Stähle, Vol. 2. Verlag Stahleisen mbH, Düsseldorf
21. Bühler H, Rose A (1969) In: Archiv Eisenhüttenwesen 40: 411–423
22. Moyer JM, Ansell GS (1975) In: Metallurg. Trans. 6A: 1785–1791
23. Abbasi F, Fletcher AJ (1985) In: Mater. Sci. Techn. 1: 830–837
24. Leblond JB, Mottet G, Devaux J, Devaux JC (1985) In: Mater. Sci. Techn. 1: 815–821
25. Greenwood GW, Johnson RH (1965) In: Proc. Roy. Soc. 283A: 403–432
26. Denis S, Gautier E, Simon A In: [2]: 393–398
27. Eckstein H-J (1971) Wärmebehandlung von Stahl. VEB Deutscher Verlag für Grundstoffindustrie, Leipzig
28. Krauss G (1980) Principles of Heat Treatment of Steel. American Society for Metals. Metals Park, Ohio
29. Hougardy H, Wildau M (1983) In: Härterei-Techn. Mitt. 38: 121–127

30. Macherauch E (1979) In: Z. Werkstofftech. 10: 97–111
31. Macherauch E, Vöhringer O (1986) In: Härterei-Techn. Mitt. 41: 71–91
32. Macherauch E, Löhe D (1987) In: Handbuch der Fertigungstechnik, Vol. 4/2, Wärmebehandeln, p. 585–648. Carl Hanser Verlag, München
33. Besserdich G, Unpublished results. Institut für Werkstoffkkunde I, Universität Karlsruhe
34. Parrish G (1980) The Influence of Microstructure on the Properties of Case-Carburized Components. American Society for Metals. Metals Park, Ohio
35. Ebert LJ (1978) In: Metallurg. Trans. 9A: 1537–1551
36. Chatterjee-Fischer R (1973) In: Härterei-Techn. Mitt. 28: 276–284
37. Hildenwall B (1979) Dissertation Nr. 39. Linköping
38. Ericsson T, Hildenwall B In: [8]: 19–38
39. Burnett JA (1985) In: Mat. Sci. Techn. 1: 863–871
40. Koch M In: [10]: 112–121
41. Parr T (1980) Report FVA 34/11. Institut für Werkstoffkunde I, Universität Karslruhe
42. Parr T, Scholtes B, Müller H, Vöhringer O In: [7], Vol. 1, p. 279–292
43. Macherauch E (1988) In: Kloos, Grosch. Proc. Induktives Randschichthärten, p. 164–208. AWT-Tagung Darmstadt
44. Bollenrath F, Domke W (1956) Forschungs-Berichte des Wirtschafts- und Verkehrsministeriums NRW, Nr. 322
45. Frank DA (1964) Dr.-Ing. Dissertation, TU Berlin
46. Brückner U, Schuler W, Walter H In: [7], Vol. 1: 293–308
47. Melander M (1985) In: J. Heat Treating 4: 145–166
48. Ericsson T In: [5]: 87–113
49. Solina A et al. (1984) In: J. Heat Treating 3: 193–204
50. Chatterjee-Fischer R, Rothe R, Becker R (1984) In: Härterei-Techn. Mitt. 39: 91–98
51. Yu HJ (1977) Dr.-Ing. Dissertation. Universität Karlsruhe
52. Ericsson T, Yao SC, Melander M (1987) In: Härterei-Techn. Mitt. 42: 280–287
53. Scholtes B Unpublished results. Institut für Werkstoffkunde I, Universität Karlsruhe
54. Lamb M, West DRF, Steen WM (1986) In: Materials Science and Technology 2: 974–980
55. Scholtes B, Mordike BL In: [1], Vol. 2: 655–662

7 Effect of Workpiece Surface Properties on Cooling Behaviour

F. Moreaux and G. Beck

Quenching specialists have often noticed unaccountable variations among the results of that meaningful phase of heat treatment. Those at times epidemic-like variations entail poor hardness penetration and/or distortions.

One of the major causes of such crises is due to inappropriate mastery of the processed part surface properties but it is also brought about by the intricate liquid vaporization mechanisms in nonsteady heat transfer.

In the case of a steady state for a liquid used at saturation point temperature and unagitated, the various vaporization types are defined in a curve called Nukiyama's (see Chap. 2). On such a curve, the transition temperature between film boiling and nucleate boiling called Leidenfrost's temperature is well defined for a specific liquid.

In the case of nonsteady state heat transfer, e.g. for quenching, a curve may be defined thus providing appropriate description of the various vaporization types met during the cooling process of a metallic part in a vaporizable liquid.

This chapter is a description of the effect of surface properties on quenching in vaporizable liquids.

This provides with better quenching control and besides offers new techniques.

7.1 Effect of Quenching Conditions on Liquid Vaporization Types

The determination principle of the cooling curves is detailed in Chap. 7. To define the various vaporization types in the course of the quenching process a standardized metallic test-piece [1] is used. It is fitted with a thermoelectric couple. It is heated to a high temperature and then dipped into the liquid. How the test-piece temperature varies depending on time is recorded, thus providing a quenching curve which is specific for the liquid and the test-piece material. Generally such a curve looks like the one in Fig. 7.1. There are three areas: film boiling, nucleate boiling and natural convection.

Two temperatures are meaningful on this curve: the transition temperature between film-boiling and nucleate boiling θ_1 and temperature at the beginning of natural convection θ_2. By taking the difference between those two temperatures

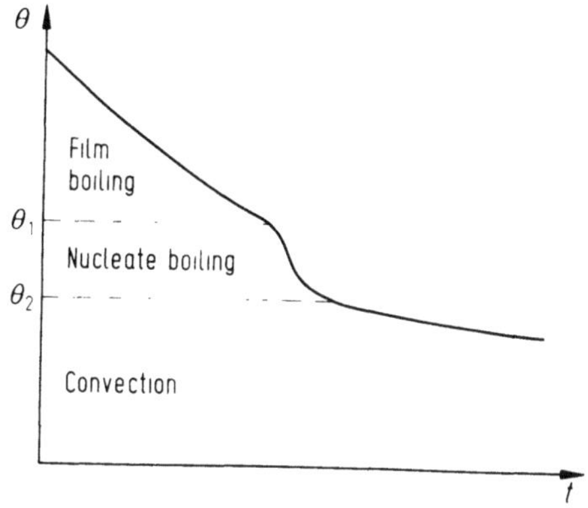

Fig. 7.1. Cooling curve. Definition of θ_1 and θ_2

into account the metallurgist can define the most appropriate cooling scheme to process a specific metallic alloy part.

7.1.1 Transition Between Film-Boiling and Nucleate-Boiling

Whether quenching in a vaporizable liquid is reproducible or not depends on the stability of the film-boiling and hence on the temperature at which the transition between film-boiling and nucleate boiling occurs.

Liquids Used at Saturation-Point Temperature

For boiling liquids, for example, water at 100 °C, liquid nitrogen, the transition between film boiling and nucleate boiling always occurs at the same temperature for a specific test-piece. This temperature can be modified only by the superficial properties of the test-piece. The liquid agitation has no effect on the transition: the film boiling is perfectly stable [2]. The vapour blanket perfectly isolates the solid from the liquid; no contact what so ever may occur before transition.

Subcooled Liquids

Such a liquid is taken at a temperature which is beneath its saturation-point temperature. Under such circumstances, film boiling does not achieve stability and the surface is sometimes wetted transiently by the liquid before nucleate boiling [3] or there are explosive transitions at various wall temperature [4–6].

Under-Vacuum Liquids.

The pressure above the liquid alters the transition temperature between film and nucleate boiling. Thanks to a specific device [7] it is possible to assess the cooling power of a quenching oil under a primary vacuum around 0.13 Pa. The Fig. 7.2 enables to compare the cooling power of a paraffinic oil (viscosity = 31 mm^2/s at 50 °C) in atmospheric pressure and then under a 0.13 Pa vacuum, the oil being unagitated. Vacuum quenching dramatically lengthens film boiling.

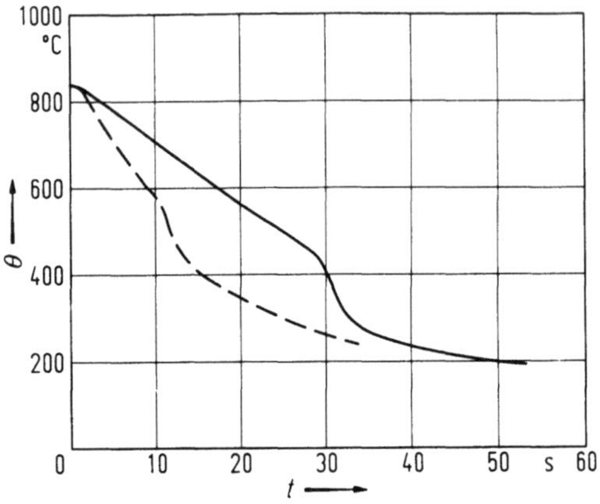

Fig. 7.2. Cooling of a silver sample ($d = 16$, $h = 48$ mm) quenched from 850 °C in a paraffinic oil at 25 °C. ——— under primary vacuum; - - - - - atmospheric pressure

θ_1 temperature drops from 555 °C to 460 °C, θ_2 temperature drops from 420 °C to 280 °C. This shows how difficult it is to perform efficient vacuum quenching. Yet it is possible to reduce the time of film boiling by agitating the liquid or by introducing specific additives [7].

7.1.2 Instability of Film-Boiling in Sub-Cooled Water

In Fig. 7.3 one can notice the various cooling laws that may be found when quenching a silver test piece 16 mm in diameter, 48 mm high with a hemispheric bottom from 850 °C into pure 40 °C water. The conditions for the experiment are seemingly similar: unagitated liquid, regular input speed into water. This set of curves shows that film boiling is not reproducible: it can be destabilized at a high temperature, then start again, it can also be stable and be destabilized only at a low temperature, when nucleate boiling starts.

What is the cause of this haphazard instability of film-boiling? In the course of the cooling process, when the film boiling is on, even a slight agitation of the water in the quenching tank induces the process. Thus it seems that in the case of sub-cooled water, the agitation of the liquid may influence the stability of film-boiling.

Experimental Study

It is possible thanks to a device to destabilize film-boiling: it consists of a 0.5 mm in diameter hypodermic needle along with a pressure system for the water; the whole device enables water at the same temperature as that of the tank to be introduced 2 mm away from the test-piece surface with a variable speed during the cooling process [8]. The location with a maximum destabilization effect is the bottom of the test-piece. The perturbation which may be controlled both for power and direction, has an effect near the liquid–vapor interface and it will destabilize the film-boiling.

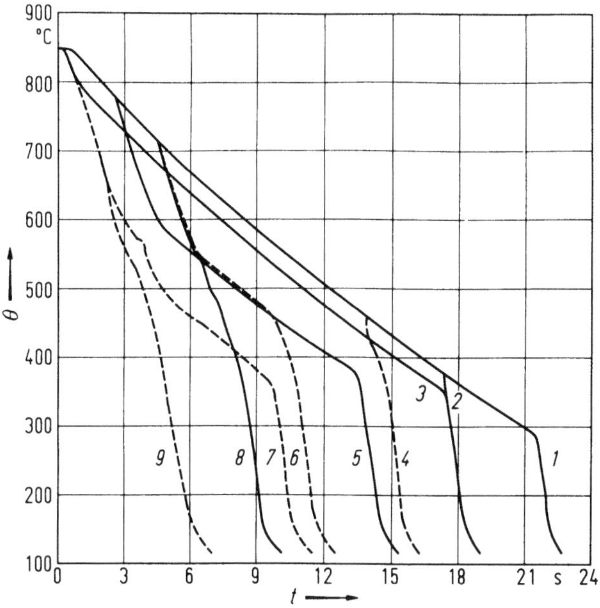

Fig. 7.3. Possible cooling curves in still water at 40 °C. Silver sample quenched from 850 °C

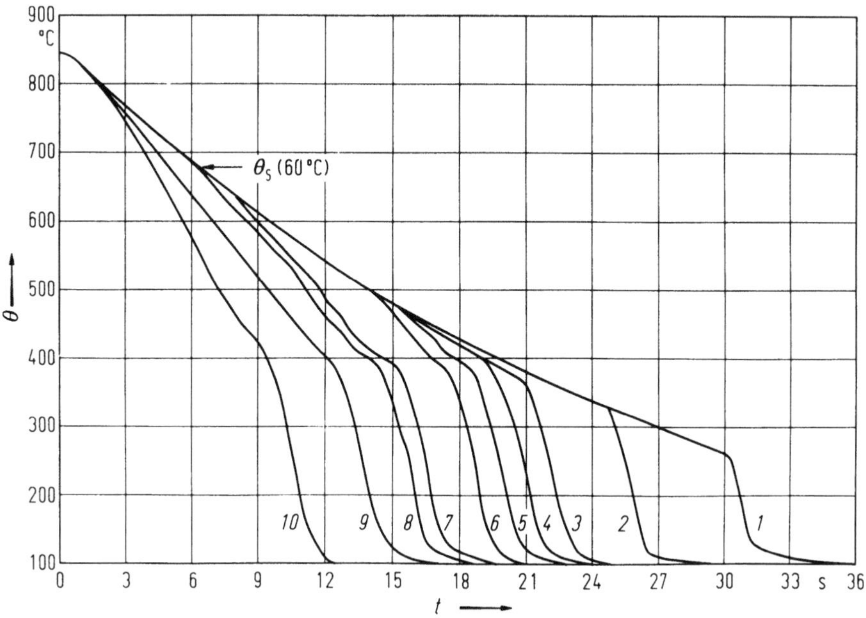

Fig. 7.4. Quenching of a silver sample in water at 60 °C with injection: (*1*) Ui = 0; (*2*) Ui = 0.13; (*3*) Ui = 0.26; (*4*) Ui = 0.34; (*5*) Ui = 0.40; (*6*) Ui = 0.46; (*7*) Ui = 0.63; (*8*) Ui = 1.05; (*9*) Ui = 1.40; (*10*) Ui = 2 m/S

Figure 7.4 shows the results with water at a 60 °C temperature. A higher imput speed, Ui, destabilizes film boiling up to a maximum above which there is no further propagation of the destabilization. In that case it is $Ui = 1.05$ m/s and for $\theta_s = 680$ °C. Above this value of $\theta_s(\theta_L) = \theta_s(60\ °C) = 680$ °C film boiling is stable [9].

For other values of water temperature, the higher stability temperature for film boiling may be assessed in the same way. When $\theta_L > 60$ °C the maximum temperature above which film boiling is stable, decreases. When $\theta_L = 85$ °C this temperature $\theta_s(\theta_L) = \theta_s(85\ °C)$ is equal to the conventional transition temperature $\theta_1(\theta_L) = \theta_1(85\ °C)$: film boiling is stable.

When θ_L decreases, destabilization temperature $\theta_s(\theta_L)$ increases sometimes up to the initial quenching temperature i.e. 850 °C. The speed Ui necessary to destabilized film boiling for a specific surface temperature decreases with θ_L. It should particularly be noticed that when $\theta_L = 30$ °C a speed $Ui = 0.10$ m/s is enough to destabilize the film boiling as soon as the test piece is introduced. This may account for spontaneous destabilization occurrences when water temperature is 40 °C. Natural disorders that may happen in the tank are enough to spontaneously destabilize the film boiling process: liquid motion due to the introduction of the test-piece, for instance.

Electric contact measurements [9, 10] show that film boiling destabilization entails a real wetting of the warm surface even with an 850 °C temperature. This means that water-surface contacts may occur at temperatures that are far higher than that of homogeneous water nucleate boiling [11].

The destabilization is often explosive, the wetting surface moving at around 2 m/s [10].

Film Boiling Stability Diagram

The complete set of results may be summarized in a diagram $\theta_s = f(\theta_L)$ Fig. 7.5.

The curve picturing $\theta_s(\theta_L)$ variations versus θ_L is nearly a straight line. On one side there is the stable film boiling area while on the other side there is unstable film-boiling. The lowest temperature for the film-boiling - nucleate boiling transition $\theta_1(\theta_L)$ is also mentioned on the diagram; that temperature is not easy to determine particularly when θ_L is low and these are average measures.

Film-boiling stability increases along with θ_s and θ_L but however, farther with θ_L than with θ_s. The liquid temperature gradient due to the sub-cooling $(100 - \theta_L)$ is parameter than matters most [8, 9].

That film boiling stability diagram is drawn, in the case of water, for a polished silver testpiece. Film boiling instability occurs whenever a metallic test-piece is quenched into a sub-cooled liquid. The stability and unstability areas obviously may vary.

Some parameters may influence film-boiling stability:

Influence of Solid Surface Roughness

On Figs. 7.3 and 7.6 the regular line curves are achieved with a very well polished silver test-piece, the dotted line curves are achieved with the same test-piece

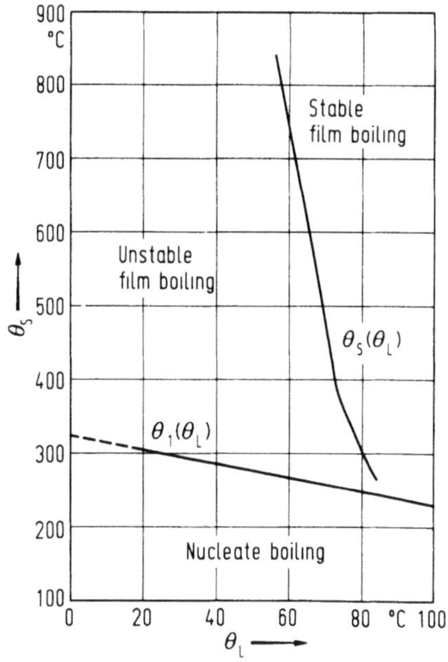

Fig. 7.5. Stability diagram of film boiling. Case of water (silver sample). θ_s = surface temperature of the sample, θ_L = water temperature

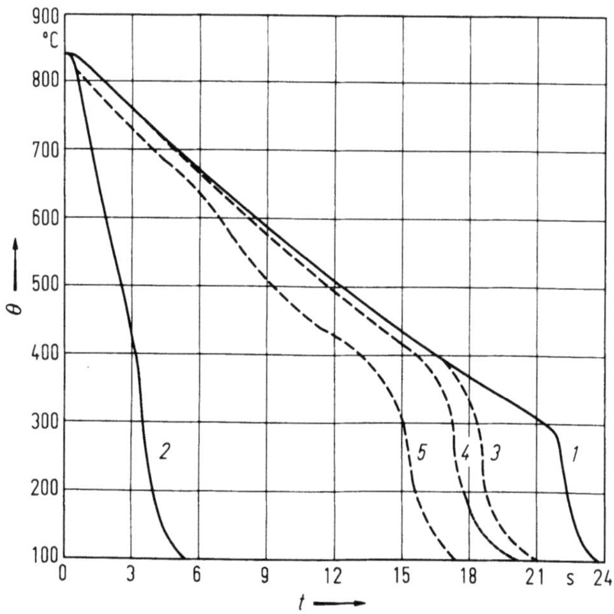

Fig. 7.6. Quenching of a silver sample in water at 30 °C with injection: (*1*) Ui = 0; (*2*) Ui = 0.10; (*3*) Ui = 0; (*4*) Ui = 0.80; (*5*) Ui = 0.92 m/s. ——— polished sample; ----- rough sample

but its surface has been abraded with abrasive paper (n°80). In the latter case film-boiling stability is dramatically boosted. For $Ui = 0.8$ m/s (curve 4) there is no destabilization. A speed $Ui = 0.92$ m/s (curve 5) is necessary for a beginning of instability to occur at $\theta_s = 625$ °C but this happens but once and the film boiling starts again and stays until "normal" transition, which shows its great stability.

In order to destabilize film boiling, the speed Ui, necessary in the case of curve 5 Fig. 7.6 is nine times higher than in the case of curve 2, corresponding to the same but well-polished test-piece. The great stability of film-boiling on a rough test-piece entails an excellent reproducibility of the curves found even for $\theta_L < 30$ °C.

In addition to this great stability, a typical film-boiling-nucleate boiling transition appears earlier with a rough test-piece. The surface roughness due to abrasive paper creates as many sites that promote micro-contacts between the liquid and the solid at the end of the film-boiling process [3].

Physico-Chemical Surface Properties Influence

Sub-cooled water film-boiling instability occurs as well, when quenching an aluminium-alloy, steel or nickel test-piece since the liquid agitation (global or injection) entails the destabilization of a film-boiling that has started.

However, if there is no liquid agitation whatsoever, a silver surface is particularly unstable. To separate the surface from bulk material properties

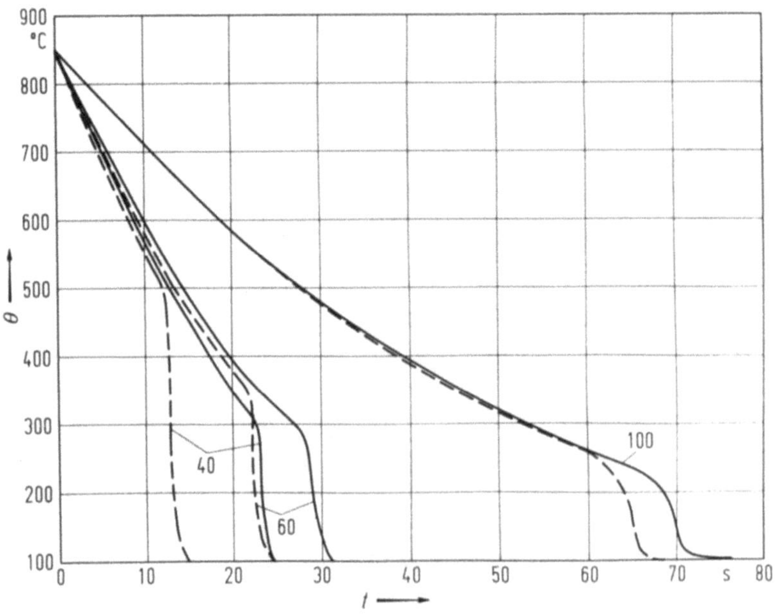

Fig. 7.7. Cooling curves for quenching in water at: 40 °C, 60 °C, 100 °C. ——— silver sample; ----- silver sample coated with Nickel (thickness 10 μm)

electrolytic deposits are made on similar test pieces (cylinder of 16 mm diameter and 48 mm high) a nickel test piece and a copper one with a 10 μm silver coating are quenched without injection into sub-cooled water. Random film-boiling destabilization occurs although it pratically never happens with nickel or copper test-pieces. Besides the film boiling is longer with a silver coating than for non silver-plated nickel or copper.

Inversely if a 10 μm nickel coating is made on a silver test piece the film-boiling process is more stable but transition appears sooner if the liquid is not agitated [9]. Thus, film-boiling stability may depend on the surface properties of the test-piece since a metal coating A on the metal test piece B gives the whole thing the properties of A and conversely (Fig. 7.7).

7.1.3 Cooling Law Calculation

During the quenching for a part with specific features, the heat transfer process at the liquid–solid contact surface is conditioned by the properties of the liquid used.

The data about that surface heat transfer, along with conduction within the solid and initial quenching conditions, fully determine a unique cooling law at every point in the solid. This law must be known by the metallurgist [12].

Liquid–solid contact surface temperature measurements are difficult during the quenching process. The only accurate data available are the cooling laws recorded by thermocouples set in the solid, particularly in its geometrical center. The heat transfer to the surface of the solid can only be studied through heat conduction within this solid, which is analysed numerically.

The boundary conditions necessary to solve the heat equation numerically are chosen a priori and they are included in a curve $q = f(\theta_s)$ which shows the heat flow density variations in the heat carried off the surface by the liquid, depending on the solid temperature, θ_s, at that place.

How temperature profiles evolve with time in the piece is determined particularly thanks to a two-dimensional calculation method with finite differences [13].

The cooling law $\theta = f(t)$ thus calculated for the test-piece center is compared to the experimental curve, which makes it possible to check a posteriori the validity of the boundary condition $q = f(\theta_s)$ supposed for the calculation.

In the case of quenching into 100 °C water, where coolings are quite reproducible, the boundary condition $q = f(\theta_s)$ is determined for three metals: silver, nickel, aluminium used for 16 mm diameter, 48 mm high cylindrical test-pieces.

Figure 7.8 shows the three curves $q = f(\theta_s)$ along with a curve for the steady state heat transfer (Nukiyama's curve). Figures 7.9 and 7.10 allow to compare experimental curves and the calculated ones.

The numerical method allows to simulate very well the cooling curves, but the curves $q = f(\theta_s)$ are quite different for the three metals even if the experimental

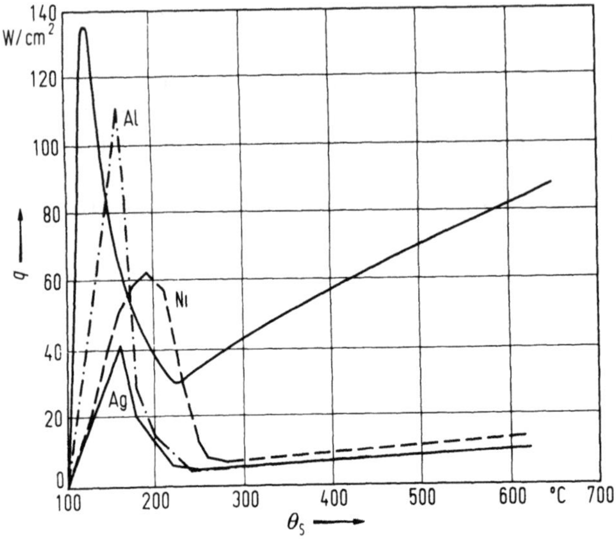

Fig. 7.8. Curves $q = f(\theta_s)$ for aluminum, silver, nickel samples quenched in water at 100 °C. Comparison with steady state curve

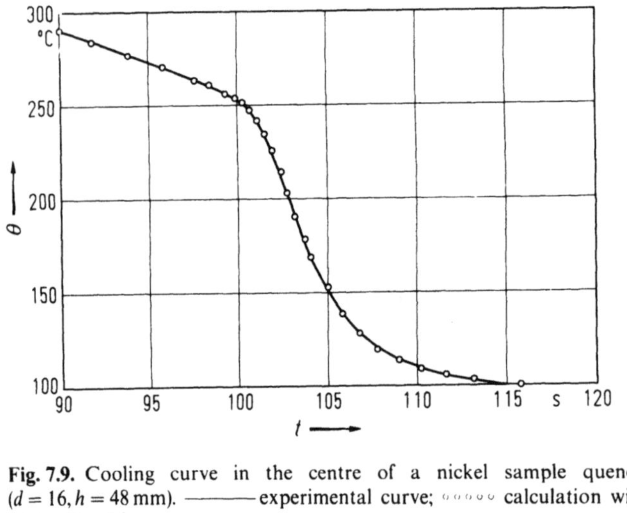

Fig. 7.9. Cooling curve in the centre of a nickel sample quenched in water at 100 °C ($d = 16, h = 48$ mm). ——— experimental curve; ○○○○○ calculation with the boundary condition of the Fig. 7.8

conditions are the same. The values of heat flow maximum vary between one and three for silver and aluminum, the temperature of the end of film boiling are not the same for the three test-pieces so the heat transfer coefficient [14].

Hence it appears clearly that it is impossible to define a universal curve $q = f(\theta_s)$ that would apply only to the liquid. There is a strong relationship

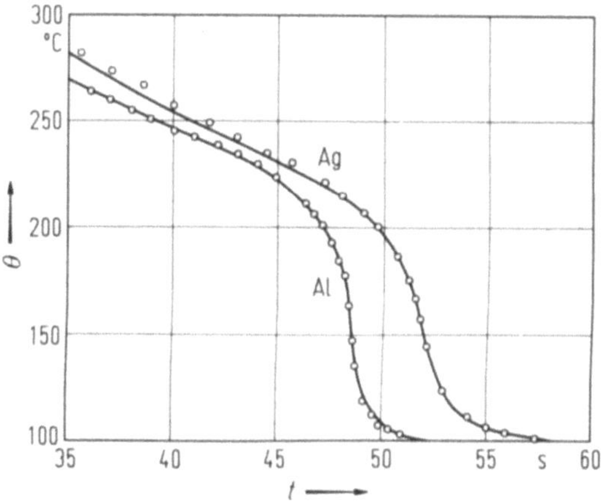

Fig. 7.10. Cooling curves in the centre of aluminum and silver samples quenched in water at 100 °C ($d = 16$, $h = 48$ mm). ——— Experimental curves; ooooo calculation with boundary condition of the Fig. 7.8

between the reaction of the solid and that of the liquid. As a result there is no studying two phase heat transfer phenomena without taking the heat providing solid properties into account.

So, the cooling of a metallic part by quenching in a vaporizable liquid cannot be fully foreseen by calculation, it will always be necessary to experiment to know the temperature at a given point of the part at least in order to define the boundary condition $q = f(\theta_s)$.

7.2 Influence of the Workpiece Surface's Thermophysical Properties

The surface of quenched parts rarely shows thermophysical features similar to those of the inside metal either on account of oxidization or on account of various intentional or unintentional deposits that may coat them. The mechanical state of the surface may also vary depending on whether it is polished or uneven for various reasons.

The addition of all those superficial factors may dramatically alter the cooling process. Thus, it is possible to make use of these factors to further improve quenching control.

7.2.1 Influence of the Initial Workpiece-Liquid Contact on the Cooling Process

Just as the heated (θ_s) solid meets the quenching liquid (θ_L), a temperature (θ_c) appears on the surface contact. This temperature is common for the two matters. Its value determines the whole cooling process. This temperature may either cause or not the liquid vaporization. This temperature θ_c cannot be measured but thanks to simplifying hypotheses it may be calculated.

Case When θ_c Does not Cause Vaporization

Supposing the contact between the test-piece and the liquid is instantaneous, the approximation to consider both matters as infinite is physically speaking justifiable as the thermal signal that they have met has not reached the limits of both matters in any measurable way. In that case θ_c is calculated from the heat equation.

$$\theta_c = \frac{\theta_S E_S + \theta_L E_L}{E_S + E_L}$$

$E = \sqrt{\lambda c \rho}$: Effusivity ($\lambda$ = thermal conductivity, c = specific heat capacity, ρ = mass density)

This temperature remains steady as long as the thermal signal cannot be noticed, for example at the center of the test-piece for which the thermal way to the surface is the shortest.

Such a relation only requires the initial temperatures of the matters and their effusivity. The initial quenching regime in a non vaporizable liquid is when temperature θ_c does not noticeably vary [15].

If θ_c does not depend on the geometry of the piece, the duration of the initial regime is dependent on how long the piece is supposed to be infinite, thus on its dimensions, as is shown in Fig. 7.11. The measure of the duration of the initial regime first of all enables to check that θ_c really is the temperature forced on to the test-piece surface by the quenching process. This is the case of quenching made in salts or molten alloys.

Case When θ_c Causes Liquid Vaporization

If at that temperature θ_c the quenching liquid may be vaporized, the initial contact may be interrupted before the contact surface between the piece and

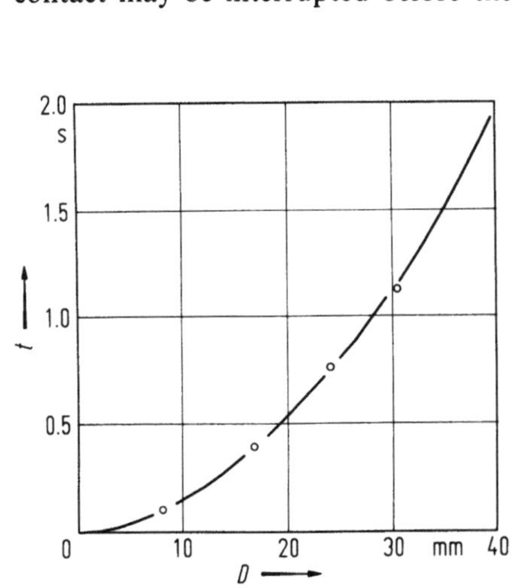

Fig. 7.11. Variations of the time of the initial quenching regime in a non vaporizable liquid versus nickel samples diameter (quenching from 900 °C in melted tin at 300 °C)

the liquid has reached θ_c, on account of the vapor phase. Thus it is the value of the superficial effusivity of the piece which determine the vaporization process (film or nucleate boiling) and then the whole cooling process [15].

Critical Vaporization Temperatures

Let us take the case of boiling water. A nickel cylinder oxidized in the air, 24 mm in diameter and 72 mm high is quenched from 200 °C and 220 °C respectively. Then the difference of temperature $\Delta\theta$ between the surface and the centre is measured, which enables to detect the appearance of film boiling in a sensitive way. Figure 7.12 shows that above $\theta = 220$ °C the contact temperature triggers film boiling.

This result may be generalized, first in function of the surface effusivity and secondly for various water temperatures (Fig. 7.13). To perform quick quenching processes, the conditions must avoid film boiling: the water must be as-cold as possible and the superficial effusivity of the piece must be low. The figure shows it is possible from 1000 °C and without film-boiling to quench a test-piece, with a superficial effusivity around 2 (SI) in iced water. Such conditions are rather restrictive and hardly possible in the case of industrial facilities.

So, the cooling process may be noticeably accelerated by coating the test-piece surface with a deposit meeting the E requirements, which avoids film-boiling. To quench into 20 °C water from 800 °C, for instance, a deposit of Na_2SiO_3 is satisfying, whatever the properties of the piece. This can be checked on the Fig. 7.14 where an insulating, with weak E, but very thin film dramatically boosts the cooling process by preventing the occurrence of film-boiling [15, 16].

Figure 7.15 shows that the quenching of a test-piece coated with a weak E deposit is faster and steadier, but that the maximum of the temperature difference between centre and surface is the same that for the bare test-piece.

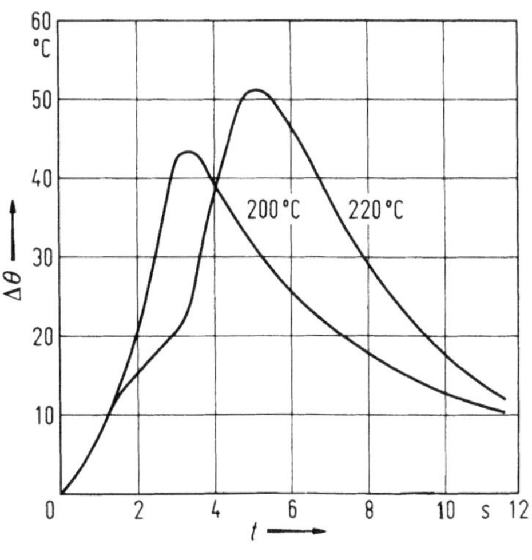

Fig. 7.12. Quenching of a nickel sample ($d = 24$, $h = 72$ mm) in water at 100 °C from 200 °C and 220 °C. Variations of the temperature difference between the centre and the surface versus time

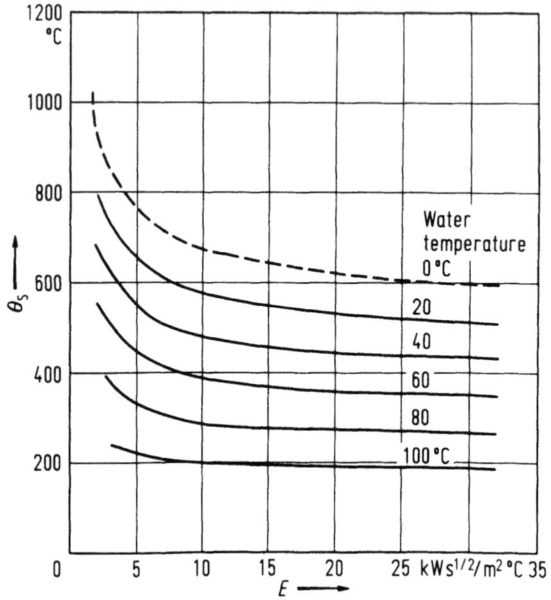

Fig. 7.13. Variations of the initial critical temperature of quenching versus the superficial effusivity of the solid (water)

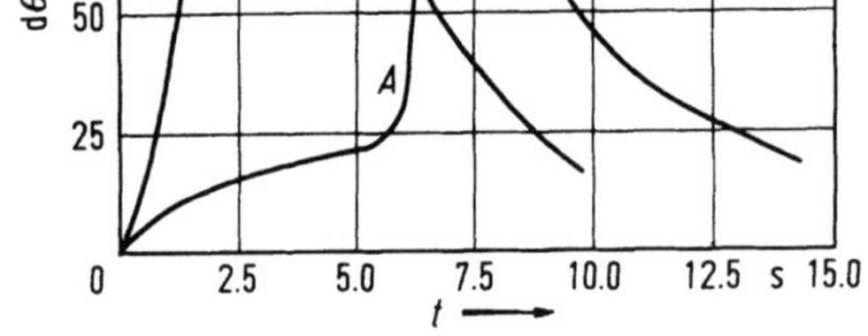

Fig. 7.14. Cooling rate (at 1.5 mm of the surface) versus time of a nickel sample ($d = 24, h = 72$ mm) quenched from 800 °C in water at 20 °C. *A* NiO; *B* Na_2SiO_3 (no film boiling)

Preventing any film-boiling is useful from two practical points of view:

—high temperature cooling is both faster and steadier
—there is no distortion due to the progressive elimination of the film boiling, which triggers serious temperature gradients between film-boiling and nucleate boiling areas.

For instance water quenched carbon steel files are, on account of their dissymetry, test-pieces which are very sensitive to distortion. Figure 7.16 shows that the file with a low effusivity surface (thanks to a very thin sodium silicate

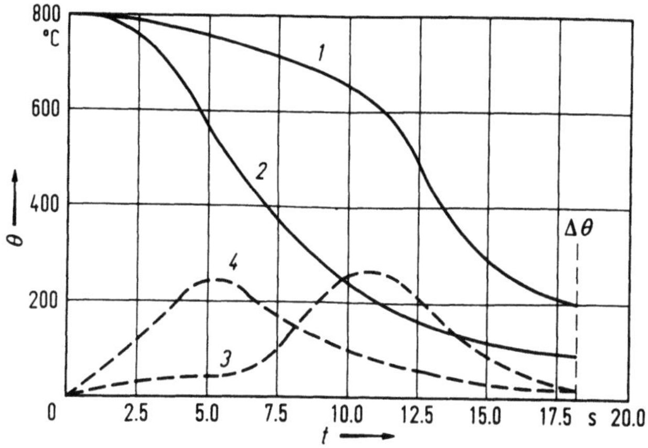

Fig. 7.15. Influence of the superficial effusivity on the quench cooling in water at 20 °C (nickel sample $d = 24$, $h = 72$ mm). *1* bare; *2* coated with 10 μm Na_2SiO_3 (centre temperature); *3* bare; *4* coated with 10 μM Na_2SiO_3 (temperature difference between centre and surface)

Fig. 7.16. Comparison between two files (0.45% C steel) quenched in water with and without Na_2SiO_3 coating

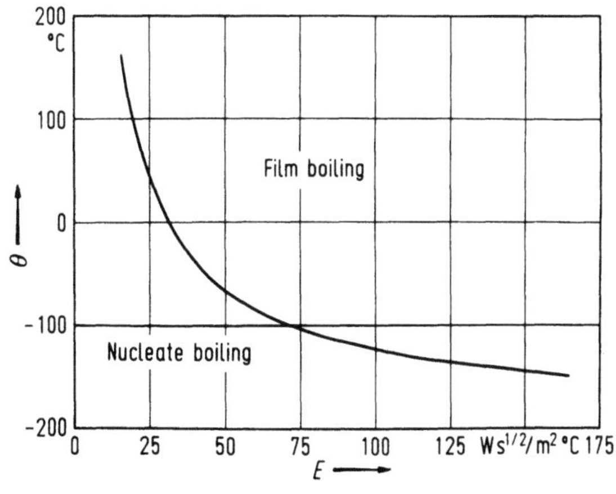

Fig. 7.17. Variations of the initial critical temperature of quenching versus superficial effusivity in the case of liquid nitrogen

coating) is cooled quickly and evenly. Its post-quenching structure is really martensitic but, above all, it has not been distorted like the other file quenched in similar conditions but without the coating.

These results are valid for other vaporizable liquids such as, for instance, liquid nitrogen [17] with which it is very difficult to avoid film-boiling (Fig. 7.17).

In what was previously said the piece surface temperature was supposed to be uniform.

7.2.2 Surface Thermal Resistance Effect on the Cooling Process

Up to now only the surface contact properties have been taken into account through effusivity, by considering the piece superficial thermal resistance as negligible. It appeared that the quenching conditions to avoid film boiling are very restrictive.

Now the effect of fins on the surface to eliminate film boiling thanks to the lower temperature at their ends will be analysed.

The thermal resistance R between the bottom and the top of the fin is written:

$$R = L/\lambda S$$

(L = length of the fin, S = section of the fin).

If the fin conductivity is lower, the same effect may be achieved with a shorter fin, but it may be wider. If the fins are made of an thermal insulator, the fins may even be replaced by a continuous coating which lowers the apparent superficial temperature of the piece. Figure 7.18 shows this reasoning. The advantage of an insulating coating is that it avoids film-boiling on the whole surface whereas the fins are only wetted at their ends [17].

In the case of liquid nitrogen, although it is not easy to avoid film-boiling, it is possible to do so by coating a test-piece quenched from normal atmosphere temperature with synthetic rubber. The result is a noticeable increase of the heat flux out of the test-piece, as its is shown in Fig. 7.19. The cooling time seems to be clearly influenced by the thickness of the coating, hence its thermal resistance (Fig. 7.20) [17].

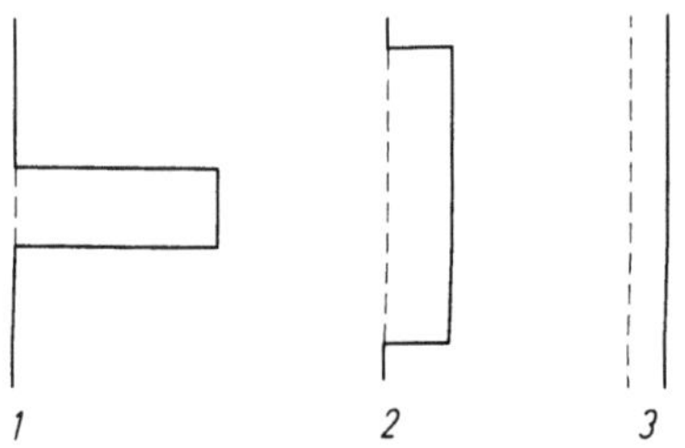

Fig. 7.18. Scheme of fins with the same thermal resistance. *1* high λ (metal); *2* average λ (oxide); *3* small λ (insulation)

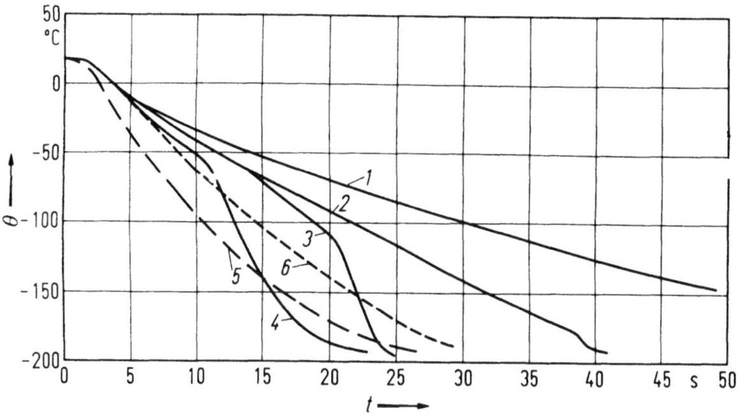

Fig. 7.19. Nickel samples ($d = 10, h = 10$ mm) quenched from 20 °C in liquid nitrogen with coatings of different thickness. *1* bare; *2* e = 2 μm; *3* e = 60 μm; *4* e = 90 μm; *5* e = 200 μm; *6* e = 500 μm

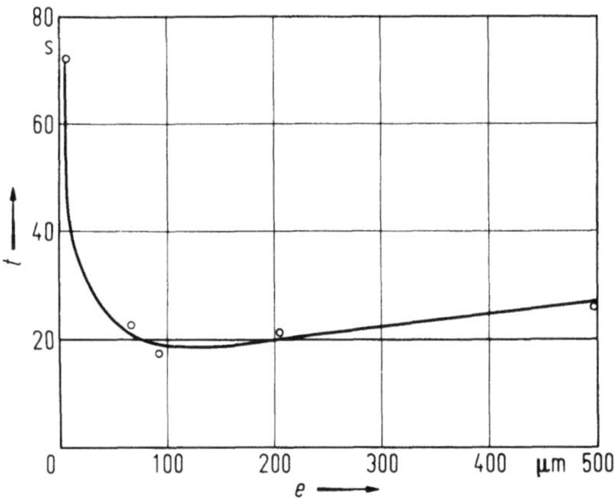

Fig. 7.20. Influence of the coating thickness on the cooling time of the sample between 20 °C and −180 °C. (nickel sample $d = 10$, $h = 10$ mm)

The analysis of the phenomena, which induces to replace film-boiling by a special vaporization process, larvate boiling, was proposed [16, 18]. It implies very fast surface temperature changes in connection with direct periodic liquid-surface contacts [19].

It should be noticed that if larvate boiling starts thanks to a thermal resistance on the test-piece surface, there is still a condition of low superficial effusivity which is necessary to ensure the process [18]. This can be checked on Fig. 7.21. The thermal resistance effect is eliminated thanks to a high-effusivity silver coating but without any influence on thermal resistance. It is allowed again by another low effusivity Na_2SiO_3 thin coating.

This experiment shows the critical effect of the superficial thermal properties of the workpieces on their cooling by quenching in a vaporizable liquid.

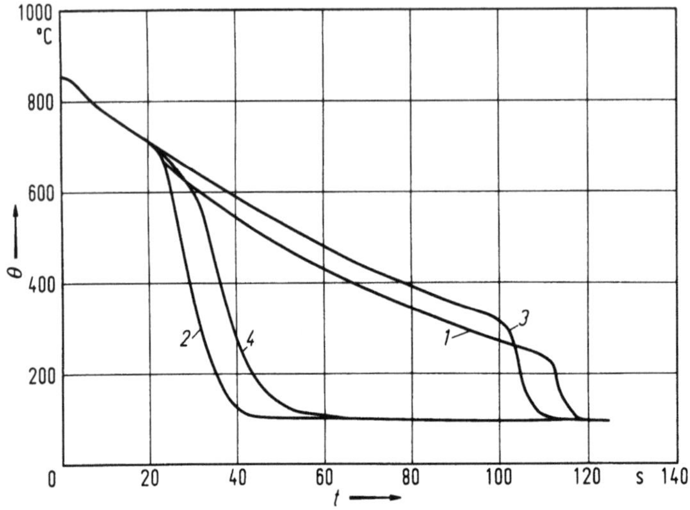

Fig. 7.21. Cooling curves in the centre of a cylindrical nickel sample ($d = 16, h = 48$ mm) quenched from 850 °C in water at 100 °C. *1* bare; *2* coated with 200 μm ZrO_2; *3* coated with 200 μm ZrO_2 + 50 μm Ag; *4* coated with 200 μm ZrO_2 + 50 μm Ag + 10 μm Na_2SiO_3

How this Can Really Be Used

Interesting results were found when using coatings which provide quenched pieces with larvate boiling cooling. Both of these examples are very clear:

Water Quenching of an Aluminum AZ5GU (7075) *Type Alloy*

A cylindric 50 mm diameter 7075 piece was coated so as to create larvate boiling by quenching from 465 °C in 100 °C water.

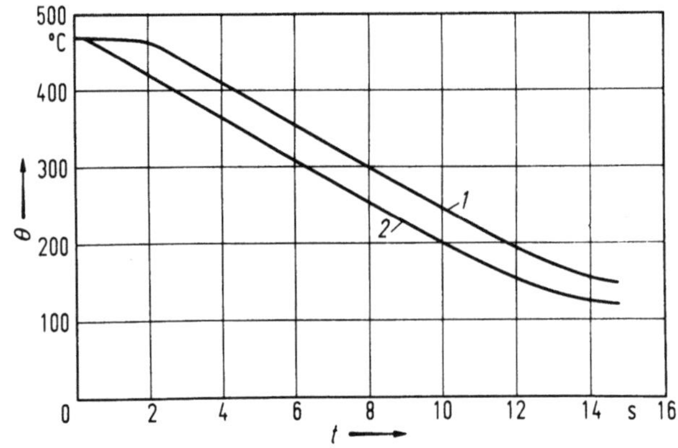

Fig. 7.22. AZ5GU (7075) sample ($d = 50, h = 150$ mm) coated with insulating material and quenched from 465 °C in water at 100 °C. *1* centre; *2* surface

Figure 7.22 shows the cooling process which is both fast and steady for the test piece surface as well as for its centre. The resulting metallurgic characteristics are similar to those of conventional water quenching but the fact there is no nucleate-film boiling transition noticeably reduces the amount of residual stresses; this is shown by the table below where the surface residual stresses are compared. It should be noted that the quenching with a coating in 100 °C water provides faster cooling than the usual quenching in 70 °C water [20].

Cylinder 7075	$d = 20$–$h = 60$ mm			$d = 50$–$h = 50$ mm			$d = 80$–$h = 240$ mm			$d = 120$–$h = 360$ mm		
Water temperature °C	20	70	100 with coating	20	70	100 with coating	20	70	100 with coating	20	70	100 with coating
Superficial stresses MPa	184	143	79	179	152	112	182	136	118	221	168	156

Spring Steel Cylinder Quenching in Mineral Oil

This is the case as it happens in industrial facilities of whether quenching is possible or not. We put the stress on the utmost importance of the influence of the type of surface for the workpieces to be quenched [21].

The oil used for the quenching of cylindric mechanical parts made of silicon steel 55S7 (9260) type, previously heated in a furnace with a neutral atmosphere

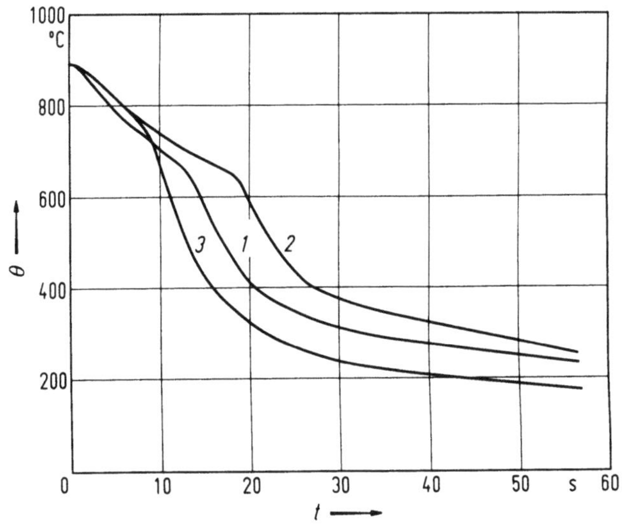

Fig. 7.23. Cooling curves in the centre of a spring steel sample 55S7 (9260) ($d = 20$, $h = 60$ mm) quenched from 900 °C in mineral oils at 50 °C. *1* oxidized sample—mineral oil A; *2* non oxidized sample—mineral oil A; *3* non oxidized sample—mineral oil B

may cause cracks to appear on the quenched pieces. This means a lot of pieces have to be rejected. However, as soon as the austenitizing is performed without a protecting atmosphere, the cracks disappear.

The cooling power of the quenching bath has been determined. The test-piece has been quenched in 50 °C oil *A*, with two different surface aspects (Fig. 7.23).

1. Oxidized test piece after air-heating (curve 1).
2. Non oxidized test piece after vacuum heating (curve 2).

In the latter case, cracks can often be noticed, which implies a systematic control of each piece. If the austenitizing is performed in an oxidizing atmosphere, the oxide layer alters the cooling process and this is interesting from two points of view:

— better quenching penetration due to faster cooling
— softer film-boiling-nucleate boiling transition thus eliminating all cracks. A new quenching oil (oil *B*) was designed to provide non-oxidized pieces with a cooling process close to cooling type 1 (curve 3).

7.2.3 Influence of the Surface Condition on the Cooling Process

In the case of vaporizable liquid quenching, however fine the test-piece surface polishing may be, it has no effect on the film-boiling process. Yet deeper scratches have a more noticeable effect as soon as they are 0.6 mm deep. For instance 1 mm deep scratches entail a twice faster film-boiling cooling speed along with a noticeably higher nucleate boiling-film boiling transition temperatures. In the case of boiling water quenching, this transition temperature increases for instance from 220 °C for a polished surface to 400 °C for a surface with 1 mm deep scratches.

7.3 Quenching Control by Adding a Solute to the Water

The alteration of the workpiece surface properties entails a cooling rate change. In what was mentioned above the coatings were made prior to quenching. By adding solutes into the quenching water, it is possible to alter either the surface of the cooling piece in order to disturb the type of liquid vaporization or directly the liquid vaporization processes.

7.3.1 Aqueous Solutions of Inorganic Solutes

The particular properties of saline solutions have been known for a long time by metallurgists under the name of "salt effect" [22, 23].

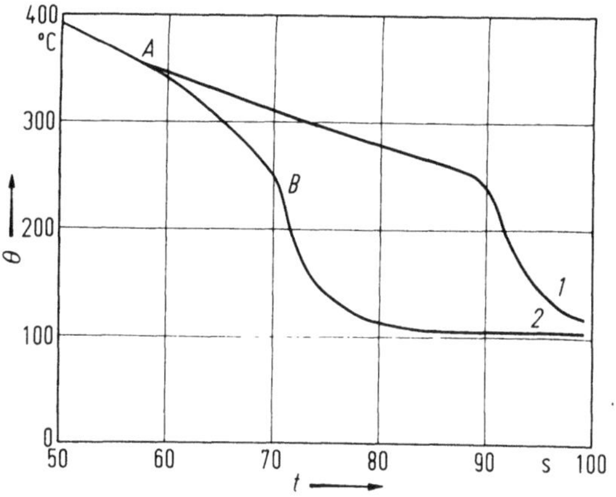

Fig. 7.24. Quenching curves of a nickel sample ($d = 16$, $h = 48$ mm). *1* water at 100 °C; *2* aqueous solution of 10% NaCl at 101 °C

The addition of sodium chloride for instance, provides for an earlier start of the water film-nucleate boiling transition temperature, and thus reduces film-boiling (Fig. 7.24). Besides, it increases the cooling rate whatever the bath temperature.

When a metallic test-piece is quenched into a 10% NaCl aqueous solution at 100 °C, at the end of the film-boiling a new type of vaporization occurs before nucleate boiling. This is shown by the AB location on the cooling curve (Fig. 7.24).

After film-boiling a continuous salt film covers the test-piece surface. This coating creates a thermal resistance on the test-piece surface along with larvate boiling, which was already met with the previous paragraph. During the film-boiling process there is water vaporization and the liquid-vapor interface is enriched with salt; when it is saturated, the salt precipitates in the test-piece surface [23].

The heat transfer coefficient does not vary during the film-boiling process when thermally stable salt is added into the water. Otherwise it may be altered by the salt decomposition (the vapor composition differs from that of the liquid).

The maximum cooling rate during the nucleate boiling process, which was noticed during the quenching in stable saline solutions may be accounted for by the Van Stralen [24] model. Bubbles, which are more numerous but smaller than in the case of pure water, appear on the hot surface, which avoids their coalescence. Heat flux evacuation is then made easier thanks to more numerous surface wettings and also thanks to the microconvection created by the agitation of those tiny bubbles near the surface. During that process, the salt coating dissolves again within the solution.

The cooling laws of a nickel test-piece quenched from 700 °C into NaCl, KCl, LiCl and NaOH aqueous solutions brought to boiling point show that for a given concentration the more soluble in water the solute is, the shorter the film-boiling process is (Fig. 7.25).

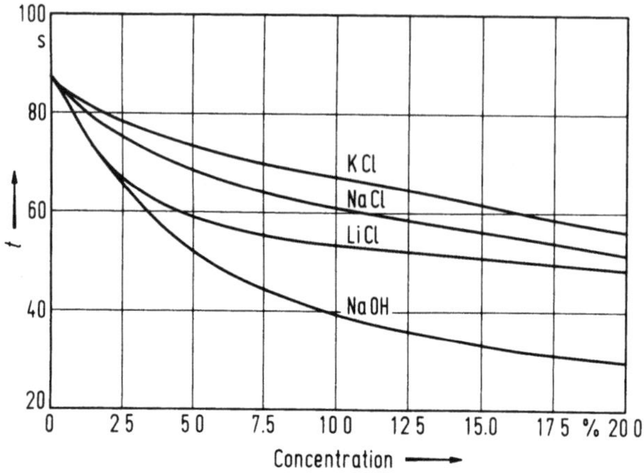

Fig. 7.25. Variations of film boiling duration versus the aqueous solution concentration. Nickel sample ($d = 16$, $h = 48$ mm) quenched from 700 °C in boiling solutions

The more soluble the solute is and the higher its concentration is, the higher the nucleate-film boiling transition temperature is.

7.3.2 Organic Polymer Aqueous Solution

There are organic polymers which are soluble in water. Their solutions may be used as quenching liquids. Among those polymers some are worth quoting: polyvinyl alcohol, polyethylene oxide and its by-products, polyacrylic or polymethacrylic acid salts or polyvinylpyrrolidone. Adding a polymer into the water will alter its vaporization properties and entail a change in its cooling power by processes which are similar to those of the salt effect but also by the viscosity effect (Chap. 8).

The most frequently used organic polymers belong to the polyalkyleneglycols (PAG). Their solubility is inverted, which means that their water solubility decreases when the temperature increases.

Organic polymer aqueous solutions are generally used at atmosphere temperature; like in the case of sub-cooled pure water, there are unstable film-boiling occurrences, which means that the results are not reproducible [25].

Figure 7.26 shows the cooling laws for polymer quenchant *A* (temperature 20 °C) when motionless (curves 1, 2, 3). In that case the film-boiling duration varies from one quenching to another. However, thanks to $Ui = 5$ m/s injection (curve 4) it is possible to suppress film-boiling and thus to make the cooling reproducible. Here again liquid agitation is a very important parameter [25].

Figure 7.27 shows the stability diagram of polymer quenchant *A*. This diagram shows that the film-boiling unstability area is quite limited. In the case of pure water, this area is far more important for similar destabilization condition

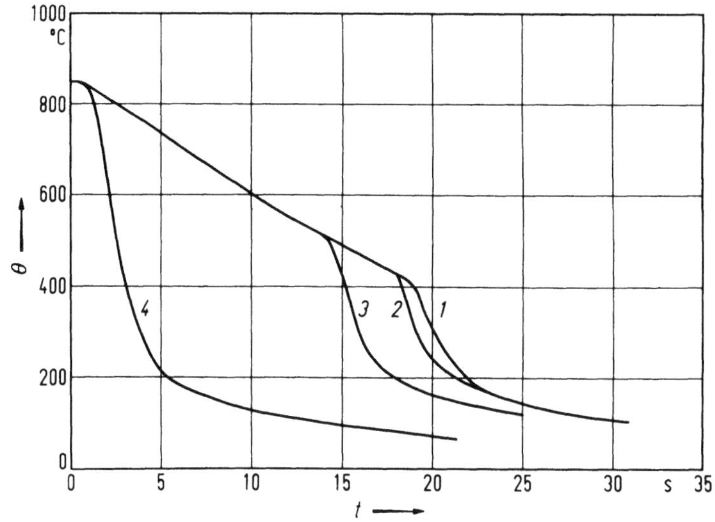

Fig. 7.26. Cooling curves of a silver sample quenched in Polymer quenchant *A* (5%) at 20 °C (*1–3*) without agitation (*4*) injection Ui = 5 m/s

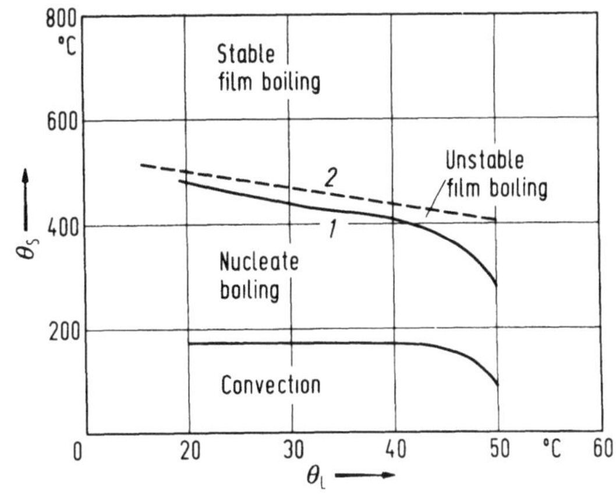

Fig. 7.27. Stability diagram of film boiling. Case of Polymer quenchant *A* (5%) *1* Ui = 0 m/s; *2* Ui = 2 m/s

($Ui = 2$ m/s). The polymer has a stabilizing effect on the film-boiling process, which makes its use possible for Aluminum alloy quenching instead of pure water [26]. The liquid speed variation allows either to suppress film-boiling or to control its duration in a reproducible way.

Steels Quenching

Mineral oils present some drawbacks among which their inflammability, their toxicity, their aging may be quoted. However, in the case of steel-quenching, they provide the best cooling power-quenching power compromise.

Thanks to its high convection temperature (θ_2) an oil enables to quench a steel piece as slowly as possible around the martensitic transformation temperature. So, the temperature gradient in the piece is limited and the transformation process is as regular as possible.

However, at high temperature, as there is still a film-boiling process, the oil quenching power is limited.

Preparing efficient aqueous solutions implies the following method:

—One: the film boiling stability diagram is determined for different polymer concentrations. This enables to determine the concentrations for which the film-boiling stability area is wide [25].
—Two: the liquid agitation influence is studied in order to suppress the unstable film-boiling area and by replacing it with nucleate boiling. The non-reproducibility problems are thus solved [27].
—Three: the specific additive influences are studied to see which ones may help raise the natural convection beginning temperature to a value close to that of a mineral oil [28].

Optimizing those three parameters helps prepare an aqueous solution made of organic polymers plus specific additives which gives a cooling curve from 850 °C and whose natural convection temperature is similar to that of a mineral oil.

Figure 7.28 shows the results found in the case of the quenching of a wedge test-piece ($d = 30$ mm, $h = 105$ mm, top-angle: 20°) in 38C2 (5038) steel [29]. On the one hand, one may notice the influence of liquid agitation and on the other hand one remarks that hardness penetration is systematically better in the case of polymer quenchant *A* than in that of mineral oil. An example of industrial achievement is shown on Fig. 7.29 [29].

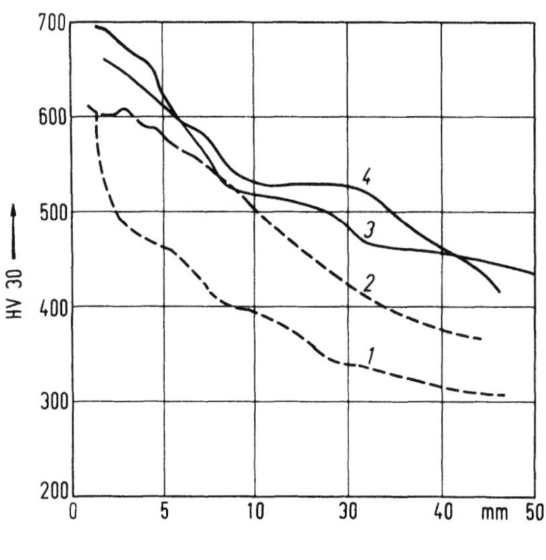

Fig. 7.28. Hardness penetration in the wedge sample 38C2 (5038) steel influence of the velocity of the liquid: (*1*, *3*) $U = 0.25$ m/s; (*2*, *4*) $U = 1.2$ m/s. ——— Polymer quenchant *A* 20 °C; - - - - - Mineral oil 150 °C

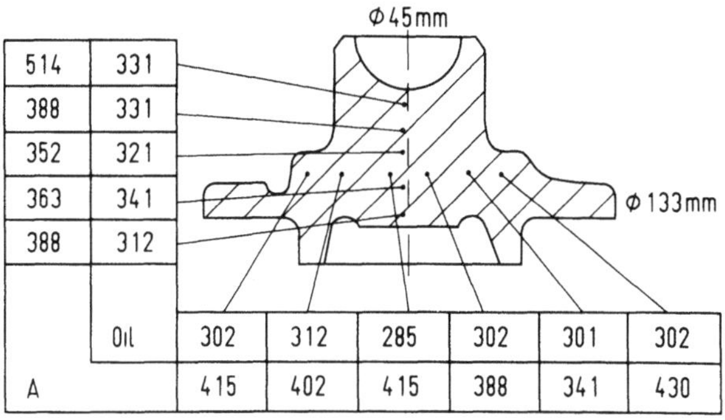

Fig. 7.29. Hardness penetration (HB 5/750) in a 46C2 (5046) steel. Polymer quenchant *A* at 20 °C. Mineral oil at 70 °C

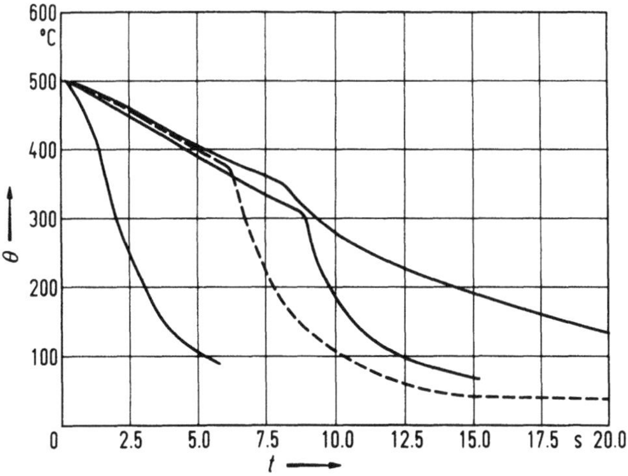

Fig. 7.30. Possible cooling curves in still water at 60 °C. Silver sample

Aluminum Alloys Quenching

As the best, cooling rate must be moderate at high temperatures, but higher at intermediate temperatures between 400 °C and 250 °C, the occurrence, in the case of water quenching, of a short film-boiling process between 500 °C and 400 °C is therefore looked forward to. The case of quenching in 60 °C water meets this requirement (Fig. 7.30, dotted line).

However quenching in 60 °C water is not reproducible enough, as is shown in Fig. 7.30. In fact it is possible to find quite different cooling laws even though the operational conditions were seemingly identical.

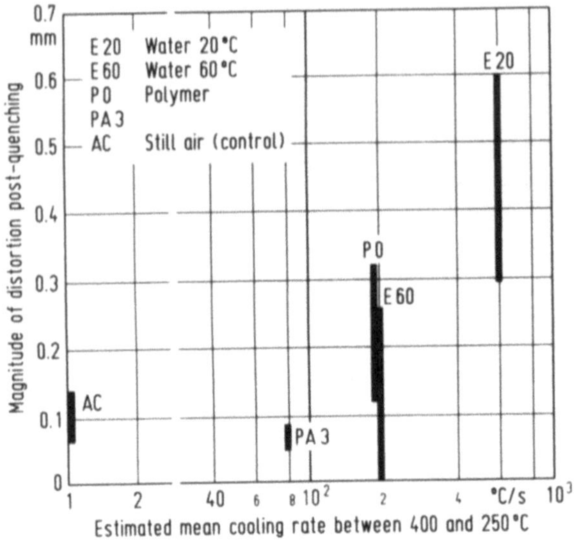

Fig. 7.31. Distortion of thin (4 mm) cylindrical aluminum alloy test-pieces quenched in various media

Thanks to additives, it is possible to make an aqueous solution of PAG (PA3) with stable film-boiling and which gives a reproductible cooling curve similar to the dotted line in Fig. 7.30.

In order to show the distortions, a tubular, split, aluminium alloy test-piece (4 mm thick) is quenched in different liquids. Four test-pieces are quenched at the same time so as to provide with statistics [26, 30].

Figure 7.31 shows the results depending on the average quenching rate in different liquids.

The distortions are important and scattered in the case of 20 °C (E20) and 60 °C (E60) water quenching, as well as in the case of quenching in a polymer solution (PO) available in shops. However, in the case of polymer quenchant PA3, the distortions are not significantly different from those due to still air quenching. Besides the mechanical properties, although they are inferior to those found with 20 °C water, are still satisfactory.

References

1. Dumont C et al. (1975) Principes de conception et de choix d'une huile de trempe. In: Traitement Thermique 94: 81–90; Traitement Thermique 95: 67–75
2. Moreaux F, Chevrier JC, Beck G (1976) Hydrodynamic and thermal study of the stability of the boundary layer in the case of film boiling. Proceedings of the I.C.H.M.T 29.8–4.9. Dubrovnik (Yougoslavie) 615–623
3. Bradfield WS (1966) Solid–liquid contact in stable film boiling. In: I/E.C Fundamentals 5: 200–204
4. Stevens JW, Witte LC (1971) Transient film and transition boiling from a sphere. In: Int. J. Heat and Mass Transfer 14: 443–450

5. Stevens JW, Witte LC (1973) Destabilization of vapor film boiling around spheres. In: Int. J. Heat and Mass Transfer 14: 669–678
6. Hoeje OC et al. (1975) An investigation of the collapse and surface rewet in film boiling in forced vertical flow. In: J. Heat Transfer: 166–172
7. Vigneron B et al. (1982) Determination du pouvoir refroidissant d'huiles de trempe sous vide. In: Traitement Thermique 169: 37–39
8. Flament G, Moreaux F, Beck G (1978) Instabilité de la caléfaction au cours du refroidissement par trempe dans l'eau sous-refroidie. C.R. Ac. Sc. 287B, 321–323
9. Flament G, Moreaux F, Beck G (1979) Instabilité de la caléfaction à haute température sur an cylindre vertical trempé dans un liquide sousrefroidi. In: Int. J. Heat and Mass Transfer 22: 1059–1067
10. Flament G, Moreaux F, Martin M (1979) Déstabilisation de la caléfaction: propagation du front de mouillage. In: Letters in Heat and Mass Transfer 6: 208–216
11. Spiegler P et al. (1963) Onset on film boiling and the foam limit. In: Int. J. Heat and Mass Transfer 6: 987–989
12. Beck G, Chevrier JC (1971) Comparaison des données de trempe, déterminées à l'aide d'une méthode numérique, à celles du régime permanent. In: Int. J. Heat and Mass Transfer 14: 1731–1735
13. Archambault P, Chevrier JC (1977) Distribution de la température au sein d'un cylindre trempé dans un liquide vaporisable. In: Int. J. Heat and Mass Transfer 20: 1–6
14. Flament G, Chevrier JC, Moreaux F (1977) Transfert de chaleur en mode transitoire entre un solide à haute température et l'eau bouillante. In: C. R. Ac. Sc. 284B: 339–342
15. Beck G (1971) Contribution à l'étude thermocinétrique du régime initial de trempe des métaux et alliages. In: Revue Générale de Thermique 109: 37–34
16. Chevrier JC, Moreaux F, Beck G (1972) L'effusivité et la résistance thermique des zones superficielles du solide déterminent le processus de vaporisation du liquide en régime de trempe. In: Int. J. Heat and Mass Transfer 15: 1631–1645
17. Chevrier JC, Beck G (1971) Influence d'un dépôt de faible conductivité thermique sur le mécanisme de refroidissement par trempe d'une éprouvette métallique dans l'azote liquide. Application à la trempe à l'eau. In: Mémoires Scientifiques de la Revue de Métallurgie LXVIII 6: 391–400
18. Moreaux F, Chevrier JC, Beck G (1975) Destabilization of film boiling by means of a thermal resistance. In: International Journal of Multiphase Flow 2, 183–190
19. Kikuchi Y et al. (1986) The effect of thin insulating layer on heat transfer characteristics during quenching of hot metals in satured water. In: Transactions ISIJ 26: 576–581
20. Archambault P et al. (1980) A contribution to the optimization of the 7075 heat treatment. In: Material Science and Engineering 43: 1–6
21. Moreaux F, Simon A, Beck G (1980) Relations between quenching process, hardness depth and quench defects in steels. In: Heat Treating 1: 50–56
22. Zavarine IN (1935) Quenching in water, brine and oil. In: Metal Progress 27: 43–46
23. Moreaux F, Beck G (1970) Influence sur le pouvoir de refroidissement par trempe dans l'eau, de la mise en solution d'un sel thermiquement stable. In: Mémoires Scientifiques de la Revue de Métallurgie LXVII 4: 285–293
24. Van Stralen SJD (1968) The growth rate of vapour bubbles in superheated pure liquids and binary mixtures. Int. J. Heat and Mass Transfer 11: 1467–1489
25. Moreaux F et al. (1984) New polymer quenchants for steels and their characterisation by cooling curves and hardness penetration Heat Treatment' 84. 2–4 Mai Londres (Angleterre) Proceedings 18.1–18.5
26. Moreaux F et al. (1986) Nouvelles solutions aqueuses de polymères organiques pour la trempe des aciers et des alliages d'aluminium. Proceedings vol. III. 5éme ICHTM. 20–24 Octobre Budapest (Hongrie) p. 1799–1805
27. Olivier J et al. (1986) Brassage des fluides de trempe: conception et conséquences métallurgiques. In: Traitement Thermique 206: 29–42
28. Moreaux F et al. Procédé de modification du pouvoir refroidissant de milieux aqueux destinés à la trempe d'alliages métalliques. Brevet français n° 87 18109. Brevet européen n° 88 420422.3
29. Moreaux F et al. (1988) New aqueous solutions for metallic alloys quenching. Importance of agitation. Proceedings 167–170. 6ème ICHTM. 28–30 September, Chigago (U.S.A)
30. Archambault P, Chevrier JC, Beck G (1978) Sur l'existence de refroidissements optimaux des pièces en alliage d'aluminium. C. R. Ac. Sc. 286: 139–142

8 Determination of Quenching Power of Various Fluids

8.1 Methods and Standards for Laboratory Tests of Liquid Quenchants

Hans M. Tensi

The rapid cooling of hardenable steels or age-hardenable aluminum alloys is usually achieved by their immersion in, or their spraying with, fluids. The consequent changes in the mechanical properties of the materials – designated as hardness values in the simplest case – are determined by the temperature/time profiles attained during quenching.

Component parts have usually undergone the majority of processing steps and thus claimed a large percentage of manufacturing costs prior to quenching. For this reason, a failure of hardening due to incorrect cooling is generally a great loss. Researchers are therefore committed to describe the cooling process in different fluids as accurately as possible, so that hardening failures, which, for example, might be caused by the use of an unsuitable fluid, are avoided. To this end, the cooling performance of different fluids has to be measured as precisely as possible. The requirements to be met by the temperature/time profile during cooling depend upon the desired mechanical properties of the material and upon its alloy composition (cf. Chap. 1). Since the metallographic background of the property changes occurring during cooling is well known and modern data processing provides convenient ways to link hardness acceptance with the material and thermal properties of the component part, it is expedient to ascertain the quenching effect of different fluids in accordance with internationally standardized guidelines. The practical reasons for this are as follows:

- Examination of possible modifications in the quenching effect of currently used quenching liquids,
- Selection of quenching liquids on the basis of the desired mechanical properties, with given metallographic and thermal properties of the component part, and
- Computer-assisted hardness prediction for the physical and cost-related optimization of the quenching process, with given metallographic and thermal properties of the component part.

At present various methods of testing the cooling effect of fluids are in use around the world. The technical committee "Scientific and Technological Aspects of Quenching" in the IFHT[1], has therefore taken efforts to prepare internationally standardized proposals.

[1] IFHT: International Federation for Heat Treatment

A report on two proposals is given in the following. The Laboratory Test for Quenching Oils has already been presented to the ISO as an international standard [1]; a corresponding proposal concerning aqueous polymer solutions is to be adopted in the near future [2].

8.1.1 Laboratory Test for Industrial Quenching Oils

This laboratory test has been available as

DRAFT INTERNATIONAL STANDARD ISO/DIS 9950

since 28 October, 1988 [1]. The testing methods applied in several countries were checked in the course of its preparation. The results attained in different laboratories extensively agreed when the test conditions described in the proposal were observed.

The most important aspects of this ISO/DIS Standard will be addressed in the following. If the laboratory test is to be carried out conforming to the standards, it is imperative to use the complete text of the proposal.

The fact that, in laboratory tests, there is no forced convection in the oil baths, places considerable restrictions on the information yielded by the standard. It must therefore be explicitly stated that the results attained in the laboratory cannot be directly transferred to the cooling performance of the corresponding oils in commercial quenching baths. This international test standard thus only meets the first of the three application requirements listed in Chap. 7.1.

The sample immersed in the bath is a reusable nickel base alloy cylinder. Temperature is measured in the cylinder's gravity centre with a NiCr/Ni shielded thermocouple (details of the sample are shown in Fig. 8.1.1). An unalloyed, highly viscous paraffinic mineral oil is prescribed as a reference quenchant. The properties of this oil have to be secured by careful storage when not in use (e.g.,

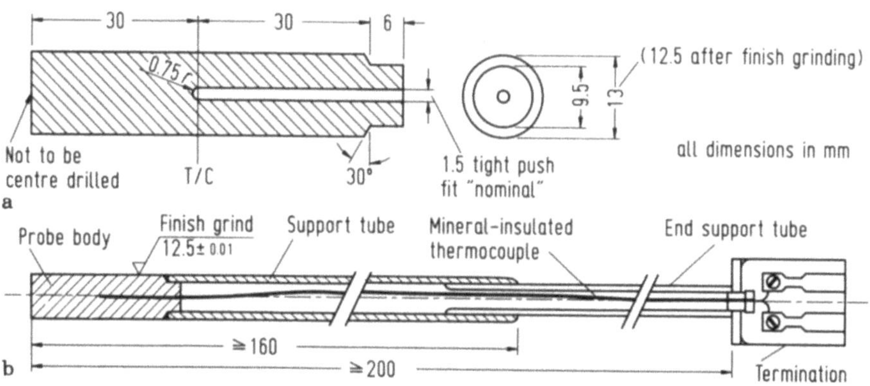

Fig. 8.1.1 a, b. Thermal sample [1]. **a** sample details; **b** general essembly

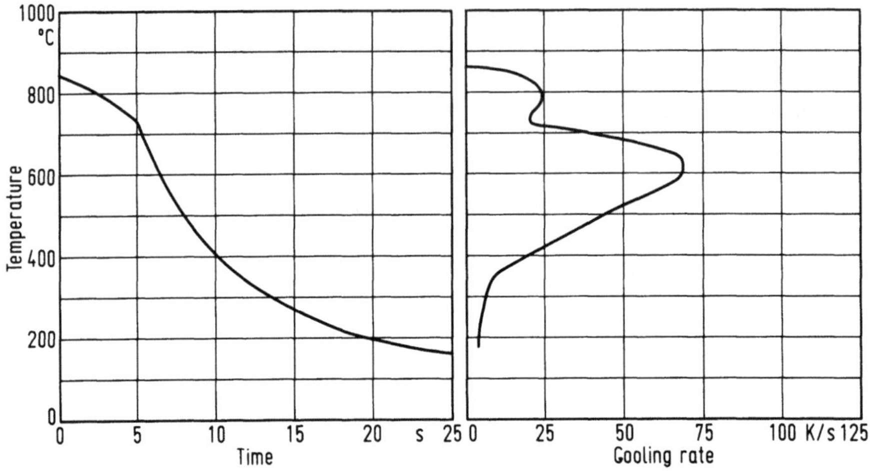

Fig. 8.1.2. Typical temperature/time and temperature/cooling rate plots for test probe cooled in a quenching oil [1]

in a closed container; maximum storage time no longer than two years). In addition to the listed minimum and maximum values of the reference oil's physical properties, the minimum and maximum values of the cooling performance measured for the standardized sample under the given conditions are defined (e.g., max. cooling rate, sample temperature at max. cooling rate, time elapsed before given temperatures are reached, etc.).

Furthermore, for carrying out the cooling test conforming to standards, the surface of the sample must be passivated in an unprotected furnace atmosphere. This requires at least 6 preliminary experiments (quenching the sample from 850 °C). The cleaning and reconditioning of the sample surface is precisely prescribed. The quenching fluid is used in a quantity of 2 liters. The sample should be lowered into the bath by an automatic apparatus, since irregularities of immersion may also have an effect on the characteristic cooling values. In the case of digital measuring, frequencies of $>20\,s^{-1}$ are recommended.

The test results are represented by the cooling curves $T(t)$ and $\dot{T}(T)$ (cf. Figs. 8.1.2a, b) and by the numerical statement of the cooling times elapsed before given temperatures and cooling rates have been reached.

8.1.2 Laboratory Test for Industrial Polymer Quenchants

The preparative works leading to this tentative standard were more complex: aqueous polymer solutions are subject to much greater influences by variations of physical and chemical parameters than oils, and polymeric quenching media also exhibit extremely great differences in their wetting behaviour.

It was thus essential to take forced bath convection into account within the framework of this proposal. In addition, further measuring rules for registering the wetting process had to be adopted.

Apparatus for the generation of bath convection:

Forced convection in quenching baths generally exerts a very significant influence on the characteristics of cooling. Convection not only raises the heat transfer coefficients during the different phases of boiling in oils, and increasingly in aqueous polymer solutions, but also strongly modifies the wetting process, which in turn has considerable effects on the cooling process inside the sample (cf. Chap. 5.2, pages 93–97).

If convection is to be included as a quantifiable cooling parameter of the quenching process, it is necessary to generate in the quenching bath convection conditions of physically precisely defined intensities and, subsequently, to determine the influence exerted by these on the wetting process and the cooling in the sample.

A complex convection apparatus has been developed for basic experiments (Fig. 8.1.3). It produces an irrotational and low-turbulence flow in the quenching area of a closed fluid circuit, with a fairly regular flow profile over the cross-section. This is true for flow velocities of approx. < 1.5 m/s (Fig. 8.1.4). The

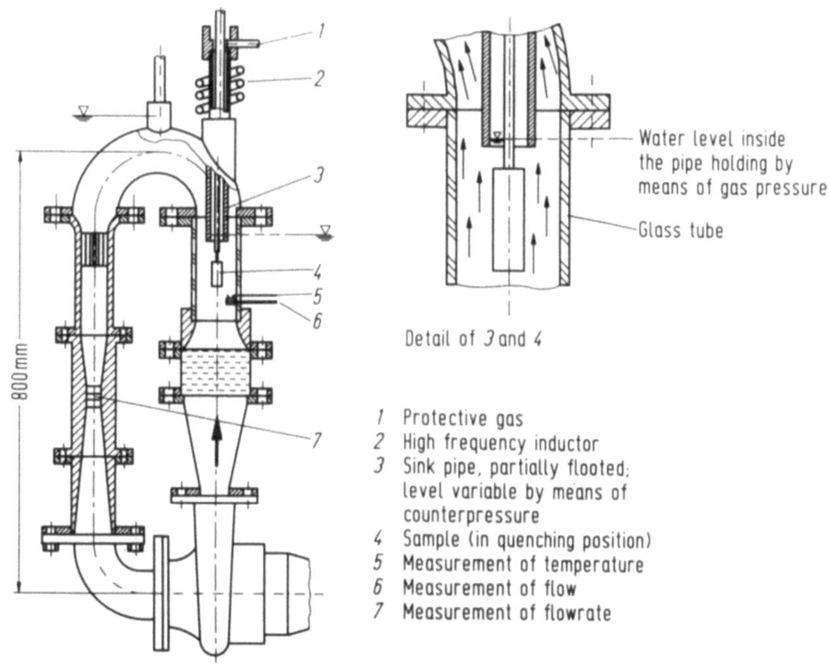

Fig. 8.1.3. Convection system for generation of exactly defined fluid flow, permitting measurement of kinematics of wetting and cooling curve (in centre of sample) during quenching; inductively heated sample ($\phi 15 \times 45$ mm) is lowered into flowing quenchant (volume of bath 9 dm^3) [3, 4]

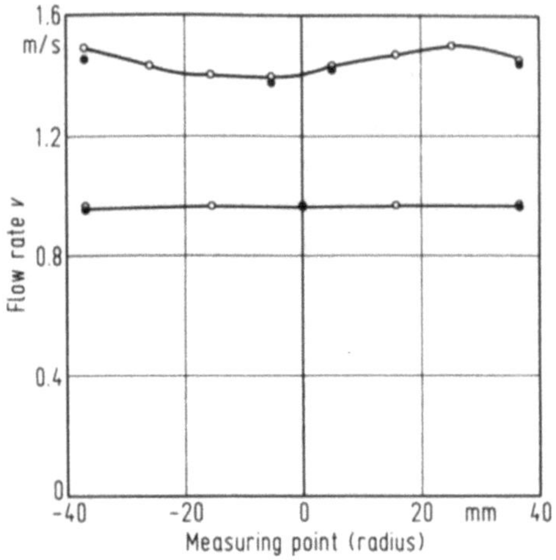

Fig. 8.1.4. Distribution of velocity of flow across critical zone for two series of measurements taken at point 6 in Fig. 8.1.3 using destilled water at 25 °C

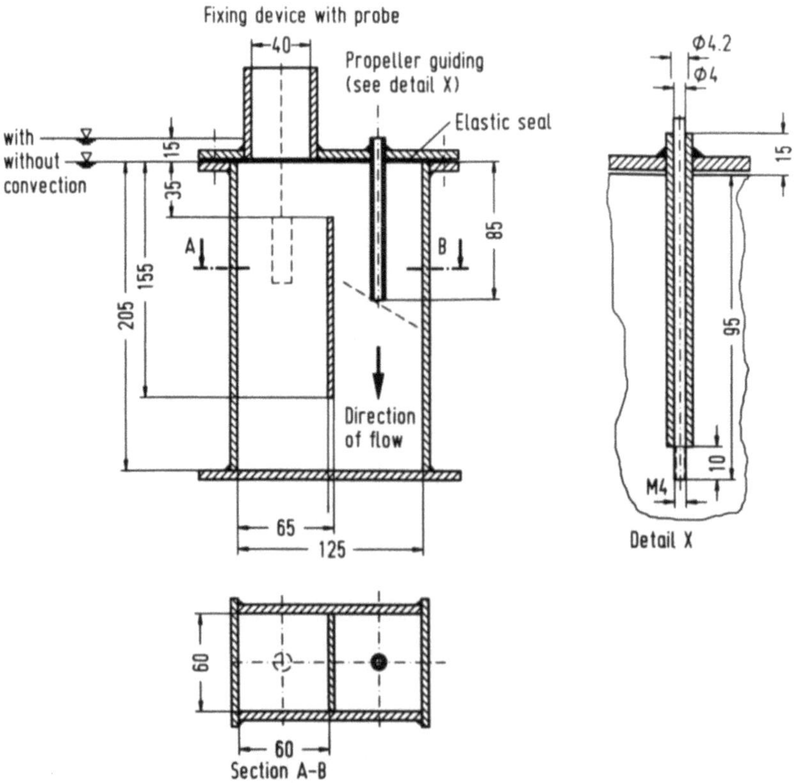

Fig. 8.1.5. Simplified convection system for generation of fluid flow (all dimensions in mm, volume of bath 2 dm^3) [2]

volume of the fluid is in this case 9 liters. The probes 5 and 6 (Fig. 8.1.3) are used for measuring the temperature and flow velocity in the reaction area. The sample, 4, is inductively heated under air or protective gas in a guiding pipe (2); it is lowered through the fluid level in the pipe into the fully developed flow in the reaction area of the apparatus (the fluid level in the pipe is set at approx. 1 cm above its lower end by the gas pressure adjusted to the flow velocity; cf. details 4–3 in Fig. 8.1.3).

Wetting kinematics is the cooling parameter most strongly influenced by convection (cf. Chap. 5.5 pages 102–109).

In order to develop a simple convection apparatus for in-house construction at justifiable costs, the most sensitive cooling parameters were measured in the complex and in the simple convection apparatus, with identical flow velocities in the two systems. When, for example, the temperatures at the onset of wetting measured in the optimized version of the simple apparatus (cf. Fig. 8.1.5) are compared with those measured in the complex apparatus, the values extensively agree up to flow velocities of <0.6 m/s. In Figs. 8.1.6a and b, strings can be seen that indicate an almost laminar flow without gas bubbles (single-phase flow) up to 0.6 m/s. Figures 8.17a and b show that the effects of the flow in the two systems are almost identical. It becomes apparent that the T_s values are excellent in both types of apparatus, when water is used as a quenching medium;

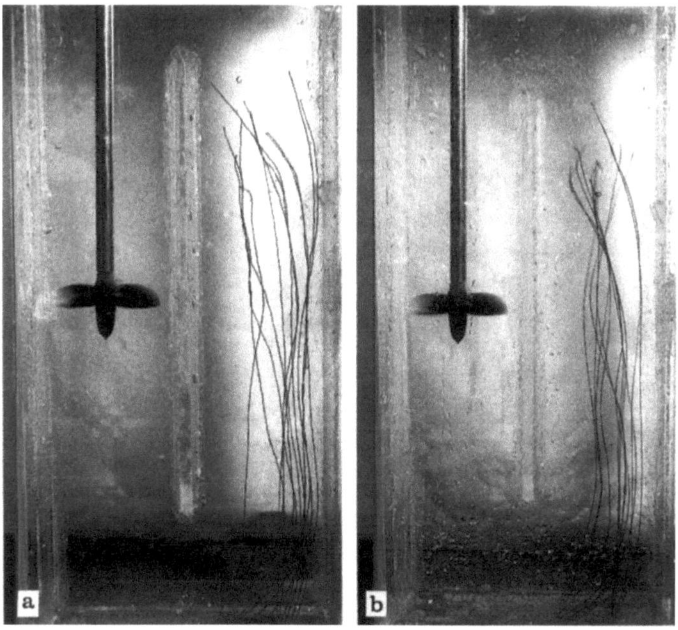

Fig. 8.1.6 a, b. Thin "flow threads" to illustrate flow without bubbles and without twist: bubble formation in low concentrated watersolution polymers begins at about $v > 0.6\ ms^{-1}$. **a** $v = 0.25\ ms^{-1}$; **b** $v = 0.6\ ms^{-1}$

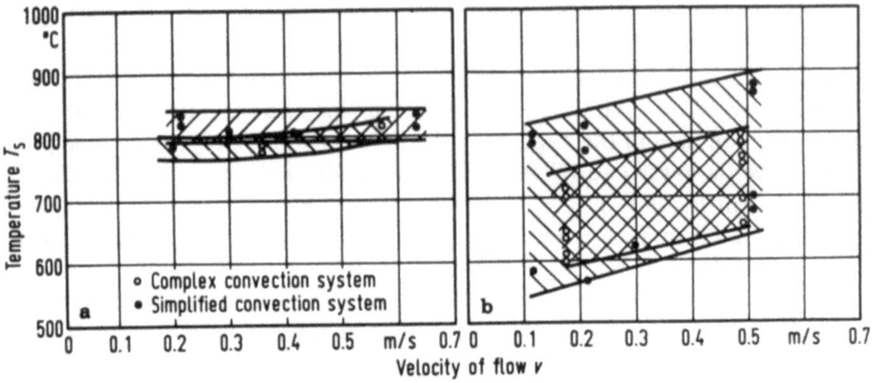

Fig. 8.1.7 a, b. Starting Temperature T_s of wetting the sample surface (measured in the centre of the sample) influenced by the velocity of flow v. Comparing the measurements in the complex (cf. Fig. 8.1.3) and the simplified convection systems (cf. Fig. 8.1.5) **a** destilled water; **b** water solution of polymer (Fluid H) with 20 vol.%

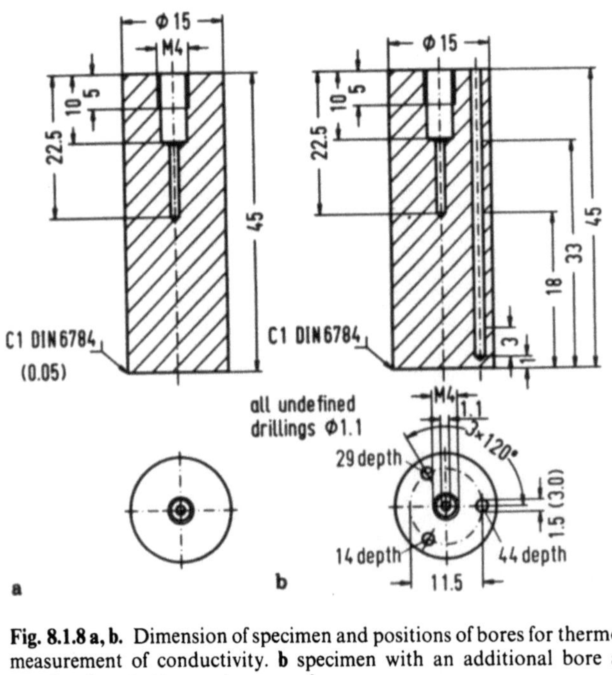

Fig. 8.1.8 a, b. Dimension of specimen and positions of bores for thermocouples. **a** Specimen used for measurement of conductivity. **b** specimen with an additional bore accomodating three thermocouples close to the specimens surface

in the case of aqueous polymer solutions (here: Polymer "*H*" with a volume percentage of 20), the dispersion of the T_s values is greater in the simple system (here, with increased viscosity, the less effective smoothing of the flow in the case of decreased viscosity can be noted). Agreement, however, is good, when one considers the simplicity of the required apparatus ("simple system").

The designe specifications for the system are included in the test proposal of the IFHT. The impeller is available on the market as a finished product[1], to mention just one constructional detail.

Measuring principles for the ascertainment of the cooling performance of aqueous polymer solutions:

CrNi cylinders with the dimensions stated in Fig. 8.1.8 are used as immersion samples. The temperature in the sample is registered by a NiCr/Ni shielded thermocouple, just as in the draft standard for quenching oils. Since the wetting process cannot be reliably measured by a temperature probe positioned in the centre of the sample, additional measuring devices are used. Two fundamentally different methods are available.

A) Ascertainment of the Wetting Process by Conductance Measurements

The wetting state is ascertained by measuring the electrical conductance between the sample and a counter electrode of arbitrary shape [3, 4, 5] (Fig. 8.1.9). When, during the vapour blanket phase, the hot sample is largely insulated, the conductance between the sample and the counter electrode is low (Fig. 8.1.10). When the vapour blanket breaks down over an arbitrary position of the sample surface (i.e., beginning of local wetting), conductance increases. The wetted portion of the sample surface approximately corresponds with the amount of the increase in conductance. When the sample surface is completely wetted, conductance is at its highest value (this statement, too, is only true in approximation, since the formation of bubbles in the fluid frequently causes the con-

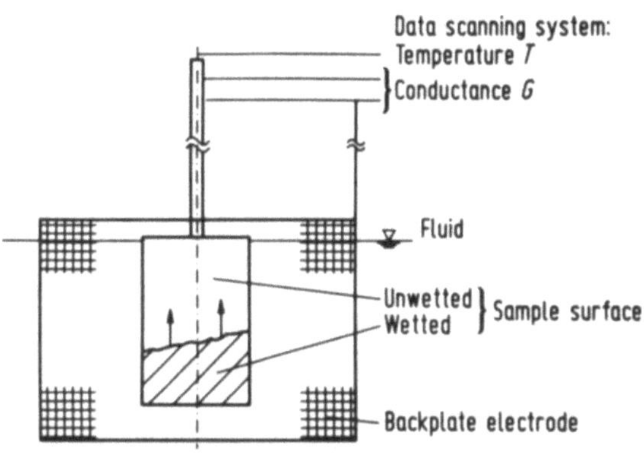

Fig. 8.1.9. Measuring principle for determination of percentage of sample surface wetted during cooling process of immersion cooled samples

[1] The code of the impeller: Diameter 50 mm; number of blades 3; pitch setting 42 mm ($\phi \times 0.85$)

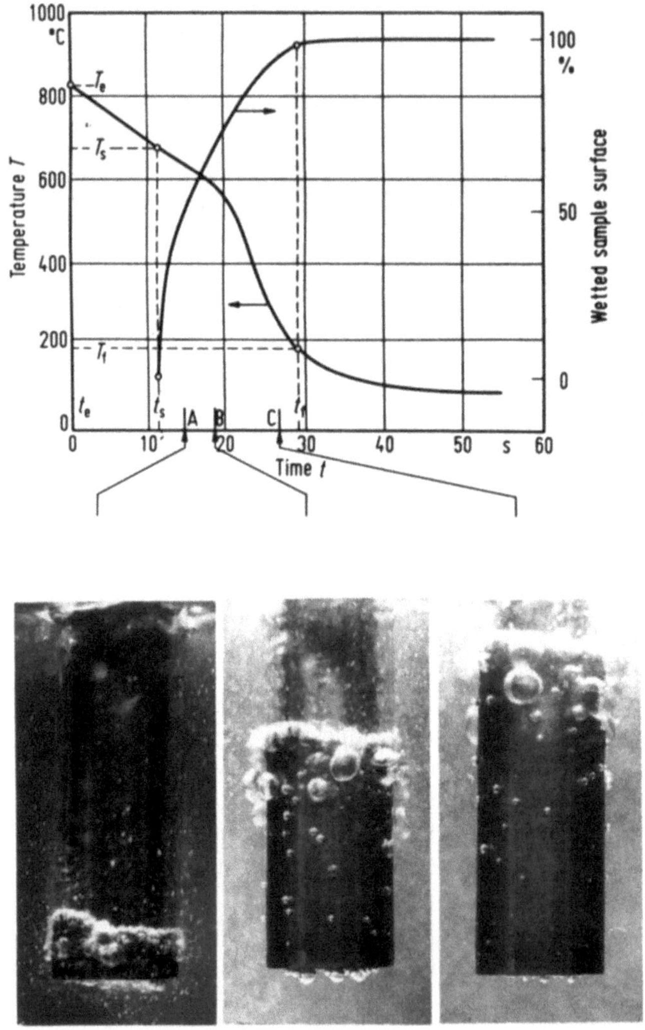

Fig. 8.1.10. Temperature in centre of sample and surface area wetted as a function of time during cooling of cylindrical Cr–Ni steel sample (cf. Fig. 8.1.8a). *A*, *B*, *C* time of taking photographs during cooling; the wetting front runs from the lower edge to the upper one; T_s, t_s temperature resp. time of starting wetting; T_f, t_f temperature resp. time of finishing wetting

ductance to drop during the wetting process and/or immediately after it; thus, for example, an asymptotic approximation to the limit of conductance can be observed after the conclusion of wetting). The change in conductance, set in direct proportion to the wetted area of the sample, and the temperature course in the sample centre during immersion cooling are schematically shown in Fig. 8.1.10. The parameters "temperature in sample centre at onset of wetting",

T_s, the associated "time of onset of wetting", t_s, and the "temperature/time value of the conclusion of wetting", T_f and t_f respectively, are the

- most important parameters for the characterization of the cooling performance of aqueous polymer solutions, and
- important parameters for the computation of the time-dependent temperature distribution in the sample and for the prediction of the hardness distribution to be expected.

Thus, in conjunction with the measured temperature drop, all the requirements formulated in Chap. 8.2 which should be met by a method of testing the quenching effect of fluids, are met.

The change in conductance during the progression of wetting in all known polymer solutions and also in quenching oils can be ascertained by means of a conductance meter with a measuring range of μS to 1 S. If a conductance meter is not available, the progression of wetting can be ascertained with reasonable accuracy by means of thermocouples positioned inside the sample close to its surface:

B) Ascertainment of the Wetting Process with Three Near-Surface Probes

The preliminary test proposal includes two variations for the realization of the experiment: Insertion of three bundled thermocouples (three individual probes packed together or a single shop-made triple thermocouple) into a near-surface bore of the sample, or insertion of a single industrially produced shielded triple cover thermocouple into the bore (cf. Fig. 8.1.8b). Figure 8.1.11 shows in basic form the information yielded by the three temperature probes on the progression of the wetting edge from the lower end of the sample to its top. This method

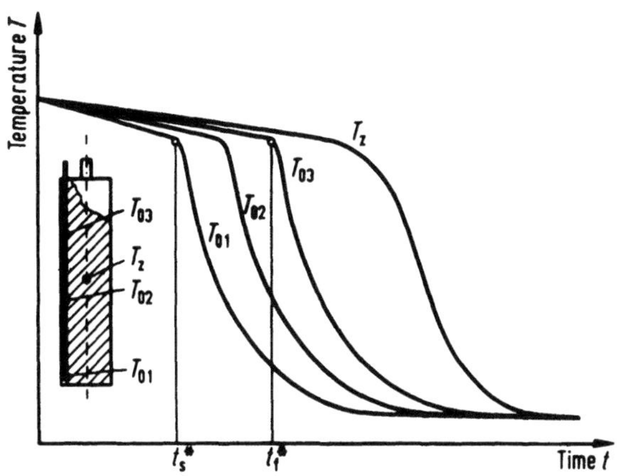

Fig. 8.1.11. Determination the kinematics of wetting by means of three thermocouples positioned close to its surface ($T_{01} - T_{03}$) and one in the specimens centre (T_c); the calculation of the total wetting of the specimen is based on the extrapolation of the time interval ($t_{f*} - t_{s*}$)

also – though not with the same resolution achieved by the conductance measurements – permits the ascertainment of the onset and conclusion of wetting and of the velocity of the advancing wetting edge. Photographs of the wetting process are depicted in Fig. 4.10b (and 4.4 Chap. 4.3). Figure 4.13 in Chap. 4.4 shows basically differing wetting processes during immersion cooling.

8.1.3 Representation of Results

The increase in conductance is graphically represented in addition to the temperature drop in the sample centre, shown as functions $T(t)$ and $\dot{T}(T)$. When three temperature probes close to the sample's surface are used, the three functions $T_{01-03}(t)$ are also represented in the diagram.

The insufficient information yielded by the "transition temperature" T_{trans} on the wetting process becomes particularly apparent in all the events in which wetting does not take place in an "explosive" manner.

In addition to the numerical values to be computed on the basis of the temperature drop in the centre (similarly as in 8.1), the temperature and time values at the onset and conclusion of wetting are ascertained as the actual central parameters.

The most important experimental parameters are: quenching bath at two temperatures, without convection and with convection of two different intensities (0.3 and 0.6 m/s); distilled water at 60 °C is used as a calibration liquid in the three convection conditions. The temperature of the sample is 850 °C at immersion. Conditioning of the sample surface should be carried out similarly to the way described in the proposal concerning quenching oils.

At present, the proposed testing method for aqueous polymer solutions is still in the international trial stage, which is to be concluded in 1991.

References

1. Draft Int. Standard ISO/DIS 9950 from 1988 (VDC 621.78.084: 621.78.065.2) Industrial Quenching Oils – Determination of Cooling Characteristics – Laboratory Test Method; S. Segeberg
2. A Proposal for Standardization "Measurement of the Cooling Characteristics of Aqueous Polymer Solutions" to the Int. Organization for Standardization ISO/DIS by Tensi HM, Künzel Th. and Stitzelberger-Jakob P. Technical University Munich
3. Tensi HM and Stitzelberger-Jakob P. Influence of Wetting Cinematics on Quenching and Hardening in Water Based Polymers with Forced Convection World Materials Congress, 1988, Chicago (USA) Published by ASM Int., L. of Cong. Nr. 88-071657 ISBN: 0-87170-334-2, 171–176
4. Tensi HM and Stitzelberger-Jakob P. Evaluation of Apparatus for Assessing Effect of Forced Convection on Quenching Characteristics Materials Science and Technology 5 (1989) 718–724
5. Tensi HM, Welzel G and Künzel Th. Problems of Getting Characteristics out of Quenching Experiments useful to control the Heat Treatment of Metallic Materials 8th Int. Heat Transfer Conf., San Francisco, 1986, p. 3031–3035
6. Tensi HM and Stitzelberger-Jakob P. Influence of Rewetting on the Quenching Behaviour of Aluminium Specimens HTM 43 (1988) 148–151

List of Symbols

G electrical conductance, S
T_b temperature of bath, °C
T_c temperature in centre sample, °C
T_f temperature in sample centre at conclusion of wetting (final temperature of wetting), °C
T_s temperature in sample centre at beginning of wetting (starting temperature of wetting), °C
t_f time at which wetting is concluded
t_s time at which wetting starts
v velocity of flow of cooling fluid, $m\,s^{-1}$

8.2 Concept of the Grossmann's H-Value and Its Shortcomings

B. Liščić

8.2.1 Theoretical Background and Definition of the "Quenching Severity H"

The Grossmann's Quench Severity Factor "H" is based on the Newton's Law of cooling:

$$dQ = \alpha \cdot A(T_s - T_\infty) \cdot dt \tag{8.2.1a}$$

for a cylinder

$$dQ = \frac{\lambda}{r} \cdot A(T_c - T_s) \cdot dt \tag{8.2.1b}$$

where:

Q = heat transferred from the surface of a body to the environment [J]
α = heat transfer coefficient [$W \cdot m^{-2} \cdot K^{-1}$]
A = surface-area [m^2]
T_s = surface temperature of the body [K]
T_∞ = ambient temperature [K]
T_c = temperature in the centerline of the cylinder [K]
r = radius of the cylinder [m]
d = diameter of the cylinder [m]
λ = thermal conductivity [$W\,m^{-1}\,K^{-1}$]
t = time [s]

According to this law the heat transferred from the surface of a body to its environment is proportional to the temperature difference between the surface temperature and the ambient temperature, and is characterized by the heat transfer coefficient. During cooling of a body the same amount of heat is transferred by heat conduction from the center to the surface of the body (in this case cylinder). This heat conduction is proportional to the temperature difference between the body centerline and its surface, and is characterized by the thermal conductivity of the body's material.

Using the mentioned law the assumption is taken into account that the heat transfer coefficient α is constant during the whole quenching process i.e. that the quenching medium behaved uniformly throughout the entire temperature range of cooling, what the quenchant, of course, does not.

From Eqs. (8.2.1a) and (b) follows: if $(T_s - T_\infty) = K_1$ and $(T_c - T_s) = K_2$

$$\frac{\alpha}{\lambda} = \frac{1}{r} \cdot \frac{K_2}{K_1} \quad [\mathrm{m}^{-1}] \tag{8.2.2}$$

This shows, that the cooling process depends on the ratio: $\frac{\alpha}{\lambda}$. It was in 1925 when Gröber [1] named the ratio $\frac{\alpha}{\lambda}$ as "relative heat transfer coefficient" ("relative Wärmeübergangszahl") giving it a designation "h" [m^{-1}]. Later Russel [2] in 1936 designated the cooling intensity also with "h". Grossmann and colaborators [3] have later in 1939, (because of more practical application in their nomograms for cooling intensity) used the value $\frac{h}{2}$ giving it the designation "H" as it is still used today. "H" should have the dimension [m^{-1}] although it is often used as a dimensionless number.

The common used characteristic value in the heat transfer is the dimensionless BIOT-number: $\mathrm{Bi} = (\alpha/\lambda) \cdot r$. Grossmann and Asimov used instead of Bi-number the dimensionless value: $h \cdot r$ or $H \cdot d$. This product is important, because every body with the same $h \cdot r$-value shows the same heat flux.

There is an American definition of the severity of quench [4] that reads: "the severity of quench is the ability of a quenching medium to extract heat from a hot workpiece which can be expressed in terms of the "H" value".

If we take the λ-value of the cooling steel body as a constant (i.e. not dependent on temperature and possible structural changes), than it is possible by using the "h"-value to define (as a first approximation) the "cooling capacity" of a quenching medium, provided that the body is axially symmetric.

The "H" value of still water (of 18 °C) is conventionally taken as 1.0. The H-values of oil, water and brine, taking into account different degrees of agitation are given in Table 8.2.1. This table was intentionally compiled from two different sources [4] and [5]. It may therefore be read only in the way that the data from one source correspond to the degree of agitation taken from the same source. This comparison has been made to show two facts:

a) degrees of agitation are not uniformly and not precisely enough defined.
b) even for comparable degrees of agitation there are substantial differences between the corresponding H-values.

Table 8.2.1

Method of quenching		*H*-value for: [m^{-1}] Oil		Water		Brine	
According to: [5]	According to: [4]	Accord. to [5]	Accord. to [4]	Accord. to [5]	Accord. to [4]	Accord. to [5]	Accord. to [4]
The workpiece and the quenchant-motionless	No agitation	0.3	0.25—0.30	1.0	1.0	2.2	2.0
—	Mild agitation	—	0.3—0.35	—	1.0—1.1	—	2.0—2.2
Moderate movement of the workpiece in still quenchant	Moderate agitation	0.4—0.6	0.35—0.40	1.5—3.0	1.2—1.3		
—	Good agitation	—	0.4—0.5	—	1.4—1.5		
Strong movement of the workpiece in still quenchant	Strong agitation	0.6—0.8	0.5—0.8	3.0—6.0	1.6—2.0	7.5	
Strong or violent agitation of the quenchant or spraying of the quenchant	Violent agitation	1.0—1.7	0.8—1.1	6.0—12.0	4.0		5.0

8.2.2 The Use and Evaluation of *H*-Values

There are several different methods of how to evaluate the "*H*" value of a quenching medium:

a) The most simple and actually very often used method is *to take the "H" value from a Table* (see Table 8.2.1). Using this method one has to be aware that in such tables usually only "*H*" values for water, oil (in general terms) and brine can be found. Another problem associated with this method is, as it is mentioned above, that the degree of agitation is not specified precisely enough and no data about quenchant temperature exist.

   All this speaks, that this method can give just a rough comparison of the quenching intensity.

b) A more precise evaluation of the "*H*" value is possible according to [6] by using the relevant Lamont diagram, see Fig. 8.2.1. This diagram is valid for

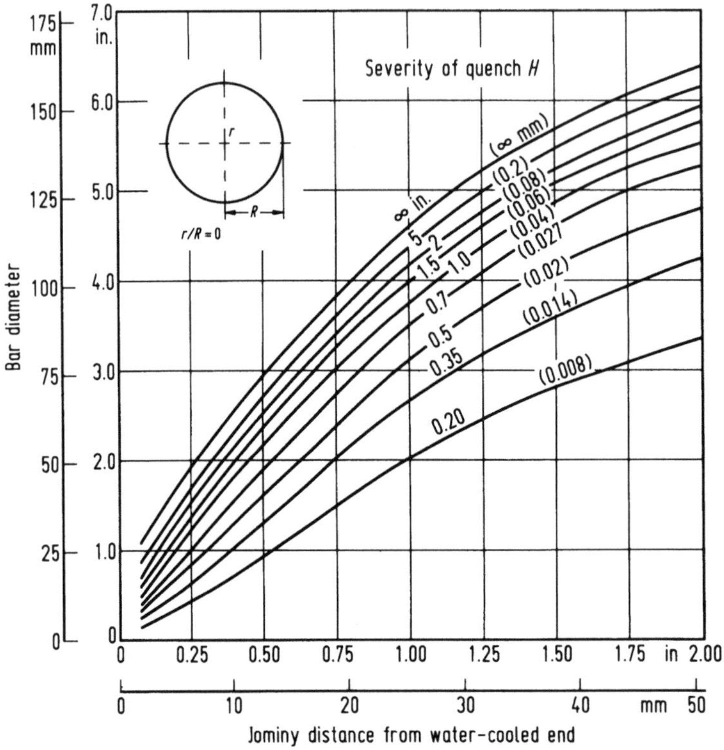

Fig. 8.2.1. A Lamont diagram, for the center of round bars ($r/R = 0$), showing the relation between bar diameter and the Jominy distance from the quenched end, that can be used to determine the quench severity factor "H"

the center of round bars, i.e. for:

$$\frac{r}{R} = 0$$

r = radius of unhardened portion of the cross-section
R = radius of the bar

Lamont has developed such diagrams for different ratios in the range of

$$0 \leqq \frac{r}{R} \leqq 1,0 \quad \text{in steps of } 0,1\ldots \tag{8.2.3}$$

the last one being valid for the surface of the bar. By knowing the hardness required at a specified point of the bar cross-section, and having the relevant Jominy curve for the steel grade in question, one takes the corresponding Jominy distance for the hardness required (as a vertical) and the relevant bar diameter (as a horizontal) and on the intersection of both values reads the adequate "H" value.

For this method same comments apply in respect of agitation rate and quenchant temperature as under a) because the "*H*" values assigned to different curves are the same as in tables.

c) There is also a diagram referred to as the 'Grossmann chart", see Fig. 8.2.2 relating bar diameter, hardenability of the steel and the severity of quench in terms of "*H*" values. If e.g. a bar of 25 mm Dia is quenched in a strong oil quench, reading the chart in Fig. 8.2.2 gives the intersection of the horizontal for 25 mm Dia, and the line for $H = 0.70$, and reads a value of 4/16″ (6.35 mm) distance from the quenched end of the Jominy hardenability specimen. This means that *the center* of the bar having 25 mm Dia will have the same hardness as that shown at 6.35 mm from the quenched end of the standard Jominy test specimen made of the steel being quenched.

According to [7] the application of this chart is limited because the quenchants involved are described in general terms (e.g. "very good oil quench—good agitation"). An additional complication is the fact that what appears to be good agitation in a quenching bath with no parts in it, may be poor agitation when a load of parts is in the bath. As could be seen from

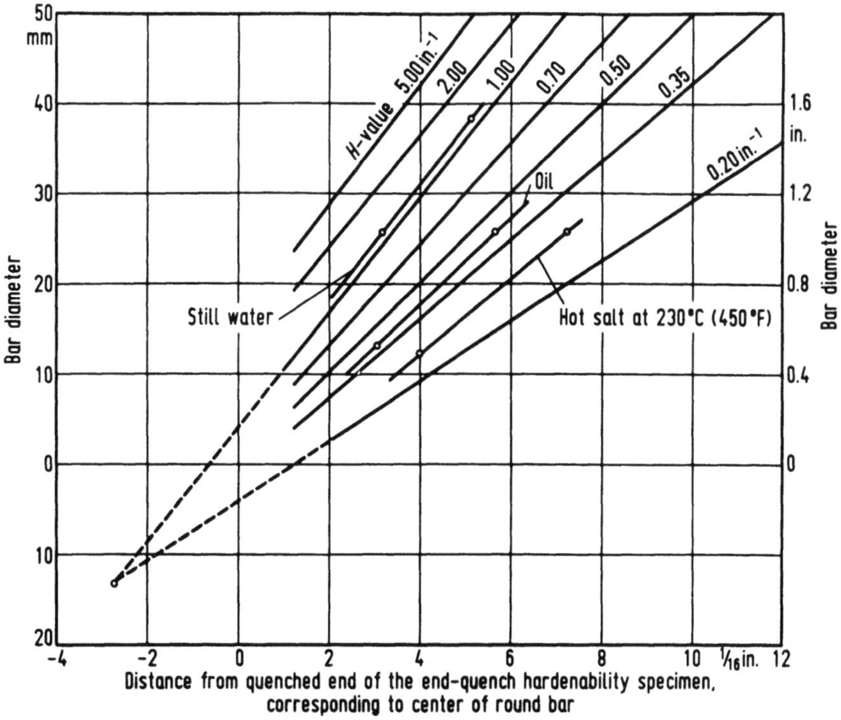

Fig. 8.2.2. "Grossmann chart," relating bar diameter, hardenability of steel and severity of quench [7]. Explanation of *H*-values: 5.00 = strong brine quench, voilent agitation; 2.00—poor brine quench, no agitation; 1.00 = poor water quench no agitation; 0.70 = strong oil quench, good agitation; 0.50 = good oil quench, good agitation; 0.35 = good oil quench moderate agitation,. 0.20 = poor oil quench, no agitation

Fig. 8.2.2, the lines indicating various severities of quench on the Grossmann chart converge at a point outside of the chart. Actually if one looks strictly on the diagram in Fig. 8.2.2, this applies only for H values $\leqq 1.0$. This common point of intersection can be used as a second point in drawing any new quench severity lines. Usually two bars of different diameters are to be quenched, when a new quench—severity line *valid for a certain quenchant and specified quenching conditions* is to be additionally drawn into the Grossmann chart. The common intersection point outside the chart provides upto some extent the possibility of checking the results.

In such a case of drawing additional new quench severity lines, the specimens can be quenched also along with a batch of parts thus taking into account the conditions of temperature and agitation that actually prevail in production quenching of the relevant batch.

d) *The various diameter rounds test.* This experimental procedure provides a very useful method of determining the "H" value of certain unknown quenchant under specified conditions. This is actually based on the Grossmann's method to estimate the critical diameter "D_k", see Fig. 8.2.3. As it could be seen: quenching rounds of bigger diameters, the unhardened portions of the cross-section (dashed areas) become bigger. Naturally, the depth of the hardened portions is bigger when quenching in a more severe quenchant (compare between oil quenching and water quenching of the same diameter in Fig. 8.2.3). Besides, the unhardened portion becomes bigger even with small increase in diameter, when quenched in a mild quenchant (oil), but it doesn't so when quenched in a severe quenchant (water). Grossmann found that on the boundaries between hardened and unhardened portions a structure of 50% martensite is present, and the hardness gradient in this bordering region is the steepest. That means that all points on the cross-section marked by "X" i.e. at boundaries between hardened and unhardened portion have followed the same or very similar cooling law

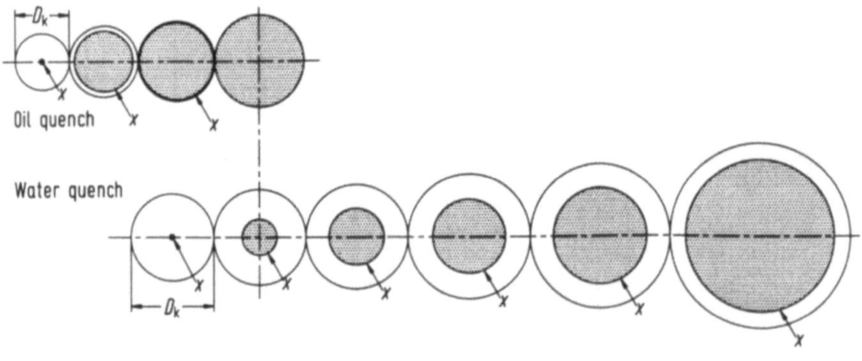

Fig. 8.2.3. Schematic presentation of the Grossmann's method to estimate the critical diameter "D_k" by quenching round bars of different diameters in oil and water. Dashed areas represent unhardened portions of the cross-section. Points (or boundaries) marked with "X", mark 50% martensitic structure

(same "H" value). The diameter at which the 50% martensite structure appears exactly in the center of the cross-section, with $r = 0$, when quenched with a certain severity factor "H", Grossmann named as critical diameter "D_k". There are diagrams which correlate the critical diameter "D_k" to the ideal critical diameter "D_I" (which can be calculated from the steel composition), and the severity of quench ""H". [22]

The experimental various diameter rounds test takes considerable time to prepare several specimens (especially cutting of hardened specimens may be a problem), and to search for the diameter size of the steel to be used which is sensitive enough to the severity of the quenchant to be applied. Using this experimental method one has to take into account, according to [8], that "the hardening in the center of a round bar is disturbed by the mechanical stresses produced in the quenching process, and therefore H-values are not a precise measure of quenching effect".

e) Another experimental method to quantitatively evaluate the "H" value of an unknown quenchant has been shown in principle on Fig. 8.2.4. Using this method all conditions of the cooling process are kept constant, and the only factor that is changed is the influence of the quenchant. In this way comparative H-values to an agreed standard condition are obtained. A steel cylinder of 1/2″ (12.7 mm) Dia is used having a thermocouple in its geometric centre. The temperature vs. time (cooling) curve is first taken for standard condition i.e. for still water of 18 °C (see curve 1 in Fig. 8.2.4), for which, by convention $H_w = 1.0$. Then the other cooling curve 2 by quenching the cylinder in the unknown quenchant is taken. The average cooling rate within a temperature region (e.g. 700 °C to 300 °C) in K/s is than compared. The ratio of the cooling rate of the unknown quenchant (v_q) to the cooling rate

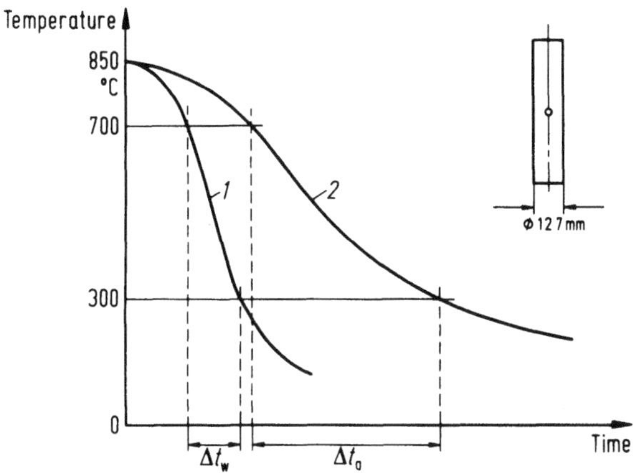

Fig. 8.2.4. Principle of experimental evaluation of the H-value relative to the H-value of still water ($H = 1.0$). 1 Still water of 18 °C 2 Unknown quenchant

of still water (v_w) gives H-value of the unknown quenchant:

$$H_q = \frac{v_q}{v_w} = \frac{\Delta t_w}{\Delta t_q} = \frac{1}{\Delta t_q} \ldots \tag{8.2.4}$$

This procedure has been described in [9] [10]. Recently efforts have been made to extend the use of H values also for polymer—solution quenchants, including mathematical and statistical calculations.

f) Monroe and Bates [11] have performed an investigation of three different quenchants: a fast oil, water and a 16% solution of a polyglycol in water. Instrumented probes of 25, 38 and 50 mm diameters and a length of four times the diameter, machined from austenitic steel AISI 304 with a thermocouple placed in the geometric centre, were used.

The H-value for various experimental conditions was first determined by solving the Fourier's equation with a finite element heat transfer program which allowed specific H-values to be used as input values.

Cooling curves at the center of various sized bars were calculated and the cooling rate at 700 °C determined.

With a polynomial least squares fit, relation between the cooling rate at 700 °C and the H-value was established.

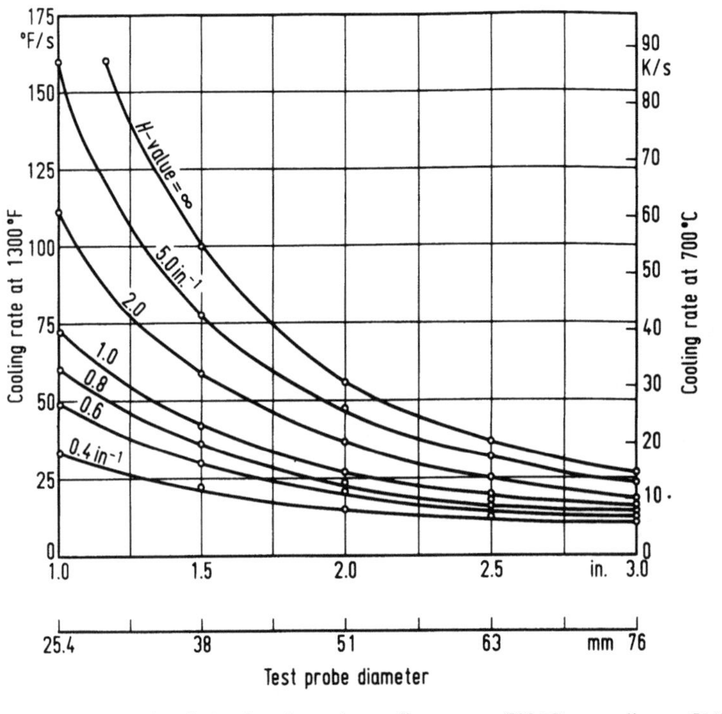

Fig. 8.2.5. Calculated H-values from the cooling rate at 700 °C, according to [11]

By experiments H-values have been determined by recording the cooling curve in the center of cylindrical instrumented probes, determining the 700 °C cooling rate from the cooling curve, putting this value in the polynomial expression relating the H-value to the 700 °C cooling rate, and solving for the H-value. Figure 8.2.5 shows calculated H-value from the cooling rate at 700 °C for different test probe diameters.

This procedure allowed to evaluate H-values of the above mentioned quenchants under different conditions (agitation velocity and temperature).

Two objections can be raised to this method. The first one is that the H-value which should characterise the quenching intensity during the whole quenching process has been related to one point on the cooling curve only (700 °C). The other is that measuring the cooling curves *in the center* of bars of one to two inches diameters, due to damping effect, cannot reflect the real heat flux on the surface.

g) In another recent publication [12], Totten, Dakins and Jarvis have dealt only with a specific "Oil like" polymer quenchant (UCON Quenchant E), designing the experiments by using a 2^3 factorial in order to study the interrelationship of the quench bath variables

	Units	Low	Middle	High
Circulation rate (CR)	ft/min	50	75	100
Polymer concentration (C)	%	15	20	25
Bath temperature (T)	°F	100	120	140

and the H-value.

Using the experimental cooling curve data, the following statistical formula has been developed:

$$H\text{-Value} = 1.954 + 0.006\,\mathrm{CR} - 0.02650\,\mathrm{C} - 0.00975\,\mathrm{T} \qquad (8.2.5)$$

The standard deviation for this statistical model is ± 0.09073. Figure 8.2.6 shows a plot illustrating the above mentioned statistical model for limits of all three variables given above. This data show, that H-values for less than 0.4 to greater than 1.0 are achievable by varying: the quenchant concentration bath temperature and circulation rate. This plot also illustrates that H-values:

—increase with increasing circulation rate
—decrease with increasing quenchant concentration, and
—decrease with increasing quenchant bath temperature.

8.2.3 Shortcomings of the *H*-Value

a) The most serious objection to the H-value is, that as a constant value it cannot really describe the heat transfer taking place at the surface of a workpiece, when quenched in an evaporable fluid.

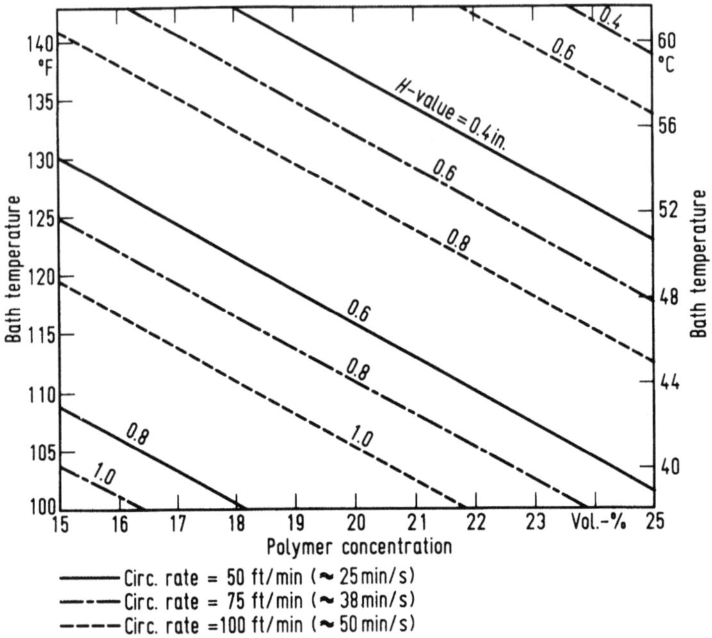

Fig. 8.2.6. *H*-value for different concentrations, circulation rates and bath temperatures, of the polymer-solution UCON quenchant *E*, according to [12]

As it is well known the heat transfer coefficient "α" changes very much during different stages of the quenching process. According to [14] e.g. for a quenching oil of 150 °C, the heat transfer coefficient during the boiling phase is about:

$\alpha_B = 5000\,[\mathrm{W/m^2K}]$, while during the convection phase it is only $\alpha_K = 350\,[\mathrm{W/m^2K}]$.

Figure 8.2.7 shows according to [13] the temperature-time plot (a) and the temperature vs. heat flux density plot (b) when a 8 mm Dia. silver cylinder is quenched in oil. As the full line in Fig. 8.2.7a shows, really a steep temperature fall is taking place during the boiling phase. The dashed cooling curve calculated from constant *H*-value doesn't show this fall, and intersects the real cooling curve as per definition, as the half temperature T_H. In Fig. 8.2.7b the full line shows the real heat flux density as function of temperature. The dashed line shows the heat flux density calculated according to Newton's law with a constant heat transfer coefficient $\alpha = 840\ [\mathrm{W/m^2K}]$, that corresponds to the value $H = 0.42$ if the heat conductivity $\lambda = 25\ [\mathrm{W/m \cdot K}]$ (for steel) is taken, according to the formula:

$$H = \frac{1\alpha}{2\lambda} \cdot 0.025 \quad (H \text{ in } 1/\text{inch}) \qquad (8.2.6)$$

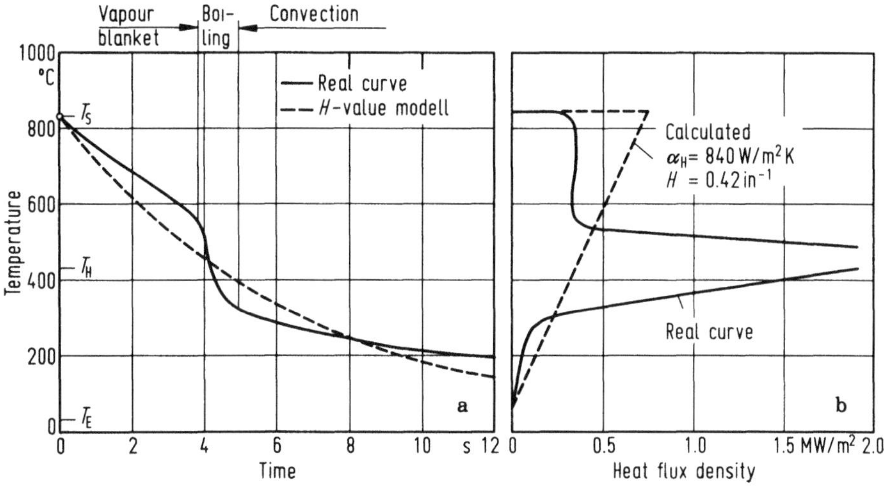

Fig. 8.2.7 a, b. Temperature—time plot (**a**) and temperature-heat flux density plot (**b**), when quenching a silver cylinder of 8 mm Dia in oil, according to [13]. ——— real curves; ----- calculated using an H-value of 0.42. T_s = temperature at start of cooling; T_E = temperature on the end of cooling; $T_H = T_S - T_E/2 + T_E$ (half temperature)

The differences between reality and calculations using a constant H-value are evident.

It should also be taken into account, that the heat conductivity "λ" itself is a temperature dependent value.

b) When using H-value taken from the tables, there is no possibility to take into account neither the:

—specific characteristics e.g. of an oil (its composition; viscosity), nor the temperature of the quenchant itself.

Figure 8.2.8 shows, according to [10] how the H-values, experimentally established in different temperature regions, depend on the temperature of the quenchant, in this case a plain mineral oil, Fig. 8.2.8a, and a blended mineral oil Fig. 8.2.8b.

Not knowing to which specific conditions an H-value pertains creates the real situation that the same H-value can be attached to different quenchants and quenching conditions. That is even more so when taking into account different agitation rates.

c) When using H-values taken from the tables (see Table 8.2.1), it is evident, that the agitation rate of the quenchant hasn't been defined neither uniformly, nor precisely enough which is one of the major factors why different H-values can be found in tables for similar quenching conditions.

d) When measuring the H-values of a quenchant experimentally by comparing it to the $H = 1.0$ value of plain still water of 18 °C (as described in 8.2.2), the fact has not been taken into account, that plain water of different provenance

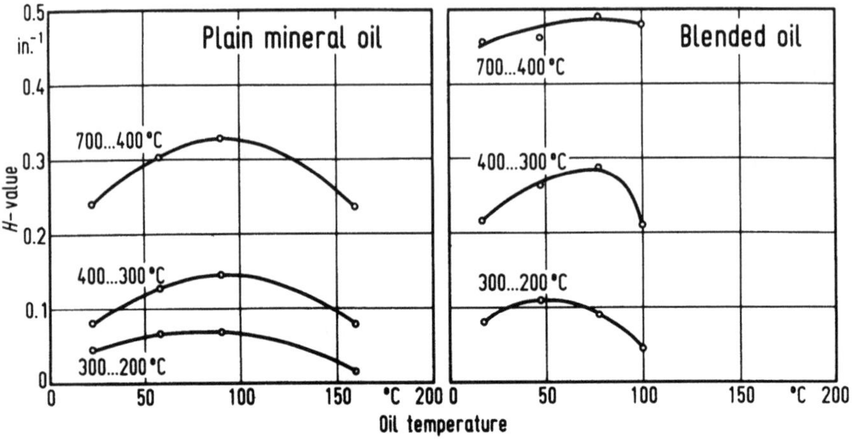

Fig. 8.2.8 a, b. Dependence of the H-values on the quenchant's temperature in different temperature regions for a plain oil (**a**) and for a blended oil (**b**) according to [10]

may contain different quantities of gases as: N_2, O_2, CO_2 as well as different quantities of dissolved carbonates. The gases present in water can influence the thickness of the vapour film in the first phase of quenching as well as the duration of this phase. Figure 8.2.9 shows, according to [15] three cooling rate vs. temperature curves, measured experimentally by a silver ball method

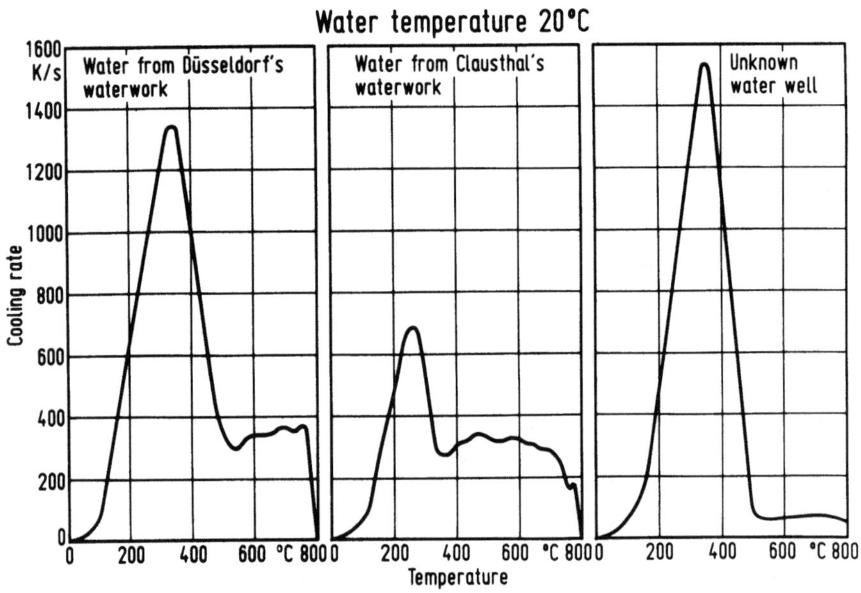

Fig. 8.2.9. Cooling rate vs. temperature curves measured experimentally by a silver ball method for three samples of plain water of 20 °C of different provenance, according to [15]

for three samples of plain water of 20 °C of different provenance. As it could be seen remarkable differences exist among water samples of different provenance not only in respect of the cooling rate maximum but also in respect of temperature at which the cooling rate maximum occurs.

e) Experimental measuring of the H-value of an unknown quenchant is usually done by using a small diameter (12.7 mm) specimen which cools down to below 300 °C in some 10 to 20 seconds. Contrary to that, quenching of a real part (e.g. a cylinder of 60 mm Dia) takes much longer time-see Fig. 8.2.10, according to [16], especially when considering core temperatures. Comparing the temperature scales on the left hand side and the right hand side diagrams in Fig. 8.2.10 one can see that after 28 seconds when the boiling phase measured with the cylinder of small diameter is over, and the convection phase starts, the surface temperature of the 60 mm Dia cylinder has dropped only to about 450 °C, but the temperature in its core is still about 770 °C. This is a proof that actually the heat extraction from the core of this cylinder takes place during the convection stage of cooling in this oil, which lasts much longer than previous two stages. During convection cooling stage, as it is known the factors that mostly influence the cooling

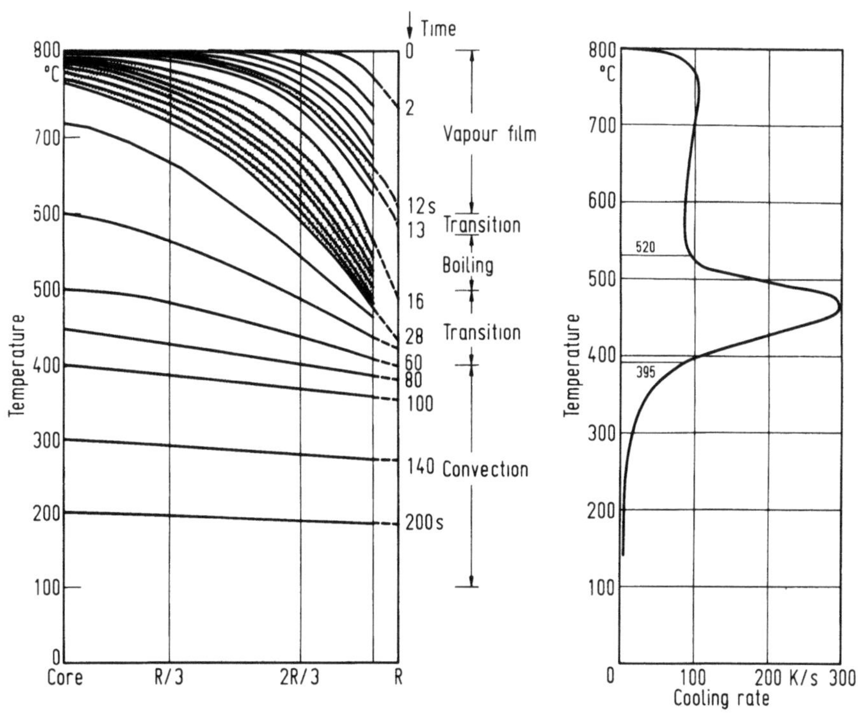

Fig. 8.2.10. Comparison of thermal gradients during quenching of a 60 mm dia cylinder in oil with correspondent cooling rate vs temperature curve for the same oil, taken with a specimen of small diameter, according to [16]

intensity are: the temperature difference between surface of the part and quenchant itself and the agitation rate. Because the small diameter cylinder has already cooled down before the bulk of the heat has been extracted from the core of the real part, it cannot give any information about cooling the real part during its longest period of convection cooling. Because the hardness distribution on the cross section, after quenching, considerably depends on the heat extraction process during convection stage, predictions of the hardness distribution on cross sections of bigger dimensions using an experimentally established H-value are in-accurate or may even be misleading.

Finally, one has to be aware, that the H-value concept cannot be used for all those quenching processes during which two cooling media (e.g. interrupted quenching or martempering), or changed cooling conditions (e.g. change of the agitation rate), are used.

8.3 Practical Problems when Measuring Temperature Within Quenching Specimens

While for laboratory tests, specimens of small diameter are used, when measuring and recording temperature vs. time (Cooling) curves in practice, usually instrumented probes of bigger diameters and bigger mass are used. When using a cylindrical probe for such measurements in practical conditions, usually the following precautions are taken in order to assure a high reproducibility of results:

—The length of the probe is always $\geqq 3$ Dia in order to neglect the heat extraction through the both ends of the cylinder thus assuring the supposition that in the half length cross-section only a radial—symmetrical heat flow exists.
—The probe is made of a metal without structural transformations when heated to high temperature in order to avoid additional heat input resulting from structural transformations.
—The surface condition of the probe is kept standard.
—The response time of the thermocouples used is adequately short in order to register fast changing temperatures.

In spite of all this precautions, one has to be aware of the following facts and problems which can influence the accuracy and reproducibility of this temperature measurement when an unsteady heat transfer with transient fast changing temperatures on the probe surface takes place:

a) *Damping of phenomena on the probe's surface:* We are interested to register all specific phenomena that occur during quenching on the surface of the probe, especially in evaporable quenchants (vapour film, boiling). The accuracy of measurement of temperatures depends on the position of the thermocouple sensing tip as well as on the response-time of the thermocouple. The faster the temperature change, the higher is the dependence. Damping of this fenomena is well known. If the thermocouple sensing tip is shifted only little to the interior of the workpiece, the phenomena occurring on the surface would be greatly damped. The bigger the distance of the thermocouple sensing tip position from the surface, the greater will be the damping effect.

b) *Wetting of the surface:* The local differences in heat transfer on the surface and hence the local temperature differences result from the local isolation effects coming from vapour bubbles which stick to the surface during transition from the vapour blanket to the boiling stage. This transition does not occur simultaneously on the whole surface but rather starts on the sharp edges and spreads from there later onto the other parts of the surface. This wetting kinematics which is comprehensively described in Chap. 5.0, depends on the wetting properties of the quenchant as well as on the wetting properties of the workpiece surface itself.

The phenomenon has an effect on the accuracy and reproducibility of the temperature measurement that is opposite to the damping effect. The nearer the thermocouple sensing tip is to the surface, the more will be temperature measurement dependent on the local temperature differences at the surface.

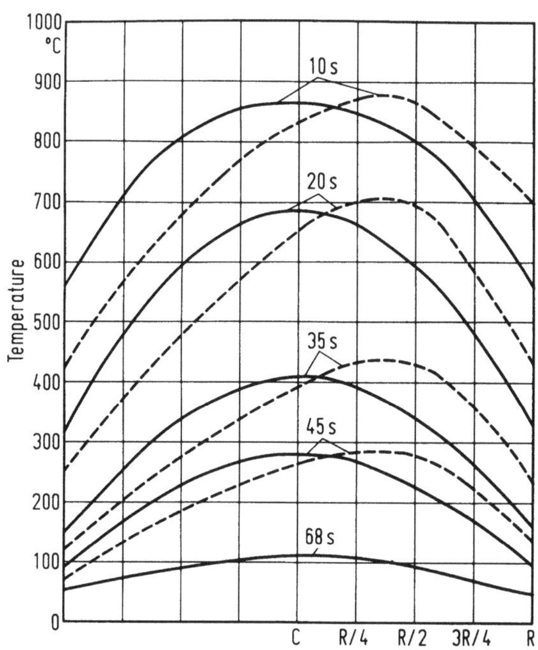

Fig. 8.3.1. Symmetric (—) and unsymmetric (---) cooling of a 35 mm Dia cylinder, quenched in water

c) Another reason which can impair the reproducibility of the temperature-time curves measured by the probe is an unsymmetric heat extraction, see Fig. 8.3.1. In this case the supposition of a radially symmetric heat flow has not been fulfilled (due to e.g. an unsymmetric flow of the quenching fluid), resulting in unsymmetric temperature fields.

8.4.1 Concept and Aims of the Temperature Gradient Method in the Workshop Practice Based on Heat-Flux Density

8.4.1 Concept and Aims of the Temperature Gradient Method

In designing the method for practical measurement and recording of the quenching intensity in workshop conditions, a special technology of temperature measurement of transient fast changing temperatures, as well as an adequate microcomputer software for: collecting, storing, calculating, graphical display and retrieval of data and results, have been used.

Contrary to the Grossmann's value concept, expressing the quenching intensity by a sole number only, the main idea was to express the quenching intensity by continuous change of adequate thermodynamic functions during the whole quenching process, taking into account that the method should be sufficiently sensitive to reflect changes in each of important quenching parameters (specific character of the quenchant, its temperature and mode and degree of agitation.)

The main aims of the method are:

1. It should enable a real comparison of the quenching intensity among different quenchants, quenching conditions and quenching techniques.
2. It should provide an unambiguous relation between measured and evaluated quenching intensity and the depth of hardening (expressed by hardness distribution over the cross-section).
3. It should furnish as much as possible information about thermal stresses and possible superposition of structural transformation stresses that will occur at a quenching process.
4. It should provide the basis for automatic control of the quenching intensity during the quenching process.

The essential feature of the method is a cylindrical instrumented probe of 50 mm Dia × 200 mm length having two (or in the latest version three) thermocouples displaced radialy on the half-length cross-section in characteristic points.

This instrumented probe, having adequate mass, and heat capacity when heated to starting temperature (usually 850 °C) is capable of following the whole quenching process of real parts (in an oil bath e.g. upto 500–600 seconds or in a vacuum furnace upto 1000–3000 seconds), contrary to the small mass laboratory specimens which quenching in oil is finished in 15 to 30 seconds.

This makes possible a more precise prediction of the hardness profile upon quenching using the own established data base and a computer aided method which is described in Chap. 11.2.8. The position of the thermocouple sensing tips within the cross-section of the probe have been selected as follows, because of the following reasons:

—The outher thermocouple on the very surface measures the surface temperature of the probe (T_s), which is important to register all the phenomena that are taking place on the surface during quenching.
—The intermediate thermocouple ($T_{-1.5}$) measures the temperature at a point radialy displaced at 1.5 mm below the surface. The readings of T_s and $T_{-1.5}$ in each moment of cooling enable to easily calculate the temperature gradient near the surface of the probe which is taken as an indicator of the real quenching intensity, see [17].
—The central thermocouple (T_c), which is provided with the new version of the probe is placed in the center of the cross-section itself. The purpose is to indicate how long does it take to extract heat from the core, and to provide in every moment the temperature difference between the surface and the core which is essential for calculation of thermal stresses.

8.4.2 Description of the Method [18] [19]

When a trial is carried out in practice, the data of the relevant quenching conditions (kind of quenchant, its temperature and its degree of agitation) should be specified and electronically stored.

The method for measuring and recording the quenching intensity is based on the known physical rule, that the heat flux at the surface of a body is directly proportional to the temperature gradient at the surface multiplied by the thermal conductivity of the body material, as per the equation expressing the heat flux density:

$$q = \lambda \frac{\partial T}{\partial x}$$

q = heat flux density (i.e. quantity of heat which is transferred in a unit of time through a surface unit, perpendicular to the surface)..... W/m^2

λ = thermal conductivity of the body material...... W/Km

$\frac{\partial T}{\partial x}$ = temperature gradient at the probe surface, perpendicular to the surface..... K/m.

Since the thermal conductivity of low alloy steels changes in the temperature range between 800 °C and 100 °C, this temperature range was divided in three temperature regions and the conductivity was taken as a separate value in each of these three regions for thermal calculations.

The essential feature of the method lies in measuring and recording the temperature gradient near the surface of a special cylindrical probe during the entire quenching process.

Temperature measurements should be made on the true surface in order to record all phenomena that are taking place during quenching (e.g. vapour blanket, boiling). It is well known, that if the point of temperature measurement is shifted even a little from the surface into the interior of the specimen, the temperature drop on the temperature v.s. time curve will be greatly damped.

In order to determine the temperature gradient near the surface, the temperature on the surface itself, and the temperature at a point just underneath the surface should be measured. Surface temperature measurement of a specimen has, until now, been considered to be rather inaccurate, if at all possible. However, using a new concept of fast-response thermocouples for transient thermal measurements, measurement of the surface temperature becomes possible and reproducible.

For measurement of the temperature on the true surface of the probe, the thermocouple must fulfil the following requirements:

—it should be two-dimensional instead of three-dimensional,
—it should be flat with the surface,
—it should, with its own mass, disturb the heat flux as little as possible.

Figure 8.4.1 shows the instrumented cylindrical probe made of austenitic stainless steel AISI 304, developed in cooperation with the NANMAC Corp., Framingham Centre Massachusetts (USA). The outher thermocouple, that

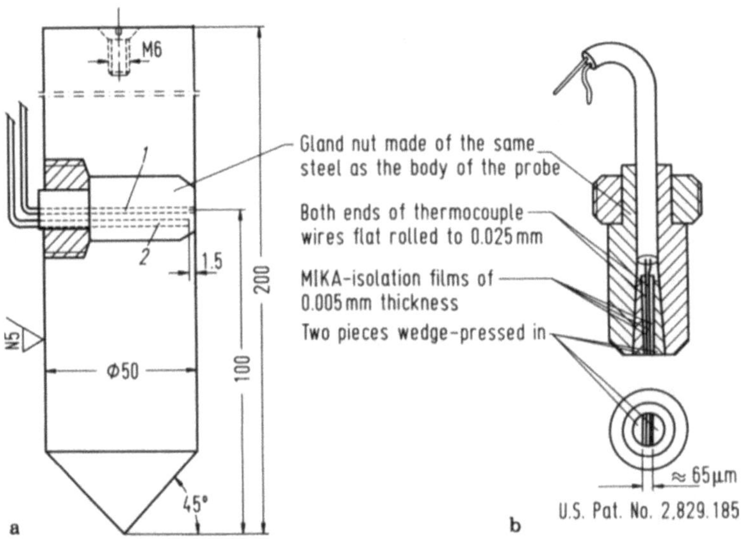

Fig. 8.4.1 a, b. Liščić-Nanmac Probe for measurement of the temperature gradient on the surface (**a**) Detail of the Nanmac—Thermocouple (**b**)

meets all the above mentioned requirements, measures the temperature on the surface of the probe and the second one measures the temperature on the same radius but at the point 1.5 mm below the surface.

The hot junction of the outher thermocouple is formed on the probe surface itself by polishing the sensing tip region. During the polishing action,

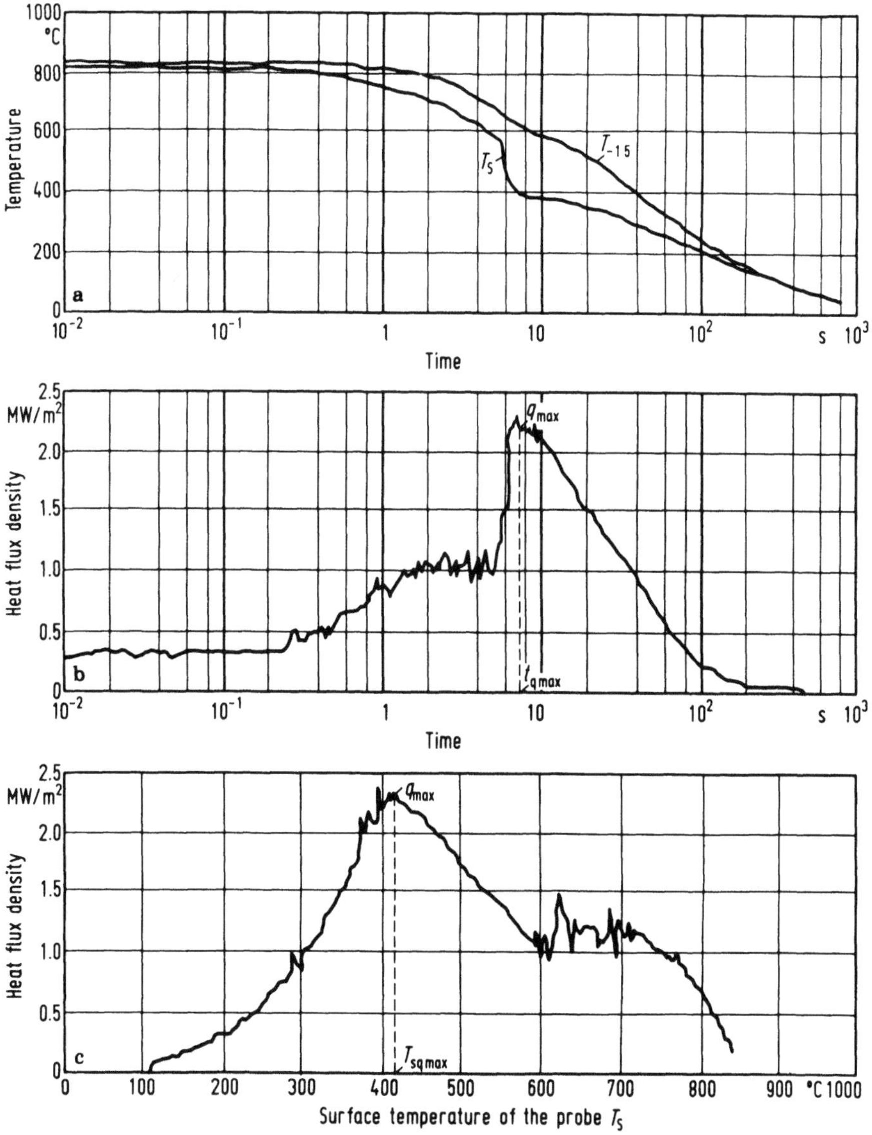

Fig. 8.4.2 a–c. Recorded temperature vs time (cooling) curves for the surface and the point 1.5 mm below the surface (**a**); calculated heat flux density vs. time (**b**); calculated heat flux density vs. surface temperature (**c**); obtained by quenching the probe in still mineral oil of 20 °C

microscopic particles of both metals that form the thermocouple bridge the central mica layer and join together. The specific features of the cylindrical probe are:

—The response time of the thermocouple is 10^{-5} s: the fastest temperature changes can be recorded.

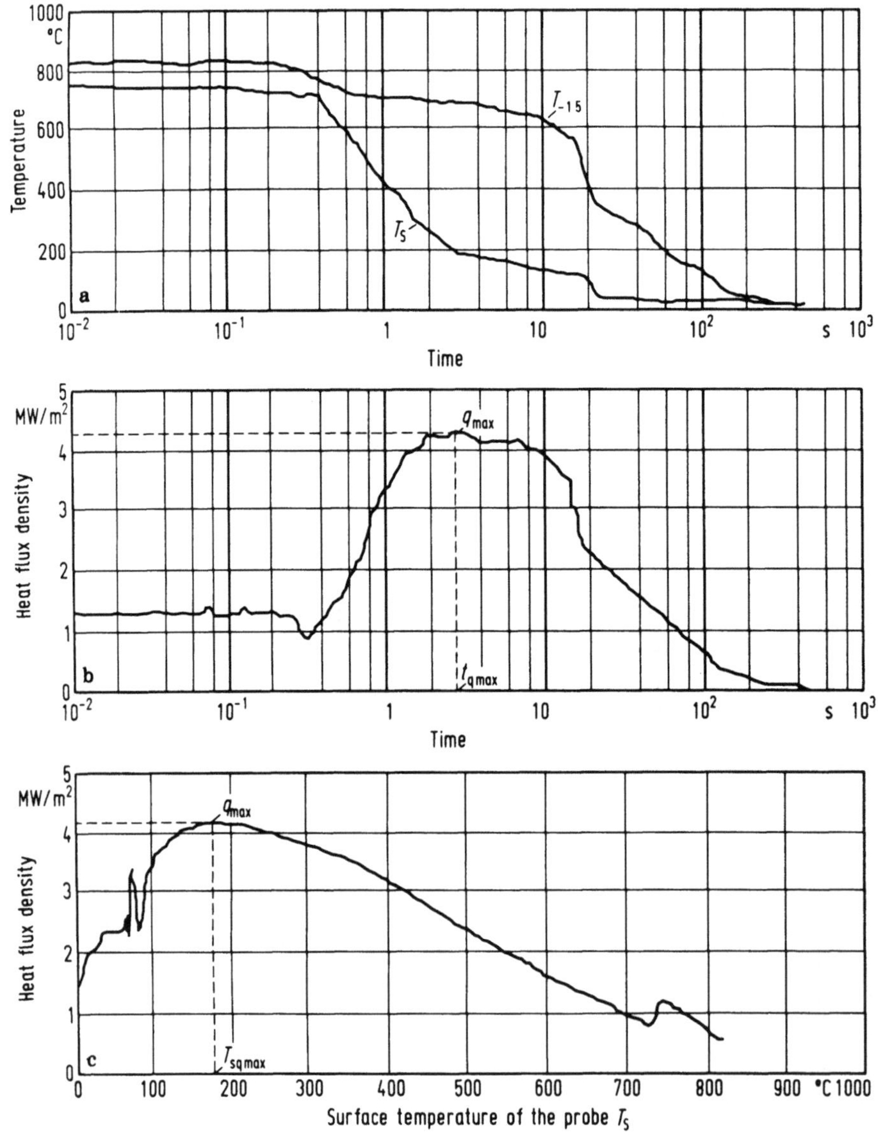

Fig. 8.4.3 a–c. Recorded temperature vs time (cooling curves) for the surface and the point 1.5 mm below the surface (**a**); calculated heat flux density vs time (**b**); calculated heat flux density vs. surface temperature (**c**); obtained by quenching the probe in still water 20 °C

—The internal thermocouple can be positioned with an accuracy of ± 0.025 mm.
—The surface condition of the probe can be maintained by polishing the sensing tip before each measurement.
—The body of the probe, made of an austenitic stainless steel, does not change in structure during the heating-quenching process, nor does it evolve of absorb heat because of structural transformations.
—The size of the probe and its mass ensure a sufficient heat capacity and a symmetrical radial heat flow in the cross-section plane where the thermocouples are located.
—According to [20] the average heat transfer coefficient during the boiling stage, generally depends on the specimen diameter, but for diameters >50 mm. This dependence is not remarkable.

When a test of the quenching intensity is performed, the probe is heated to 850 °C in a suitable furnace and transferred quickly to the quenching bath and immersed. The probe is connected to a microcomputer. Adequate software enables to store the temperature v.s. time data from both thermocouples and to calculate and graphically display the relevant functions.

Figure 8.4.2 shows the recorded cooling curves for the surface of the probe and for the point 1.5 mm below the surface (a), as well as relevant heat flux density versus time (b), and heat flux density versus surface temperature (c) curves, obtained by quenching the probe in a 20 °C mineral oil without agitation.

Figure 8.4.3 shows the same curves obtained by quenching the probe in water of 20 °C, without agitation.

Higher heat flux density values are obtained by quenching the probe in water than by quenching it in oil, as can be seen by comparing Fig. 8.4.2 and Fig. 8.4.3. Greater quenching intensity of water can also be seen from the fact, that the heat flux density maximum (q_{max}) is reached in third second after immersion, when quenching the probe in water, and in eighth second after immersion, when quenching it in oil. During the critical time interval between

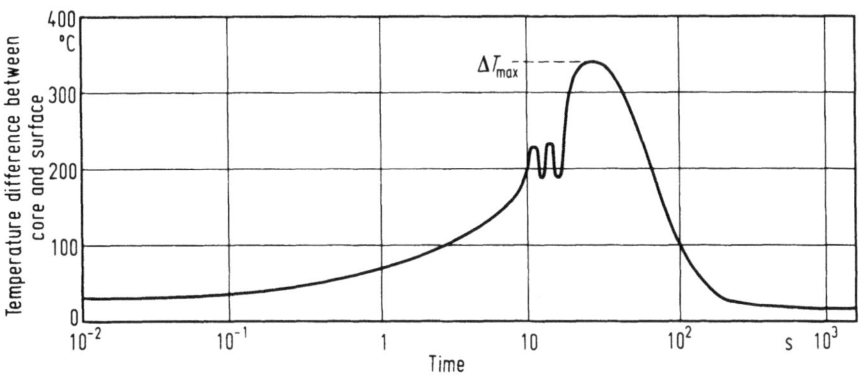

Fig. 8.4.4. Temperature differences $\Delta T = T_c - T_s$ between core and surface of the probe, when quenched in still mineral oil of 20 °C

one and ten seconds after immersion, the heat flux density values are much higher for water-quenching than for oil-quenching.

It is a well known fact, that the cooling rate in the low temperature region is much higher in water quenching than in oil quenching. This is also clearly confirmed using the method described. Namely, comparing Fig. 8.4.2c and Fig. 8.4.3c (heat flux density v.s. surface temperature) shows that the heat flux density maximum ($T_s q_{max}$) occurred at 420 °C for oil quenching and at 170 °C for water quenching.

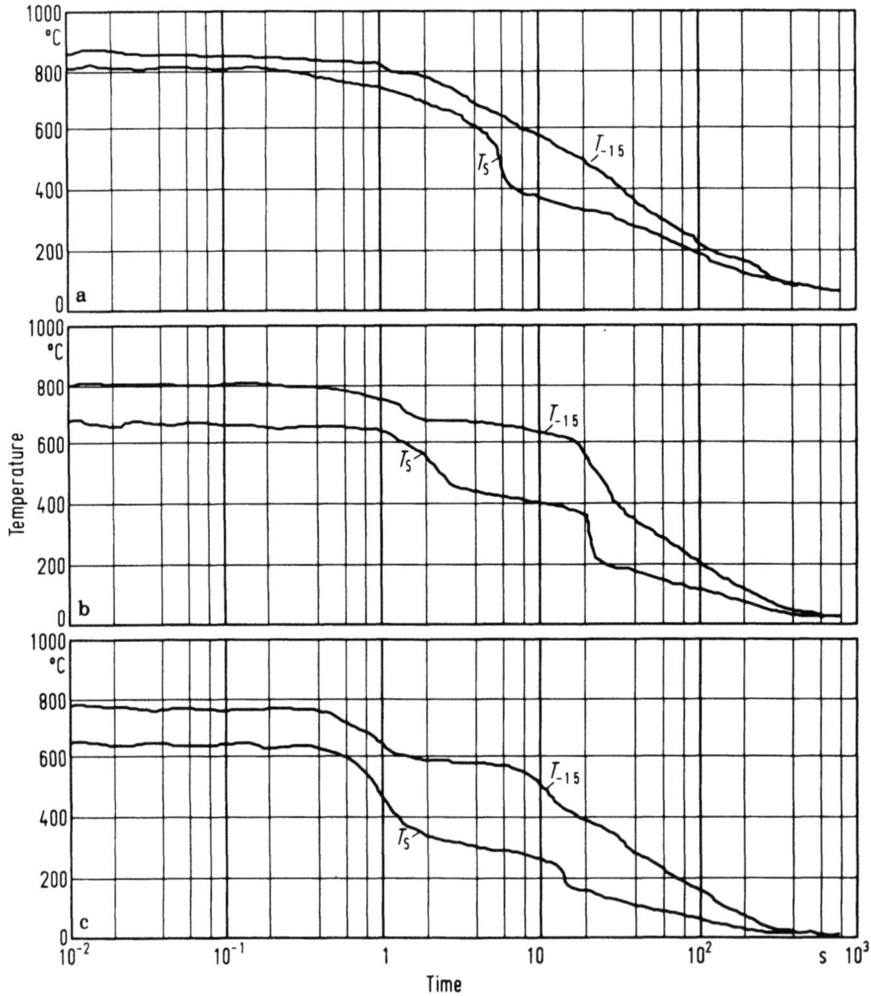

Fig. 8.4.5 a–c. Temperature vs. time (cooling) curves recorded by the Liščić-Nanmac probe during quenching from 850 °C in a mineral oil of 20 °C with following agitation rates **a.** 0 m/s; **b.** 1.0 m/s; **c.** 1.8 m/s

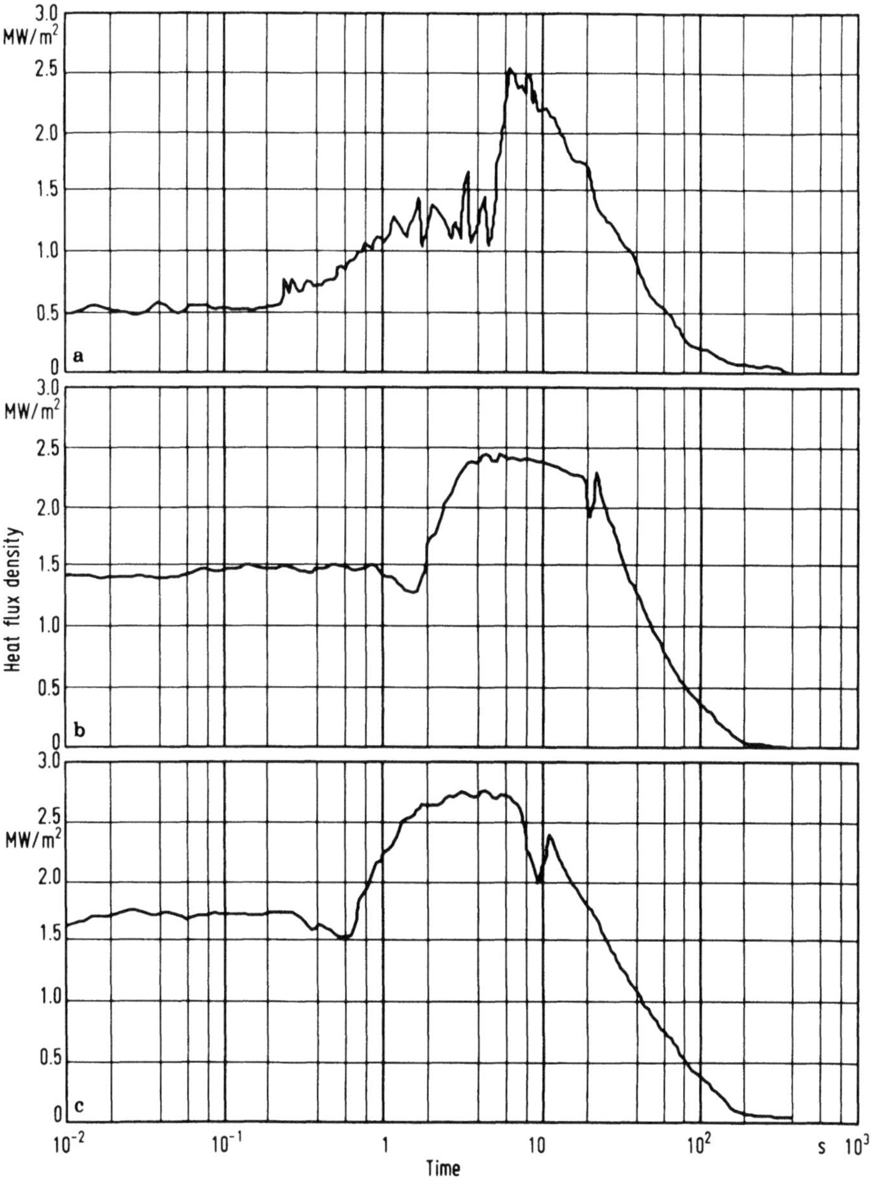

Fig. 8.4.6 a–c. Heat flux density vs. time curves obtained by computer calculation from measured cooling curves from Fig. 8.4.5, when quenching from 850 °C in a mineral oil of 20 °C with agitation rates. **a** 0 m/s; **b** 1.0 m/s; **c** 1.8 m/s

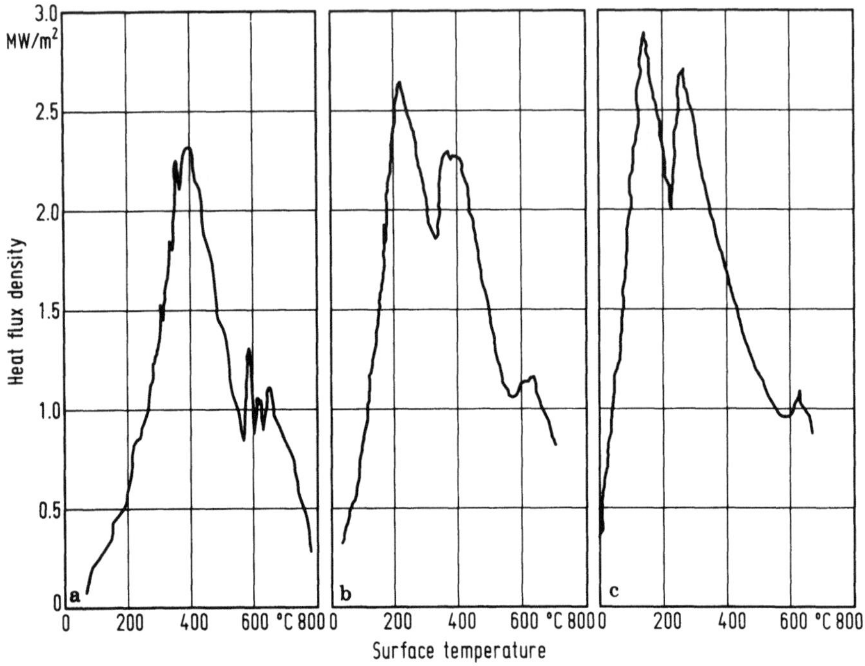

Fig. 8.4.7 a–c. Heat flux density vs. surface temperature curves obtained by computer-calculation from measured cooling curves on Fig. 8.4.5, when quenching from 850 °C in a mineral oil of 20 °C with agitation rates **a** 0 m/s; **b** 1.0 m/s; **c** 1.8 m/s

The new version of the probe has also a third thermocouple in the centre of the cross-section, and enables to record the temperature differences between the core and surface during the quenching process. Figure 8.4.4 shows a plot of these temperature differences when quenching the probe in still mineral oil of 20 °C. As it can be seen the maximum temperature difference in this case occurred in about 23rd second after immersion and amounted to 340 °C. High temperature differences (>200 °C) are present between ten and sixty seconds after immersion. Using the temperature v.s. time curves for the same mineral oil of 20 °C—Fig. 8.4.2 a—one can see that during this period the surface temperature (T_s) has dropped from 400 °C to 270 °C.

If a steel is quenched, the martensite start temperature (M_s) of which is below 270 °C, no superposition of *high* thermal stresses and those resulting from structural transformations will take place, but if the M_s temperature of the steel being quenched is higher than 270 °C, such a superposition of stresses is likely.

The described method is enough sensible to record changes in quenching parameters as e.g. the agitation rate of the quenchant.

Figure 8.4.5 shows temperature v.s. time (cooling) curves recorded during quenching from 850 °C in a mineral oil of 20 °C with different agitation rates.

Figure 8.4.6 shows corresponding heat flux density v.s. time curves, and Fig. 8.4.7 shows corresponding heat flux density v.s. surface temperature curves, for the same quenching conditions.

8.5 Definition and Evaluation of the Quenching Intensity

In order to define and evaluate the quenching intensity in the workshop practice from the cooling curves recorded by the probe and from the heat flux density v.s. time diagram—the following method of evaluation was adopted.

The quenching intensity (I) is evaluated by the four following criteria:

—The maximum heat flux density	q_{max}	[MW/m^2]
—The time (from the immersion) when the maximum heat flux density occurs	$t_{q_{max}}$	[s]
—Integral under the heat flux density curve from immersion until t_x (proportional to the quantity of heat extracted)—t_x depends on quenching conditions	$\int_{t_0}^{t_x} q\,dt$	[MJ/m^2]
—The time from immersion up to the moment when the core. temperature (T_c) drops to 300 °C (represents the rate of heat extraction from the core).	$t_{T_{c\,300}}$	[s]

The quenching intensity (I) is higher if the values q_{max} and $\int_{t_0}^{t_x} q dt$ are bigger, and the time intervals until $t_{q_{max}}$ and $t_{T_{c\,300}}$ are shorter. This can be proved by comparing the heat flux density v.s. time diagrams in Figs. 8.4.2. and 8.4.3. Water, namely, as it is well known has a higher quenching intensity than oil and will obviously produce deeper through-hardening.

Generally there is a good coorelation between the quenching intensity evaluated as described, and the resulting dept of hardening. Higher quenching intensity evaluated according to above described criteria produces a bigger depth of hardening.

The value $T_s\ q_{max}$ which can be taken from the heat flux density v.s. surface temperature diagrams (see Figs. 8.4.2c and 8.4.3c), gives the surface temperature of the probe in the moment when the maximum heat flux density occurs. This value is of importance when juding the quenchants in respect of possible stresses, deformation and cracks. The lower the value of $T_s\ q_{max}$ is, the greater is the risk of deformation and cracking.

As it is well known quenching in water brings always greater risk of cracking than quenching in oil.

8.6 Possibilities of Automatic Control of the Quenching Process

The aim of every optimalization and control of the quenching process is the following: to realize a relevant cooling curve in selected point of the cross-section which is adequate to produce the required hardness, but avoids to fast cooling in order to minimize the thermal stresses, especially, in the region of martensite formation.

The first step to realize this aim is the optimalization of quenching conditions giving the specification for all quenching parameters of the most suitable quenching intensity for the case in question. The second step is the automatic control of this quenching process which has to answer the following question: what can be done *during the quenching process* by controling the quenching intensity in order to minimize the thermal stresses. Therefore we have to know which influential factors we can change generally for a certain quenchant in order to change the course of the quenching intensity, and especially which of these factors can be changed *during the quenching process*. Figure 8.6.1 gives a list of this factors for different quenchants. From this list we can see that for

Kind of quenchant	Possibilities of changing the quenching intensity by changing of:	Influential factors which can be changed during quenching process
Water and unorganic water solutions	Concentration of the solution. Bath temperature Agitation rate	Agitation rate
Polymer-solutions in water	Concentration of the polymer Bath temperature Agitation rate	Agitation rate
Oil	Bath temperature Agitation rate	Agitation rate
Hot salt baths	Bath temperature Agitation rate Additions of some percentage of water	Agitation rate
Fluid bed	Bath temperature Flow rate of the gas Addition of liquid N_2 Addition of dispersed water	Flow rate of the gas Addition of liquid N_2 Addition of dispersed water
Gas	Pressure of the gas Flow rate of the gas	Pressure of the gas Flow rate of the gas
Spray-quenching with water or emulsion	Pressure of water Concentration of the emulsion	Pressure of water (emulsion)
Spray quenching with mixture of pressurized air and water	Pressure of water Pressure of air	Pressure of water Pressure of air

Fig. 8.6.1. Possibilities of changing the quenching intensity of different quenchants and influential factors which could be changed during the quenching process

immersion-quenching in fluids the only factor which can be changed during the quenching process is the agitation rate. The best possibilities for an automatic (and computerized) control of the quenching intensity exists with water-air spray quenching [21].

The automatic control of the quenching intensity in a quenching tank is possible only for workpieces having a cross-section thickness bigger than critical. As "critical cross-section thickness" in this case we consider the thickness, the core of which is cooled to about 400 °C in less time than it is required for changing the regime of agitation. This means that with thin parts, the automatic control of the quenching intensity by changing the regime of agitation is not possible.

The most important element of the control system is the instrumented probe, which has to be heated through to the same temperature as the treated parts and immersed simultaneously with them into the quenching tank. For the purpose of automatic control the probe should have thermocouples on the cross-section radially displaced in the following points: *S* (surface), 3/4 *R*; 1/2 *R*; 1/4 *R*, and *C* (center).

Additionally, continuous cooling transformation diagrams (CCT) for the relevant steel grades should also be stored in the own data bank. The main task of the probe itself during automatic control is to register the "turning-point of the automatic control" (TAC), and to change the regime of agitation.

If e.g. a certain hardness value after quenching is required at the point 1/2 R of a bar, the regime of agitation can be changed to a milder one at the moment when the relevant cooling curve for this point has dropped so close

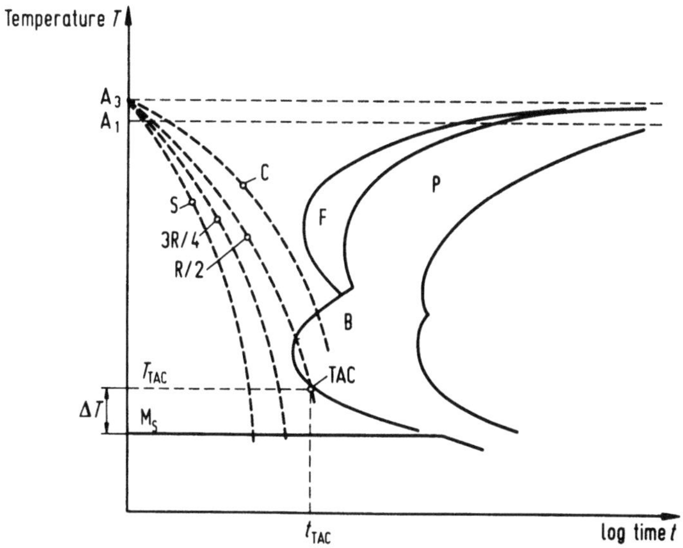

Fig. 8.6.2. "Turning point of the automatic control" (TAC) during quenching of a round bar if certain hardness is required at point 1/2 R on the cross-section

to the M_s temperature that the further cooling can not more influence the structure in this point. Then "the turning point of the automatic control" (TAC) is reached (see Fig. 8.6.2).

It is necessary to preprogramme the pobe by the following simulation:

—The adequate CCT diagram (which has to have the same coordinates as the diagrams of cooling curves), has to be called to the computer's screen and the cooling curve in the required point (stored previously in the memory for selected quenching conditions), has to be superimposed.
—The "turning point of the automatic control" (TAC) with its coordinates T_{TAC} and t_{TAC}, has to be fixed.
—The relevant thermoelement of the probe (in this case in point 1/2 R), has to be preprogrammed so that, when during cooling the temperature T_{TAC} is reached, the probe switches of the agitation automatically.

These procedure is valid, of course, only for bar diameter of 50 mm, because the probe is of this diameter.

To enable the use of the same procedure for automatic control of the quenching intensity, when quenching bars of other diameters, it is necessary to establish an auxiliary diagram shown in Fig. 8.6.3 *for every steel grade and every specified quenching conditions.* This diagram (which can be obtained by measuring experimental hardness values), shows the correlation between hardness in characteristic points of the cross section and bar diameter after quenching the relevant steel grade under specified quenching conditions. As it could be seen from Fig. 8.6.3 the same hardness e.g. will be found in point 3/4 R of a bar of 70 mm Dia as in the point 1/4 R of a bar of 50 mm Dia, if both are made of AISI 4140 and quenched under same quenching conditions.

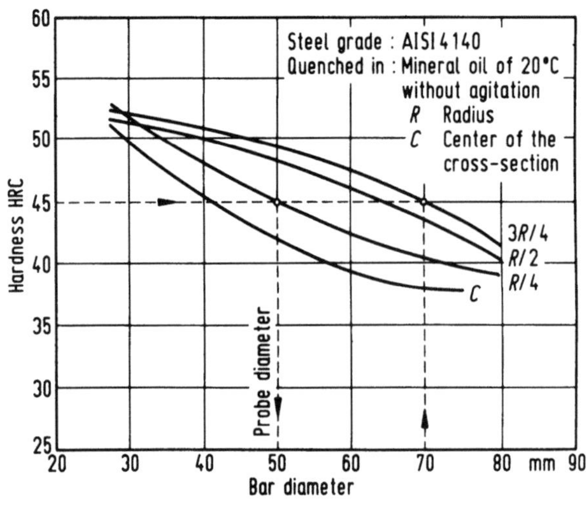

Fig. 8.6.3. Measured hardness values on round bars of different diameters made of AISI-4140, when quenching them in mineral oil of 20 °C, without agitation

References

1. Gröber H (1925) "Die Erwärmung und Abkühlung einfacher geometrischer Körper", Zeitschrift des Vereins Deutscher Ingenieure 69: 705–711
2. Russel TF (1936) Iron and Steel Institute—Spec. Report No. 14: 139–187
3. Grossman MA, Asimow M and Urban SF (1939) "Hardenability of Alloy Steels" - American Society for Metals, Cleveland, pp. 124–190
4. Metals Handbook (1964) 8th Edition vol. 2, American Society for Metals, Ohio, p. 18
5. Grossman MA, Asimow M (1940) "Hardenability and Quenching" The Iron Age, May 2
6. Lamont JL (1943) "How to Estimate Hardening Depth on Bars". Iron Age, October 14, vol. 152: 64–70
7. Boyer HE, Cary PR (1988) "Quenching and Control of Distortion", ASM International, Metals Park, Ohio
8. Carry PE (1964) Technical Report No. P-18-2-64, American Society for Metals, Metals Park, Ohio
9. Grossmann, Asimow and Urban (1939) "Hardenability, its Relation to Quenching, and Some Quantitative Data in Hardenability of Alloy Steel", American Society for Metals, Cleveland, pp. 124–126
10. Adalbert Rühenbeck (1955) "Das Abkühlvermögen der Härteöle und ihre Einordnung in die Reihe der gebrauchlichen Abkühlmittel", HTM, Band 10, Heft 2, S9–27
11. Monroe RW and Bates CE (1983) "Evaluating Quenchants and Facilities for Hardening Steel", Journal of Heat Treating, vol. 3, No. 2, Dec., pp. 83–99
12. Totten GE, Dakins ME and Jarvis LM (1989) "How *H*-factors can be used to characterize polymers", Heat Treating, Dec, pp. 28–29
13. Wünning (1981) "Berechnung und Steuerung des Temperaturverlaufs beim Abschrecken von Stahl in Wasser und Öl", HTM 36, 5: 231–241
14. Wünning J und Liedtke D (1983) "Versuche zum Ermitteln der Wärmestromdichte beim Abschrecken von Stahl in flüssigen Abschreckmitteln nach der QTA—Methode," HTM, Band 38, Heft 4: S149–196
15. Peter W (1952) "Über den Abkühlungsvorgang in flüssigen Härtemitteln und seine Beeinflussung durch die Oberflächenbeschaffenheit des Härtegutes, "HTM, Band 5: S65–96
16. Sayetat—Bournicon: "Les huiles de trempe" CETIM—France
17. Liščić B (1978) "The Temperature Gradient at the Surface as an Indicator of the Real Quenching Intensity during Hardening, HTM 33, No. 4: 179–191
18. Liščić B "Practical Measurement of the Quenching Intensity by the Temperature Gradient on the Surface of a Special Cylindrical Probe", Proceedings of the 18th International Conference on Heat Treatment of Materials, May 1980, Detroit MI, American Society for Metals, pp. 51–63
19. Liščić B and Filetin T (1988) "Computer—Aided Evaluation of Quenching Intensity and Prediction of Hardness Distribution", J. Heat Treating, vol. 5, No. 2: 115–124
20. Kobasko NI (1968) "Teplovie procesi pri zakalke stali" - Metallovedenie i termičeskaja obrabotka metallov, No. 3
21. Archambault P, Didier G, Moreaux F and Beck G (1984) "Computer Controlled Spray Quenching", Metal Progress, Oct, pp. 67–72
22. Thelning KE (1975) "Steel and Its Heat Treatment", Bofors Handbook, Butterworths, London and Boston

9 Types of Cooling Media and Their Properties

Wacław Luty

9.1 Required Properties

The primary requirement to be met by all quenchants is that they should extract heat from at least the surface layer, or in many cases from a considerable portion or even from the entire cross-section of the quenched workpiece, at a rate not less than the so-called critical cooling rate. If this requirement is fulfilled the overcooled austenite will not undergo a pearlitic transformation and thus the structure of the as-quenched workpiece will mostly be martensitic or bainitic. This requirement is necessary but not sufficient. Modern quenchants should in addition fulfill several other requirements such as:

—they should show a proper cooling kinetics within the entire range of temperatures applied, in particular their cooling rate at temperatures below M_s should be small enough to prevent inadmissible quenching deformations;
—they should be stable during their service life, in particular they should be resistant to thermal decomposition and oxidation;
—they should not interact with the surfaces of the quenched workpieces and of the quench tanks nor, when quenching is conducted in a protective atmosphere, with this atmosphere;
—they should conform to the standard requirements concerning toxicity, inflammability, and the protection of the natural and working environment;
—residues of the quenchants should easily be removable from the surfaces of quenched workpieces by washing, evaporating etc.;
—their cost per unit production should be low.

The thermal properties of quenchants and their mechanism and kinetics of heat transfer affect the surface hardness of the as-quenched workpiece, the distribution of this hardness over the workpiece cross-section, and the magnitudes and distributions of the thermal and structural stresses induced within it which in turn determine the magnitude and character of quench distortions. The requirement that within the martensitic transformation temperature range the cooling rate should be small follows from the fact that the small cooling rate produce a small temperature gradient so that the transformation proceeds almost simultaneously in the entire volume of the load, even when the quenched parts are considerably varied in size. This permits minimizing structural stresses.

The service life i.e. the time during which the quenching characteristics do not undergo larger adverse changes should be long so as to ensure the stability of quenching processes and the uniformity of hardening results. The quenchant consumption and thus the cost of the quenchant per unit production depends on the amount of the quenchant lost during a given quenching operation, and thus on its specified service life and on its losses by drug out and evaporation. These two latter effects depend on the quenchant viscosity and vapour pressure at the temperature at which the bath is operated. All these factors are involved in the production costs.

The ease of removing the quenchant residues from the surface of quenched parts is especially important when quenching is conducted in modern aggregates under a protective atmosphere, since then the quenched part surfaces are often not subjected to any mechanical finishing treatment after quenching and tempering.

9.2 General Classification and Comparison of Quenchants

A classification of quenchants may be based on their state of aggregation. Most quenchants are liquid media. Gaseous quenchants are less numerous. According to this criterion, fluidized media, composed of powdered solids fluidized by a stream of gases, and water-air spray or water mist quenchants may be classified as mixed media. The group of liquid quenchants is divided into three subgroups comprising water quenchants, oil quenchants and quenchants containing molten salts and metals.

Some authors classify water-based and oil quenchants depending on whether or not boiling occurs at quenching temperatures. Using this criterion all quenchants may be divided into two groups: evaporating and non-evaporating. The latter group comprises molten salts and metals, gases and fluidized media. The use of this criterion is fully justified since these two types of quenchants differ in their mechanism of heat transfer during quenching and thus in the relationships betwen the kinetics of quenching and the kinetics of phase transformations. This difference results from the fact that the boiling process, involving absorption of considerable amounts of heat, accelerates the cooling process in a certain temperature range which may overlap, at least partially, the temperature range in which overcooled austenite is unstable. Such an overlapping is desirable since it facilitates the quench kinetics to be matched with the kinetics of phase transformations.

Several specifically different characteristics of the individual groups of quenchants define their range of applications. Table 9.1 lists the most often used quenchants classified according to the above described criteria. The table also indicates the conditions of the contact between the individual quenchants and the quenched part surface, and the conditions of the quenchant movement, since these factors considerably affect the cooling rate.

Table 9.1. General classification of quenching media including possible modes of their circulation and of bringing them in contact with the quenched surface

Quenchant group/ state of aggregation	Sub-group	Quenchant type	Conditions of contact with quenched surface/natural convection				
			natural convection	laminar flow	stream of controlled direction	turbulent flow	spray
Liquid quenchants	Water based quenchants	Water	+	+	+	+	+
		Solutions of in-organic compounds	+	+	+	+	
		Solutions of organic compounds	+	+	+	+	+
		Water-oil emulsions			+	+	+
	Oil quenchants	Vegetable oils	+	+		+	
		Conventional minerals oils	+	+	+	+	
		High-performance mineral oil	+	+	+	+	
		Fast mineral oils	+	+	+	+	
		Hot mineral oils	+	+		+	
	Molten salts and metals	Saltpetre baths	+	+	+	+	
		Saltpetre-water baths	+	+	+	+	
		Salt-alkally baths	+	+		+	
		Molten metals					
Gaseous quenchants		Technical vacuum					
		Air	+	+	+	+	
		Inert-reducing gas mixtures	+				
		Protective atmospheres	+				
		High-pressure gases	+	+	+		
Solid quenchants		Plate, moulds, dies [possibly water-cooled]	+				
		Self-cooling by heat conduction[1]	+				
Mixed media		Cold fluidized beds				+	
		Hot fluidized beds				+	
		Water-fog					
		Water-air stream					

[1] This mode of cooling is employed in surface hardening of thin layers after austenitizing by a concentrated stream of energy (high frequency induction heating, plasma heating, electron beam and laser heating)

This chapter has chiefly been designed to describe individual quenchants in detail. However in order to emphasise important differences between quenchants as regards their most important technological property, namely, the quenching power, we shall start by comparing the quenching power of typical quenchants.

The quenching power may be described by means of various coefficients or diagrams (see Chap. 6) such as

—Grossmann's coefficient H,
—heat transfer coefficient α in $W/(Km^2)$,
—time taken for the quenched part to be cooled from 800 to 500 °C,
—the cooling curves representing the variation of temperature with time or with cooling rate.

Comparing the coefficients H and α and the time required for the quenched part to be cooled from 800 to 500 °C, we may quantitatively compare the average quenching powers of various quenchants, but most information about their thermokinetic properties can be obtained from the cooling curves. Table 9.2 gives values of H for typical quenchants and shows how strongly this coefficient is affected by a relative movement of the quenchant with respect to the quenched parts. In Table 9.2, the values of H are given according to Grossmann [1], who however presented them in terms of reciprocals of inch (which were sometimes treated as non-dimensional numbers) whereas those given in Table 9.2 have been converted to mm^{-1} since the contemporary literature uses the S.I. units. We can see that the coefficient H is not a very precise measure of the quenching power since it has almost the same value (about 1.0) for quiescent water and for intensively agitated oil, whereas in practice the quenching results obtained with these two madia may differ considerably (which can also be inferred from the cooling curves). Figure 9.1 shows the data of Table 9.2 in a more demonstrative graphical form.

Table 9.3 compares the values of the heat transfer coefficient α for various quenchants; the values are listed in increasing order.

Table 9.2. Severity of quench after Grossmann [1]

Agitation of fluid or piece	Oil		Water		Brine	
	in^{-1}	mm^{-1}	in^{-1}	mm^{-1}	in^{-1}	mm^{-1}
No circulation	0.25	0.01	0.9	0.035	2	0.079
	0.30	0.012	1.0	0.039		
Mild circulation (or agitation)	0.30	0.012	1.0	0.039	2	0.079
	0.35	0.014	1.1	0.043	2.2	0.086
Moderate circulation	0.35	0.014	1.2	0.047	—	—
	0.40	0.016	1.3	0.051		
Good circulation	0.40	0.016	1.4	0.055	—	—
	0.50	0.020	1.5	0.059		
Strong circulation	0.40	0.020	1.6	0.063	—	—
	0.80	0.030	2.0	0.079		
Violent circulation	0.80	0.030	4.0	0.15	5.0	0.20
	1.10	0.043				

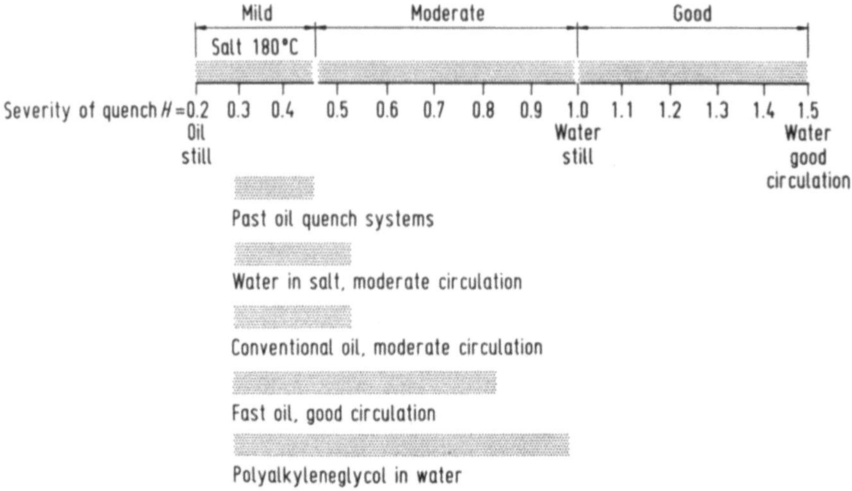

Fig. 9.1. Quenching severity coefficients H of typical quenchants [3]

Table 9.3. Heat transfer coefficients of various quenching media [2]

Quenchants and their circulation	Heat transfer coefficients, W/(K·m²)
Stream of nitrogen at 1000 hPa	100– 150
Strong circulation of nitrogen at 5000 hPa	300– 400
Salt bath (GS-430) at 550 °C	350– 450
Fluidized bed	400– 500
Salt bath (AS-140) at 180 °C	600– 800
Oil no circulation (20–80 °C)	1000–1500
Oil good circulation (20–80 °C)	1800–2200
Water (15–25 °C)	3000–3500

Cooling curves (shown in Fig. 9.2 for various quenching media) provide more information than the coefficients H and α. They permit us to estimate how the cooling rate varies with the temperature of the quenched surface within the entire range of quenching temperatures. This information is very important for predicting the performance of a quenchant. The higher the inclination angle of the cooling curve with respect to the time axis, the greater is the cooling rate within the given temperature range. It can clearly be seen from Fig. 9.2 that between the cooling curve for conventional oil and that for water a wide range of cooling rates exists which is only partially covered by the curve for a fast oil. The cooling rates falling within this range, and even exceeding it, can be obtained with water-based solutions of polymeric compounds (Fig. 9.3). It is

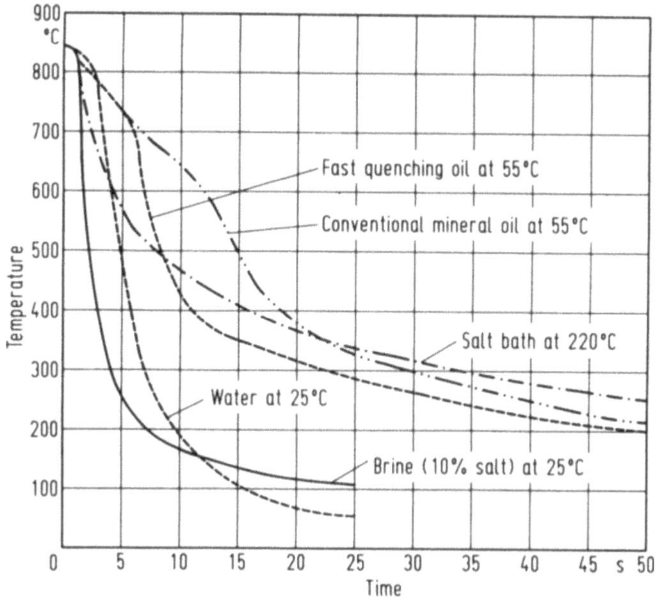

Fig. 9.2. Center cooling curves of an 18/8 steel specimen $\phi 12.7 \times 37.5$ mm, quenched in conventional cooling media

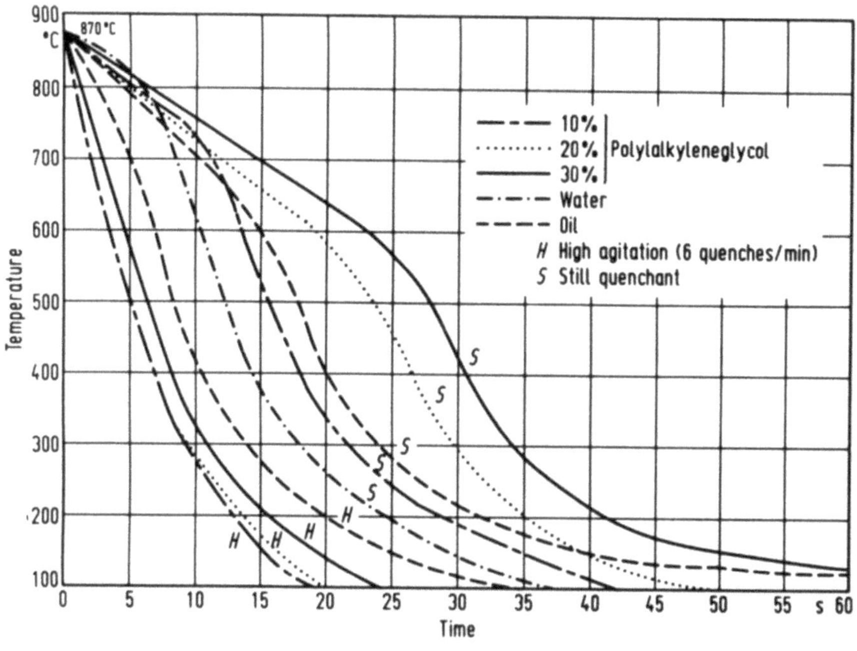

Fig. 9.3. Comparative cooling curves for quenchants operated within a temperature range of 38–49 °C, after [3]

for this reason that the use of these solutions has recently been growing. Their cooling rate may be adjusted within a very wide range, starting from rates smaller than those obtained with oils up to the rates greater than those of water, by controlling the concentration and the intensity of agitation. This possibility provides great money savings in industrial applications since it enables a proper selection of the grade of steel and the kind of the quenching medium.

9.3 Water as a Quenching Medium

9.3.1 General Characteristic

In the time when carbon steels were the only steels known, water was almost the only quenchant employed. Nowadays, plain water, usually tap water, is also successfully used for quenching carbon steel or low-alloyed steel of low hardenability, wherever the products, workpieces or tools made of such steel have simple shapes, are not liable to cracking, and the temperature and the circulation conditions of the available water conform to adopted specifications.

Before proceeding to describe the advantages and disadvantages of water when used as a quenching medium, we should consider its specific physical (Table 9.4) and chemical properties as compared with those of most other fluids. The most characteristic feature of water as regards its quenching properties is the exceptionally high quenching power (the heat transfer coefficient) due chiefly to the high specific heat of vaporization and the high specific heat capacity of water. The thermal conductivity of water, on the other hand, is very small in comparison with that of metals. The fact that the boiling temperature of water is much lower than that of oil quenchants is very unfavourable. All these thermal properties of water are jointly represented by the diagram of the heat transfer coefficient plotted as a function of the temperature of the quenched part surface. Figure 9.4 shows such diagrams for various surface conditions, as given in [4]

Table 9.4. Thermal properties of water

Property	Symbol	SI Units	Value
Specific latent heat of vaporization at 0 °C at 100 °C	Δh_v	kJ/kg	2495.4 2253.0
Specific heat capacity at 15 °C	C	kJ/(K·kg)	4.18
Thermal conductivity at 20 °C	λ	W/(K·m)	0.68
Medium heat transfer coefficient	α	W/(K·m²)	3000–3500

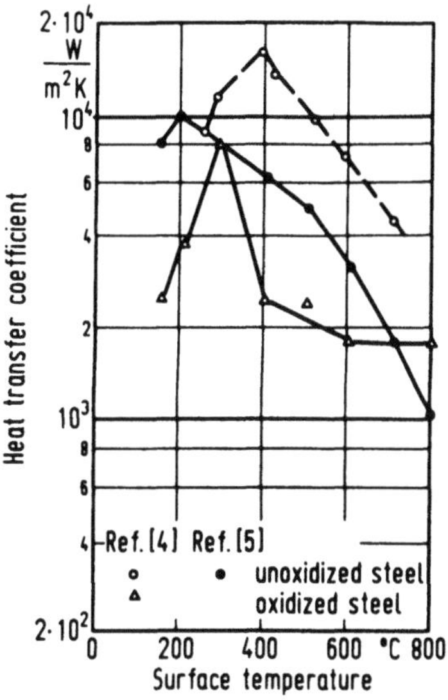

Fig. 9.4. Heat transfer coefficients for immersion cooling in water at 20 °C after [4] and [5]

and [5]. As can be seen from the figure, the maximum values of α occur at relatively low temperatures. Another important physical property of water is its ability to dissolve numerous inorganic compounds (salts, alcalis, acids), organic compounds and gases. As regards the chemical properties of water, it has the advantage of being stable in contact with the quenched parts heated up to the hardening temperatures, and the disadvantage in that it easily corrodes the surfaces of steel products.

Other advantages of water, important from the point of view of the production practice, are

—non-flammability,
—low cost,
—no hazards to health,
—easy scale removal, and
—no damage to the natural environment when drained

to the wastes.

Water used as a quenching medium has however several essential disadvantages. The most important is that the variation of its cooling rate as a function of temperature does not correspond with the kinetics of phase transformations. The range of the quenched surface temperatures (380–100 °C) at which water boils and thus the cooling rate is at a maximum coincides with the range of martensitic transformation temperatures for most grades of steel. This results in strong structural stresses and distortions which may causes cracking. The vapour blanket occurring at temperatures above the boiling point is very unstable and the vapour persists in holes, cavities and corners of the quenched parts even when the temperature of the bath decreases. In effect, the cooling rate at this stage oscillates, thereby producing thermal stresses, distortion and soft spots. Another disadvantage of water is that its quenching power strongly depends on the temperature, on the intensity of its agitation and on the impurities contained in it. How these factors affect the quenching power will be discussed below.

9.3.2 Effect of the Temperature of Quenching Water Upon its Quenching Power

Figure 9.5 shows the cooling rate curves for a silver ball of 20 mm diameter quenched in quiescent soft water of various temperatures (a), and the water flowing at a velocity of 0.25 m/s (b). In a quiescent water of temperature 20 °C, the phase of vapour blanket occurs at temperatures between 800 and 380 °C, and the cooling rate during this phase is relatively high (about 180 K/s), though much smaller than that in the boiling phase. This high cooling rate occurring during both the phases makes it difficult or even impossible to harden parts of varied cross-sections.

At temperatures close to the boiling point, an increase in the temperature of the quench water causes the vapour blanket temperature range to widen to about 200 °C. In the vapour blanket stage as well as in the boiling stage the cooling rate rapidly decreases with increasing temperature and its maximum shifts towards lower temperatures. The variation of the average relative cooling

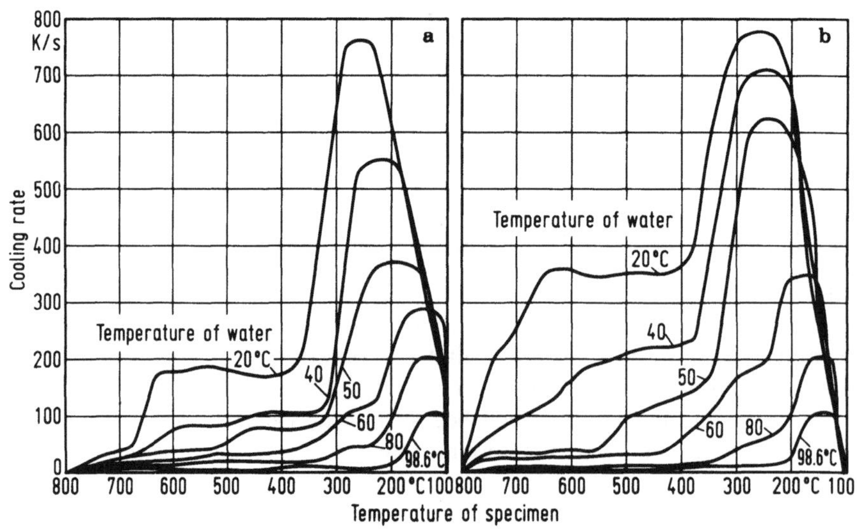

Fig. 9.5 a, b. Cooling rate of a silver ball $\phi 20$ mm. **a** in still water, **b** in a stream of water at a velocity of 0.25 m/s [6]

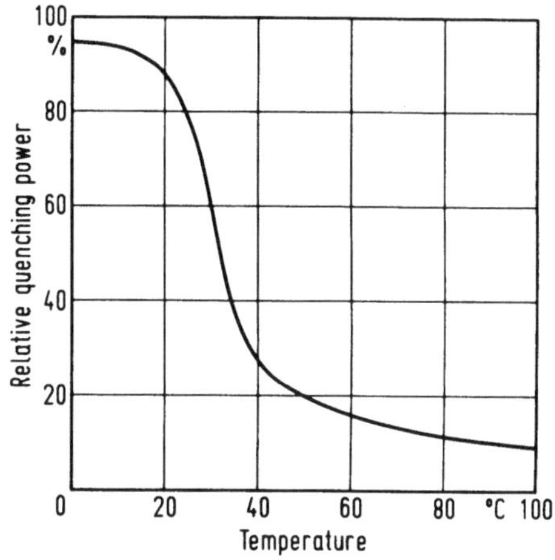

Fig. 9.6. Relative quenching power of water as a function of temperature [7]

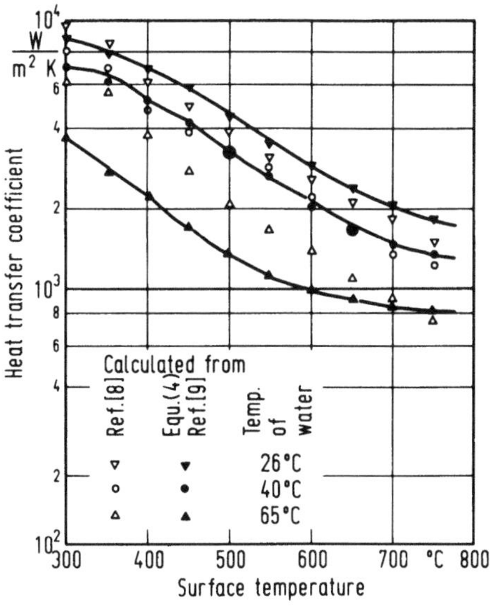

Fig. 9.7. Heat transfer coefficients during immersion in water at 26, 40 and 65 °C, calculated using Eq. [4] from [9] and the formula given in [8]

rate at the quenched workpiece surface as a function of water temperature is shown in Fig. 9.6, and the variation of the heat transfer coefficient in Fig. 9.7.

It follows from the above that, as the water temperature increases, the cooling rate and thermal stresses decrease but the structural stresses increase. This together with the instability, already mentioned, of the vapour blanket, lead to cracking, the effect confirmed by hardening practice. Moreover, when the quench water is circulating (Fig. 9.5b), an increase in temperature during the vapour blanket stage results in considerable differences, usually local, in the cooling rate, producing soft spots and distortion. This is why in practice the cooling

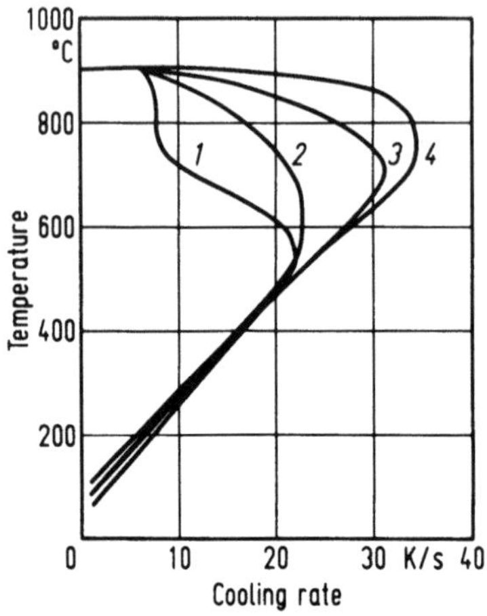

Fig. 9.8. Effect of water temperature on the cooling rate at flow velocity of 0.45 m/s: 1–65 °C, 2–53 °C, 3–41 °C, 4–19 °C [10]

rate of water should not be controlled by adjusting its temperature. In most practical cases the quench water has a temperature between 15 and 25 °C, although slightly lower or higher temperatures may occasionally appear to be favourable. If the quench water is not well agitated with respect to the quenched part, or the quenched part has a sophisticated shape and rough surface, the water temperature near this surface may locally increase even up to 100 °C.

The effect of the quench water temperature upon the cooling rate and upon the value of the coefficient of quench severity has also been studied by R.W. Monroe and C.E. Bates [10]. They examined this effect at the centre of 38 mm diameter bars made of austenitic steel. Their results (Fig. 9.8) not only confirm the observation described above that the cooling power of the quench water considerably decreases with increasing temperature, but also indicate that at temperatures exceeding 65 °C the presence of the vapour blanket affects the quenching process even at the bar centre.

9.3.3 Effect of Agitation Rate

Comparing the portions *a* and *b* of Fig. 9.5 we can see that if in the vapour blanket stage the quench water is agitated with respect to the quenched part surface, the cooling rate increases and, thus, the quenching process proceeds more uniformly. This effect may be attributed to the increased rate of vapour condensation, resulting in the vapour blanket becoming thinner so that more water can evaporate. However at temperatures used in practice the effect of this agitation upon the maximum cooling rate is insignificant. On the other hand, experience indicates that when the quench medium is agitated with respect to the quenched part, or vice versa, the uniformity of the hardnesses developed is improved (Table 9.5) and quench distortions and cracking are reduced.

Table 9.5. Effect of water velocity on the hardness after quenching of steel specimens ϕ 12 mm [6]

Water velocity m/s	Rockwell C hardness of steel grade specimens		
	SAE-1144	SAE-5140	SAE-W1-1, 2C Extra
0	55–60	52–56	48–65
0.5	57–60	52–57	60–64
1.0	59–62	58–60	63–66

The effect of the quenchant temperature and agitation upon the quenching power and the stability of the vapour blanket is stronger in water than in oil quenchants. This is the reason why water should not be used for hardening workpieces of sophisticated shapes or containing closed holes. The obstructions hinder the quenchant movement so that the temperature may locally increase and the vapour blanket stage may shift towards lower temperatures. Obviously, all these effects produce distortions, cracking, non-uniform hardness and soft spots. Similar effects may also occur with parts of relatively simple shape, if they are packed too tightly in the supporting basket or if the basket design is incorrect.

9.3.4 Effect of Water Contamination

If the quench water contains substances, even in small amounts, dissolved or forming emulsions in it, the cooling kinetics is considerably altered as compared with that observed in distilled or rain water. Since in many localities tap water, and also groundwater, contain small amounts of soluble salts, gases and solid contaminants, the quenching powers of waters available in different localities may differ from one another and also from the quenching power of distilled water. The most important effect of contaminants is that exerted upon the persistence of the vapour blanket. Depending on this effect, contaminants may be divided into two groups: increasing or decreasing the vapour blanket persistence.

The first group comprises solid, liquid and gaseous substances that are insoluble or poorly soluble in water. The solid (e.g. ceramic particles, soot) and liquid (fats, soaps, gelatines) substances forming suspensions or emulsions, which are prone to surface reactions, promote the formation of vapour and hinder its condensation. As a result, the duration of the vapour blanket stage increases and the vapour blanket vanishes at lower temperatures. Oils and soaps are the most active agents in this respect. At the same time however these agents increase non-uniformity of cooling and enhance the adverse effects associated with this non-uniformity, such as soft spots and distortion. Gases act in a similar way but their action is weaker. CO_2 evidently increases the duration of the vapour blanket stage and thereby decreases the cooling rate. This is why boiled water, and water that has long been used for quenching, provides more reproducible hardening results (more uniform hardness) than those obtained with fresh water. For the same reason compressed air is not recommended for agitating the quench water, especially in small tanks. Fresh tap water should not be used for quenching. In order to maintain the quench water at a constant temperature, it should be circulated in a closed system.

Salts, acids and alkali dissolved in water increase its cooling power by reducing the durability of the vapour blanket; when the concentration of these compounds is sufficiently high, the vapour blanket stage does not occur at all. How the electrolyte-forming non-organic compounds affect the kinetics of quenching will be explained in Sect. 9.4.

9.4 Water Solutions of Non-Organic Salts and Alkali

The fact that even a small amount of salt, alkali or acid dissolved in water markedly changes its thermokinetic properties has been utilised for reducing or eliminating several disadvantages of pure water by adding some salt or alkali to it. Such solutions have been called 'brines'.

The most widely used brine is one containing an admixture of sodium chloride (kitchen salt) added alone or in conjunction with other salts. Figure 9.9 shows how the concentration of sodium chloride in a water solution affects the cooling rate-temperature curve of this solution. We can see from the figure that as the concentration of sodium chloride increases to 10%, the width of the vapour blanket temperature range decreases (this effect begins at a concentration of 1%) until it is reduced to zero, whereas the cooling rate rapidly increases

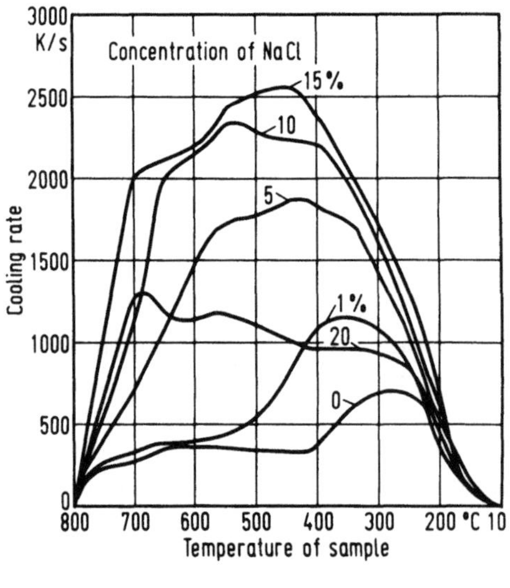

Fig. 9.9. Effect of NaCl concentration in a water solution on the cooling rate of a silver ball ϕ20 mm—as a function of temperature [6]

within the entire range of quenching temperatures. As the concentration of sodium chloride further increases, the cooling rate decreases, but its variation with the quenched surface temperature is relatively uniform.

The influence of other salts and alkali is similar, whereas the action of gases is weaker. This can be seen from data given in Table 9.6. Sodium alkali solutions, even of small 5–20% concentrations, provide the greatest cooling rates, but sodium chloride is cheaper and less hazardous in use. Calcium chloride solutions in the range from 5 to 20% provide practically the same cooling rates as sodium chloride solutions; sodium carbonate solutions are slightly slower.

The major advantages of brines in comparison with water are:

- —within the entire quenching temperature range, the temperature of the quenching solution affects the cooling rate less than it does in water (Fig. 9.10) and thus the quenching operation may be conducted at a higher temperature, e.g. at 50 °C or, with steels of greater hardenability, even at 70 °C;
- —soft spots, distortion and cracking occur more rarely;
- —agitation of the quenchant affects less the quenching power;
- —thanks to the reduction or elimination of the vapour blanket stage, quenching proceeds more uniformly even when the load comprises fine and tightly packed workpieces, thereby improving the uniformity of as-quenched hardness.

This latter effect can be explained by the fact that, when the small saturated salt crystals deposited on the metal surface are heated up by the metal, water escapes from them by microexplosions thereby breaking the vapour blanket and facilitating the nucleation of vapour bubbles. As a result, the duration of the vapour blanket stage is shortened or sometimes this stage does not occur at all and is replaced by the boiling stage. Moreover, this effect prevents the vapour bubbles from being trapped in the concavities present on the surfaces of

Table 9.6. Cooling rate in the range 880–730 °C of θ 12.7 mm diameter probe 0.95% C steel, with T.C. at center, in water solutions [7]

Water solution	Concentration, %	Cooling rate, K/s	Water solution	Concentration, %	Cooling rate, K/s
Water	0	102–120	NaOH	2.5	195
NaCl	5	170		5.0	202
	10	195		11.5	202
$CaCl_2$	5	170		16.5	207
	10	193	HCl	5.0	153
	20	170		20.0	100
Na_2CO_3	10	170	H_2SO_4	5–20	143–150

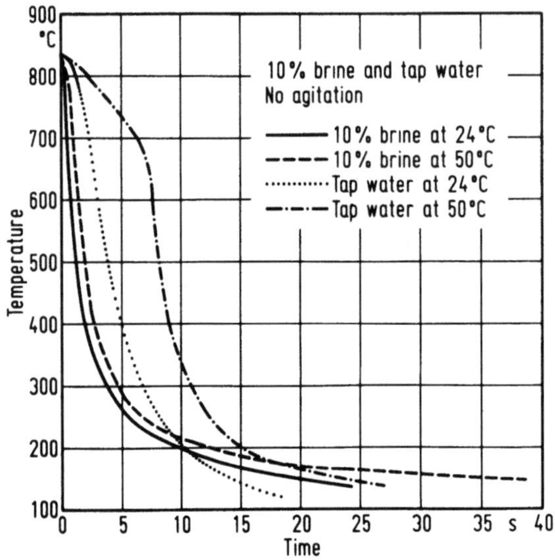

Fig. 9.10. Center cooling curves showing the effect of temperature on the quenching power of tap water and a 10% NaCl solution when quenching 18/8 stainless steel specimens $\phi 12.5 \times 63.5$ mm [7]

quenched parts or in the centre of the load if this is composed of small, closely packed workpieces.

Brines have however several disadvantages such as

—the quench tank and the whole equipment (pipes, pumps) must be protected from corrosion or made of stainless steel (this is not required when using brines containing corrosion inhibitors);
—a hood is usually needed over the quench tank to protect the operating personnel and the equipment from the harmful attack of the brine vapours;
—as compared with water, the material and labour costs involved are higher (the preparation of the solution and systematic control of its concentration require additional labour);
—with certain solutions, such as e.g. sodium hydroxide, nitrate salts, or when cyanide is added to the solution, the toxication and fire hazards incrase, and the wastes must be neutralised.

Apart from the solutions listed in Table 9.6, other solutions containing mixtures of salts and various additives, such as corrosion inhibitors, are also used. However sodium chloride solutions of concentration from 2 to 20% by weight (the concentration being most conveniently controlled by measuring the density of the solution) are in most common used. In the USSR, an aqueous solution containing 5–7% of $KMnO_4$ is also recommended. Its quenching power lies between that of water and that of oil, and it has a particular advantage of being relatively slow within the range of low quenching temperatures. The Chinese use aqueous solutions of nonorganic salts, such as the oversaturated nitrate salts solution (25% $NaNO_3$ + 20% $NaNO_2$ + 20% KNO_3 + 35% H_2O),

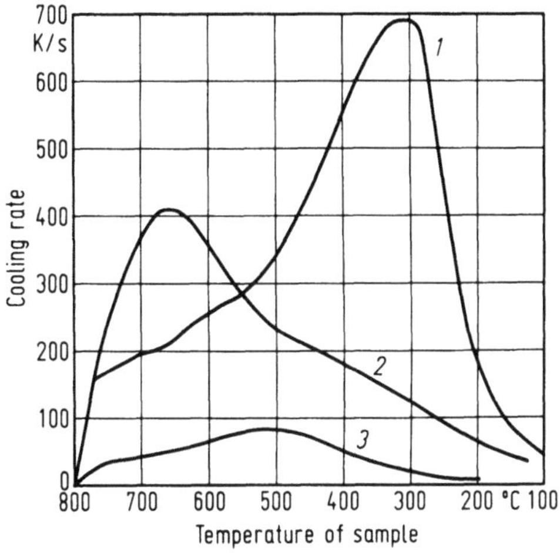

Fig. 9.11. Cooling curves obtained in quenching a silver ball ϕ20 mm in water (curve 1), oil (curve 2) and in supersaturated saltpetre salt solution (curve 3) [11]

the $CaCl_2$ solution of density ranging from 1.38 to 1.46 (38–60% $CaCl_2$) and ca. 30% solution of a mixture composed of 49% $ZnCl_2$ + 49% NaOH + 2% soap powder. Figure 9.11 shows the cooling curves for the nitrate salt solution in comparison with those for water and oil. This solution, and also other solutions containing oxidizing compounds, must not of course be used for quenching parts that have previously been heated in salt baths with an admixture of cyanides, since this would threaten with explosion. The calcium chloride solution of composition as given above is recommended for quenching workpieces made of structural or tool carbon steels and having varied cross-sections and sophisticated shapes. Such workpieces may also be quenched in a solution containing zinc chloride, sodium alkali and soap powder (the composition as given above), which is a milk-like emulsion. This latter quenchant provides high hardnesses and little distortion; the surface of the quenched workpiece is clean and silver-gray in colour.

In countries where metallurgical industry is highly developed, various mixtures of salts containing inhibitors of various trade names are commercially available; their chemical compositions are selected to fit the applications for which they have been designed.

9.5 Water–Oil Emulsions

Several tens years ago, in search for filling the wide gap between the cooling powers of water and oil, a conception emerged of using an emulsion of oil in water to obtain a quenchant of intermediate quenching power. In order to prepare such an emulsion, an appropriate type of oil is of course required and also some small amount of an additive that will reduce the surface tension and thereby stabilize the emulsion, i.e. prevent the two mutually insoluble liquids from separating. However several studies on the thermo-kinetic properties of water–oil emulsions have indicated that their use is limited.

Figure 9.12 shows how the cooling rate depends upon the oil content of these emulsions, according to A. Rose [12]. As can be seen, in emulsions containing more than 10% oil, the vapour blanket stage, during which the cooling rate is much slower than that in water during this stage, persists until very low temperatures are reached. Moreover, the immersion hardening practice has shown that the vapour blanket is very unstable and thus generates the adverse effects discussed before (soft spots, distortion, cracking). At moderate and low temperatures, the cooling rate of these emulsions is similar to that of water. As the temperature of an emulsion increases, the cooling rate, and which is even worse, the uniformity of cooling, decrease in a greater degree than they do in water baths (Fig. 9.13).

It follows from the above that water–oil emulsions cannot be used for immersion hardening, except in special cases. They are occasionally used for cooling steel workpieces, susceptible to temper brittleness, after these workpieces had been tempered at a high temperature. In the past, before polymer quenchants were developed, water–oil emulsions were quite widely used for spray quenching in induction or flame surface hardening operations, since with this technique the vapour blanket is mechanically broken by the compressed quenchant streams. Brines are not suitable for these hardening techniques since the crystallizing salt chocks up the sprayer orifices.

A comparison of thermokinetic properties of various water-based quenchants, including water–oil emulsions, when used for immersion and spray

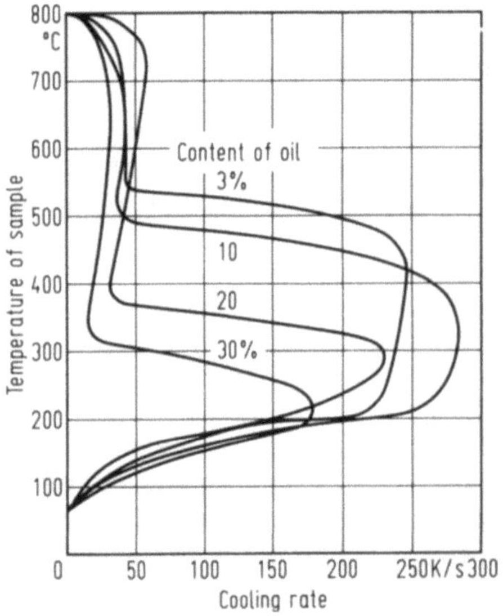

Fig. 9.12. Cooling curves obtained in quenching a silver ball $\phi 20\,mm$ in water-oil emulsions containing 3–30% soluble oil [12]

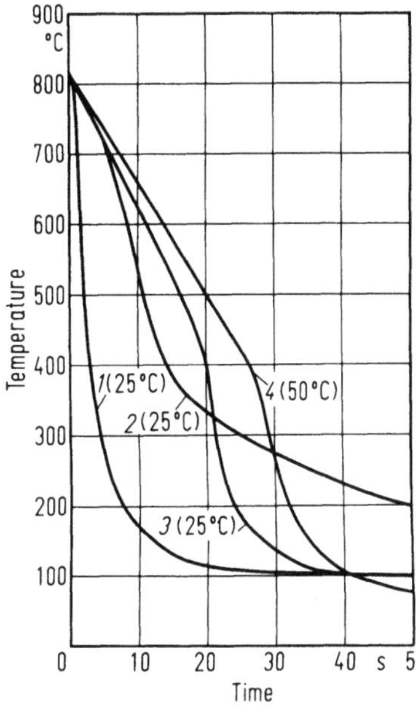

Fig. 9.13. Center cooling curves for a 18/8 stainless steel specimen ϕ12.7 × 38 mm; quenched without agitation in: *1* tap water at 25 °C; *2* conventional oil at 25 °C; *3* 10% soluble oil +90% water at 25 °C; *4* 10% soluble oil +90% water at 50 °C [7]

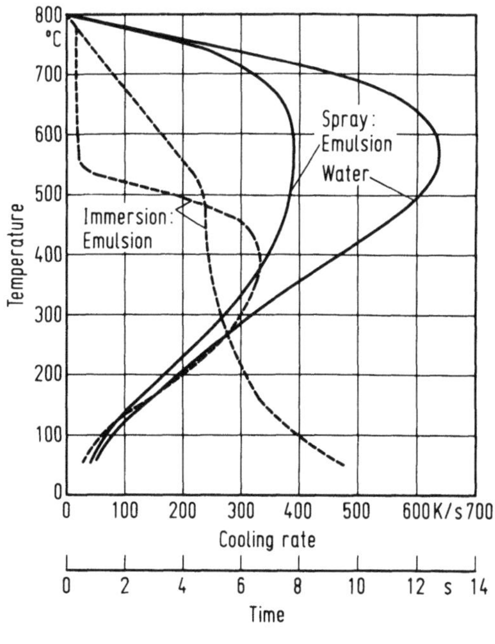

Fig. 9.14. Comparison of cooling rate curves for immersion and spray quenching of an Inconel 600 alloy specimen ϕ8 × 24 mm in water and in 5% emulsion of a soluble oil [13]

quenching is given in [13]. As can be seen from Fig. 9.14, in spray quenching, water–oil emulsions are much slower than water, but in immersion quenching their behaviour is similar to that of water, since boiling, and thus a high quenching rate, unfavourable occur within the range of martensitic transformation temperatures. For this reason, and also because the oils used for preparing emulsions easily decompose at higher temperatures, water–oil emulsions have recently largely been replaced by synthetic quenchants. Moreover due to the degradation of these oils under thermal shocks, deposits form locally on the quenched part surface, leading to nonuniform quenching.

9.6 Aqueous Polymer Solutions

Since several decades, water and mineral oils have been the most widely used quenchants. These two types of quenchants differ in their quenching characteristics so widely that they meet modern quenching requirements only in certain specific applications. As described in Sect. 9.3 and 9.4, the range of application of brines is also very limited. The quenching characteristics of oils, which are still predominantly used for quenching, will be discussed in Sect. 9.6. Here we shall only mention their major disadvantages, justifying the need for seeking new quenchants, other than water and oil. These disadvantages are:

—the cooling rate in the range of pearlitic transformation temperatures is insufficiently high when quenching greater parts made of certain grades of steel such as carbon and low alloy steels:
—oils are flammable and evoke fumes and vapours harmfull to the operating personnel and to the natural environment and thus require a provision of safety measures which are however often ineffective;
—quenched parts must be washed in alkali solutions, which raise problems in the waste management.

Because of the disadvantages of water and oil quenchants and the limitations on their use, new quenchants have long been sought to meet the following requirements:

—the cooling rate within the range of pearlitic transformation temperatures should be greater than that of mineral oils;
—the cooling rate in the range of martensitic transformation temperatures should be similar to that of oils;
—the quenchants should be non-flammable and harmless to the operating personnel and the natural environment.

Numerous studies carried out in many countries during the last decades have shown that the above requirements may be fulfilled by aqueous solutions of certain polymer compounds. Thus, a new group, comprising numerous quenchants of widely varied properties, have came into use. Below, we shall discuss the properties of these quenchants in detail.

9.6.1 General Characteristic

Polymer quenchants were first used in the USA and France in the 1950s. They were based on polyvinyl alcohol which was the major constituent stimulating the quenching power, and contained defoamers, bactericidals and corrosion inhibitors.

In the next years, several other quenchants containing high-molecular polymer compounds were developed. K.H. Kopietz [14] classified all the polymer quenchants commercially available in the USA in 1978 into 5 types (Table 9.7). All these quenchants have been designed either to replace quench oils in some applications and thus to avoid their disadvantages, or to fill the gap between the cooling rates of water and oil so that steels of lower hardenability could be quenched (which is important from the economic point of view).

Polymers dissolved in water essentially alter the mechanism and the kinetics of heat transfer during quenching, although in principle the three known stages of this transfer (the vapour blanket, the boiling and the convection stages) still occur. During the first stage, the mechanism of the quenching action is the same in all kinds of polymer quenchants. This mechanism is determined by the increased viscosity of polymer solutions in comparison with that of water and by the polymer-enriched film formed at the cooled surface, which isolates and stabilizes the vapour blanket. As a result, the vapour blanket stage is prolonged depending upon the kind of the polymer employed, upon its concentration, and upon the temperature and the degree of agitation of the quenchant.

The increased density of the solution near the quenched surface obviously results from water evaporating from this surface when wetted. This increased density in turn promotes polymerization processes and causes a thin film to form at the quenched surface; if PVA, PVP or ACR quenchants are used, the film is composed of a gel or, if the quenching is conducted in a PAG quenchant, it is composed of the polymer since at temperatures exceeding the critical temperature (usually 74 °C) polyalkylene glycols do not dissolve in water.

In polymer solutions, when at a certain temperature the vapour blanket is broken, the boiling stage begins just as in other quenchants. But as the viscosity of the solution increases and the rate of its motion decreases, the intensity of the boiling process decreases. When the temperature of the quenched surface decreases below the boiling temperature of the quenchant, the heat transfer chiefly proceeds by convection. During this stage, the cooling rate also depends on the physical properties of the solution, especially on its viscosity.

Quenchants such as the PVA or the PVA + GLY (commercially known as e.g. Aquaplast and Aquaquench *A*) are used for both surface spray hardening and immersion volume hardening. Even when the polymer concentrations are small, ranging between 2 to 4%, and thus low costs are involved, the cooling rate of these quenchants within the temperature range from 800 to 300 °C is slower than that of water and markedly faster than that of oils. Unfortunately, the cooling rate continues to be greater than that of oils at temperatures below 300 °C, i.e., in the range of martensitic transformation. This fundamental

Table 9.7. Polymer quenchants available on the US market (1978) after [14]

Polymer type	Polyvinyl alcohol	Polyvinyl alcohol + poly oxyalkylene glycols	Poly-oxyalkylene glycols	Polyvinyl pyrrolidone	Alkali polyacrylate
Abbreviation	PVA	PVA&GLY(PAG)	GLY(PAG) (Range of some different products)	PVP	ACR
Polymer concentration (weight)	16%	12%	40%–60%	10%	10%
Solubility in water	Normal	Normal	Inverse	Normal	Normal
Dry residue	Brittle solids	Elastic solids	Viscous fluid	Brittle solids	Brittle solids
Viscosity of neat concentrate, mm^2/s (cSt) at 37.8 °C	10,000	890	300–550	10.6	555
Viscosity of aqueous solutions, mm^2/s (cSt) at 37.8 °C 30%	25.5	5.2	8.7–11.5	2.6	51
20%	8.4	2.7	4.4–5.3	1.7	31
10%	2.4	1.4	1.9–2.2	1.1	16
5%	1.3	1.0	1.1–1.3	0.9	9

disadvantage is exhibited not only by PVA quenchants but also, to a greater or smaller extent, by all water-based quenchants, since their boiling temperature is considerably lower than that of oils. This is why polymer quenchants have until recently been considered inferior to oils as regards their effect upon distortions and cracking, and thus unable to replace the latter. It was only in 1983 that quenchants of the ACR type, whose cooling curves in the lower temperature range are similar to those of oils, have been developed.

The PVA type quenchants have however other essential disadvantages and for this reason their use has recently been limited; in some countries their production has even been stopped. In addition to the high cooling rate at low temperatures, already mentioned, the other disadvantages are:

—a film, difficult to remove, of rubber-like compounds is formed on the surfaces of the quenched workpieces, pipes and quench tanks;
—as a result of the thermal decomposition of these quenchants, acetic acid is generated which because of its corrosive action [15] and uncontrolled effect upon the cooling power must be neutralized.

Quenchants of the GLY type, briefly called PAG (commercially known as e.g., Aquaquench 251 and 1250, Osmanil E2, Ucon) have been in most common use since twenty years. The most characteristic feature, distinguishing polyglycols from among all other polymers, is what is called the inverse solubility in water. This term denotes the polymers' property of being insoluble in water at temperatures exceeding 63–85 °C, due to which during the initial phase of cooling the surface of quenched workpieces is covered with a polymer film that restrains heat transfer.

The PVP type quenchants, whose main constituent is polyvinyl pyrrolidone, have been developed in the USA (examples of commercially available quenchants of this type are Parquench 60 and 90). In comparison with the PAG quenchants, their major advantage is that their cooling rates in the upper temperature range are faster while, in the lower temperature range, slower. This is shown in Table 9.8, reprinted from R.W. Foreman [16]. We can see that the

Table 9.8. Quench Data for polyglycol [PAG], polivinyl pyrrolidone [PVP], and fast quench oil (25.4 mm diameter probe, either 304 or 4140 steel, with T.C. at center. Austenitizing temp. was 870–900 °C) [16]

Type and concentration of quenchant	Bath temp. °C	Velocity m/s	Cooling rate at 700 °C K/s	Cooling rate at 200 °C, K/s
Fast oil	60	0.5	46.1	5.5
16% PVP [Parquench]	52	0	32.8	5.5
		0.75	57.3	12.5
25% PVP [Parquench]	55	0.75	47.3	12.5
16% Polyglycol	52	0	16.1	10.6
	52	1.08	34.8	15.1

thermokinetic properties of PVP quenchants are similar to those of fast oils. In its Technical Bulletin the Parquench's manufacturer (Park Chemical Co, Detroit) reports on tests in which 50 mm diameter shafts made of 0.5% steel when quenched in water developed hardnesses of 64–65 HRC but were prone to cracking, whereas when quenched in a 10% Parquench they acquire the same hardness without cracking. According to data given by the manufacturer, the 15–20% Parquench may be considered to be fully interchangeable with oil.

In 1978, another polymer quenchant was developed in the USA. Its main active constituent is sodium polyacrylan (In Table 9.7, designated as the ACR type). The properties of this quenchant essentially differ from those of all the polymer quenchants developed earlier, and this fact determines its applications. As seen in Table 9.7, the viscosity of the ACR type quenchants is much greater than that of all other water-based quenchants, and at concentrations in excess of 10%, even greater than that of mineral oils. Thanks to this increased viscosity and to other physical and chemical properties of the ACR polymers, the vapour blanket stage is considerably stabilized. At concentrations of about 20% or more, this stage is prolonged to cover the entire quenching temperature range.

Figure 9.15 shows the cooling curves for the 2.5–20% ACR solutions, as compared with the cooling curves for water and a fast oil. As can be seen, the ACR solutions of concentration exceeding 5% cool much more slowly than the oil until a temperature of about 200 °C is reached, whereas at temperatures below 200 °C, the oil is markedly slower. Up to about 200 °C, the cooling rate remains almost constant, except at very small concentrations (below 5%) when it

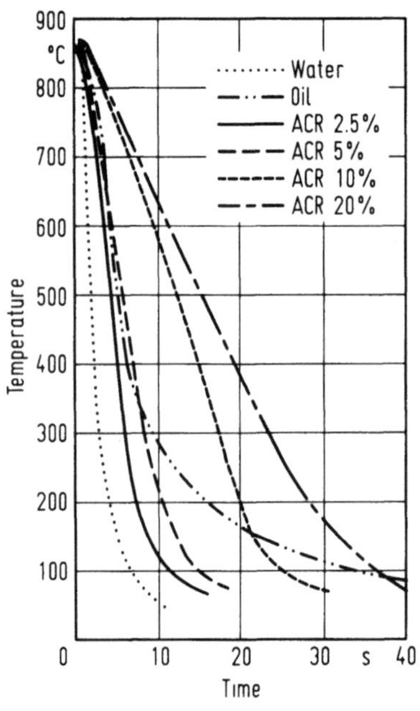

Fig. 9.15. Quenching characteristics of ACR solutions for various concentrations at 26.7 °C. Test specimen 10 mm diam. × 60 mm, austenitic steel, agitation 0.2 m/s [14]

varies in the same way as the cooling rate of other water based quenchants. A quenchant of this type is manufactured in the USA and other West-European countries under the commercial name of Aquaquench 110.

From the cooling curves for ACR solutions, shown in Fig. 9.15, we can infer the range of their application. They are primarily used when fine pearlite structures, in some cases including a small amount of bainite, are required. Therefore, they are chiefly employed in non-martensitic quenching processes, although in high-hardenability steel, quenchants of this type with concentrations of 5% or less can also produce martensitic structures. However, in meeting these requirements, the ACR quenchants do not prove to be obviously superior over other polymer quenchants and, because of their small cooling rate within the higher temperature range, they may even be considered less suitable for such quenching operations.

Numerous practical industrial applications of the ACR quenchants and the benefits resulting from their use are described in [14]. Here we only mention some of these applications, such as e.g. the quenching of hot automobile forgings directly from the forging temperature, the quenching of spring plates, torsion rolls and gears of certain types, the inter-operational quenching of deeply carburised parts (e.g., parts of large-size bearings), patenting of wire, etc.

A rough cmparison of the quenching powers of the 20% polymer quenchants described above with the cooling powers of water and oil is given in Fig. 9.16. It can be seen, though when the curves are plotted in the temperature-time coordinates this effect is not clearly marked, that in the higher temperature

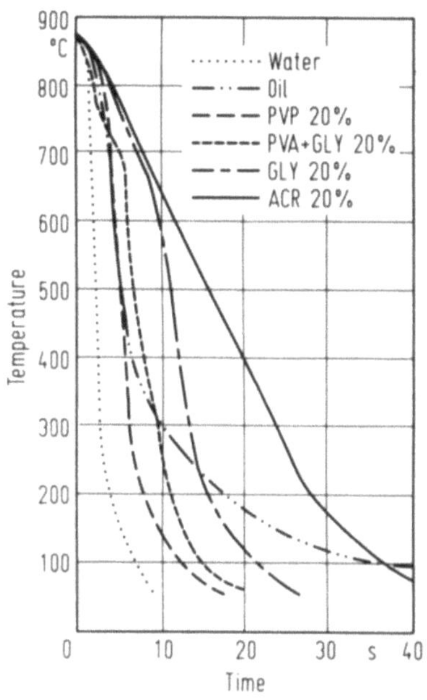

Fig. 9.16. Quenching characteristics of different polymer solutions of 20% concentration, at 26.7 °C [14]

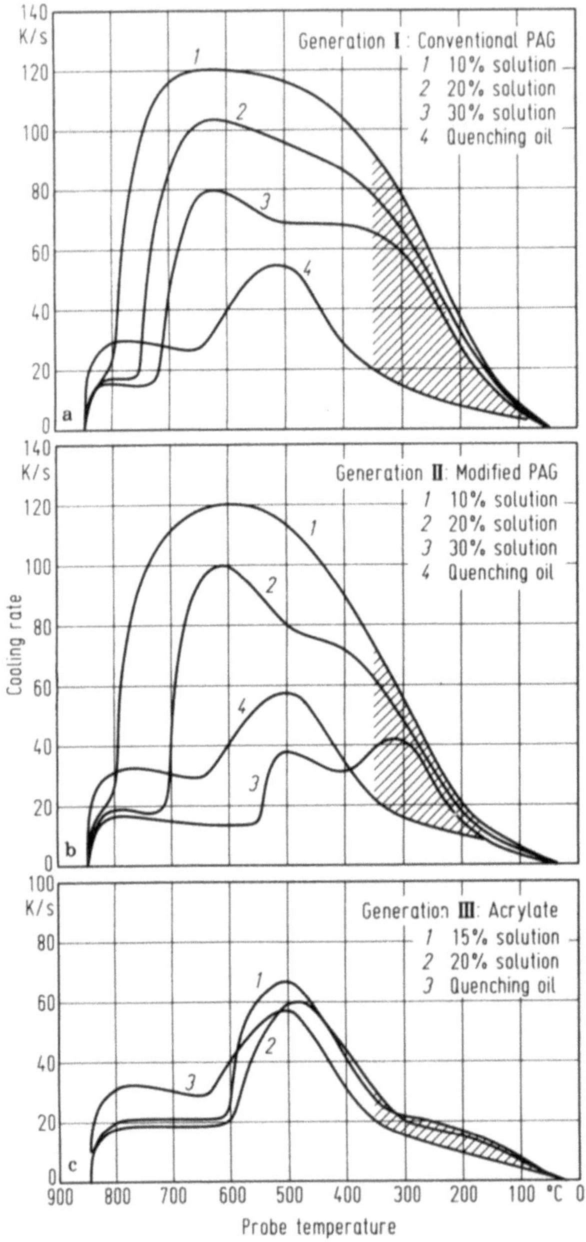

Fig. 9.17. Comparative cooling curves for centre of an Inconel 600 alloy specimen $\phi 12.5 \times 60$ mm for three generations of polymer quenchants and a normal-speed oil. Tested under agitated conditions at 40 °C [17]

range the cooling curves of the PVP, and to some extent, of the PVA + GLY quenchants resemble those of the oil more than those of the other quenchants examined. The 20% ACR quenchant, on the other hand, provides an almost constant cooling rate throughout the entire quenching temperature range, markedly slower than that of the other quenchants. Therefore, when treated with this quenchant, most steel grades will undergo a pearlitic or bainitic transformation.

R.T. Bergen [17] has classified the contemporary polymer quenchants, available on the European market, and described their development differently to Kopietz. Omitting the outdated PVA quenchants and the PVP quenchants as little used in Europe, he divides the remaining quenchants into three generations: the first based on polyalkylene glycols (PAG), the second also based on polyglycols but with modified molecular structure and viscosity (PAG mod.) and the third based on polyacrylates (ACR). Figure 9.17 compares the thermokinetic characteristics of these three types of quenchants. The first and the third generations differ most in their martensitic transformation temperature ranges (represented by the cross-hatched area positioned below 350 °C in the figure). The ACR quenchants provide a further considerable improvement in this regard. In view of the common belief, prevailing until recently, that no quenchant can compare with oils as regards their technological advantages, the development of ACR quenchants has been a considerable achievement. It should however be noted that within the higher temperature range the ACR quenchants do not show the desired intermediate cooling power that would fall half-way between the cooling power of water and that of oil.

Bergen reports on his experiments in which using a specimen 12.5 mm in diameter and 60 mm long at a temperature of 300 °C he obtained the cooling rate of 60–80 K/s with the first generation of quenchants, 40–50 K/s with the second generation, 15–25 K/s with the third, and 10–20 K/s with a conventional oil.

A classification of modern polymer quenchants into three categories, similar to that of Bergen, has been given by N.A. Hilder [18]. His classification however

Table 9.9. Some physical properties of the polymer concentrate tested [18]

Polymer type	PAG	PVP	Polyacr. [ACR]
Kinematic viscosity at 40 °C, mm^2/s	430	75	600
Solubility in water	Inverse (inversion temperature 74 °C)	Normal	Normal
Solids content, wt %	60	10	23
Dry residue	Viscous liquid	Brittle solid	Brittle solid
Electrolytic character	Nonionic	Nonionic	Anionic polyelectrolyte

differs from the Bergen's in that PAG quenchants are not divided into two generations and that PVP quenchants are included. Hilder's classification together with the major physical properties of the concentrates is given in Table 9.9. Based on this classification, we shall compare below the useful properties of the individual quenchants.

9.6.2 Performance Characteristics

The basic performance characteristic is certainly the quenching power, most fully represented by the plot of the cooling rate as a function of the temperature of the quenched workpiece (or of the test specimen). For the sake of simplicity, the quenching powers of various quenchants are often compared by comparing only certain characteristic points on the cooling curves, such as e.g. the maximum cooling rate and the cooling rate at a temperature of 300 °C. Knowing the maximum cooling rate we can estimate whether the pearlitic transformation can be avoided (and thus we can specify the severity of quench). The cooling rate at 300 °C provides information about the influence of the quenchant upon hardening distortions.

The quenching power of polymer media may easily be adjusted between wide limits; this is one of the major advantages of quenchants of this type. The parameters of this control may be the concentration of the solution, its temperature and rate of agitation. By way of example, Fig. 9.18 shows, after Hilder [18], the cooling curves for a 15% polymer used for quenching a specimen,

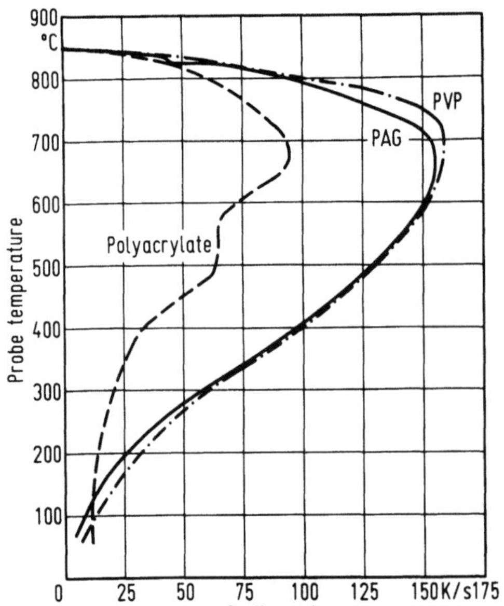

Fig. 9.18. Centre cooling data for an Inconel 600 alloy specimen $\phi 12.5 \times 60$ mm in 15% polymer solutions, tested at 30 °C and 0.5 m/s agitation [18]

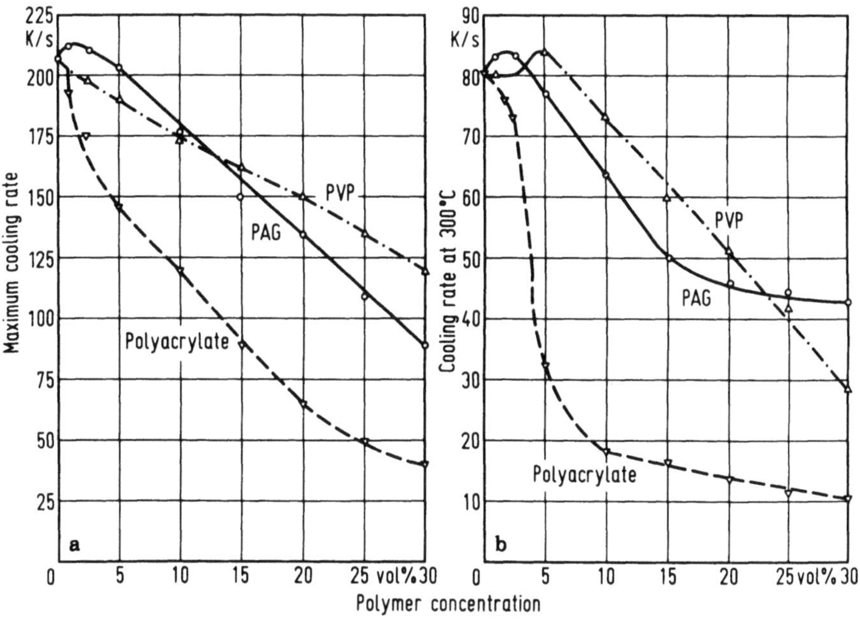

Fig. 9.19 a, b. Effect of polymer concentration on the maximum cooling rate (**a**) and on the cooling rate at 300 °C (**b**). Tested at solution temperature 30 °C, 0.5 m/s fluid velocity, specimen $\phi 12.5 \times 60$ mm [18]

12.5 mm in diameter and 60 mm long, made of Inconel 600 alloy, with the quenchant flowing in a laminar way at a rate of 0.5 m/s. We can see that the PAG and PVP quenchants behave in a similar way, and their quenching powers are greater than the quenching power of the ACR solution within the entire temperature range. Within the lower temperature range (below 350 °C) this low cooling rate is advantageous, but at temperatures of pearlitic transformation this restricts the range of application of this quenchant.

Figure 9.19 gives a more thorough characteristic of the influence that the quenchant concentration has upon the cooling rate within both the higher and the lower temperature ranges. For all the quenchant types, the maximum cooling rate decreases almost in proportion to increasing concentration, whereas at 300 °C, this advantageous proportionality is only maintained by the PVP quenchants.

The variation of the cooling rate with the quenchant temperature within the temperature range of diffusion transformations is for all the quenchant types similar to the variation of the cooling rate with concentration, that is, as the quenchant temperature increases, the maximum cooling rate decreases. However at 300 °C, this relationship is true for the PAG quenchants only (Fig. 9.20).

The third of the parameters controlling the cooling rate, the rate of the quenchant movement (agitation), has a reverse effect to that of temperature but

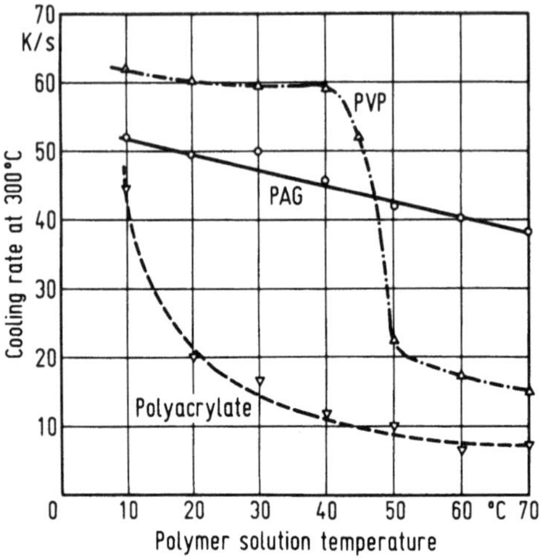

Fig. 9.20. Effect of the polymer 15% solution temperature on the cooling rate at 300 °C temperature of specimen [18]

affects the cooling rate only when that is at its maximum. In PAG and ACR quenchants the effect of the agitation rate upon the cooling rate at 300 °C is insignificant, whereas in PVP quenchants this cooling rate increases with increasing agitation rate but only by about 0.2 m/s. The differences in the behaviour of the individual quenchants result from their different heat transfer mechanisms.

In hardening practice it is very useful to estimate the global effect of these three parameters upon the cooling rate. Hilder [18] has determined these relationships experimentally and presented them in the form of formulae. The notation used in these formulae is:

MR—maximum cooling rate (K/s)
R_{300}—cooling rate at 300 °C (K/s)
C—polymer concentration (volume %)
T—quenchant temperature (°C)
V—quenchant velocity (m/s)

and the formulae are:
for PAG quenchants:

$$MR = 244.6 - 4.3C - 1.7T + 47.2V$$
$$R_{300} = 83 - 1.6C - 0.36T + 14.1V$$

for PVP quenchants:

$$MR = 223.7 - 2.94C - 1.9T + 71.9V$$
$$R_{300} = 96.3 - 1.8C - 1.02T + 37.68V$$

for ACR quenchants:

$$MR = 160.59 - 5.63C - 1.10T + 114.8V$$
$$R_{300} = 58.6 - 2.5C - 0.57T + 43.4V$$

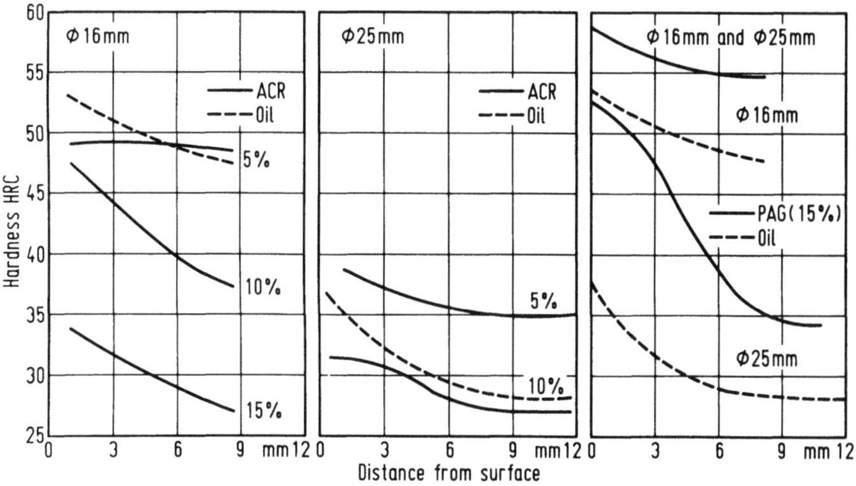

Fig. 9.21. Hardness of a 0.46% C steel specimen after hardening in Aquaquench ACR 5–15%. Aquaquench 251 (PAG) 15% and in a fast oil (Gulf Super Quench U) [19]

Obviously, the choice of the parameters stimulating the quenching power of polymer quenchants considerably affects the hardening results, especially the distribution of hardness over the cross-section of the quenched workpiece. This may be well illustrated by the results obtained by S. Segerberg [19], shown in Fig. 9.21. As we can see, 16 mm diameter bars made of 0.46%C carbon steel show hardness of 58 HRC when quenched in a 15% PAG solution (Aquaquench 251), 54 HRC in a fast oil, and only 49.5 HRC in the 5% ACR Aquaquench. Data for 25 mm diameter bars quenched in higher-concentration ACR solutions are also given in the figure. Segerberg [19] also gives data illustrating how the temperature of aqueous PAG and ACR solutions affects the distribution of hardness in a 0.46%C steel and an alloy steel containing 0.24%C, 1%Cr and 0.17% Mo (SAE4130 grade).

Apart form the quenching power, other quenchant characteristics, important from the practical and economic points of view, are: the service life, the dragout loss, the sensitivity to contamination, and toxicity. The results of studies on these characteristics can be found in [18, 19] and in other works. The service life of a bath is defined by the dragout of its active constituents, their thermal decomposition and the growth of microorganisms within it. The smallest dragout occur in PVP quenchants, slightly greater in PAG, and the greatest in ACR. Nevertheless, polymer quenchants show much less dragout loss than quench oils.

9.6.3 How to Use Polymer Quenchants

From the point of view of performance, the ease of adjusting the quenching power of polymer madia constitutes their considerable advantage, but when controlling actual quenching processes it is this ease that makes the handling operations more troublesome than those required when using oil baths. The manufacturers of various commercial polymer solutions give detailed instructions on their use. We shall discuss below some general requirements that apply to most if not to all polymer quenchants.

From the point of view of the reproducibility of hardening results, the most important requirement is that the bath concentration and temperature should systematically be monitored, since these factors have an essential influence upon the quenching power of all water based quenchants. If the mass of the bath is sufficiently greater than the mass of the quenched load and the bath is moderately agitated, the temperature can relatively easily be maintained contant within $\pm 5\,°C$. The necessary mass of the bath may approximately be taken to be 12–15 litres per 1 kg gross load, and even more when the load consists of fine parts or of parts with a high surface-to-volume ratio [15]. If these requirements cannot be fulfilled, the use of heat exchangers is recommended. As to the rate of bath agitation, it should not be too high since this could cause adverse physical and chemical effects to occur on the quenched part surface. If pumps are used for the agitation, their efficiency per hour is required to be 5–10 times as great as the volume of the quenchant contained in the quench tank.

The control of concentration which, when quenching parts or tools susceptible to cracking, should be maintained at its optimum value within an accuracy of $+1\%$, may be effected by measuring one of the three parameters:

—the density (measured by a hydrometer),
—the kinematic viscosity, or
—the light refraction coefficient (measured by a refractometer).

The latter method is the simplest, since the measurement takes no more than a few seconds. A droplet of the solution is placed on he prism of a small hand-operated optical device, from the division of which, when the device is directed against light, one can read the related value with the naked eye. The concentration sought is then found from an appropriate standard diagram or table. The check should be repeated every week. When the solution is contaminated with sludge (deposited due e.g. to the use of hard water) or with oil, the value of the refraction coefficient may change. If this occurs, the solution should be filtered, or at least its viscosity, which is the most precise measure of the concentration, whould be checked out at regular intervals. Figure 9.22 shows approximately, according to Hilder [18], how the viscosity varies with concentration.

The concentration of hydrogen ions (the pH) should also be checked at least once a week and, if necessary, corrected by adding inhibitors.

In closed systems, such as e.g. the automatic induction hardening apparatus, the bath is often infested with bacteria. To remedy this, sometimes a new bath must be prepared after a few months of operation. The service life of the baths used in immersion hardening is longer, ranging from one to several years. To

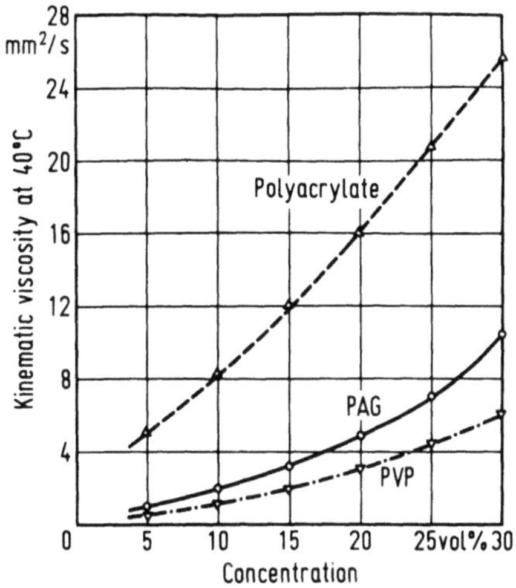

Fig. 9.22. Kinematic viscosity of polymer solutions as a function of concentration [18]

counteract the bacterial infections, the bath should not long be kept still but periodically agitated or blown through with compressed air.

Generally, polymer quenchants are not toxic and their contact with skin is safe. Nevertheless, it is recommended that rubber gloves should be used in handling them since some of them contain inhibitors (e.g. sodium nitrite) which

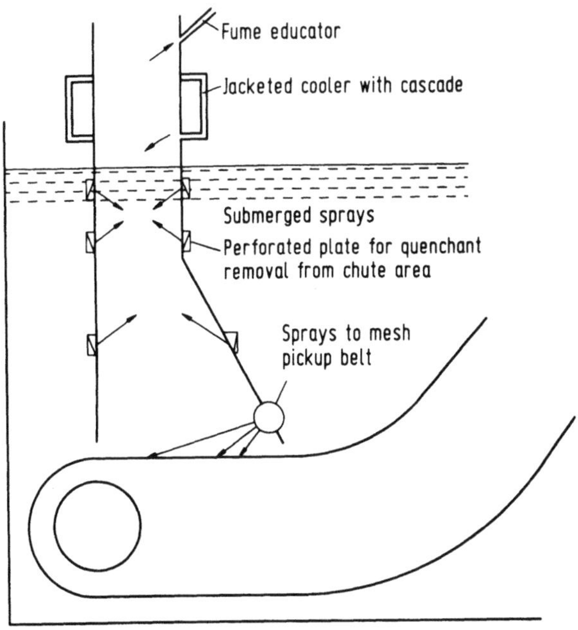

Fig. 9.23. Chute quench with submerged sprays [20]

may be harmful to the skin. Before being drained from the quench tank to a sediment collector, the bath should be diluted in the ratio about 1:50. The solution should not be drained directly to rivers or to water reservoirs.

Polymer baths may be used in heat treatment equipment of almost any type. For example, their use in continuous furnaces designed for quench-hardening treatment under a protective atmosphere does not require any redesign or additional equipment. When water based solutions are used as substitutes for oil in continuous carburising or carbonitriding furnaces, only a simplecooling system providing a cooler cascade in the chute need to be installed (Fig. 9.23) so as to prevent water vapours from entering the furnace atmosphere and thereby altering the carbon and nitrogen potentials. In multi-purpose controlled atmosphere over-type furnaces a tight curtain is required between the heating and pretreatment chambers. Polymer baths may also without any restrictions replace oil in tanks of pit or chamber batch-type furnaces.

9.7 Mineral Quenching Oils

When alloy steels came into use and appeared to undergo distortion and cracking when quenched in water, other quenchants, cooling less intensively, were sought. First, various kinds of oil and vegetable or animal fats were attempted. When fresh, they have very favourable quenching characteristics but also show several disadvantages. The most important one is that their life is short since they are prone to oxidation and decomposition. Moreover, they are expensive, malodorous, and when used for quenching parts that have previously been heated in salts, soaps of alcali metals precipitate to form sludge on the quenched surfaces.

When at the end of the 19th century several cheap products of petroleum appeared on the market, mineral oils almost entirely superseded the vegetable and animal oils for quenching. The latter are yet sometimes added to mineral oils to improve their quenching characteristics such as e.g. the wettability.

As the mechanism and the kinetics of phase transformations that proceed in overcooled austeninte, and the heat transfer effects occurring in quenching media were better known, the properties of quench oils were improved and their assortment was extended. Despite their well known disadvantages (fire hazard, adverse effects on the environment and the quenching power insufficient for certain applications), mineral oils are still the most commonly used quenchants, since they permit achiving the optimum combination of high hardness with the minimum quenching distortion. As compared with other quenchants (except water), oils are less troublesome to control and monitor. Their application range is very wide, although one should not ignore fire hazard when quenching heavy loads with high surface area-to-mass ratios and when partial or spray quenching is to be performed.

9.7.1 Composition of Quenching Oils

Mineral oils are mixtures of very many organic compounds. For this reason their boiling temperature is not constant, as is the case with single-component fluids, but varies within a certain range depending on several factors. The position and width of this range for a given oil, considerably affect its thermokinetic properties and, to a certain extent, depends on the origin and the method of refining of the petroleum of which the oil has been manufactured. However, the major factors determining the quenching power of a given type of oil are:

—the fraction ratio of the so-called base oil, and
—various substances added to stimulate the performance characteristics of the oil, such as its quenching power.

The influence of these two factors should always be taken into account when designing a particular quench oil to match its characteristics with the hardening technique to be employed.

The importance of the fraction ratio of base oil was examined by G. Beck et al. [21] and T.J. Tkaczuk et al. [22], among several authors. The former authors identified the fractions according to the major hydrocarbon compounds contained in them (paraffinic, naphthenic and aromatic oils) and the latter according to their distillation temperature range: 300–350 °C (No1), 350–400° (No2), 400–450°C (No3) and 450–500 °C (No4).

With paraffinic oils, the critical temperature at which the vapour blanket phase passes into the boiling phase (Leidenfrost's point) hardly depends on the bath temperature if the latter ranges between 25° and 120 °C, whereas it markedly depends on the oil viscosity. At a viscosity of 12 mm^2/s at 50 °C the critical temperature is about 500 °C, at 60 mm^2/s 50 °C about 600 °C and at 228 mm^2/s at 50 °C about 700 °C. This explains the fact, well known in practice, that the quenching power of oils hardly depends on the oil temperature. This characteristic property of oils clearly distinguishes them from water and all water based quenchants, for which the temperature is one of the most important parameters. Despite this property, however, the temperature of quench oils is usually maintained between 50 and 90 °C but this is aimed at slowing the martensitic transformation and at facilitating agitation by decreasing the oil viscosity. The boiling temperature range, i.e., the difference between the temperatures at which boiling starts and ceases, is in paraffinic oils relatively narrow (130–200 °C).

The viscosity of naphthenic oils is smaller than that of paraffinic oils at the same temperature, but the thermokinetic properties of these two groups of oils are very similar.

Aromatic oils show a poor thermal resistance and a low viscosity. They however improve the wettability of the quenched surface and prolong the boiling stage. Base oils are as a rule largely composed of paraffin or naphthene fractions. Figure 9.24 shows the differences between the thermokinetic properties of the individual oil fractions.

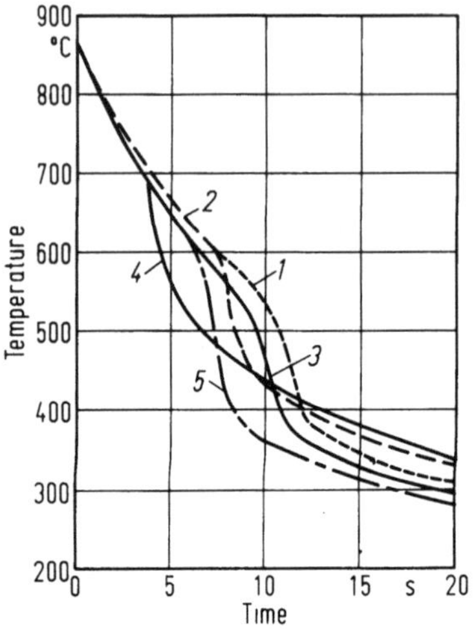

Fig. 9.24. Cooling curves of a silver specimen $\phi 16 \times 48$ mm in different fractions of oils at a temperature of 80 °C [21]: *1* naphthenic oil 4.5 °E/50 °C; *2* naphthenic oil 18 °E/50 °C; *3* paraffinic oil 1.5 °E/50 °C; *4* paraffinic oil 30 °E/50 °C; *5* aromatic oil

The properties of the individual fractions according to Tkaczuk et al. [22] are shown in Fig. 9.25 and in Table 9.10. From the results of their study these authors conclude that

—the fraction ratio of base oils considerably affects the quenching power of quench oils;
—fractions whose distillation temperatures are low (300–400 °C) have a low quenching power within the upper range of quenching temperatures;
—fractions whose distillation temperature ranges from 400 to 500 °C show a high quenching power at temperatures higher than 500 °C and insufficiently low below this temperatures (since boiling proceeds in them within a narrow temperature range);

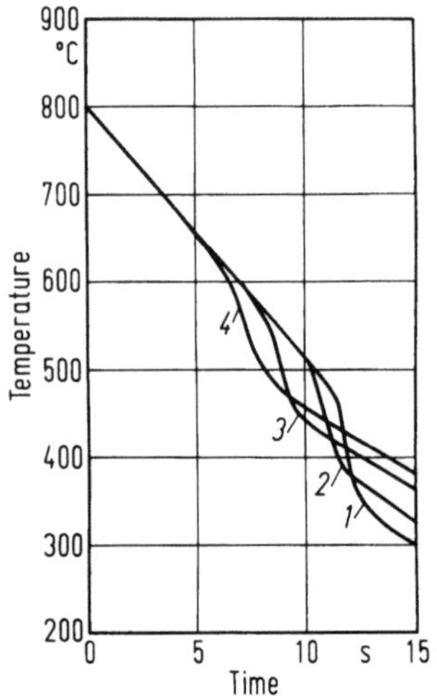

Fig. 9.25. Centre cooling curves for a silver ball $\phi 20$ mm in base oils, after [22]. Fractions of distillation temperature °C: *1* 300/350; *2* 350/400; *3* 400/450; *4* 450/500

Table 9.10. Characteristics of base oils fractions after [22]. Cooling of silver ball φ 20 mm

Index	Fractions of distillation temperature °C			
	300–350 No. 1	350–400 No. 2	400–450 No. 3	450–500 No. 4
Viscosity at 50 °C mm²/s	6.31	10.43	32.19	61.09
Viscosity at 50 °C, °E	1.5	1.85	4.35	8.05
Leidenfrost point, °C	462	484	534	589
Temperature range of boiling, °C	117	95	86	84
Cooling time in vapour blanket stage, s	11.8	10.8	8.8	6.8
Cooling time in boiling stage, s	1.23	1.08	0.92	0.79

—the thermokinetic properties of base oils can to a certain extent be modified by mixing the oil fractions in various proportions.

A different classification of base oils has been given by A. Bashford and A.J. Mills [23]. Table 9.11 compares the properties of three typical selectively refined base oils whose viscosity, (100, 150 and 500 SUS), measured by Saybolt's method, has been taken as the criterion for their classification. The three oils are solvent-refined paraffinics highly resistant to oxidation and thermally stable, the features deciding about the service life of quenchants. The table also includes the spindle oil, since this may be used as a base oil.

The quenching power of quench oils can be increased and the temperature range of the individual cooling stages adjusted not only by modifying the fraction ratio of the base oil but also, even in a greater degree, by adding special substances, such as metal sulphonates, nickel-zinc dithiophosphorate, various petrochemical compounds, and also natural fats of wegetable or animal origin. These special additives can affect the cooling curves by either

—lowering the temperature at which the convection stage begins (an addition of a few percent surface active substances reduces this temperature by about 100 °C);
—increasing the temperature at which the vapour blanket phase ends, without shifting the convection temperature.

Table. 9.11. Typical characteristics of base oils [23]

Base oil	Spindle oil	100 Solvent Neutral	150 Solvent Neutral	500 Solvent Neutral
Relative density	0.873	0.865	0.875	0.885
Flash point (open), °C	200	200	210	260
Viscosity at 100 °C, cSt	3.0	4.0	5.2	11.0
Viscosity at 50 °C, °E	1.8	2.3	3.0	7.9
Viscosity index	65–90	99	95	95

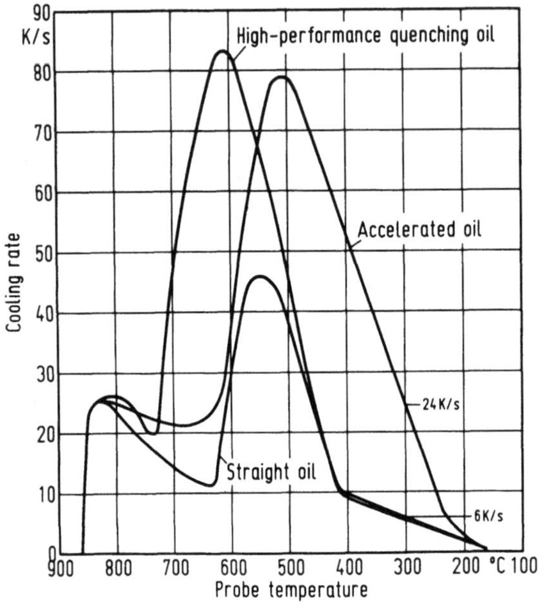

Fig. 9.26. Comparison of cooling rate for a straight oil, a conventionally-accelerated oil and a high-performance oil. Test specimen ϕ 12.5 × 60 mm of Inconel 600 [23]

The additives chiefly affect the properties of quenchants by changing the surface tension and thereby the ability of the quenchant to wet the quenched surface.

Some additives accelerate the cooling process not only within the upper temperature range by also within the lower; this latter effect is undesirable as it favours distortions and cracking. Other additives, such as e.g. nickel-zink dithiophosphorate, because of their susceptibility to thermal decomposition, may increase the need for cleaning the quenched surface after quenching. In the recent years, however, several fast oils which do not suffer from these disadvantages have been developed. These oils have been called 'high performance" oils in contrast to "accelerated" oils known thus far. The thermo-kinetic properties of these two groups of oils are compared in Fig. 9.26.

Apart from the additives stimulating the cooling kinetics, modern oils also contain several other additives, such as oxidation inhibitors, detergents, sludge inhibitors, defoamers, and emulgators facilitating the removal of oil residuses with cold water.

The heat-flux density during cooling in oils is on average [21]:

—about 2.1 MW/m^2 ($50\,cal/s\cdot cm^2$) during the vapour blanket stage (at 600 °C)
—about 8.37 MW/m^2 ($200\,cal/s\cdot cm^2$) during the boiling stage (at 450 °C), and
—about 0.85 MW/m^2 ($20\,cal/s\cdot cm^2$) during the convection stage.

It should however be remembered that these values have been determined under certain definite cooling conditions, such as the type of quenchant, its temperature and rate of agitation, the condition of the surface that participate in the heat exchange, etc.

9.7.2 Classification and General Description of Quenching Oils

Usually, the specialised producers of modern quenching oils offer several types of them. As regards the operating temperature, all the commercially available oils may be divided into two groups, (low- and high temperature oils), whereas taking into account their quenching power, as is usually done, into three groups (slow and fast low-temperature oils and high-temperature oils). Table 9.12 lists the oils according to these criteria, in addition divided into subgroups generally based on their chemical formulation. Depending on the base oil used by the manufacturer and on the number and kind of additives, oils belonging to the same group or subgroup may considerably differ in their physical, chemical and technological properties. It is therefore justified not only by commercial but also by technical reasons that various trade names are given to individual quenching oils.

Conventional oils (group 1) are usually composed of paraffinic and naphthenic fractions of viscosity ranging from 5.5 to $50\,mm^2/s$ at 50 °C and contain additives that increase their resistance to ageing. Parts of bright-coloured surface quenched in such oils remain bright-coloured after the treatment. After the oil has however been used for some time, the products of its ageing make this surface darker. Sme conventional oils of viscosity ranging from 5.5 to $14\,mm^2/s$ at 50 °C contain additives that prevent the quenched part surface from darkening even if the oil has long been used. The quenching power of this group of oils is approximately determined by the oil viscosity. Oils of low viscosity cool more quickly than those of high viscosity.

Fast oils (group 2) are composed of distillation fractions of low viscosity and contain additives that enhance wetting or by adhering to the quench part surface increase the quenching power. The surfaces of parts quenched in these oils usually darken during quenching.

The quenching power of fast oils, especially those of lowest viscosities, is so high that, when quenched in them, even carbon steels are hardened through, up to sections of about 25 mm. The quenching power of fast oils is therefore close to that of still water, which has been confirmed by measurements of the Grossmann cooling rate coefficients. Although their flash point lies below the flash points of other types of oils, fire hazard is not increased but even decreased. This apparent paradox is explained in Fig. 9.27. In a fast oil of small viscosity and low flash point the increase of temperature in the immediate vicinity of the cooled part is not so great since the quenchant circulates easily and almost the whole volume of the bath participates in the heat transfer. With a dense oil, on the other hand, during the initial phase of cooling only a small portion of the oil bath takes part in his process.

Table 9.12. General classification of quenching oils after [24[1]]

Classification according to the operating temperature		Classification into groups according to quenching power and applications		Subgroups according to chemical composition		Physical and chemical characteristics				
						Viscosity at 50 °C	Flash point °C	Ash content %	Coce content %	Saponification number mg KOH/g
A	Low-temperature oils (up to 90 °C)	1	Conventional (slow) oils	a	Conventional oils			0.01	0.05	2
				b	Compounded oils[2]/					2
		2	Fast and high-performance oils	a	Fast high-performance oils	6 °E/45 cSt/	230	0.01	0.10	2
				b	Fast high-performance and compounded oils					2
B	Hot oils (above 90 °C)	3	Hot oils (for mar quenching and austempering)	a	Conventional oils	6 °E/45 cSt/	230	0.01	0.10	2
				b	High-performance oils			0.01		

[1] Special quenching oils for use in vacuum aggregates are not included in this classification
[2] oils containing vegetable or animal fat additives

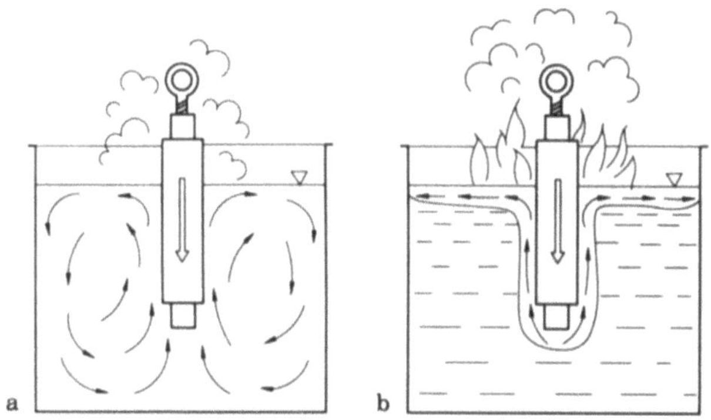

Fig. 9.27 a, b. Effect of viscosity and the flash point of oil on fire hazard. **a** small viscosity and low flash point oil; **b** great viscosity and high flash point oil

Of the group of fast oils, oils of lower viscosity are more suitable for quenching parts of greater cross-sections than are denser oils, since the boiling temperature range of the former oils is positioned lower than that of the latter. K.H. Kopietz [25] reports that when quenching a 20 mm diameter silver ball in an oil of viscosity of 2.0 °E/50 °C, the maximum cooling rate occurs at 500 °C, whereas when the oil viscosity is reduced to 1.5 °E/50 °C, the maximum cooling rate shifts to 400 °C.

High-temperature oils (group 3) have not been used for marquenching and austempering until the 1950s when appropriate blends of base oils with oxidation inhibitors were developed. Without these inhibitors, oil is very quickly oxidized at higher temperatures, becomes much denser and is no longer suitable for use. Modern high-temperature oils can be used up to 200 °C, or in some cases even to 250 °C, their long service life (up to several years) being maintained. The quenching power of these oils is much higher than that of nitrate baths. With an addition of appropriate emulgators, which practically do not alter their quenching power, they can be washed out with water like oils of groups 1 and 2.

9.7.3 Physical and Chemical Properties

To identify individual petroleum products, and thus quenching oils, several physical and chemical parameters are determined. The quenching power is obviously the basic useful characteristic of quenching oils but a knowledge of their physical and chemical properties facilitates the selection of an appropriate type of oil to fit a given application, and to evaluate its service life and the cost involved. The thermal characteristics of oils essentially differ from those of water (see Sect. 9.3). The values of the major thermal characteristics of oils are, approximately,

—specific heat capacity $-2.01\,\mathrm{kJ/K \cdot kg}$,
—specific latent heat of vaporization $-251.2\,\mathrm{kJ/kg}$ for highly viscous oils and $167.5\,\mathrm{kJ/kg}$ for oils of small viscosity,
—thermal conductivity $-0.127\,\mathrm{W/K \cdot m}$.

All the physical and chemical characteristics subject to routine checking may be divided into two groups: basic and auxiliary. The basic parameters include: viscosity, the flash point, the fire point and density. The auxiliary parameters are much more numerous; they include: the so-called viscosity index, colour, neutralisation number, saponification number, Conradson's carbon residue, resistance to oxidation, ash, asphalt and water contents, and vapour pressure.

Viscosity is a measure of internal friction which occurs in fluids when adjacent fluid layers displace themselves with respect to one another. In Newtonian fluids such as oils the tangential force T that balances the internal friction is proportional to the surface area of the layers being in contact and to the velocity gradient of their relative movement. This may be written as

$$T = \mu \cdot A \frac{dv}{dx}$$

where A is the surface area of the adjacent layers, dv is the difference in their velocities, dx is the spacing between the layers, and μ is the dynamic viscosity index expressed in pascales-seconds [Pas]. Engineers however most often use the concept of kinematic viscosity defined as the ratio of dynamic viscosity to density, and thus, in the SI system of units, expressed in $\mathrm{m^2/s}$ or $\mathrm{cm^2/s}$ (this latter unit was formerly called the stoke). The viscosity of oils is usually expressed in $\mathrm{mm^2/s}$ (formerly centistokes, $1\,\mathrm{mm^2/s} = 1\,\mathrm{cSt}$).

To simplify the measurement of viscosity, in addition to dynamic and kinematic viscosities which are absolute quantities, a relative viscosity is often measured. It is defined as the ratio of the dynamic viscosity of the fluid under test to the dynamic viscosity of a reference fluid. If we take water to be the reference fluid, and measure viscosity by measuring the time taken by 200 ml of the fluid examined to flow through a standard nozzle (Engler's viscometer), then the unit of the relative viscosity is the Engler degree (°E). The relations between the viscosity units described above can be found in the literature in the form of formulae, diagrams (Fig. 9.28) and tables.

The effect of temperature upon viscosity is different in different oils. In order to evaluate this effect quantitatively, a conventional quantity known as the viscosity index has been introduced. The value of this index is determined by comparing the viscosity-versus-temperature curve for the oil examined with similar curves for two reference oils, chosen so that their viscosity-temperature characteristics radically differ from one another. The value of the viscosity index of that reference oil which shows the smallest viscosity variations is taken to be 100, and that for the oil showing the largest viscosity variations to be 0. Therefore, the viscosity of an oil of higher viscosity index will vary with

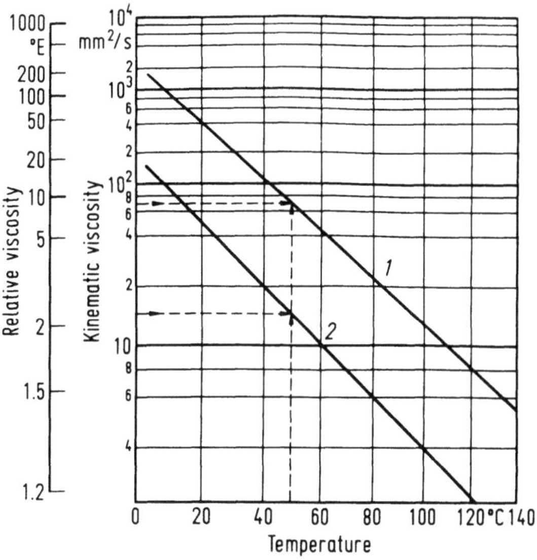

Fig. 9.28. Relative and kinematic viscosity as a function of temperature for two oils: *1* oil of viscosity of 70 mm^2/s at 50 °C; *2* oil of viscosity 14 mm^2/s at 50 °C [26]

temperature less than that of an oil of smaller viscosity index. The value of the viscosity index is determined using special monograms.

The viscosity of a fresh quenching oil not only determines to some extent its quenching power but is also an important economic criterion. Using quenching oils of lower viscosities, dragout losses are reduced and less energy is consumed for oil agitation and circulation through heat exchangers and filters. Moreover, with such oils, cleaning is greatly facilitated. By measuring the viscosity and the flash point of an oil that has already been operated for some time we can evaluate the degree of its ageing. This is so since, as is well known, an increased viscosity and a raised flash point indicate an increased oxidation degree of the oil, a decreased viscosity and a lowered flash point indicate its chemical decomposition (degradation), and an increased viscosity occurring in conjunction with a lowered flash point indicate that the oil has undergone both oxidation and chemical decomposition.

The flash and fire points are chiefly determined (in an open or close crucible) in order to specify the maximum operating temperature of a given quench oil bath, that is admissible for safety reasons. The fire point is defined as the temperature at which the oil educes such an amount of gases that a flame passed over it set them to fire for a short while. The operating temperature should be at least 30–50 °C lower than the fire point. The flash point, which is more often determined, differ from the fire point in that the gases set to fire above the oil surface do not extinguish at once but are on fire for at least 5s. The flash temperature may be determined in an open crucible (Markusson's method) or in a closed crucible (Martens–Pensky's method). The flash temperatures determined by the latter method are usually lower by at least 20 °C and thus this method is recommended as more safe when specifying the admissible operating temperature of the quenching oil examined. The admissible operating temperature is usually specified about 30 °C below the fire point and 60–80 °C below the flash point.

The density of an oil indicates to some extent its chemical structure. If the density is smaller than 0.9 kg/l, the oil is of the paraffin type, if higher, the oil chiefly contains naphthene fractions. Paraffinic oils are more economic than naphthenic, since a paraffinic oil of the same mass (which defines the cost) as a naphthenic one occupies a larger volume.

The neutralisation number is the amount of KOH in milligrams required for free organic acids contained in 1 g of the oil examined to be neutralised. Knowing the neutralisation number of a given oil we can guess what additives are contained in it. An increase of the neutralisation number with increasing exploitation time indicates that some ageing processes have been proceeding in the oil. Usually, the neutralisation number of 1 mg KOH/g is considered to be admissible.

The saponification number is the amount of KOH in milligrams required for free acids and esters to be saponified. The value of this number depends on how much animal and vegetable fats the oil contains and should not exceed 2–3 mg KOH/g.

The Conradson number indicates how much the oil is susceptible to carbonisation. It is determined by vaporising the oil in an air-free atmosphere and then measuring the residues in weight percents. This test permits comparing the sludging properties of various oils.

The resistance to oxidation of an oil may be treated as a conventional criterion for evaluating its service life. This parameter is of particular importance when examining high-temperature oils. As in other physical and chemical tests, the test conditions are standardised. Quenching oils are tested for their resistance to oxidation by heating them up to a specified temperature (e.g. to 130, 150 or 180 °C), then passing, at a specified flow intensity (e.g., 3 l/h), air or oxygen through them for a specified period of time (e.g., for 24 h), with a copper plate immersed in the oil to act as a catalyst, and then determining certain physical and chemical properties of the oil and comparing them with those determined prior to the oxidation test. Usually, the properties determined after this test are: the viscosity, the coke content and the neutralisation number. An inspection of the appearance of the oil before and after the oxidation test also permits a qualitative estimate of the processes that have been proceeding in the oil during the oxidation.

Apart from the properties discussed above, other properties, such as the vaporisation power, the solidification temperature and the contents of ashes, water and other contaminants may also be determined. The water content is of particular importance; it should not exceed 0.05%. The effect of water contamination upon the quenching power of oils and upon fire hazard will be discussed later in Sect. 9.7.6.

Different oil types widely differ in their physical and chemical properties (Table 9.13). Table 9.14, prepared using data given in [27], lists all the basic

Table 9.13. Changeability range of physico-chemical properties for quenching oils [24]

Property	Unit	Changeability range
Flash point	°C	160–310
Fire point	°C	172–320
Density at 15 °C	kg/l	0.849–0.933
Kinematic viscosity at 50 °C	mm^2/s (cSt)	9.44–284.5
Viscosity index	—	0–152
Water	%	0–0.1
Neutralization number	mg KOH/g	0.02–3.60
Saponification number	mg KOH/g	0.1–54
Conradson carbon residue	%	0–3.10
Ash	%	0–3.30

and auxiliary physical and chemical properties of quenching oils and compares their practical usefulness in evaluating various quenching oils.

9.7.4 Quenching Power of Oils

The most important performance characteristic of a quenching oil is the variation of its cooling rate as a function of the temperature of the quenched surface. This characteristic is examined in the laboratory using methods discussed in Chap. 6 and is best represented by the specimen temperature-cooling rate curves. We can read from these curves the critical temperatures T_1 and T_2 that separate the succesive phases of heat transfer, determine the width of the boiling temperature range ($\Delta T = T_1 - T_2$), the maximum cooling rate and the temperature at which it occurs. These parameters have been called the thermokinetic properties.

J. Mansion et al. [24] give a comprehensive and statistically representative description of the thermokinetic properties of 122 types of oils, classified as in Table 9.12 into three groups, and divided into two subgroups each. Figure 9.29 shows the results of tests performed by these authors using a silver specimen 8 mm in diameter and 24 mm long. As seen from the figure, different types of oils, even those belonging to the same group or subgroup, considerably differ in their thermokinetic properties. This fact necessitates working out effective methods for the selection of appropriate oil to fit a given application and at the same time makes us aware that the term 'quenching in oil' is very non-precise.

Although oils belonging to groups 1a and 2a differ from one another considerably, within each of these groups the differences in the critical temperature do not exceed 30–40 °C. The properties of compounded oils (groups 1b and 2b) are much more varied depending on the kind and concentration of fats added to them.

Fast oils (group 2) not only provide a much greater maximum cooling rate than do conventional oils (group 1) but also have a markedly wider boiling

Table 9.14. The significance and practical importance of testing the physical and chemical properties of mineral quenching oils

Type of quenching oil		Low-teperature oils										High-temperature (hot) oils				Significance and interpretation of property
		Conven-tional slow oils		Compoun-ded oils		Fast (high-performance) oils		Washable oils		Vacuum oils		Conven-tional oils		High-performance oils		
		oil condition: f—fresh, u—used														
Property		f	u	f	u	f	u	f	u	f	u	f	u	f	u	
basic	Quenching power (cooling curve)	+	+	+	+	+	+	+	+	+	+	+	+	+	+	Major useful property
basic	Density g/ml	+	±	+	±	+	±	+	±	+	±	+	±	+	±	Costs criterion and an approximate indication of the oil fraction ratio
basic	Flash point °C	+	+	+	+	+	+	+	+	+	+	+	+	+	+	Fire hazard criterion and indication of oil volatility
basic	Kinematic viscosity mm^2/s (cSt)	+	+	+	+	+	+	+	+	+	+	+	+	+	+	Dragout loss, mobility and application range criterion
	Colour and appearance	±	+	±	+	±	+	±	+	±	+	±	+	±	+	Approximate criterion for evaluating water contamination and sludge content
	Viscosity index	±	±	±	±	±	±	±	±	±	±	±	±	±	±	Carbon structure criterion: indicates how oil viscosity varies with temperature

auxiliary	Neutralization mg KOH/g number	+	+	+	+	+	+	+	+	+	+	+	+	+	+	Indicative of the presence of additives: helps evaluating the degree of oxidation
	Saponification number mg KOH/g			+	+											Indicative of the presence of vegetable and animal fats
	Washability							+	+							Indicative of whether the oil can be washed out with water (by forming an emulsion with it)
	Conradson carbon residue, %					+	+							+	+	Indicative of the oil susceptibility to form carbonaceous deposits
	Resistance to oxidation	±		±		±		±				+		+		Service life criterion (chiefly for hot quenching oils)
	Asphalt content, %	±	±	±	±	±	±	±	±	±	±	±	±	±	±	Criterion of cleanness of the quenched surface
	Ash content, %	+	±	+	±			+	±	+	±	+	+			Indicative of the presence of metal oxide additives and of the loss of additives; criterion of the oil quality
	Sulphate ash content, %					+	+	+	+							
	Water content, %	±	+	±	+	±	+	+	+	±	+	±	+	±	+	Safety criterion; quenching power criterion
	Evaporisation after Noak	+		+		+		+		+		+		+		Cost criterion; indicative of the oil degradation

[a] indicates a test recommended as very useful ± indicates a test occasionally useful

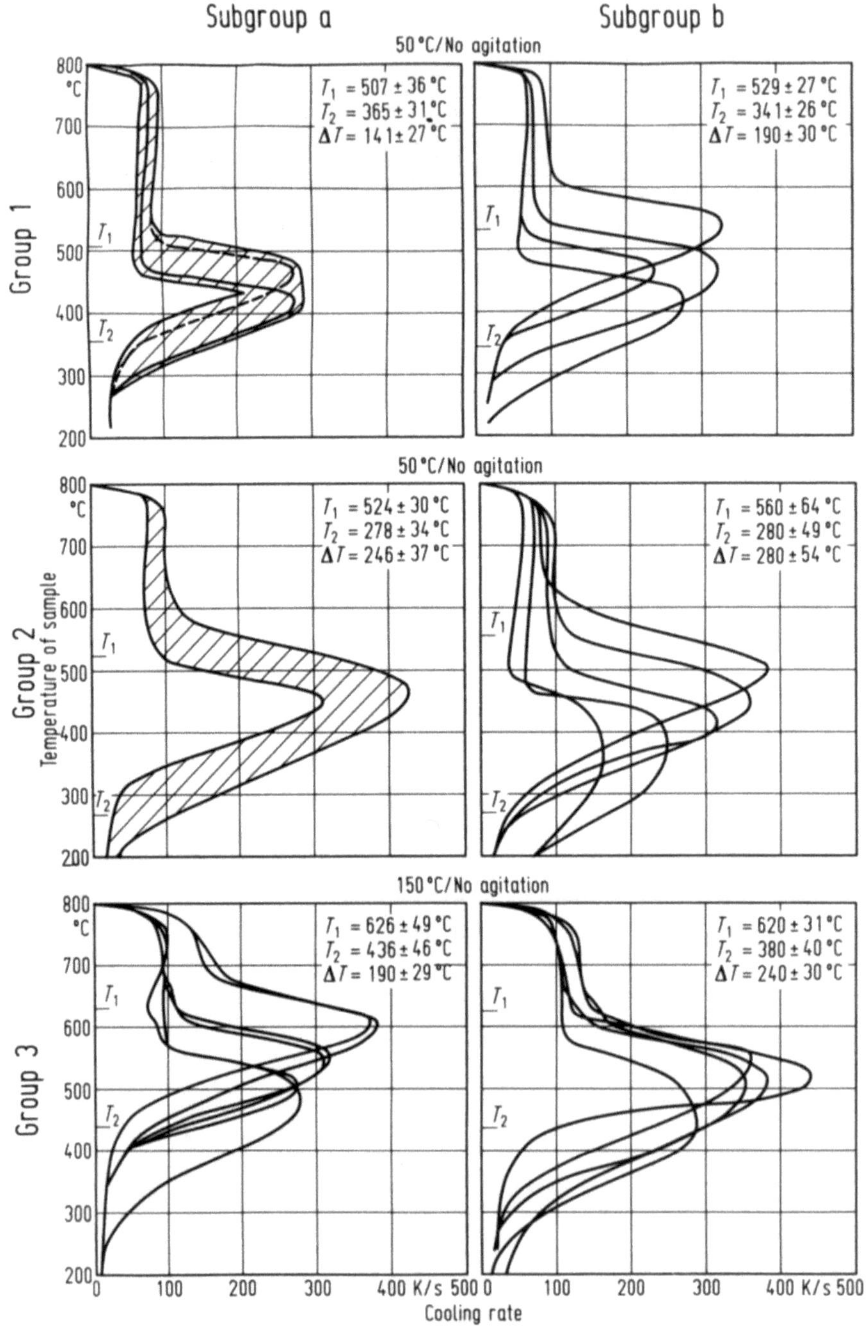

Fig. 9.29. Cooling rate scatter for quenching oils divided in separate groups according to classification given in Table 9.12 after [24]

temperature range. Tanks to these properties and also to the fact that the temperature T_2 of fast oils is shifted downwards by about 90 °C, parts quenched in them are hardened to greater depths and thus low-alloy steels, or even carbon steels, may be hardened.

The properties of high-temperature oils (group 3) are very widely varied depending on the composition and nature of the base oils employed and the kind and content of additives. Therefore, oils of this group should also be carefully selected so as to fit a given application.

9.7.5 Effects of Oil Temperature and Agitation

As mentioned in Sect. 9.7.1, one of the characteristic features of quenching oils is that their quenching power hardly depends on the bath temperature. This is illustrated in Fig. 9.30, where we can see that in the temperature range of pearlitic transformation the average cooling rate is in conventional (slow) oils practically independent of the oil temperature and in refined (fast) oils it slightly decreases after the oil temperature exceeds 70–80 °C. The figure also shows how the cooling rate increases with decreasing viscosity. In practice, when low-temperature oils are used, the quench bath is heated up to a temperature between 50 and 80 °C for reasons not connected with the quenching power. The oil is chiefly heated up in order to decrease its viscosity and thereby to facilitate agitation and to reduce dragout losses. In some cases by heating up the bath we wish to stop martensitic transformation so as to reduce distortion.

In contrast to the bath temperature, the mode and rate of oil agitation considerably affect the cooling rate and the uniformity of quenching. A forced circulation or agitation of the quenchant performs important tasks, since it

—continuously removes the portion of the fluid that is in contact with the cooled surface so as to prevent its overheating and, where oil is used as the quenchant, to prevent its degradation,

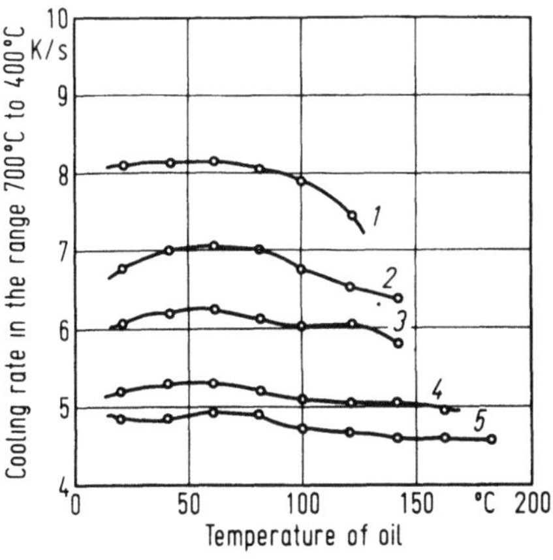

Fig. 9.30 Effect of oil temperature on the centre cooling rate within the 700–400 °C range for a steel specimen ϕ55 mm: *1* fast oil 1.4 °E/50 °C, *2* fast oil 2 °E/50 °C, *3–5* conventional oils of viscosities 1.6, 2.2 and 3.5 °E/50 °C [25]

—ensures uniform quenching of the entire surface and of all parts contained in the load so as to avoid soft spots and excessive distortion,
—equalizes temperature throughout the quenchant volume, ensuring that this temperature will nowhere exceed its admissible level (this is especially important in oils because of fire hazard).

When considering the importance of oil agitation we should first distinguish between two types of this motion and define methods or measuring its rate. In fluids of all types (liquids, gases), a laminar or turbulent flow can occur. The laminar (layered) flow occurs where particles of the fluid move in a certain definite direction, following mutually parallel paths. The rate of such a flow is expressed in units of speed, i.e., in m/s, and is easy to measure. When dealing with liquid quenchants, a laminar flow of rate about 0.2 m/s is described as mild, and that of rate 2 m/s as very intensive. Between these two extremes, we may of course conventionally specify certain intermediate degrees of the intensity of laminar flow.

The turbulent flow occurs after a certain linear velocity, defined by means of what is known as the Reynolds number, is exceeded. It is characterised by eddy currents and chaotic, as regards direction and velocity, movements of fluid particles. The measurement and specification of a turbulent flow is much more difficult than those of a laminar flow. In quenching, especially in oil quenching, turbulent agitation is however more desirable than laminar, since it improves the uniformity of quenching and thereby produces less distortion. Depending on the type of oil and on the section size of the quenched part the effect of agitation upon as-quenched hardness may be varied, but usually strong. By way of example, Table 9.15 gives hardnesses at the mid-radii of two bars, 20

Table 9.15. Effect of turbulence for various quenching oils on the hardness in half radius of steel bars SAE-510 θ 20 and 40 mm [24]

Oils group after table 7.2	Diameter of bars, mm	Hardness HV_{300N} after quenching in oil	
		no agitation	turbulence
1a	20	532	563
	40	399	444
1b	20	540	574
	40	374	449
2a	20	575	592
	40	501	512
2b	20	577	572
	40	511	536
3a	20	359	527
	40	322	378

and 40 mm in diameter, made of SAE-5140 steel, which have been developed by quenching in various oils.

American authors usually specify the quench severity by means of Grossmann's coefficient H. Using this method, the value of H for a fast quiescent oil is about 0.5, whereas for the same oil but violently agitated it increases to 0.8–1.1, which approximately corresponds to the quench severity in quiescent water. We should however remember all reservations applied to the Grossmann method.

In industrial quench tanks, both laminar and turbulent flows may occur depending on the tank design and the mode of agitation of the quenchant. Under these conditions, a quantitative estimate of the flow rate is very uncertain and is usually made by indirect methods (e.g. by determining the pump efficiency or the horsepower required for agitation divided by the quenchant volume), or the intensity of motion is only described as e.g. poor, moderate or high. If the quenchant is agitated by means of pumps and external heat exchangers, its motion is usually mild and laminar. This is sufficient for the temperature of the bath to be maintained at a desired level but does not ensure uniform quenching. The effect of this kind of agitation upon the cooling rate is noticeable but relatively insignificant.

The three purposes, mentioned above, for which the quenchant should be agitated, can much better be achieved by using motor-driven propellers installed in an appropriate position. If the power delivered by the motor that drives the propeller is sufficiently great with respect to the volume of the quenching oil, the oil agitation will be turbulent. According to Furman [28], the motor power required to induce a turbulent flow in an oil of volume $0.25\,m^3$ is 200 W, for an oil volume of $2.5\,m^3$ is about 2 kW and for a volume $10\,m^3$ about 7 kW. Water based quenchants require less power, thanks to their lower viscosity. A turbulent flow may also be induced by means of ultrasonic generators. If this is the case, the duration of the vapour blanket stage is shortened [28].

9.7.6 Contamination of Quenching Oils with Water

Even a small amount (below 0.1%) of water contained in quenching oils may produce very undesirable effects, such as soft spots, excessive distortions and cracking. Moreover, the presence of water in the quenching oil greatly increases fire or even explosion hazard. Below, based chiefly on the considerations presented in [29], we shall discuss the following aspects of this problem:

—sources of water contamination,
—the effect of water contamination upon the rate and uniformity of quenching,
—the effect of water contamination upon fire hazard,
—methods for detecting water in a quenching oil and for determining its content.

Water can get into the oil in various ways. The most common sources of water contamination are leaks in the oil cooling system, (i.e. in heat exchangers), and the condensation of atmospheric moisture on the external surfaces of various pipes or parts of the furnace (e.g., when a chamber furnace with a vestibule is installed above the quench tank). Water can also get into the oil when that is stored in contaminated barrels out of doors, when the pumps employed are contaminated with water, when the roof of the hardening plant leaks, etc.

It is obvious that water contained in a quenching oil usually in the form of fine-dispersive globules must alter the cooling characteristics of the oil, since an uncontrolled vapour blanket is formed, often locally, on the surface of quenched parts. At water concentrations as small as about 0.2%, the vapour blanket stage persists to considerably lower temperatures. Moreover, since usually this vapour blanket does not cover the entire quenched surface, quenching proceeds non-uniformly. These are typical conditions favouring the formation of soft spots and the development of distortions and cracks, especially when the quenched parts have sophisticated shapes and are made of steel of low hardenability. Water contamination also causes the boiling stage to persist to lower temperatures (Fig. 9.31), whereas it accelerates heat convection that produces structural stresses and distortions.

When using a quenching oil contaminated with water, fire hazard increases due chiefly to the formation of foam composed of oil mixed with air. At the moment of immersing parts to be quenched, the quenching oil almost always ignites but extinguishes after the parts have completely been immersed, since the gases expelled below the oil surface cannot get in touch with oxygen, whereas when they come out of the oil their temperature is already below the flash point. When, however, water is present in the oil, a great amount of steam is produced (1 l of water produces 600 l of steam) which favours the formation of inflammable foam. At higher water concentrations, the amount of foam may be so great that the foam overflows the edges of the quench tank, further

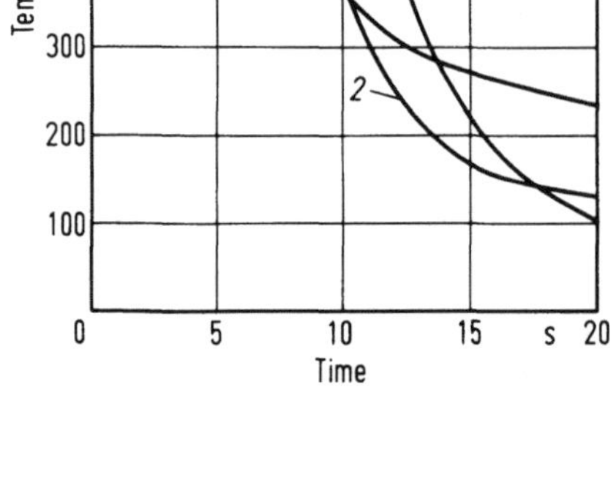

Fig. 9.31. Effect of water content in a still fast oil on the cooling curves for a steel specimen ϕ10 mm: *1* no water; *2* 0.2% water; *3* 0.5% water [29]

increasing fire hazard. If water contained in the oil is not dispersed but rests at the bottom of the quench tank or is trapped in 'dead' spaces where oil does not circulate, the pressure of violently produced steam may lead to explosive expulsions of the oil which may then be set to fire.

A very dangerous situation occurs when an oil contaminated with water is used in modern closed hardening equipment where the space above the oil is filled with a protective or carbonaceous atmosphere. Although in this case oxygen contained in the air is not accessible, the gas-oil foam may easily enter the heating chamber of the aggregate where its violent gasification may result in an explosion. An additional factor which makes the situation difficult to control is that the oil cannot be observed and the operating personnel cannot notice the first symptoms of the formation of foam.

The typical symptoms indicating the presence of water in a quenching oil are:

—intensive 'boiling' of the oil when the load is being cooled down
—sputtering and characteristic sizzling (as during frying),
—intensive foaming, especially when the load is being cooled down,
—fire does not extinguish, even after the load has been submerged completely,
—the oil becomes turbid and its colour changes to become milky-brown,
—the pressure in the hardening aggregate considerably increases.

If such symptoms occur, the quenching operation should be stopped and the oil tested for water contamination. The simplest qualitative test consists of heating a small oil specimen in a glass flask placed over a flame or in a beaker placed on an electric plate heated up to a temperature above 100 °C. Oil free of water remains quiet, while that containing water will characteristically sizzle or sputter, and foam will form. It is recommended to measure the oil temperature during this test so as not to exceed the boiling point. It would also be well to determine quantitatively the water content (using e.g. the xylol method).

Water can be removed from an oil by various methods, depending upon the design and size of the quench tank and upon the type of oil. The simplest method consists of heating the oil cautiously up to a temperature of 110–130 °C so that the water evaporate. The oil should be heated slowly and foam forming on it should be observed since the bath can 'seethe'. For this reason it is advisable to pump the oil from the quench tank to a greater tank such that the oil only fills 3/4 of its volume. While evaporating the water, the oil should be continuously circulated. The water may be considered to evaporate completely when foaming ceases.

When dealing with quench tanks of great size it is recommended that the bath be maintained quiescent for about 24 hours so as to allow water to settle in the lower portion of the tank from which it can be pumped or drained out. This operation is difficult with washable oils because they contain emulgators. Unwashable oils, fresh or little used, may be passed through a centrifuge. This may reduce the water content down to about 0.05–0.1% [29].

After dewatering by one of the above techniques, the oil should be tested for water contamination using a laboratory method.

9.7.7 Ageing Process in Quenching Oils

Conditions prevailing during quenching operations induce various complex chemical processes in the quenching oil, leading to significant changes in the useful characteristics of the oil. These processes, generally called the ageing of the oil bath, chiefly include the thermal decomposition of heavy hydrocarbons into more volatile or gaseous compounds, and oxidation of some compounds. The oxidation products undergo polymerization resulting in a considerable increase of the oil viscosity. Moreover, as a result of the oil contacting the quenched surface, heated up to a high temperature, oil cracking and secondary reactions of the decomposition products occur. Oxidized hydrocarbons are unstable and when decomposing form sludge which settles on the walls of the quench tank and pipes, chokes up the coolers and filters and contaminates the surfaces of quenched parts. Under these conditions, quenching is likely to proceed non-uniformly bringing about all well known consequences.

In conventional paraffin oils, the ageing process usually brings about an increase of the quenching power. This is because the product of oil decomposition and oxidation reduce the surface tension and thereby improve the wetting of the quenched surface, and also because the solid contaminants make the oil to boil more easily. In modern high-performance oils, the ageing only slightly

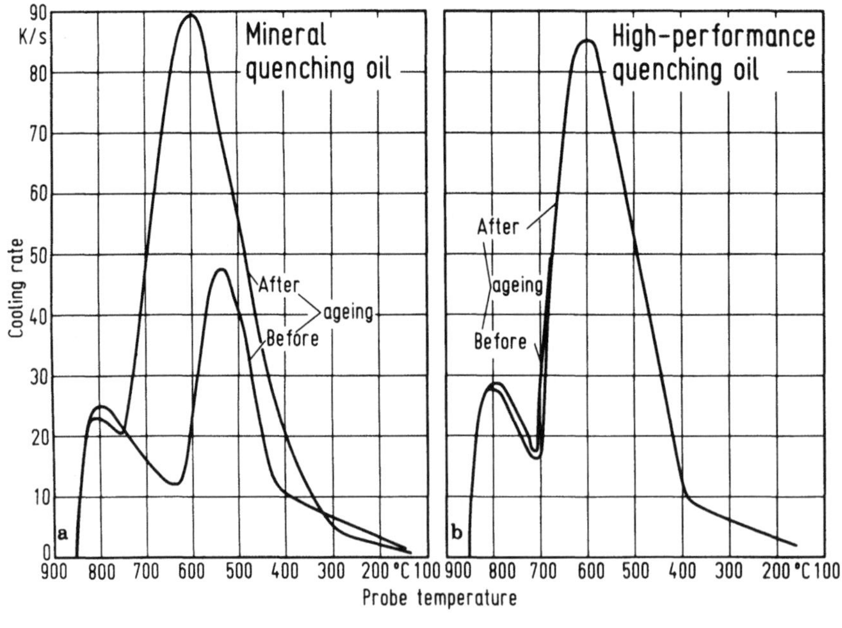

Fig. 9.32 a, b. Comparison of cooling rate profiles before and after ageing for (**a**) straight mineral oil and (**b**) a high-performance quenching oil, as determined using the Wolfson Engineering Group test, after Bashford and Mills [23]

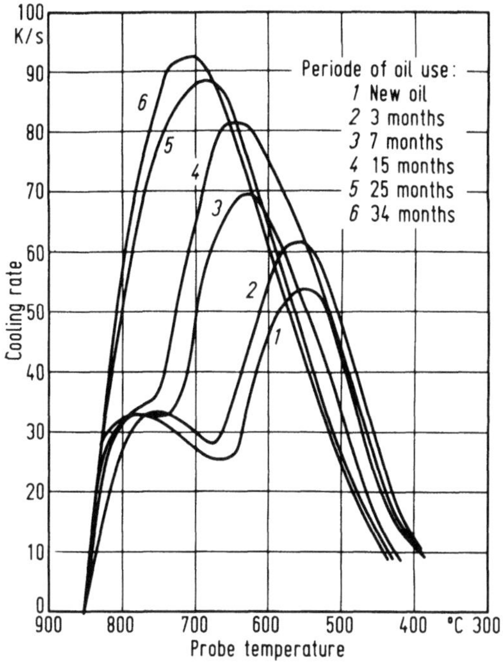

Fig. 9.33. The effect of period of use on the cooling rate temperature profile of the optimum oil at 200 °C, after J. M. Hampshire [30]

affects the thermokinetic properties, even after a several years' exploitation (Fig. 9.32). The service life of high-temperature (hot) oils is of course shorter, though it largely depends on the quality of a given oil and on the conditions under which it is operated. In the recent years, the leading manufacturers of quenching oils have made considerable achievements in improving the service life of quenching oils. As in conventional paraffinic oils, the quenching power of hot oils increases with increasing exploitation time (Fig. 9.33).

As can be inferred from the above considerations, the service life of quenching oils is not limited by adverse changes in the oil quenching power but is primarily affected by chemical processes, the products of which contaminate the oils and make them denser. Because of the chemical degradation, hot oils sometimes acquire a gelatinous consistency after they have been cooled down. Such an oil cannot be used again and it is difficult to remove it from the tank.

In practice, the quenching oil is systematically supplied with fresh oil in order to recompense drag out losses which can amount to about 10–25 kg per 1000 kg of the load. Among other factors, these supplements of fresh oil contribute to lengthening the service life of the quench oil up to several years.

9.7.8 Hot Quenching Oils

Marquenching and austempering provide essential advantages chiefly in that they minimize the dimension changes and distortion of the quenched parts, improve some of their mechanical properties, and are less expensive. Formerly, nitrate quenching baths were chiefly used for these techniques, since oils, when operated at 120–200 °C, underwent oxidation and their viscosity increased leading to shorten their service life. Since the time when efficient additives inhibiting oxidation and stimulating the quenching power have been developed, hot quenching oils have largely superseded nitrate baths, which has considerably contributed to the wide-spread use of marquenching, and for certain grades of steel also austempering.

Oils used for these quenching techniques must be able to operate at temperatures about 120–250 °C for a long time, and their viscosity should be such that permits vigorous agitation necessary for uniform quenching. They should be more resistant to ageing than are low-temperature oils and thus an addition of appropriate oxidation inhibitors is crucial for the service life of these oils. Because of the higher bath temperatures, the quenching power of hot oils is generally lower than that of low-temperature oils (Fig. 9.34), but it is markedly higher than that of nitrate baths. This permits hardening parts of greater section size, where good mechanical properties at the core are required.

According to their quenching power, hot oils may be divided into two groups: conventional and high-performance oils. Table 9.16 lists certain properties and applications of five types of the 'Martemp' oil, of which the three initial are conventional and the remaining two high-performance oils. The quenching power of conventional oils is chiefly determined by their viscosity (the quenching power decreases with increasing viscosity). In high-performance oils, the quenching power primarily depends on the relevant additives intended to enhance it.

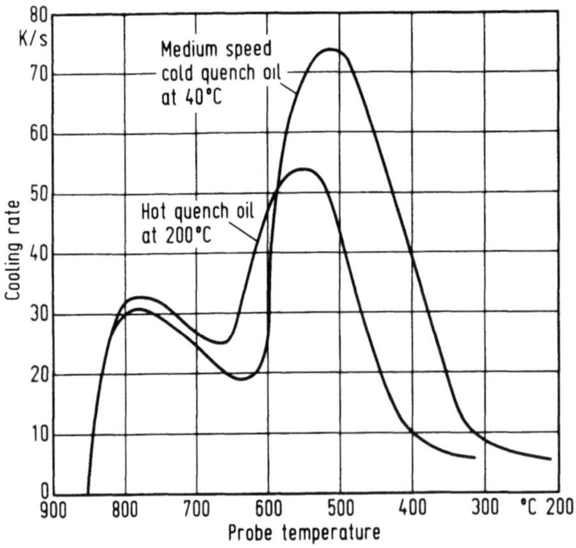

Fig. 9.34. Cooling rate curves for one of the hot oils at 200 °C and for a medium-speed cold oil at 40 °C [30]

Table 9.16. Properties and applications of high-temperature quenching oils type MARTEMP [31]

Oil type	Viscosity (mm²/s) at temperature, °C		Flash point °C	Temperature range of application[1] °C	Examples of applications
	40	150			
Martemp 600	58	3.2	261	60–150	Matempering of steel parts of
Martemp 800	105	4.3	290	80–160	moderate and high hardenability
Martemp 3500	480	9.1	360	150–250 max. 265	Austempering of nodular iron castings
Martemp 722	73	3.5	265	60–150 max. 180	Martempering of parts inclined to warping
Martemp 875	110	4.3	295	70–160	Martempering of big steel parts

[1] Upper temperature recommended for oil quenching systems applying protective atmosphere

Like low-temperature oils, all hot oils may contain emulgators, which are added in order to facilitate their washability, but do not affect the quenching power.

The service life of hot oils depends on the conditions under which they are operated, among other factors on whether the quench tank is open (i.e., whether the oil is in contact with air) or closed and the bath is protected by a special atmosphere. In order to minimize the effect of other important factors influencing the service life of hot quenching oils [32] it is recommended that

—the heat density supplied by the elements used to heat up the oil should not exceed 1–1.5 W/cm^2;
—the mode of agitation should be selected so as to minimize the amount of air entering the bath;
—violent agitation should only be used when the load is just being cooled, whereas during lay-offs the rate of agitation should be reduced and only

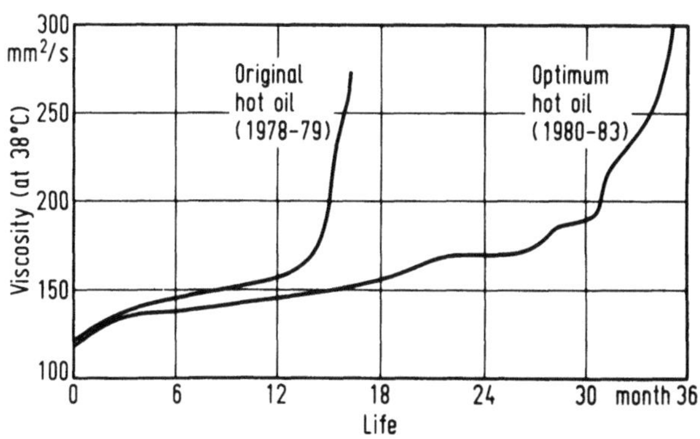

Fig. 9.35. Life of the optimum hot oil installed in 1980, compared with the originally-used products, assessed on the basis of viscosity increase in a laboratory test [30]

maintained at such a level that ensures a uniform temperature throughout the bath;

—in open quench tanks the surface area of the bath that is exposed to air should be reduced to the indispensable minimum.

If all the above rules are observed, the service life of high quality hot oils should be 2–2.5 years (Fig. 9.35). If the bath is operated in a protective or carbonaceous atmosphere, the service life of the oil may even be longer.

9.7.9 Vacuum Quenching Oils

Since the time when vacuum aggregates were first used for quenching, cooling has been one of the most difficult operation. Early aggregates were so designed that post-austenitising cooling could only be effected by means of a stream of an inert gas (usually nitrogen). This however only permitted hardening parts of small sections made of high-alloy self-hardening steels, such as stainless or high-speed steels. Therefore, the manufacturers of vacuum furnaces began to install quench oils tanks in them.

In the first experiments on oil quenching after vacuum austenitising, the oil was operated in vacuum or in nitrogen compressed to about atmospheric pressure. It appeared that in vacuum the quenching power of the oil was much lower than that achieved in the nitrogen atmosphere. Some authors attribute this to the fact that in vacuum volatile fractions easily evaporate. When quenching in nitrogen, on the other hand, parts austenitized at temperatures of about 1000 °C or higher showed structural defects such as carburazing or, in high-speed steels, incipient melting, which especially occur at corners and sharp edges. The carburazing effect can be attributed to methane evolving from the oil after the latter has come in contact with the hot quenched surface. Figure 9.36

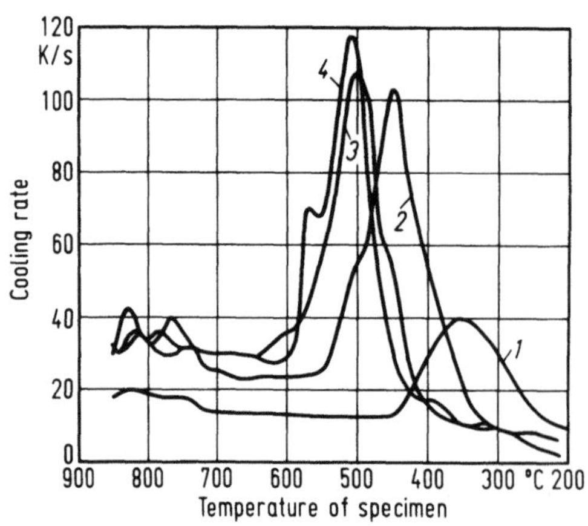

Fig. 9.36. Effect of nitrogen pressure on the cooling rate of a copper ball ϕ20 mm in a vacuum oil: *1* 6.5×10^{-3} hPa *2* 200 hPa; *3* 665 hPs; *4* 1013 hPa [33]

shows how the nitrogen pressure affects the cooling curve at the centre of a 20 mm.diam.copper ball [33]. As can be seen, this pressure affects not only the cooling rate but also the range of boiling temperatures.

Vacuum oil quenching does not produce the adverse effect upon the structure of the near-surface layer, since the gases evolving from the oil are immediately pumped out and thus the oil undergoes less decomposition and the carbon potential of the gases does not increase. The carburazing effect of quenching oils also depends on: the thermal stability of the oil, its quenching power, the rate of agitation, the size of the load and the section size of the quenched parts.

To avoid the adverse effects described above, some manufacturers of quenching oils have developed special types of oils to fit vacuum quenching. The major requirements that these oils should fulfill are: a low vapour pressure, a small viscosity, no effect upon the cleanness of the quenched surface and a sufficiently high quenching power. Some specialists however think that any conventional quenching oil providing bright hardening can be used in vacuum furnaces, provided that the quench bath is properly prepared by

—heating the oil, with appropriate precautions, under the atmospheric pressure, at a temperature slightly below the flash point for several hours so that the volatile fractions and water residues are removed, and
—properly degasing and agitating the oil during the load heating cycle performed in vacuum.

Based on experience, some manufacturers of vacuum furnaces recommend the following measures to be taken in order to avoid carburization and incipient melting when quenching parts of section size of the order of 50 mm made of high-speed steels [34]:

—until the load is cooled down to at least about 990 °C, a low vacuum should be maintained above it;
—austenitization should be performed at the lower admissible austenitizing temperatures for an appropriately prolonged time;
—in some cases, it appears sufficient to cool the load down to about 950 °C in an inert gas and then to continue the cooling in oil.

From the above considerations we can see that though the use of quenching oils in vacuum furnaces permits hardening parts of greater sections and a wider range of steels may be used, quenching in gases is easier and less expensive. This is the reason why, in modern vacuum furnaces, the tendency for quenching at greater rates but in a high-pressure gas has prevailed.

9.7.10 Fire Hazard and Safety Precautions

As it is known, the inflammability of mineral quenching oils is their primary disadvantage. The ignition of oil in a quench tank may bring about catastrophic consequences, but if the tank is properly designed and the hardening operation is correctly carried out, this happens rarely though cannot be excluded. In order to prevent oil ignition, or when this happens to extinguish the fire, it is necessary to know all the factors that increase and decrease fire hazards in quenching oils. Primarily, we should emphasize that as a rule it is not the oil itself that is set on fire but its vapours, which together with oxygen form a flammable mixture. Similarly oil form, composed of oil vapours and air, can ignite.

The major causes of the ignition of oil vapours and gases are:

—overheating of the oil when the ratio of its mass to the mass of the quenched load is not high enough;
—overheating of the oil in the immediate vicinity of the load when the oil is poorly agitated or its viscosity is too high (see Fig. 9.27),
—quenched parts are not completely submerged because of a defect occurring in the hoists or in the conveyors or because the bath is not sufficiently deep,
—quenched parts are being immersed too slowly,
—water contamination of the oil exceeds the permissible value,
—the load is removed from the bath too early.

The ratio of the oil mass to the group mass of the quenched load should be about 10:1. When the oil agitation is good or when only a few parts are to be quenched, this ratio may be reduced to 7:1.

The importance of a forced oil circulation has been discussed in detail in Sect. 9.7.5. From the point of view of fire hazard, the oil agitation ensures a quick distribution of heat throughout the bath and thus prevents an excessive local increase in its temperature. For the same reason, the oil viscosity should be limited to 2–3 °E and the temperatures, not only the upper temperature but also the lower one (between 30–80 °C) upon which the viscosity depends, should not be exceeded. After the quenching operation has been completed, the oil temperature should be at least 40–60 °C lower than the fire point and appropriately more lower than the flash point.

Incomplete submersion of the load, irrespective of its cause, always leads to fire. That is why the hoists and conveyors should be equipped with a breakdown protection device (e.g. pneumatic) for use in the event of an electric power failure during a quenching cycle.

The rate at which the load is being immersed in the oil has no greater importance in continuous furnaces, i.e., when parts to be quenched are not numerous and are lowered into the oil gravitationally. Similarly in sealed aggregates, where the whole load is submerged at once but the oil is protected by an oxygen-free atmosphere. The situation is completely different when parts are quenched in open tanks. Here, when the parts are being submerged too slowly we shall always have fire, which though usually extinguishes after the parts are completely submerged, may also turn into a long-lasting fire. It is recommended that the rate of submerging the parts should be not less than 14–22 m/min, and the depth at which the parts are immersed should be at least 20–30 cm below the bath surface [35]. The rate of submerging is especially important when dealing with tubular parts, since large amounts of vapour formed in their interior and flowing upwards may cause fountain-like ejections of the oil.

The role of water as a factor increasing fire hazard has been discussed in Sect. 9.7.6. Water contamination poses in practice one of the greatest danger of fire. In baths with water contents of the order of 0.2–0.3%, and operated at above 100 °C, oil foam is intensely formed and can overflow from the tank not only during the quenching operation.

Measures undertaken to prevent fire in oil tanks and methods of extinguishing the fire when it has already occurred are [35, 36]:

—the oils used in a given quenching operation should show a low ability to evaporate at the temperatures employed;
—the bath should be sufficiently agitated,
—the appropriate ratio of the oil mass to the gross mass of the quenched load should be maintained;
—the bath temperature and the oil level should be monitored;
—the quench tank should be furnished with a ventilation system and possibly with carbon dioxide fire extinguishers to smother a fire if set inside the vent pipes (recommended in large ventillation systems);
—an automatic system for detecting excessive water contamination (0.1%) should be installed (especially in closed tanks);
—auxiliary oil tanks connected an appropriate pipes with the quench tank should be installed outside the hardening plant so as to enable the quench oil to be quickly (within at most 10 min) drained out;
—large open tanks should be equipped with a carbon dioxide fire extinguisher or a device producing an inert gas curtain (usually nitrogen);
—appropriate fire extinguishing system should be installed in the hardening plant.

Moreover, the operating personnel of the hardening plant should be trained in fire prevention and fire extinguishing.

9.8 Saltbaths Used in Martempering and Austempering

9.8.1 General Description

Martempering and austempering require quenching media that can operate at temperatures from 150 to 550°, such as hot mineral quenching oils described in the preceding section, molten salts or alkali, some low-melting metals or their alloys, and fluidized beds. Formerly, lead and its alloys with tin were the most often used metals (pure lead was chiefly used for wire patenting). Since however lead vapours are toxic whereas its quenching power does not much differ from that of molten salts, the use of lead for quenching has almost entirely been abandoned. Thus far, fluidized beds are not widely used either (see Sect. 9.10), nor are hot mineral oils because of their narrow operating temperature range (150–200 °C, exceptionally to 250 °C). Therefore, at present, molten salts and alkali are the most widely used quenchants in marquenching and austempering.

In molten salts, metals and fluidized beds the heat transfer mechanism is totally different from that in the fluids which begin to boil before the hardening temperatures are reached. According to the Newton's law, the heat flux density and, thus, the cooling rate, are proportional to the surface temperature gradient. Hence the fastest cooling occurs in the initial phase of quenching (at about 800–700 °C) and not in the region of pearlitic transformation as required in hardening from the point of view of the kinetics of phase transformations (Fig. 9.37). This is the major reason why, using these quenchants, marquenching and austempering may only be applied to parts of small, or occasionally medium, section size.

The quenching power of salt baths is generally poor, but can be modified within relatively wide limits by changing the agitation rate of the quenchant, by adding some amount of water to it, or by adjusting these two factors at the same time which is especially efficient. Although the thermal conductivity of molten salts is smaller than that of molten metals by two orders of magnitude, salts are superior to metals in this respect. While the specific heat capacity of molten lead is about 125 kJ/K·kg, that of salt baths and alkali baths is at least ten times as great. For this reason and also because the viscosity of salt baths is much lower than that of metal baths which facilitates a quenchant motion, the quenching capacities of these two media do not much differ.

A basic disadvantage of saltpetre baths lies in that they rapidly decompose after the admissible temperature is exceeded and that they enter in exotermic reactions with inflammable materials. These reactions may result in fire or even in explosion. Therefore, when using baths of this type, safety reasons require that appropriate precautions (detailed in Sect. 9.8.4) be strictly observed. Molten alkali and alkali-chloride mixture are much safer in use, but because of their higher melting temperatures they cannot fully replace saltpetre baths.

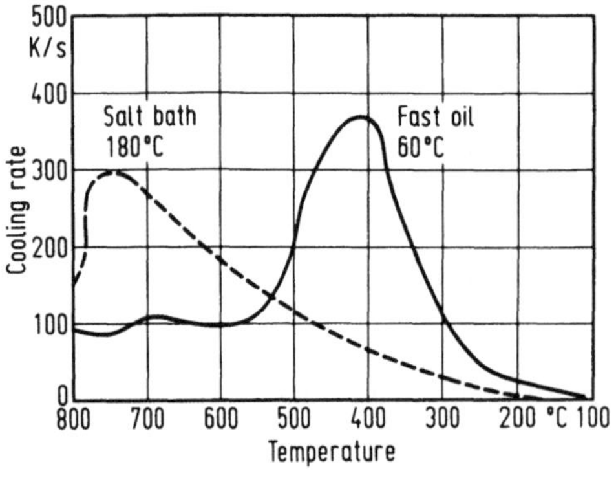

Fig. 9.37. Cooling rate of a silver ball ϕ20 mm in a molten salt and a fast oil, after E. Burgdorf

The bath temperatures employed in marquenching and austempering are imposed by the temperature (M_S) at which for specified austenitizing parameters the martensite transformation begins in a given grade of steel. Thus each of the bath types (oils, saltpetre and alkali baths) used for quenching nowadays has its specific range of application temperatures.

For oils this temperature range does not exceed 250 °C, for saltpetre baths it may be taken to be 160–500 °C, and for alkali baths is 350–600 °C. An addition of water reduces the melting temperature of salt and alkali baths and thereby widens their application range towards lower temperatures. Structural steels usually require a temperature of about 200–350 °C, while high-speed steels are marquenched at 450–600 °C.

Saltpetre salts cause slight oxidation and colouring of the marquenched surface, whereas marquenching in alkali baths provides clean light-gray surfaces. Salt residues left on the quenched surface are easily soluble in water, especially at increased temperatures.

9.8.2 Saltpetre Salts

Sodium and potassium nitrates and nitrites have relatively low melting temperatures, and the 1:1 nitrate-nitite mixtures melt at 150 °C. Figure 9.38 shows the melting isotherms for these mixtures. The sides and diagonals of this square enable us to find the boiling temperatures for binary systems, and each point positioned inside the triangle represents the composition of a triple system. Water added to saltpetre baths chiefly in order to increase their quenching power, in addition reduces their melting temperature. For example, a mixture composed of 53% KNO_3 + 40% $NaNO_2$ + 7% $NaNO_3$ but containing no water melts at 140 °C, while the same mixture containing 2–3% water melts at about 100 °C.

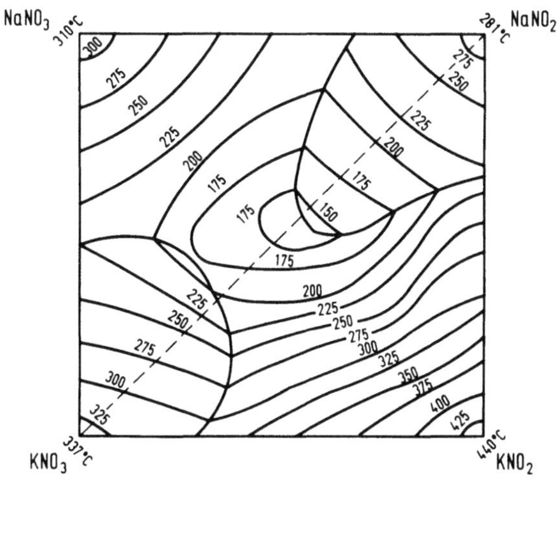

Fig. 9.38. Isothermic sections of the liquidus surface of a KNO_3–KNO_2–$NaNO_2$–$NaNO_3$ system [37]

An addition of water also reduces the mixture viscosity and thereby its cooling rate. Table 9.17 gives several examples of saltpetre baths used in practice and show their temperature application ranges. Table 9.18 lists the saltpetre baths commercially available in the Federal German Republic.

Salt baths cool relatively slowly and uniformly but, as already shown in Fig. 9.37, the highest cooling rate does not occur in the desired temperature range. The factors able to modify the cooling rate between relatively wide limits are

—bath temperature (according to the Newton's law, a decrease in the temperature increases the cooling rate), Fig. 9.39;
—viscosity, which when decreasing due e.g. to an increased difference between the melting temperature and the temperature of the bath, increases the cooling rate;
—rate of agitation, a proper selection of which is not only very important for safety reasons but also, since heat transfer occurs through convection, necessary to obtain uniform quenching;
—an addition of water in an amount depending on the bath operating temperatures; water is the most effective factor increasing the cooling rate of the bath.

Table 9.17. Chemical compositions and temperature range for applications of saltpetre salts [6]

Chemical composition, %	Melting temperature, °C	Temperature of application, °C
$NaNO_3$	308	320–550
KNO_3	334	350–550
$NaNO_2$	284	323–550
KNO_2	387	—
45% $NaNO_3$ + 55% KNO_3	220	230–550
50% $NaNO_3$ + 50% KNO_2	150	160–550
50% $NaNO_2$ + 50% KNO_3	225	280–550
25% KNO_3 + 75% $NaNO_3$	240	380–540
50% $NaNO_3$ + 50% $NaNO_2$	205	260–600

Table 9.18. Salt baths produced in the German Federal Republic and their applications [31]

Mark	Type of composition	Melting temperature °C	Applications temperature °C	Application
AS-140	Saltpetre salts	140	160–550	Marquenching
AS-220	Saltpetre salts	230	250–550	Austempering and marguenching
GS-430	Chloride salts	430	470–700	Marquenching of high-speed and tool steels

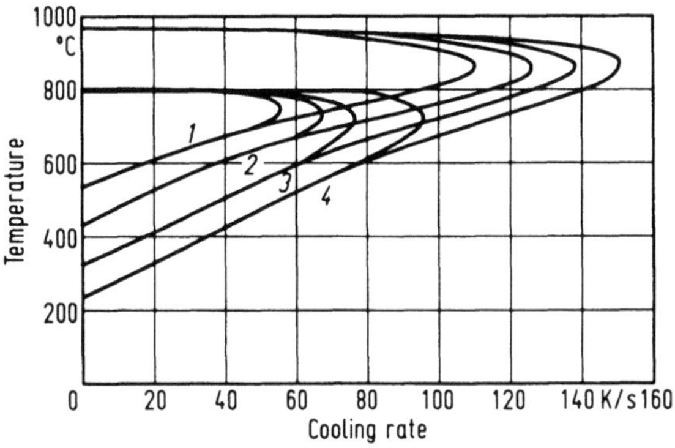

Fig. 9.39. Cooling rates of a silver ball ϕ20 mm in a molten salt 50% KNO_3 + 50% $NaNO_3$) at various temperatures after Luega and Pomp: *1* 550 °C; *2* 450 °C; *3* 350 °C; *4* 250 °C [6]

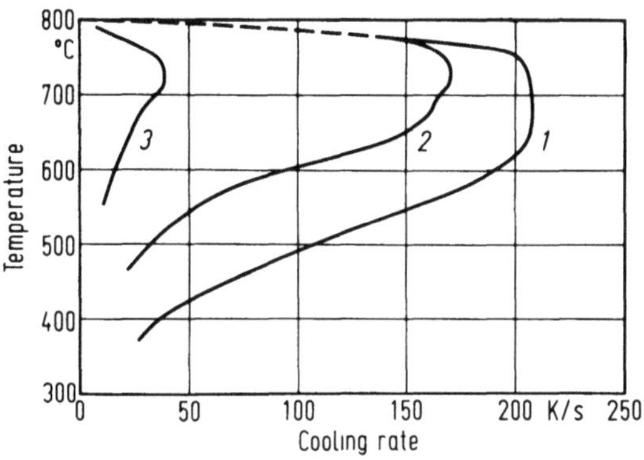

Fig. 9.40. Effect of water addition to saltpetre baths on the cooling rate of a silver ball ϕ20 mm: *1* 300 °C, 0.18% H_2O, *2* 400 °C, 0.09% H_2O, *3* 400 °C no water [38]

Approximately, the solubility of water in saltpetre salts (55% KNO_3 + 45% $NaNO_2$) is: at 200 °C about 3.8%, at 300 °C about 0.5% at 400 °C about 0.18% and at 500 °C about 0.12% [38]. The strong influence of water upon the cooling rate and upon the width of the temperature range within which the cooling rate is at maximum is illustrated in Fig. 9.40. As water evaporates, which most intensively occurs when the load is being cooled, the water content of the saltpetre bath decreases, and the water losses should be recompensed. Obviously, the rate of water evaporation depends in a considerable degree upon the bath temperature (Fig. 9.41). Too high water contents are undesirable, since excessive amounts of steam may violently agitate the bath. The recommended water contents [7], depending on the bath temperature, are

200 °C—0.5–2%
250 °C—0.5–1%
300 °C—0.25–0.5%
400 °C—0.10–0.25%.

Water should be added to melten salts and alkalis very caustiously so as to avoid spattering. While adding water, the bath should be strongly agitated.

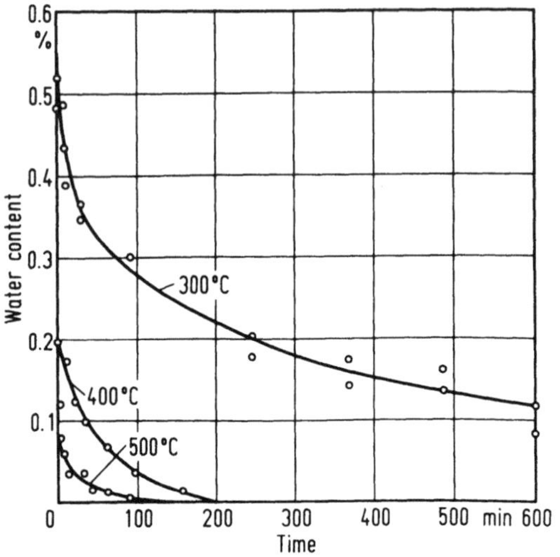

Fig. 9.41. Time variation of the water content in salt baths at temperatures: 300, 400 and 500 °C [38]

The water should be passed through a pipe of small diameter, not greater than 6 mm, and discharged, near the quenchant outflow from the pump or near the vane used for agitating the bath. An experienced operator can approximately estimate the water content of the bath by observing the intensity of evaporation when submerging parts to be quenched. The water content may also be evaluated by taking a specimen of the salt bath, weighing it, heating up to about 400 °C so that the water evaporates, and then weighing the specimen again. Another method consists of determining the solidification temperature and then reading the water content out from a diagram plotted for a salt of known composition. A device for measuring the water content based on the measurement of the electrical conductance of the bath is described in [39].

Despite their important disadvantages when used as quenchants (primarily their possible exotermic reactions and susceptibility to explosions) molten salts and alkalis are still in wide use even in highly developed countries since other quenchants able to operate at temperatures from 250 to 550 °C have not yet been developed. For safety reasons, saltpetre salts must not be used for quenching parts whose austenitizing temperature is higher than 950 °C and section sizes exceed 15–20 mm, even if a small number of parts, or a single part, are to be quenched. With larger loads, even those consisting of parts of small section size, the admissible temperature of the bath, which is about 550 °C, may be locally exceeded. This cannot happen in modern automatic equipment. Hardening plants where such equipment have not been installed prefer using alkali baths possible with some addition of various salts.

9.8.3 Martempering and Austempering in Molten Alkalis and Alkali-Salt Baths

Molten NaOH and KOH, whether or not they contain salt additives (Table 9.19), are safer in use than pure saltpetre baths and do not oxidize the quenched surface. Although NaOH melts at 328 °C and KOH even at 360 °C, mixtures of these alkalis, especially those containing water, melt at much lower temperatures (Fig. 9.42). The significance of the melting temperature lies in that the lower it is below the bath temperature, the smaller the viscosity of the bath and thereby the higher its quenching power. However the quenching power of alkalis depends on their chemical composition and their melting temperature

Table 9.19. Chemical compositions of some alkali and salts for martempering and austempering [6]

Chemical composition, %	Melting temperature, °C	Application temperature °C
$NaNO_3$	328	350–700
KOH	360	400–650
75% NaOH + 25% $NaNO_3$	280	420–540
37% NaOH + 63% KOH	160	180–350
38% KOH + 22% NaOH + 20% $NaNO_3$ + 15% $NaNO_2$ + 5% Na_3PO_4	150	160–360
70% NaOH + 15% $NaNO_3$ + 10% $NaNO_2$ + 5% Na_3PO_4	250	280–600

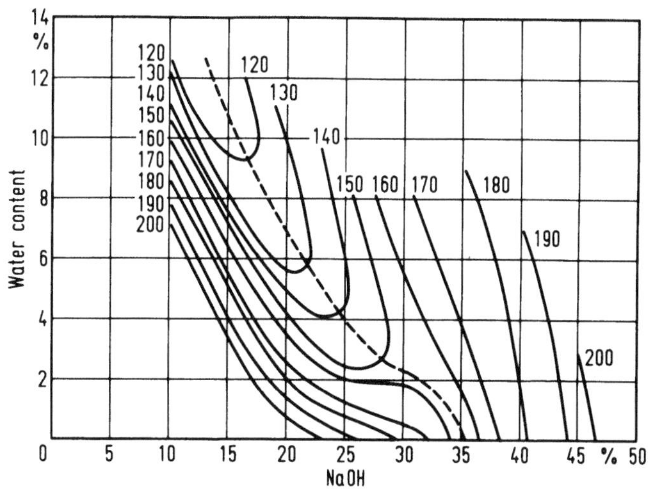

Fig. 9.42. Solidification isotherms of the NaOH–KOH–H_2O systems [40]

much less than that of saltpetre salts. The temperature of an alkali bath, on the other hand, significantly affects its cooling rate.

The most effective factor increasing the cooling rate of alkali baths is water, just as with saltpetre salts (Fig. 9.43). Water contents exceeding 10% are not however recommended since bath containing greater amounts of water can 'seethe' upon submerging the load. Besides, water should not be added to baths operated at temperatures higher than 250 °C. For baths operated at 25–300 °C, anhydrous mixtures of NaOH and KOH are most often used, whereas for those operated above 350 °C, commercially pure NaOH is recommended. A bath composed of NaOH and KOH in the ratio 20:80 and containing 6% water solidifies at 130 °C. As seen from Fig. 9.43, under the specified conditions and at a temperature of 700 °C this bath cools a 25 mm·diam. steel ball at a rate of 110 K/s. Under the same conditions, a conventional oil will cool the same ball at a rate of 55 K/s, and water at a rate of 450 K/s [40].

Parts quenched in alkali baths need not be sand-blasted or etched, but alkali residues should be course be washed out from their surface. Water added to the bath improves the quality of the as-quenched surface. However alkalis, even when carefully washed out after quenching, increase the susceptibility of the quenched part surface to atmospheric corrosion. For this reason, parts quenched in alkali baths are sometimes passivated or oxidized. An addition of Na_3PO_4 to alkali baths (see Table 9.19) increases the resistance of the quenched parts of corrosion.

As regards the service life and the stability of characteristics, alkali baths are inferior to saltpetre baths. As they adsorb carbon dioxide from the air, sodium and potassium carbonates form, and when the bath is contaminated with barium chloride derived from salt used for heating, in addition barium carbonate forms. These carbonates in conjunction with metal oxides make the alkali bath more dense, modify its composition and quenching power. To counteract these effects, it is recommended to remove the clots formed, replenish the

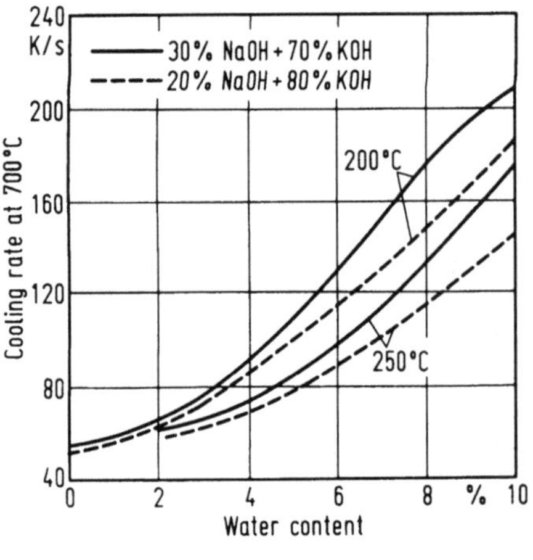

Fig. 9.43. Variation of the surface cooling rate at 700 °C as a function of water content for a steel specimen ϕ25 mm quenched in molten alkali at 200 and 250 °C [40]

bath with fresh alkali, and deoxidize it by adding $K_4Fe(CN)_6$, borax, or a mixture of the two. These operations and the necessity to monitor continuously the characteristics of alkali baths cause considerable inconvenience in practice.

9.8.4 Safety Precautions when Using Saltpetre Baths

The inflammability and the liability to explosions of saltpetre baths, the chemical aggressiveness of alkalis and the relatively high operating temperature of these baths impose rigorous safety precautions on both the design of the quench facilities and the use of these type of quenchants. Detailed instructions are usually given by the manufacturers of quench tanks and quenching salt baths. Here we shall only discuss the major general rules to be followed when using salt baths [41].

Heat treatment operations, especially when salt baths are used, should be performed in a separate, well ventilated room with two emergency exist, or at least the heat treatment facilities should be well isolated from other equipment. Quench tanks should be made of heat-resisting metal sheets lined, in the interior, with bricks or other ceramic material and should be heated by electric tubular heaters. The temperature should be controlled by at least two self-operated temperature regulators, one of which should provide an alarm signal in case the maximum admissible temperature (which for saltpetre baths is taken to be 550 °C and for the heater shields to be 660 °C) is exceeded. A safe operation of a quench tank furance largely depends on its cover which should be tight and easily actuated.

Before a quench tank furnace is brought into operation the following preparatory operations should be performed (including melting of salts):

—clean the internal walls of the tanks, heating tubes and other elements out of salt residues, scale, metal pieces etc.;
—check out the tank for defects such as fractures and pits;
—check the electric supply and measuring circuits;
—set the temperature regulator to a temperature about 50 °C higher than the melting temperature of the salt employed, switch on the heaters and gradually fill the tank with salt, powdered or broken into fine pieces; care should be taken to ensure that the heating tubes are always covered with the salt.
—heat up the salt to a specified temperature and check whether the temperature regulators operate properly.

Two carbon dioxide fire extinguishers should be installed in the room where the furnace is located.

When submerging parts to be quenched into the salt bath, the operating personnel should wear protective gloves and transparent face protectors. The load should be dry and cleaned out of oil and carbon contaminants. Inflammable materials (wood, oil, carbon, paper, rubber), cyanide salts, easily oxidizing metals (e.g., magnesium alloys), etc. should not be stored in the vicinity of the quench tank nor should they in any case be brought in contact with the salt. Parts austenitized in a salt bath containing cyanide additives should prior to quenching in saltpetre be immersed in an inert chloride bath heated to a temperature close to that at which they have been austenitized.

When working with seltpetre baths no protective coverings, especially carbonaceous, may be used. Possible fire may be smothered with carbon dioxide extinguishers or with dry sand; water must not be used for this purpose.

After a quenching operation is completed and the supply power is switched off, it is recommended that the salt, while still molten, should be poured out of the quench tank into small clean and dry containers, using an appropriate scoop. The level of salt residues left in the tank should not be higher than that of the upper heating pipes.

All tools, vessels, fixtures, etc., covered with solidified salt should be carefully cleaned with warm water. Sludge deposited on the tank bottom should be periodically removed according to the operative work-safety instructions, which should be established in accord with given working conditions and production types and made known to the personnel of the hardening plant.

9.9 Gas Quenching

9.9.1 Air Quenching

Some grades of steel show such a great hardenability that a quenching operation performed in quiescent air, applied after austenitizing, is fast enough to produce a martensitic structure, provided that the section sizes of quenched parts are not too great. The heat transfer coefficient α of quiescent air (Fig. 9.44) is however so much smaller than that of other quenchants (see Sect. 9.2) that most structural and tool steels are not hardened when quenched in this medium. A quiescent air quench applied after austenitizing usually produces a pearlitic structure; this kind of heat treatment is known as normalizing.

The cooling rate of high-pressure air is higher than that of quiescent air, but not sufficient to fit most applications. High-pressure air is sometimes used

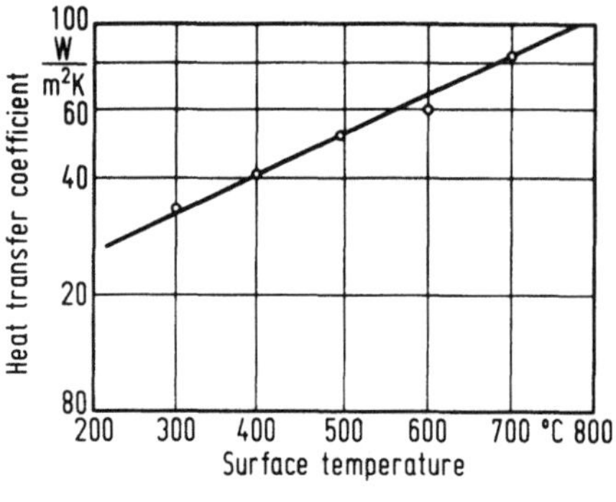

Fig. 9.44. Heat transfer coefficient for air cooled oxidized steel as a function of surface temperature [42]

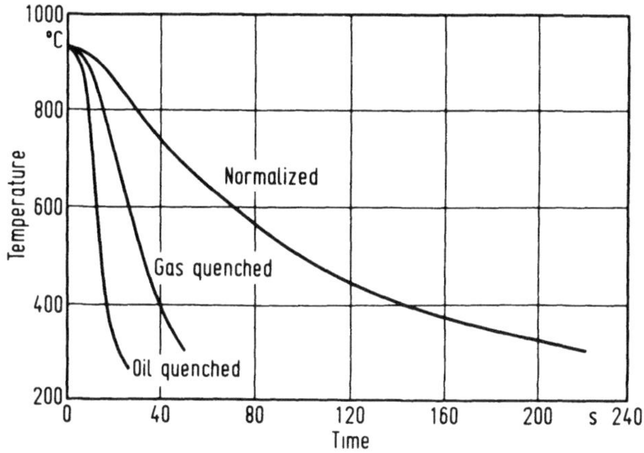

Fig. 9.45. Cooling curves obtained in quenching of 4130 steel tubing $\phi 31.7 \times 1.6$ mm in gas, oil and still air (normalizing) [7]

for quenching parts of small section size such as thin-wall pipes, metal sheet products, thin wires etc., made of steel of sufficiently high hardenability. Generally, air-stream quenching is employed when quiescent air quenching does not provide the required hardness and an oil quench may produce excessive distortions or is not recommended for other reasons. An air stream cools at an intermediate rate, falling between that of quiescent air and that of oil (Fig. 9.45). An air-stream quench is sometimes also applied to large heavy parts (e.g., rotary turbine) made of high-hardenability steels, since an oil quench may induce great thermal and structural stresses in such parts and since fire hazard is especially great in this case.

Air quenching requires an appropriate equipment, specially designed for a given part of given shape and mass so that the velocity and temperature of the air stream are sufficient and the entire quenched surface is uniformly cooled down. The air may be injected through appropriately located nozzles or fans. Typical examples of air quenching and the design principles of the relevant equipment are given in Metals Handbook [7].

In order to increase the cooling rate, the air stream is sometimes mixed with dispersed water (fog quenching). This quenching technique may for example be applied to die impressions of some type. It should however be remembered that fog quenching is very fast at low temperatures, which is a well known property of water.

9.9.2 In Situ Gas Quenching in Vacuum Furnaces

Initially, in a post-austenitizing quench performed in vacuum furnaces, the only quenchants used were circulating inert gases usually compressed to about 800 hPa.

In furnaces of this type, before proceeding to a quench operation, hatches installed in the bottom and roof of the heating chamber are lifted pneumatically and then a fan forces the cooling gas to flow through the load. After absorbing heat from the load, the gas is cooled down by passing it over the cold walls of the furnace and over water cooled heat exchangers.

Among the cooling gases that can be used in vacuum furnaces (nitrogen, argon, helium, hydrogen), nitrogen is most widely used, though the greatest quenching power is shown by hydrogen (Fig. 9.46) and helium is not much slower than hydrogen. Taking the time required for hydrogen to cool from 1000 to 500 °C to be 1, for helium this time will be 1.37, for nitrogen 1.50, and for argon as much as 1.95. As however known hydrogen is explosive and at temperature above 1000 °C shows decarburizing properties; helium, on the other hand, is very expensive and not easily available. For these reasons hydrogen was only recommended in specific cases, as an additive increasing the cooling rate of nitrogen, and its concentrations did not exceed 10%. At present, when we know other entirely safe ways of increasing the cooling rate in gas quenching, hydrogen is no longer used as an additive.

The basic factor limiting the use of gas quenching under normal pressure is the low quenching power. Even in tools made of high-speed steel this technique does not provide a sufficiently high hardness and a proper structure (the precipitation of carbides V_4C_3 cannot be avoided) when the section size exceeds 20–25 mm. It was chiefly to overcome this barrier that two- and three-chamber vacuum furnaces with provision for oil quenching were later developed. These however bring about other problems, already discussed in Sect. 9.7.9, not mentioning that furnaces of this type are much more expensive. Therefore, the designers of modern quenching installations have returned to single-chamber furnaces with gas quenching, which however is performed at considerably increased pressures (up to 0.6 MPa). The quenchant circulation paths have also been improved and more effective heat exchangers employed. The design of

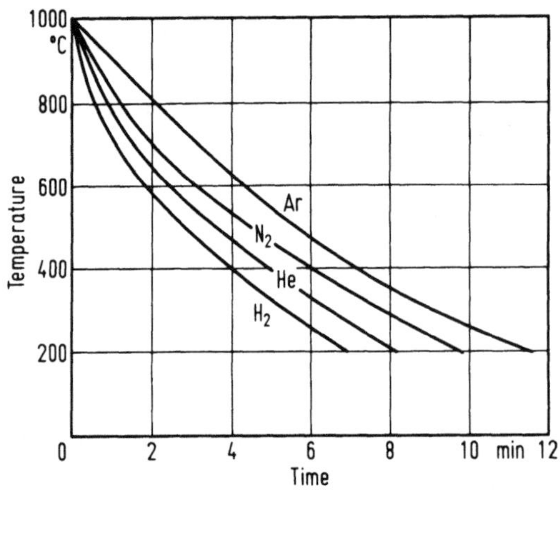

Fig. 9.46. Center cooling curves of steel specimen in 38 × 38 × 38 mm (2% C, 13, 5% Cr) various gases at a pressure of 933 hPa, after W. Cheary

these modern quenching installations has been based on theoretical studies and numerous experiments.

Using thermodynamic theory Hilpert has determined how the heat transfer coefficient α varies as a function of the pressure and flow rate of the gas and established the simplified relationship [2]

$$\alpha \approx (w \times p)^m$$

where w is the flow rate of the gas, p is the gas pressure, and m is the exponent whose value has been found empirically to be 0.6–0.8. From the relationship it follows that as the flow rate and the pressure increase, the value of the coefficient α increases and thereby the cooling rate also increases. It is physically understandable, since as the pressure and flow rate increase, the number of gas particles that come in contact with the quenched surface increase. It also follows from this formula that the cooling rate of gases does not increase in proportion to their pressure. Experiments have shown that an increase in pressure, e.g., from 1000 to 6000 hPa causes the cooling rate to rise 3.5-fold and not sixfold. R. Bauer [2] reports that in a nitrogen quench at a pressure of 1000 hPa the value of α was 100–150 W/Km^2 and at an increased pressure it was 300–400 W/Km^2. Therefore, even the values at increased gas pressures are considerably less than those for oil quenchants, but close to the values obtained with salt baths and fluidized beds (see Sect. 9.2). The values given above should however be regarded as rough estimates since the actual cooling rate depends not only on the pressure and flow rate but also on several other factors such as the design of gas supplying equipment and of heat exchangers and the arrangement and the shape (the surface area-to-mass ratio) of the load.

In the recent years, many works have been published, discussing various aspects (the conditions, possibilities and restrictions) of gas quenching in vacuum furnaces. Below we shall very briefly present some of them.

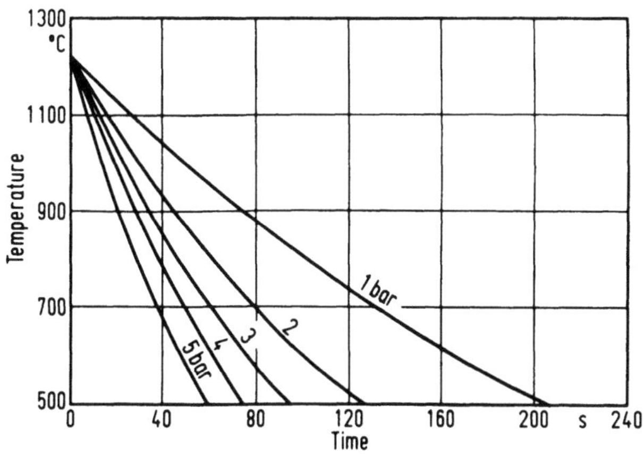

Fig. 9.47. Center cooling curves of a S6-5-2 steel specimen $\phi 25 \times 48$ mm in nitrogen at pressure 1–5 bars [44]

F. Limque and F. Bless [44] discuss the influence of nitrogen pressure, varying between 1 and 5 bars, upon the centre cooling rate of a specimen, 25 mm in diameter and 48 mm long, made of S 6–5–2 steel (Fig. 9.47). Examining a 100 mm long specimen made of the same steel, they have found how the specimen diameter affects the cooling rate and the hardness developed on the surface and at the core of the specimen after tempering it twice at 560 °C for two hours (Fig. 9.48). As seen from the figure, even with large section-sizes, gas quenching at a plessure and 5 bars provides high hardnesses.

In their next work [45], the same authors compared the cooling rates, the hardnesses developed and the service life of 25 mm·diam. by 54 mm long cutters and 6.5 mm·diam. by 13.3 mm long spiral drills, made of high-speed steels, arranged in a 100 kg net mass workload, which were quenched, after vacuum heat treatment, in a nitrogen stream under pressure of 5 bars and in a salt bath at a temperature of 550 °C. The results of the two treatments were almost identical.

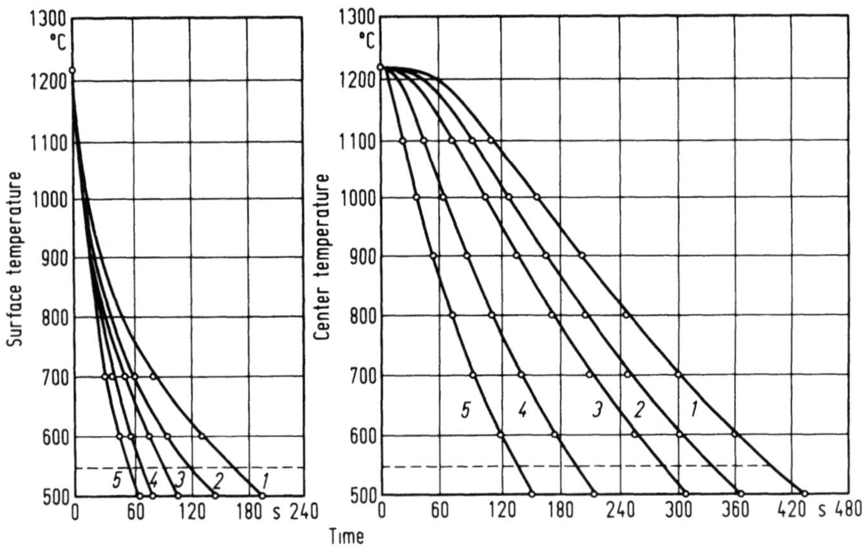

		Surface hardness	Center hardness	
1	Φ 100mm	63.5 HRC	62 HRC	after 2×tempering at 560°C/2h
2	Φ 80mm	65.0 HRC	64 HRC	
3	Φ 60mm	65.5 HRC	65 HRC	
4	Φ 40mm	65 5 HRC	65 HRC	
5	Φ 25mm	67.0 HRC	67 HRC	

Fig. 9.48. Surface and center cooling curves of S6-5-2 steel specimens ϕ25–100 mm in 5 bar nitrogen, and hardness obtained after 2x tempering at 560 °C – 2 hr [44]

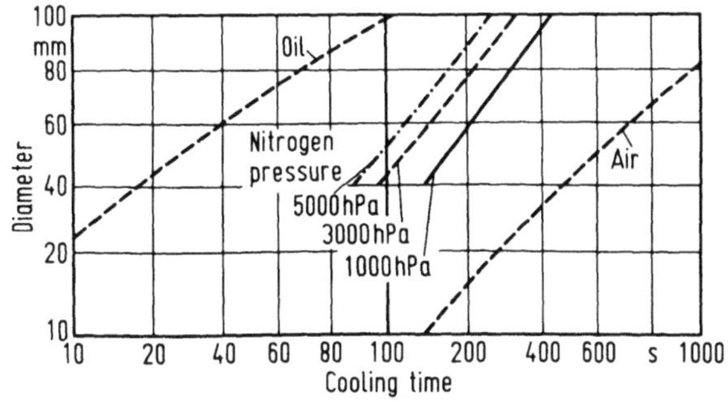

Fig. 9.49. Effect of nitrogen pressure and bar diameter S6-5-2 steel on the centre cooling time in the range 800 – 500 °C [46]

By similar tests, the same results have been obtained by the authors of [46]. Following the German school, in addition to the cooling curves shown in Fig. 9.48, as a measure of quenching power they used the cooling time, defined as the time required for the temperature to decrease from 800 to 500 °C. This parameter is very important in high-speed steels, where a rapid quench permits avoiding V_4C_3 precipitation. Figure 9.49 shows how the cooling time varies with gas pressure and bar diameter. We can also clearly seen from this figure that the quenching power of high-pressure gases falls between the quenching power of oil and that of air. The nitrogen pressure affects the as-quenched hardness of high-speed S 6–5–2 steel in a relatively insignificant degree and evidently less than does the bar diameter (Fig. 9.50). As will however later be shown, in other types of steel, the effect of the quenching gas pressure is considerable.

J.W. Bouwman [47], like Limque and Bless in [45], compres the cooling curves and hardnesses obtained by nitrogen quenching at a pressure of 5 bars and in a salt bath at a temperature of 550 °C. His tests were however performed in a vacuum furnance of improved design with provision for changing the direction of the gas stream and thereby ensuring more intensive and more uniform quenching. The gas was cooled in a heat exchanger installed outside the furnace chamber and this also improved the cooling rate. As a result, the hardening

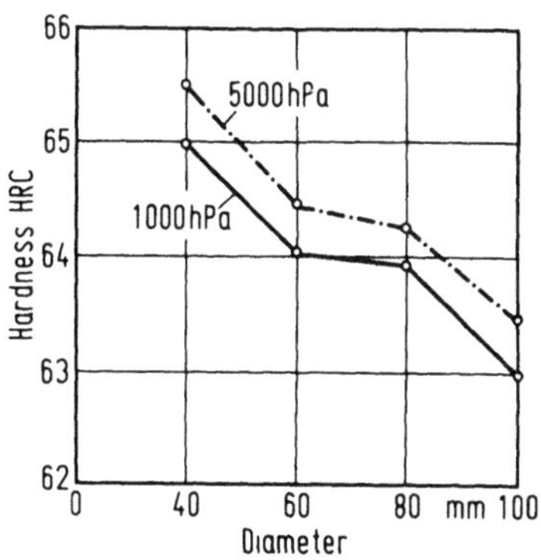

Fig. 9.50. Effect of nitrogen pressure and bar diameter S6-5-2 steel on hardness after quenching [46]

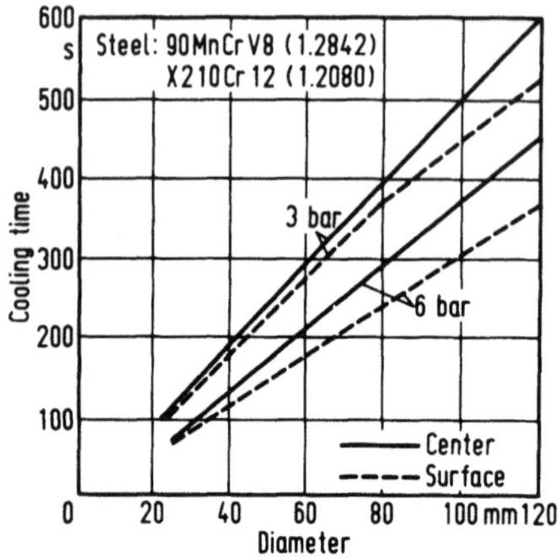

Fig. 9.51. Effect of nitrogen pressure and bar diameter on the coling time in the range 800–500 °C [48]

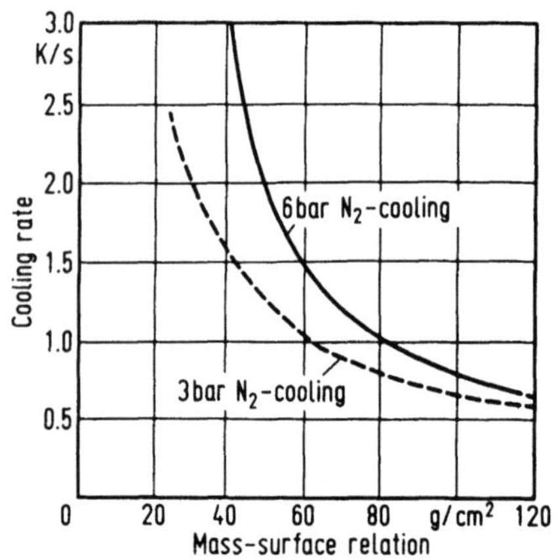

Fig. 9.52. Centre cooling rate in the range 800–500 °C for a bar ϕ40–120 mm – as a function of the mass-to surface area ratio [48]

quality was improved and the heat treatment time reduced with accompanying reductions in cost.

Apart from high-speed steels, cold work and hot work steel tools (e.g. dies) can also be in situ quenched in vacuum furnaces by a stream of high-pressure gas. H. Giesser et al. [48] examined cylindrical specimens diam. 80 and 120 mm, made of X210Cr12 steel (No 1.2080), and other cylindrical specimens diam. 40 and 80 mm, made of 90MnCrV8 steel (No 1.2842), the length of the specimens being their double diameter. The cooling curves and the as-quenched hardnesses were examined in a modern, vacuum furnace with in situ nitrogen quenching at pressures of 3 and 6 bars. Figure 9.51 shows the overall estimate of the quenching power measured in terms of the cooling time between 800 and 500 °C or in terms of the cooling time divided by 100 (the so-called λ index), just as in [46] and other studies. The curves show the evident effect of the quenching gas pressure.

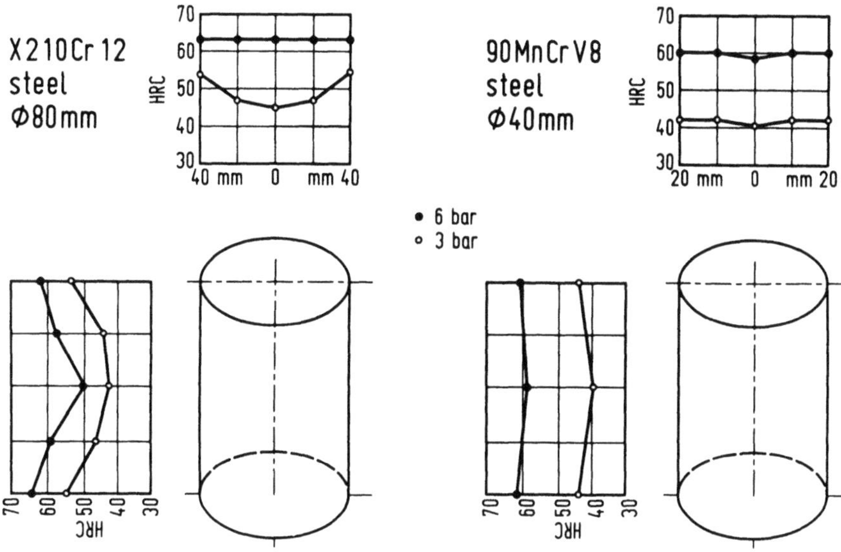

Fig. 9.53. Effect of nitrogen pressure upon the distribution of hardness after quenching [48]

Giesser et al. have also examined how the mass-to-surface area ratio affects the cooling rate within the temperature range from 800 to 500 °C (Fig. 9.52). They have found that the greater this ratio, the less the gas pressure affects the quenching process.

By way of example, Fig. 9.53 shows how the gas pressure affects the distribution of hardness over the end and side surfaces of 80 mm·diam·cylindrical specimens made of X210Cr12 steel and 40 mm·diam-specimens made of 90MnCrV8 steel. We can see that the influence of the pressure is significant. However with specimens of greater diameters, exceeding the critical diameter of a given steel grade, this influence decreases.

That the in situ gas quenching in vacuum furnaces provides required hardnesses is not its only advantage since it also minimizes distortion and ensures perfectly clean quenched surfaces so that in most cases no finishing treatment is needed after quenching. To achieve this high performance in conjunction with a high cooling rate, it is essential that the gas uniformly cool the load and that the oxygen and moisture contents in the gas be very low, not exceeding the specified admissible values. When quenching high-alloy steels, the oxygen and moisture contents should not be higher than 5 ppm each. For low-alloy steels, somewhat greater values are tolerable.

9.10 Fluidized Quenching Beds

9.10.1 Fluidization Phenomenon

If a tank whose bottom is porous (the so-called gas distributor) is filled with a fine-grained material (called the bed), such as e.g. quartz sand, up to a certain height, and a gas such as e.g. air is blown through this bottom, then with increasing flow rate v of the gas, the effects shown schematically in Fig. 9.54 occur. Initially the difference between the pressures above and below the tank bottom, Δp, increases since the bed increasingly resists the gas flow through it. The height of the bed remains first unchanged, since its gravity force F_z is greater than the sum of the buoyancy force F_w and the gas pressure F_c, which acts in the opposite direction. After a certain critical flow rate v_f (the fluidization rate), which depends on the grain size of the bed and, to a smaller degree, on the bed temperature, has been exceeded, the pressure difference Δp slightly decreases and remains at this decreased level irrespective of a further increase in the flow rate, whereas the forces F_z and the resultant force $F_w + F_c$ balance each other. After v_f is exceeded, the height of the bed increases due to its loosening. Individual grains of the bed and their clusters are set in motion, in which

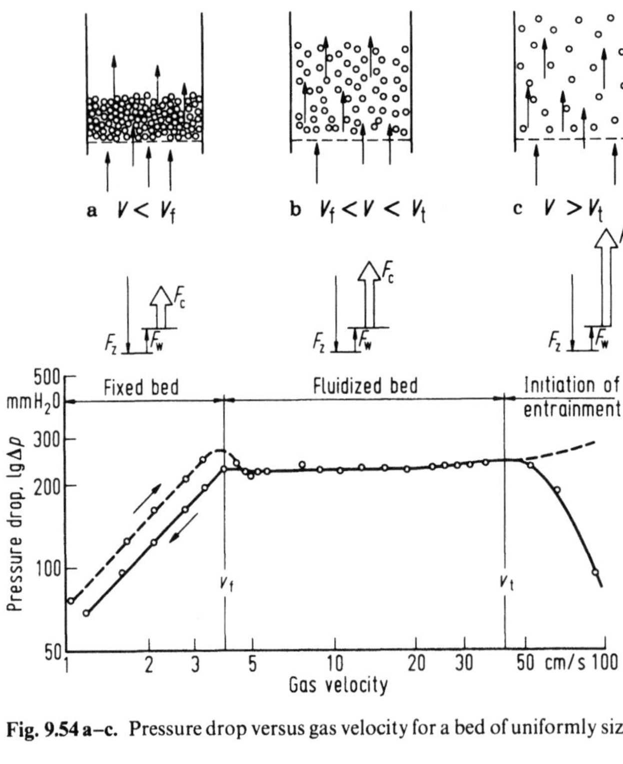

Fig. 9.54 a–c. Pressure drop versus gas velocity for a bed of uniformly sized sand particles, after [51]

they follow characteristically whirling paths. This permits the gas stream to loosen the bed further. The intensity of the motion and the degree of the bed loosening depend on the ratio of the gas flow rate to the critical rate v_f. This ratio is known as the fluidization number (λ) and is an important parameter used for controlling the fluidization process. The range of flow rates from v_f to v_t is relatively wide. Flow velocities of this range are often employed in various treatments, such as e.g., heat treatment.

The value of the fluidization number can very from about 2 to 10. In practice, values between 2 and 5 are employed since with higher values the fluidization process loses its regularity and intensive bubbling occurs. When the flow rate exceeds v_t, the buoyancy force and the gas pressure become greater than the bed gravity force, and the bed is thrown out from the tank.

Parts to be heat treated are immersed in a hot or cold fluidized bed, piece by piece or loaded all together in e.g. a basket. Such a bed has properties similar to those of fluids. The presence of the load of a given geometry affects the conditions of motion of the medium. The greater the portion of the bed cross-sectional area occupied by the load, the greater the disturbance of heat transfer. For this reason the section size of the load should not exceed 40–45% of the cross-section of the bed tank.

9.10.2 General Description

Fluidized beds show many properties, such as mobility, hydrostatic pressure, and free upper surface, characteristic of fluids. Their heat transfer coefficient is relatively high and can be adjusted between wide limits. Thanks to this property and also because of their other advantages their use as both heating and cooling media has been recently growing. In heating operations and in marquenching and martempering they are good substitutes for salt baths. The first attempts at using fluidized media for quenching were made in the 1950s, and now their use in industrial production is still growing.

The major advantages of fluidized media as employed for quenching are

—no toxic vapours and gases adversely affecting the working conditions and the environment;
—good replacement for hot oil and salt baths;
—easy temperature adjustment and small temperature variation over the bed volume;
—no need for melting operation, necessary with salts, so that the quenching installation is continuously ready to work and energy consumption is reduced;
—no boiling which adversely affects uniform quenching and causes distortion;
—almost unlimited service life and insignificantly small dragout losses;
—no cleaning, drying and passivation (needed with salt baths).

Despite these considerable advantages of fluidized quenchants, their use has so far been limited for the following reasons:

—their quenching power is not sufficiently high for quenching most grades of steel;
—dusting, difficult to avoid when quenching in them;
—a reduced cooling intensity at those places where the quenched surface is shaded from the stream of particles (the 'shade' effect);
—problems in mechanizing the quenching operations performed in fluidized media; etc.

These disadvantages of the fluidized media may of course be overcome in the future. Their use for marquenching and austempering instead of troublesome salt baths seem to have most prospects.

The value of the heat transfer coefficient α of a fluidized bed depends on many technological factors and thus can vary between wide limits. Reynoldson [52] gives values of 120–1160 W/(K·m^2), whereas according to Sommer [53] they range from 420 to 800 W/(K·m^2). The values of this coefficient as given by various authors are listed in [54]. Out of the six authors, five placed α between 397 and 600 W/(K·m^2), and only one author found it to range between 116 and 1160 W/(K·m^2). These discrepancies are due undoubtedly to differences in the test conditions.

The heat transfer coefficient is a good quantitative measure of the average quenching power of a medium, but it does not represent the variation of this power as a function of the temperature at the specimen centre, and much less at the specimen surface. This variations are to some extent represented by the centre temperature-versus-time curves (shown in Fig. 9.55 for fluidized media in comparison with the corresponding curves for other common types of quenchants) and to a much greater extent by the cooling rate plotted as a function of the specimen temperature (Fig. 9.56). We can clearly see from the latter figure

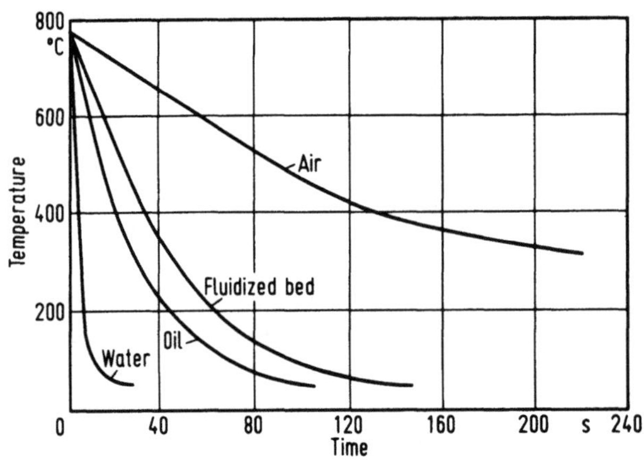

Fig. 9.55. Comparison of cooling curves for 16 mm dia. steel bars cooled in various quenching media, after E. Mitchell [52]

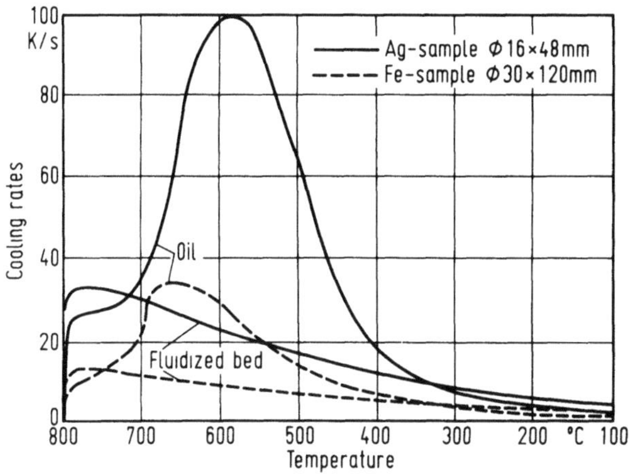

Fig. 9.56. Cooling rate of silver and low-carbon steel specimens in a mineral oil and in a fluidized bed of corundum power – as a function of the temperature at specimen center, after W. Luty [55]

that the cooling kinetics in fluidized beds is essentially different from that in oil due to the fact that the boiling process occurring in oil does not occur in fluidized beds where the heat transfer proceeds in accord with Newton's law.

9.10.3 Effect of Technological Factors on the Quenching Power

The quenching power of fluidized beds chiefly depends on the following factors:

- the grain kind and size and the grain size distribution of a given bed,
- the mode and rate of the gas flow through the bed or the fluidization number (as a measure of which we may also take the density of a given bed),
- the bed temperature,
- the ratio of the gas distributor diameter to the bed height,
- the geometry of the quenched parts, their arrangement with respect to one another and to the bed and the degree of filling the bed with the load.

The requirements that a granular material intended for a fluidized bed should fulfill are: appropriate mechanical properties, a good resistance to thermal shocks, no chemical reactivity with gases and no toxicity. It should also be easily available and cheap. The most commonly used materials are corundum (Al_2O_3), carborundum (SiC) and quartz sand (SiO_2). The effect of the kind of the bed, its grain size and the flow rate of the fluidizing gas upon the heat transfer coefficient are shown in Fig. 9.57, after [56]. It can be seen from the figure that ferro-silicon with a grain size of 80 μm has the greatest heat transfer coefficient. Because of its poor mechanical properties this material has not however found practical applications, whereas various kinds of corundum of varied grain sizes are widely used.

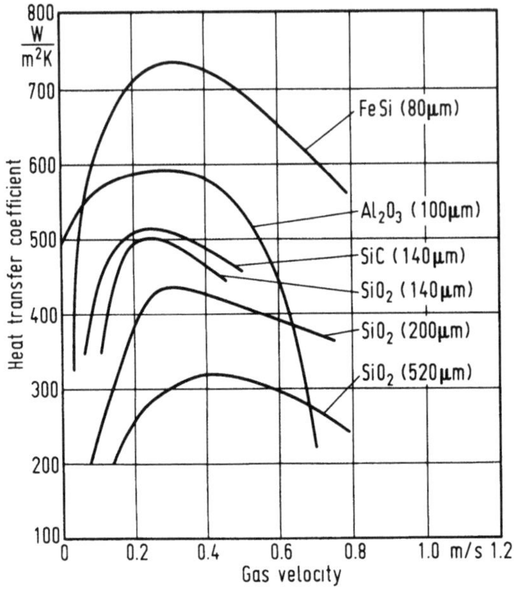

Fig. 9.57. Heat transfer coefficient of various solid particles as a function of gas velocity [56]

Tests made by Baskakov and other investigators (Table 9.20) have shown that the quenching power greatly depends on the size of the bed grains, which, as a matter of fact, has been confirmed by thermodynamic calculations. Of more recent studies those made by Sommer [57] are worth mentioning, since in his tests made on specimens 50 mm·diam by 100 mm long the heat transfer coefficient of the fluidized bed increased with increasing grain size (Fig. 9.58) which contradicted previous results. When cooling specimens 75 mm·diam by 100 mm long, in a fluidized bed with the grain size varying from 177 to 83 µm, he found no effect of the grain size upon the heat transfer coefficient. Sommer attributes his controversial results of Fig. 9.58 to the increased number of gas bubbles formed in the bed of smaller grain size.

Based on the results obtained thus far, beds of grain size 100–150 µm have been selected for practical applications. Small grains show a tendency to agglomerate due to electrostatic charges collected on their surfaces. Moreover, with such grains the fluidization process proceeds irregularly and excessive dusting occurs.

It follows from Table 9.20 that the heat transfer coefficient is also considerably affected by the bed density, which in turn depends on the density of the grains themselves and on the degree of their loosening, i.e. on the fluidization degree. As can be seen in Fig. 9.57, as the gas flow rate and, thus, the fuidization degree increase, the heat transfer coefficient initially increases, reaches a maximum and then decreases. This was also confirmed by Calperin and Einstein [58] who found that in an electrocorundum bed of grain size 150 µm the heat transfer coefficient reaches its maximum of 900 $W/K{\cdot}m^2$ when the fluidization

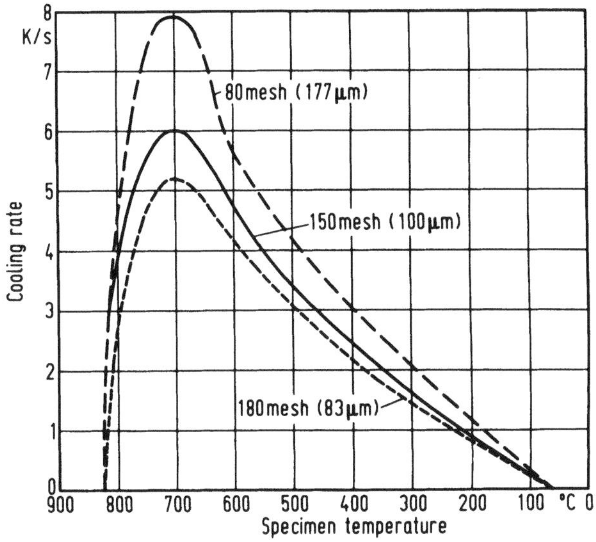

Fig. 9.58. Effect of grain diameter on the surface cooling rate of a steel specimen $\phi 50 \times 100$ mm in a fluidized bed of corundum powder, after Sommer [57]

Table 9.20. Effect of grain diameter and density of fluized bed (quartz sand) on the heat transfer coefficient [49]

Density kg/m^3	Heat transfer coefficients, W/(m^2·K) at corn diameter in µm			
	50	100	200	400
1300	700	570	400	240
650	540	390	260	160
325	410	280	200	100

degree is 3. However the bed density may be modified not only by adjusting the fluidization degree but also by modifying the grain-size distribution. This has led to the idea of using a bed of varied grain size so as to increase the bed density. Figure 9.59 shows how an addition of finer grains that fill the spaces between larger grains improves the cooling rate.

Apart from the kind of bed, the type of gas employed for fluidization may also have an effect upon the quenching power of the medium. According to Sommer [57], this effect is insignificant. However other authors, e.g. Davidson and Harrison [60] give formulas for the heat transfer coefficient that imply that this coefficient quite markedly increases with increasing thermal conductivity of the gas. The thermal conductivity of air and nitrogen is 2.4 W/(cm·K), that of hydrogen is 16.6 W/(cm·K), and that of steam 4.5 W/(cm·K). Hence, even a

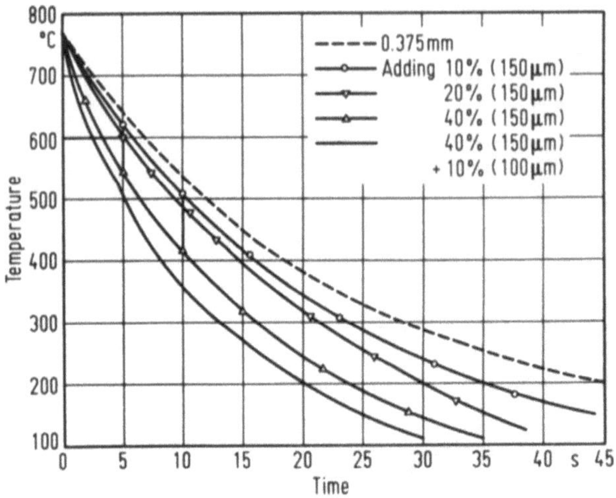

Fig. 9.59. The influence of adding fine particles on the cooling curve of a silver specimen [59]

small amount of hydrogen added to nitrogen should increase the quenching power of fluidized beds, and so should do steam in fluidized beds heated above 100 °C. Experimental studies have confirmed these predictions.

Another important factor influencing the quenching capacity of fluidized media is their temperature or, more precisely, the difference between the temperature of the cooled workpiece and that of the cooling medium. This is imposed by Newton's law. The results of experiments on this effect are given in [54, 55, 57, 61, 62] and in other works. From Fig. 9.60 it follows that in the most important temperature range from 800 to 500 °C the cooling rate markedly depends on the bed temperature. The cooling rates shown in the figure are relatively small as they have been measured at the centre of large steel sections. At the surface of the quenched parts these rates are however much greater,

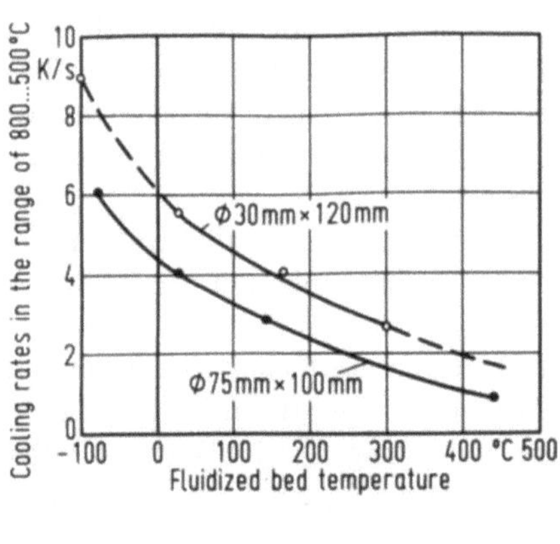

Fig. 9.60. Effect of the fluidized bed temperature on the centre cooling rate in the range 800–500 °C of a steel specimen $\phi 30 \times 120$ mm, after Luty [61.55] and of a steel specimen $\phi 75 \times 100$ mm, after Sommer [57]

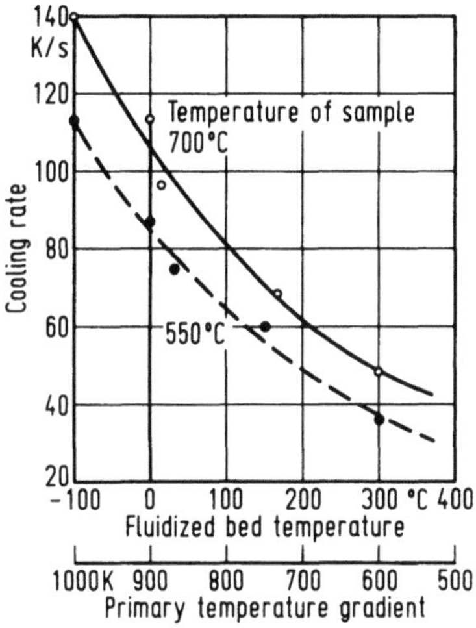

Fig. 9.61. Effect of the primary temperature gradient on the cooling rate of a silver cylindrical specimen, sized at 8 mm by 24 mm, in a fluidized bed, at the specimen temperatures of 700 °C and 500 °C after Luty [55]

which can be seen from the results of tests on a silver cylinder sized at 8 × 24 mm shown in Fig. 9.61. This figure shows the variation of the cooling rate at specimen temperatures of 700 and 550 °C as a function of the bed temperature and of the initial temperature gradient at the specimen surface. Since the specimen is cooled from a temperature of 900 °C, the maximum initial gradient at the surface is equal to the difference between this temperature and the temperature of the bed. We can see from these curves that the bed temperature and the temperature gradient at the quenched surface are very effective factors influencing the quenching power.

The effect of the temperature of a fluidized bed upon its quenching power is of particular importance in martempering and austempering, since this effect decides whether fluidized beds may or may not be used as substitutes for troublesome salt baths (see Sect. 9.8). That fluidized beds have a lower quenching power than that of salt baths is the major barrier to their wider use. Many studies [54, 62, 63] aimed to overcome this barrier have recently been performed. Below we shall discuss their most important results.

As reported in [54], nodular cast iron specimens 6 mm in diameter and 50 mm long were cooled in a fluidized bed and in a salt bath (probably a saltpetre bath with an addition of water) at temperatures of 265, 330 and 400 °C. Nodular cast iron was probably used because its thermal conductance is considerably greater than that of steel and also because castings made of it may after bainitic hardening replace much more expensive heat treated steel castings. By way of example, Fig. 9.62 shows some results of these tests from which it evidently follows that the cooling rate in the salt bath is higher than that in the fluidized bed. Time is plotted on a logarithmic scale which permits the cooling curves to be superposed on the TTT phase transformation curves in order to estimate whether required structures can be obtained.

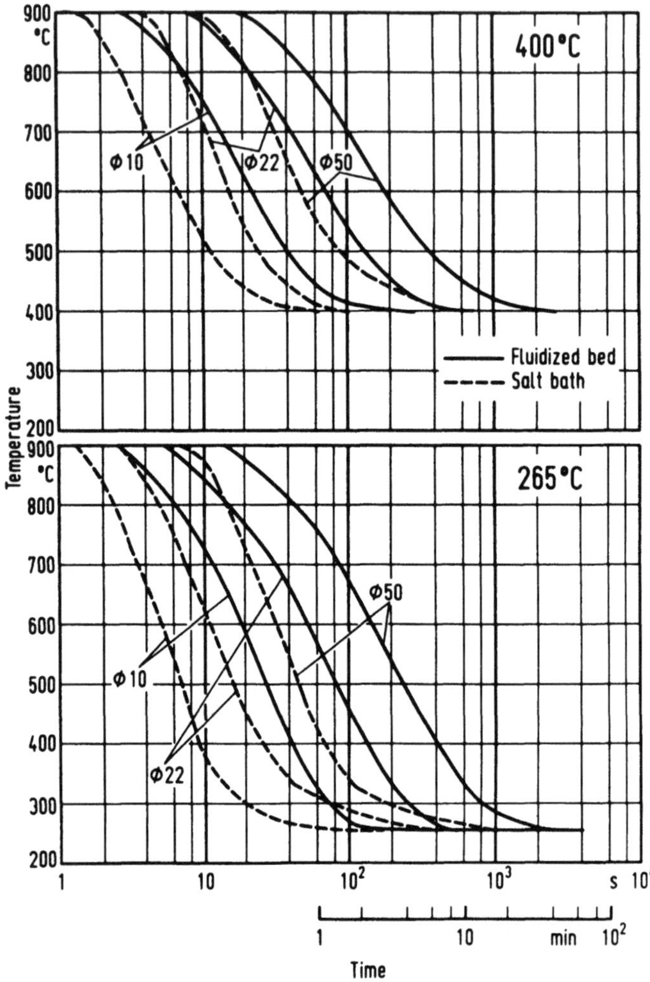

Fig. 9.62. Cooling curves for nodular cast iron specimens quenched in a fluidized bed and in a salt bath at temperatures of 265 and 400 °C [54]

in order to increase the quenching power of fluidized beds, especially those intended for austempering, the present author [62] used a mixture of electrocorundum and some sort of slag as the granular material and fluidized it with the vapours of a certain compound dissolved in water. Where the bed temperatures did not exceed 100 °C, the fluidizing gas was nitrogen containing no more than 10% hydrogen. The variation of the quenching power of these beds with temperature is shown in Fig. 9.63, together with the corresponding curves for a coventional bed and for a salt bath as used in [54]. We can see that this new bed cools much faster than the conventional one, but more slowly than the saltpetre-water bath.

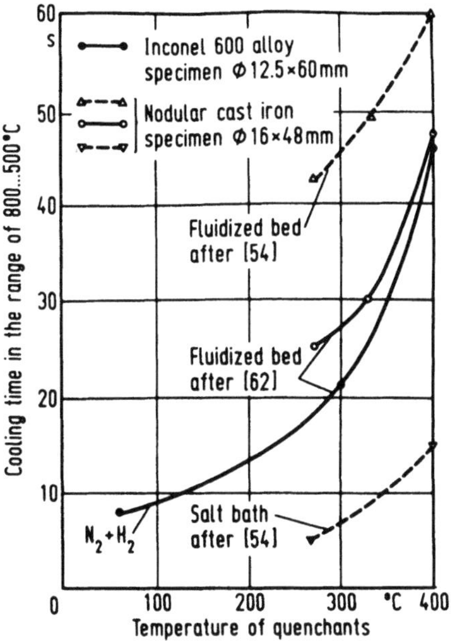

Fig. 9.63. Effect of quenchant temperature on the cooling time in the range 800–500 °C [62]

Pulkkinen and Virta [63] also attempted to increase the quenching power of fluidized beds by treating the bed surface with a water spray, by periodic water spraying of the quenched surfaces and by blowing an air-steam mixture through the bed. Only the latter technique seems to be effective and feasible in practical applications.

In all cooling media, apart from the properties of the quenchant itself, it is also the geometry of treated parts that affects the cooling rate. When cooling a single workpiece, its section size and the surface area-to-volume ratio are

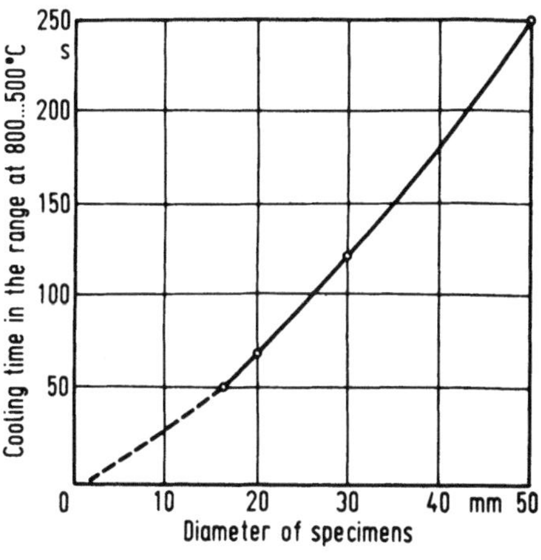

Fig. 9.64. Effect of the diameter of nodular cast iron specimens on the cooling time in the range 800–500 °C [62]

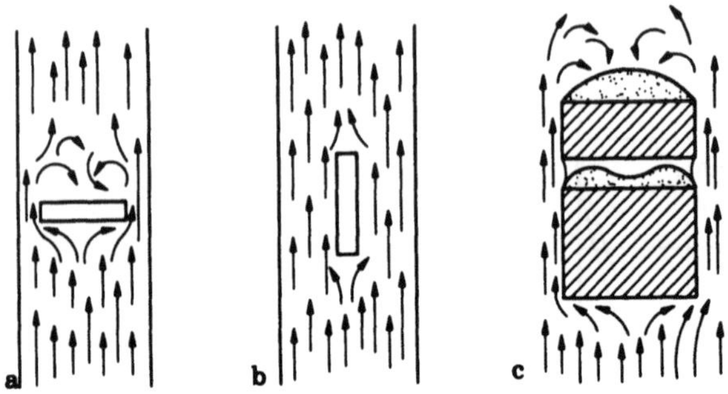

Fig. 9.65 a–c. Effect of the workpiece arrangement and of the "shade" phenomenon on the cooling conditions. a abnormal arrangement; b correct arrangement; c "shade" phenomenon

essential, whereas with loads comprised of several parts, the arrangement of the parts play the most important role. Figure 9.64 shows how the cooling time depends on the diameter of nodular cast iron bars.

The specific effect occurring when cooling in fluidized beds is the 'shade' effect, which in some cases adversely affects the uniformity of cooling and thus the uniformity of hardness developed. This effect results from the deposition of the bed material on the upper surfaces of the quenched parts and in cavities and holes present in them (Fig. 9.65). This hinders heat transfer but may quite effectively be counteracted by turning over the parts, arranging them in an appropriate manner, etc., depending on the kind and size of the load.

9.10.4 The Range of Application of Fluidized Beds

There exists a substantial literature describing numerous successful attempts at quenching in fluidized beds various machine parts and tools, made of low-, medium- and high-alloy steel, and also presenting examples of their industrial applications. To get a general opinion on the possible applications of these quenchants it is reasonable to compare the surface cooling rates obtained in fluidized beds, which amount of about 60 K/s, with the critical cooling rates for various grades of steel, which widely vary from 0.1 to 100 K/s for alloy steels and from 100 to 600 K/s for structural carbon steels. Hence we see that fluidized beds can be used for many grades of alloy steel, especially when small or moderate section sizes are in question, whereas they are not suitable for carbon steels.

When selecting a medium for quenching a given part of given section size and made of a given steel grade, it is convenient to use a diagram in which the cooling curves are superposed on the TTT phase transformation curves. Figure 9.66 shows such diagrams helpful for designing a carburizing operation on 20MoCr4 (SAE4118) steel. We can find from the figure the structures produced

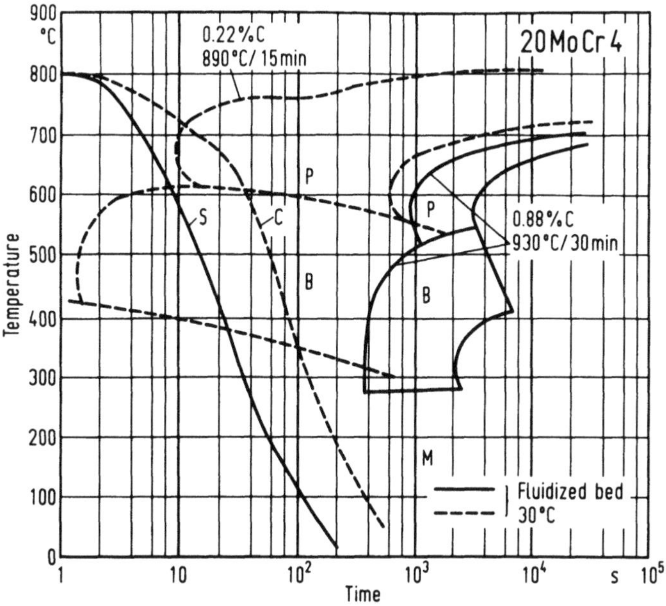

Fig. 9.66. Cooling rate curves for surface (*s*) and the center (*c*) of a low-carbon specimen of diameter 30 mm and length 120 mm in a fluidized bed of corundum powder, in comparison with a continuous TTT diagram for 2OMoCr4 steel (AISI 4118) and for the same steel after carburization to 0.88 pct C [55]

in the carburized layer (solid lines) and in the core by continuous quenching in a fluidized bed at 30 °C (dashed lines). As can be seen, the continuous quench will produce a martensitic structure in the carburized layer whereas we shall not avoid a certain amount of pearlite in the core (0.22%C).

Somer [57] also gives several examples of hardening in fluidized beds various machine parts and tools, mode of various steel grades. In 50CrV4 (SAE6150) steel parts of wall thickness 3–15 mm he obtained a hardness of 58.5 ± 1.5 HCR, and in 100Cr6 (52100) steel bushings of wall thickness 5–10 mm, a hardness of 64 HCR. Moreover, the quench distortions were smaller than those developed by oil quenching.

Ledeburitic tool steels containing about 12%Cr can very successfully be quenched in fluidized beds, even when dealing with large parts of section size up to 150 mm. Hot work tool steels can be quenched in cold fluidized beds or austempered if the parts have sophisticated shapes or widely varied section sizes. Table 9.21 compares the results of martempering ledeburitic chromium steels in a salt bath and in a fluidized bed, according to Guingand [64], whereas Table 9.22 the results of quenching several grades of tool and bearing steel in oil and in a fluidized bed, according to [61].

Numerous examples of industrial applications of fluidized beds are given in [65]. The most typical are the quenching of tools made of high-speed steel

Table 9.21. Results of hardening and isothermal quenching in salt baths and fluidized beds after [64]

Heating or cooling medium	Diameter of test pieces	Preheating temperature °C	Total time for final heating + holding at 960 °C mins.	Rockwell at surface	C hardness at centre
Salt baths	80	500	44	65.5	65
Fluidised beds[1]	80	490	51	65	65
Salt baths	40	540	36	64.5	64
Fluidised beds	40	500	41	64.5	64

[1] Small parts of the same steel but with a diameter of 8 mm were treated at the same time. Their hardness after quenching was 66 Rc

and wire patenting. Using a fluidized bed for the heating and cooling operations performed in patenting, we avoid the toxic lead or salt baths usually employed in this technique. Industrial equipment intended for this purpose is manufactured in England, France, Japan and in the Soviet Union. Figure 9.67 shows schematically the design of the equipment manufactured by the Stone-Plott Fluidfire Ltd., England. Inbetween two tanks for austenitizing and austempering at a temperature of about 500 °C, a short, water-cooled tank for fluidized bed quenching at about 100 °C is installed. The wire to be patented is placed in this tank for a short time so that its temperature only decreases to about 500 °C.

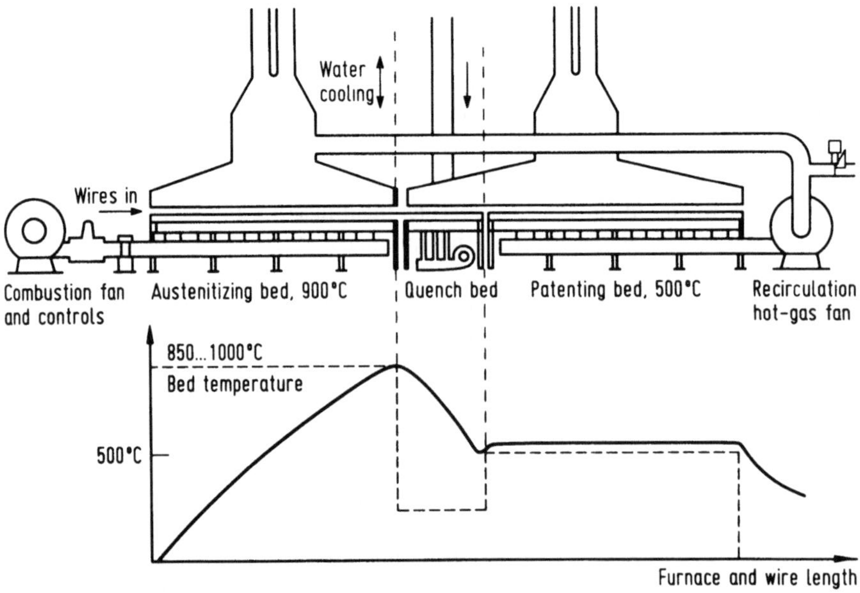

Fig. 9.67. Three zone fluidized bed arrangement and quench temperature ranges

Table 9.22. Hardness of bearing—and tool steels specimens after conventional quenching and martempering in fluidized bed and oil [106]

Grade of steel	Section		Austenitizing		Rockwell C hardnes after quenching			
					conventional in		martempering in	
	Thickness mm	Diameter mm	°C	min	oil 50 °C	fluidized bed 30 °C	oil 180 °C	fluidized bed 180 °C
100 Cr6 (52100)	5	—	860	20	64–65	65–64	64–65	64
	12	—	860	20	64–65	39	64–65	28–29
100 CrMo73	15	—	860	20	63–65	62.5–64	62.5–63	62–63
	26	—	860	25	63–65	62.5–64	62–63.5	62–63
	45	—	860	30	61–62	60–62	58–59	57–58
105WCr6	—	15	830	20	64–65	62–64	—	—
SW7M	—	20	1210	5	—	62–63	—	—

Table 9.23. Chemical composition of nodular cast iron bars

Grade	Chemical composition, %							
	C	Si	Mn	Cr	Ni	Mo	Cu	other
a	3.60	2.54	0.46	0.06	1.15	—	—	Sn = 0.036 Mg = 0.076
b	1.40	2.30	0.40	0.50	0.60	0.43	0.70	—
c	1.45	2.36	0.58	—	—	—	—	—

The successful attempts at applying fluidized beds to marquenching of high-speed steel tools and hot-work steel dies, and to wire patenting, have shown that troublesome salt baths can at least in part be eliminated. As already mentioned, in order to extend the application range of fluidized beds to include bainitic hardening of some types of steel and cast iron (especially nodular cast iron), the quenching power of these media must be increased. This has in part been achieved by the presented author as reported in [62]. To verify the results obtained, shafts made of three types of nodular cast iron (Table 9.23) were austempered and the cooling time from 800 to 500 °C was measured (see Fig. 9.63). Figure 9.68 shows the distributions of hardness measured over the surface of plates cut out at midlength of the shafts after austenitizing at 900 °C/1h and austempering in a fluidized bed at 320 ± 10 °C/2h. As can be seen, the hardnesses corresponding to a bainitic structure can be obtained for sections ranging from 16 to 50 mm in diameter, depending on the hardenability of the cast iron used.

Fluidized quenchants may also be used for accelerated cooling performed after tempering, or for heat treatment of aluminium alloys.

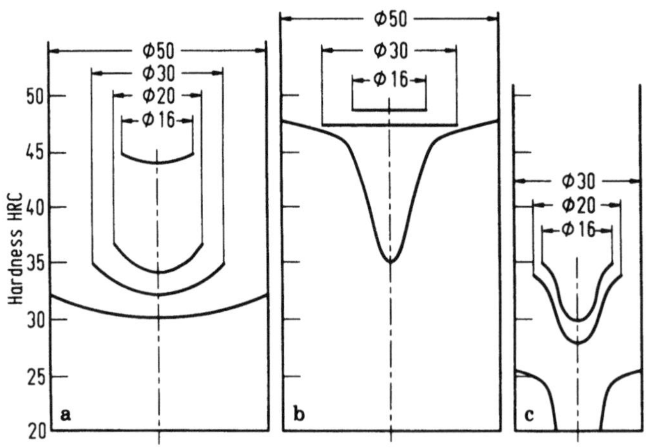

Fig. 9.68 a–c. Hardness distributions on the cross-sections of modular cast iron specimens as given in Table 9.23, after [62]

References

1. Grossman MA (1952) Elements of Hardenability. ASM, Cleveland
2. Bauer R (1985) Das Härten von hochwertigen Werkzeugen in modernen Vakuumanlagen mit Überdruck-Gasabkühlung. HMT 40: 1, s. 25–34
3. Mitchel JA (1971) Practical Utilization of Intermediate Quench Severities. Heat Treating. April, pp. 11–14
4. Lambert N, Economopulos M (1970) J. Iron Steel Inst., 208: 917–928
5. Price RF, Fletcher AI (1980) Met. Technol., 7: 203–211
6. Pietrasz ŁW (1959) Zakałocznyje sriedy. Moskwa, Maszgiz
7. Metals Handbook. Ed. 8. Ohio. ASM. Vol. 1, 1969
8. Mitsutsuka M, Fukuda K (1979) Trans. Iron Steel Inst. Jpn., 19: 162–169
9. Bamberger M, Prinz B (1986) Materials Science and Technology, April, Vol. 2, pp. 410–415
10. Monroe RW, Bates CE (1983) J. Heat Treating. Dec., Vol. 3, No. 2, pp. 83–99
11. Research Institute of Mechanical and Electrical Technology P.R.C.: Qunching Media in China. No publication
12. Rose A (1940) Das Abkühlvermögen von Stahlabschreckmittel. Archiv für das Eisenhüttenwesen., Nr. 13, s. 345–354
13. Luty W (1985) Vergleich der thermokinetischen Eigenschaften von wässrigen Abschreckmitteln bei Tauch- und Brausenabchreckung. HTM 40: 4, s. 178–182
14. Kopietz KH (1979) Controlled Quenching of Ferrous Metals in Sodium Polyacrylate Aqueous Solutions. Industrial Heating, June
15. Burgdorf E (1979) Eigenschaften und Einsatzgebiete synthetischer Abschrecklösungen. Zeitschrift für wirtschaftliche Fertigung, Bd. 74, Nr. 9, s. 431–436
16. Foreman RW (1984) Polymer Quenching Update. Industrial Heating, January, pp. 22–29
17. Bergen RT (1984) New Developments in the Technology and Application of Polymer Quenchants. Heat Treatment of Metals, No. 1, p. 7–8
18. Hilder NA (1987) The Behaviour of Polymer Quenchants. Heat Treatment of Metals, 2, pp. 31–46
19. Segerberg S (1987) Technische Eigenschaften und Gesundheitliche Verträglichkeit von Polymerabschreckmitteln. HTM 42: 1, s. 50–54
20. Lakin II (1982) The Use of Polymer Quenchants in Surface Heat Treatments. Heat Treatment of Metals, 3, pp. 73–76
21. Beck G, Dumont C, Moreaux F, Simon A (1975) Planungs- und Auswahlprinzipien für ein Abschrecköl. HTM 30: 6, s. 346–358
22. Tkaczuk TI, Rudakowa NJ, Orazowa MR (1983) Wlijanje frakcjonnogo sostawa zakałocznych masjeł na ich ochładzajuszczuju sposobnost. Mietałłowiedienje i Termiczeskaja Obrabotka Metałłow, Nr. 1, s. 16–17
23. Bashford A, Mills AI (1984) Development of Improved Additives for Quenching Oils Using Laboratory Simulations. Heat Treatment of Metals, 1, p. 9–14
24. Mansion I, Bournicon C, Sayettat C (1978) Les huiles de trempe. Centre Technique des Industries Mecaniques. Senlis
25. Kopietz KH (1963) Aufbau, Eigenschaften und Anwenung von Ölen für die Warmebehandlung von Stahl. Mineralöltechnik, Nr. 4–5
26. Keyser W (1972) Mineralöle in der Härtetechnik. Werkstatt und Betrieb, Nr. 6, s. 421–426
27. Commission mixte ATTT-SFM (1980) Caracteristique physicochimiques des huiles de trempe. Traitement Thermique, Nr. 146, s. 34–38
28. Furman GR (1971) Quenching-Metallurgical Aspects. Lubrication, v. 57, Nr. 2, pp. 13–36
29. Beitz H, Rettberg U (1979) Wasser in Abschrecköl bädern. Zeitschrift für wirtschaftliche Fertigung, Nr. 7, s. 373–377
30. Hampshire IM (1984) User Experience of Hot Oil Quenching. Heat Treatment of Metals, 1, pp. 15–19
31. Leaflets of quenchants of the company HOUGHTON-Hildesheim
32. Beitz H, Rettberg U (1975) Ölwarmbäder für moderne Ofenanlagen. Fachberichte für Oberflächentechnik. Nr. 9, s. 642–651
33. Marmier EN, Hrugije I (1983) Ochładzajuszczaja sposobnost niekotorych wakuumnych masieł pri razlicznych razrieżeniach. Mietałłowiedienie i Tiermiczeskaja Obrabotka Mietałłow. No. 1, s. 14–16

34. Luty W (1975) Sprawozdanie z podróży służbowej do Ipsen-Industries w Rockford Illinois. USA, grudzień
35. Beitz H. Ölbrände in Härtereien und ihre Vermeidung. Houghton-Hildesheim Bericht
36. Glück T (1984) Metody ochrony hartowni zwiększające bezpieczeństwo pracy. Metaloznawstwo i Obróbka Cieplna, Nr. 67, s. 35–37
37. Gulajew AP (1960) Tiermiczeskaja obrabotka stali. Moskwa, Maszgiz
38. Pietrikow WG (1977) Swojstwa goriaczich rozpławow sielitr s wodoj. Mietałłowiedienie i Tiermiczeskaja Obrabotka Mietałłow, No. 8, s. 4–7
39. Babicz AP, Drugije I (1982) Stabilizacja tiempieratury i ochłażdajuszczej sposobnosti raspława solej. Mietałłowiedienie i Tiermiczeskaja Obrabotka Mietałłow, No. 1, s. 51–52
40. Gulajew AP, i in (1953) Stupienczataja zakałka w raspławach jedkich szczełoczej. Wiestnik Maszinostrojenia, No. 8
41. Luty W (1986) Chłodziwa hartownicze. Wydawnictwa Naukowo-Techniczne. Warszawa
42. Lambert N, Economopoulos M (1970) J. Iron Steel Inst., 208, pp. 917–928
43. Ruffle TW, Byrnes ER (1979) Quenching in Vacuum Furnaces. Heat Treatment of Metals, 4, pp. 81–87
44. Limque F, Bless F (1980) Erfahrungen mit Überdruck-Gasabschrecken sowie Aufkohlen in Vakuumöfen. HTM 35: 5, s. 1–6
45. Limque F, Bless F (1982) Härten von Werkzeugen in Vakuumöfen mit Hochdruck-Gesabschreckung. Zeitschrift für wirtschaftliche Fertigung, Nr. 9, s. 1–4
46. Merlini N (1983) Trempe sous gaz à haute pression d'aciers rapides traites sous vide. Traitement Thermique, No. 171, s. 44–48
47. Bauwman JW (1984) Erfahrungen mit dem Hochdruck-Gasabschrecken in Vakuumöfen und dessen Weiterentwicklung. HTM 39: 3, s. 1–5
48. Giesser H, Bauer RE, Birzer F (1987) Möglichkeiten und Grenzen der Überdruck-Gaskühlung abkühlempfindlicher Stähle im Vakuumofen. HTM 42: 5, s. 301–308
49. Mickley HS, Thrilling CA (1949) Heat Transfer Characteristics of Fluidized Beds. Ing. Eng. Chem., No. 41, s. 1195–1197
50. Kühn F (1980) Die Abkühlung in der Wirbelschicht Gas-Wärme-International, Nr. 9, s. 481–485
51. Shirai T (1958) Fluidized beds. Kanazawa
52. Reynoldson R.W.: The development of controlled atmosphere fluidised bed furnaces for the heat treatment of metals. Gas–Wärme International. 26 (1977), Nr. 4, s. 175–186
53. Sommer P: Wärmebehandlung in der Wirbelschicht. Gas–Wärme Internationa, 33 (1984), Nr 6/7, s
54. Fargues I, Rezk A Sh, Hellio H: Utilisation du four a lit fluidise pour le traitement de fontes modernes. Traitement Thermique 194–85, p. 53–59
55. Luty W: Effect of Temperature Gradient on the Quenchig Power in Fluidized Beds. Journal of Heat Treating. ASM, December 1983, p. 108–113
56. Baskakov AP: Skorostnyj bezokislitelnyj nagrev i termiczeskaja obrabotka w kipiaščem sloe. Metalurgia, 1968
57. Sommer P: Einsatz von Wirbelschichten als Abkühlmedium. Zeitschrift für wirtschaftliche Fertigung 77 (1982) 9, s. 411–416
58. Calperin NJ, Einstein VG: Heat Transfer in Fluidized Beds, w Davidson I.E., Harrison D. "Fluidization". London Academic Press 1971
59. Rulin Z, Genlian H: The Cooling Characteristics of Fluidized Bed and its Application for Quenching of Steels and Aluminium Alloy. International 4-th Congress on Heat Treatment of Materials. Berlin 1985
60. Davidson IF, Harrisan D: Fluidization. Academic Press. London and New York 1971
61. Luty W: Untersuchung der thermokinetischen Eigenschaften und des Anwendungsbereiches einer Wirbelschicht als Abschreckmedium. HTM 36 (1981) 4, s. 194–199
62. Luty W: Fluidalne chłodziwo hartownicze, w. szczególności do hartowania izotermicznego. Metaloznawstowi i Obróbka Cieplna, 1990, No. 103–105, s. 44–48
63. Pulkkinen R, Virta J: Beschleunigtes Abkühlen von Stahl und Gusseisen in einem Wirbelbett. HTM 40 (1985) 5, s. 195–197
64. Gungand I: Heat treatment in fluidised beds. Heurtey Bulletin d'Information v. 47, Sept. 1969
65. Zawarow AS, Baskakow AP, Graczew SW: Termiczeskaja obrabotka w kipiaszczem słoje. Metallurgia, Moskwa 1981

10 Techniques of Quenching

10.1 Immersion Cooling (Direct Quenching)

HOWARD E. BOYER

Immersion cooling, refers to direct cooling (usually by a liquid quench) of the metal from its heat austenitizing temperature to, or at least to near, ambient temperature. This is the most commonly used method in the quenching of quench-hardenable steels. It is also used for rapid cooling of metals that have been solution treated at elevated temperature.

In quench hardening of steels, (as well as cast irons), certain complex transformations occur that are controlled primarily by variations in quenching mediums and techniques. In contrast, when quenching austenitic steels and some nonferrous materials prior to age hardening, transformation does not occur. In fact, rapid cooling is used solely to prevent transformation during the cooling cycles. As a rule, for quenching austenitic stainless steels and nonferrous alloys, the fastest possible cooling rate is one that is preferred for metallurgical reasons. However, practical considerations, such as distortion, frequently indicate the use of a less drastic cooling practice.

The effectiveness of quenching may be varied by changing the type of quenching medium its temperature and agitation rate. The design of the quenching system and the thoroughness with which the system is maintained also contribute to the success of the process. Design of the part likewise contributes to the mechanical properties and distortion resulting from a particular quench.

The rate of heat extraction that a quenching medium is capable of achieving is modified significantly by the manner or condition in which the quenching medium is used. These modifications have resulted in the arbitrary assignment of specific names to various quenching methods, such as direct quenching, fog quenching, gas quenching, and interrupted quenching.

Mechanism of Immersion Quenching

If all workpieces were symmetrical and alike in shape (no odd configurations) and were of the same size and/or weight, obtaining the desired hardness would be simple. In practice, however, such conditions rarely exist. Workpiece configuration, cooling power of the quenching medium, temperature of the quenching medium, and degree of agitation are the principal variables that must be considered when selecting the proper quenching procedure.

Workpiece Configuration. Several factors are involved in the mechanism of immersion quenching: (a) internal conditions of the workpiece that affect the supply of heat to the surface, (b) surface and external conditions that affect the removal of heat, (c) the heat extracting potential of the quenching fluid in the quiescent state at normal fluid temperatures and pressures ("standard" conditions), and (d) changes in the heat extracting potential of the fluid brought about by "nonstandard" conditions of agitation, temperature, or pressure. These factors are illustrated in for a furnace-heated gear that is quenched edgewise in a quiescent volatile liquid.

Note A in Fig. 10.1.1 indicates how the irregular configuration of the gear affects the flow of heat from within the gear to the quenching area. Note that high temperature persists near the surface at the roots of the teeth where large vapor bubbles are trapped. If the gear were induction or flame heated (and thus had a uniformly thin heated layer conforming to the irregular contour of the gear), heat supply to the quenching area would be more consistent, and quenching would progress more rapidly because heat also would flow simultaneously to the cold metal underlying the heated exterior. Recesses, holes (through or blind), and relatively large areas on the bottoms of workpieces all have a marked effect on how workpiece surfaces and quenching mediums interact.

A quiescent liquid experiences an unavoidable movement due to the action of immersion, the turbulence of boiling, and convection currents. This minimum agitation will eventually dissipate the accumulated heat to the surrounding large

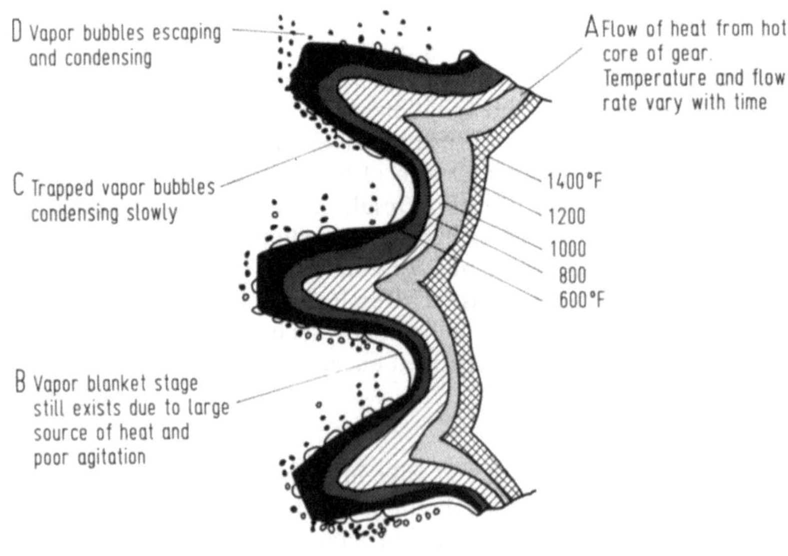

Fig. 10.1.1. Temperature gradients and other factors affecting the edgewise quenching of a gear in a quiescent volatile liquid. Source: MEI Course 6, Heat Processing Technology, Lesson 15, "Quenching Media and Equipment," ASM, 1977, p. 2

body of liquid, but local volumes of liquid will become heated, or may even vaporize, which may affect the quenching action.

Volatile quenching liquids produce some vapor at all operating temperatures. Above the boiling point, the supply of vapor becomes so plentiful that an envelope of gas is formed around the surface of the workpiece. This envelope, or "vapor blanket," is maintained by radiated heat for as long as the heat is available (note B in Fig. 10.1.1). The temperature above which a total vapor blanket is maintained is called the "characteristic temperature" of the liquid.

At lower temperatures, the vapor consists of bubbles, that vary in size depending on the relation of boundary tension of the liquid, the gas, and the solid at the temperatures concerned. In one liquid, numerous small, easily detached bubbles may be formed (note D in Fig. 10.1.1), where large, adherent bubbles, fewer in number, may be formed in another (note C in Fig. 10.1.1). This phenomenon is referred to as the "bubble size characteristic" of the liquid. For any volatile liquid, the mechanical trapping of vapor bubbles (note C in Fig. 10.1.1) will greatly retard the transfer of heat at the affected location. It thus becomes obvious that minimum volatility is a highly desirable characteristic of any quenching medium.

Various factors in addition of workpiece configuration that may affect heat extracting potential of a quenching liquid include: (a) solids that are deposited on the surface of some workpieces when they are quenched in some oils, brines, or water-base solutions; (b) gels that may form at the liquid-gas interface of the vapor blanket in polymer or other solutions; (c) deposits that form or changes that occur (for example, cracking of oil) in the liquid itself, thus affecting its viscosity; and (d) permanent loss of the volatile constituents of a quenching liquid due to low boiling point.

Cooling Curves

The most useful way of accurately describing the complex mechanism of quenching is to develop a "cooling curve" for the quenching medium under controlled conditions. A cooling curve test is sensitive to the factors listed above that may affect the cooling ability of the quenchant, because the test simulates conditions of actual practice.

Cooling curves are developed by quenching, from an elevated temperature, a test piece of the same metal of which the parts are made in a sample of the quenching medium. Sometimes an austenitic stainless steel specimen is used to avoid scaling or the necessity for a protective atmosphere. A high-speed recorder is used for plotting temperature changes, as measured by one or more thermocouples embedded in the test piece. The resulting time-temperature curve indicates the heat transfer characteristics of the quenching fluid. The typical surface and center cooling curves shown in Fig. 10.1.2 graphically describe the four stages of heat transfer from a hot solid to a cold liquid.

Although austenitic steel specimens are useful for determining heat transfer characteristics, in the quenching of ferritic steels another variable must be controlled in shop practice. This is the heat evolved from the workpiece (heat of transformation) as a phase change occurs from austenite to ferrite.

Stage A[1] in the upper left corner of Fig. 10.1.2 illustrates the first effects of immersion. Sometimes called the "initial liquid contact stage," this period is characterized by the formation of vapor bubbles that precedes the establishment of an enveloping vapor blanket. Stage A[1] lasts for only about 0.1 s (longer times result from higher liquid temperatures an parts with greater cross sections) and is relatively unimportant in the evaluation of heat transfer characteristics. It is detectable only when extremely sensitive equipment is used, and it cannot be detected when the quenching medium is viscous or contains undissolved gases, or when the quenching bath is operated near the boiling point of the liquid.

Stage A, called the "vapor blanket cooling stage," is characterized by the Leidenfrost phenomenon—namely, the formation of an unbroken vapor blanket that surrounds the workpiece. It occurs when the supply of heat from the surface of the test piece exceeds the amount of heat needed to form the maximum vapor per unit area of the piece. This stage is one of slow cooling, because the vapor envelope acts as an insulator, and cooling occurs principally by radiation through the vapor film, this stage is not detectable in cold aqueous solutions of nonvolatile solutes (at about 5% concentration) such as sodium chloride, potassium chloride, lithium chloride, sodium hydroxide, or sulfuric acid. Cooling curves for these solutions start almost immediately with stage B (Fig. 10.1.2).

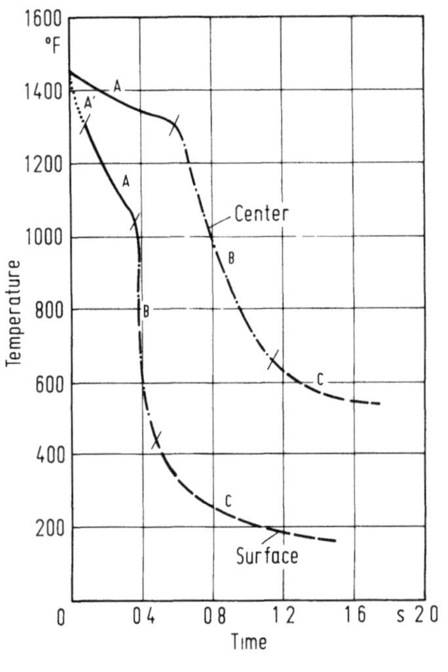

Fig. 10.1.2. Stages of heat transfer from a hot solid to a cold liquid shown on typical surface and center cooling curves. Source: MEI Course 6, Heat Processing Technology, Lesson 15, "Quenching Media and Equipment," ASM, 1977, p. 3

When saturated solutions of barium hydroxide, calcium hydroxide, or other slightly soluble materials, solutions containing finely dispersed solids, or colloidal solutions in water are used, films are deposited on the test piece during State A, which results in the prolongation of both Stages A and C. This condition usually causes a more violent action in Stage B. Solutions of some colloids or gels, such as polyvinyl alcohol, gelation, soap, and starch form an envelope of gel outside the vapor blanket formed in Stage A. The presence of this gel envelope prolongs Stage A and the succeeding stages.

Stage B, the "vapor transport cooling stage," which produces the highest rates of heat transfer, begins when the temperature of the surface metal has been reduced somewhat and the continuous vapor film collapses. Violent boiling of the quenching liquid then occurs, and heat is removed from the metal at a very rapid rate, largely as heat of vaporization. The boiling point of the quenchant determines the conclusion of this stage. Size and shape of the vapor bubbles are important in controlling the duration of Stage B, as well as the cooling rate developed within it.

Stage C is called the "liquid cooling stage." The cooling rate in this stage is slower than that developed in Stage B. Stage C begins when the temperature of the metal surface is reduced to the boiling point (or boiling range) of the quenching liquid. Below this temperature, boiling stops and slow cooling takes place thereafter by conduction and convection (see Fig. 10.1.2). The difference in temperature between the boiling point of the liquid and the bath temperature is a major factor influencing the rate of heat transfer in liquid quenchants. Viscosity of the quenchant also affects cooling rate in Stage C.

Significance of Cooling Curves. The same mechanism is involved in cooling a test piece to evaluate a quenchant as is involved in quenching an actual part in heat treating operation. For all or any selected portions of the curve, cooling curve information can be translated into cooling rate (in degrees per second), if desired. Although a cooling curve relates only to the size and material of the test piece, the thermocouple location, and the conditions of the quenching medium under which a test was performed, cooling curve data developed under one set of conditions can be translated to other conditions, by the application of heat transfer formulas.

Effect of Agitation

Agitation, that is, externally produced movement of the quenching liquid, has an extremely important influence on the heat transfer characteristics of the quenching liquid. It causes an earlier mechanical disruption of the vapor blanket in Stage A and produces smaller, more frequently detached vapor bubbles during the vapor transport cooling stage (Stage B). It mechanically disrupts or dislodges gels and solids, whether they are on the surface of the test piece or suspended at the edge of the vapor blanket, thus producing faster heat transfer in liquid cooling (Stage C). In addition to the above effects, agitation also circulates cool liquid to replace the heated liquid surrounding the workpiece.

Effect of Temperature

The temperature of the quenching medium markedly affects its ability to extract heat from a workpiece. This does not necessarily mean, however, that the lower the temperature of the medium, the faster heat will be extracted. The specific rate of heat extraction depends on the characteristics of the specific quenching medium. In general, higher temperatures lower the "characteristic temperature" and thus lengthen the duration of Stage A for aqueous liquids. Oils are more fluid at elevated temperatures, and cooling rates are thus increased. However, the boiling point is not changed. Higher liquid temperatures may decrease viscosity, affect the bubble size, or influence the breakdown and flash point of certain oils or compounds. Other factors being equal, higher temperatures of the quenchant decrease the rate of heat transfer in Stage C.

Increasing the temperature of the workpiece has relatively little effect on its ability to transfer heat to the quenching medium. The rate of heat transfer may be increased simply because a greater temperature difference exists. The most noticeable change in ability to transfer heat probably comes from the more rapid oxidation of the surface of the test piece at higher temperatures. This can either increase or decrease the heat transfer capability, depending on the thickness of the oxide developed.

Media for Immersion Quenching

Many different media have been used for quenching. The most commonly used are included in the list below, including some that are used only to a very limited extent for specific applications:

- Water
- Brine solutions (aqueous)
- Caustic solutions (aqueous)
- Polymer solutions
- Oils
- Molten salts
- Molten metals.

Gas quenching, self quenching, fog quenching and dry die quenching are not considered as immersion quenching, but these techniques will be covered later in this treatise.

10.2 Quenching Techniques

H.E. Boyer

In general, the quenching techniques in common use are:

- Immersion in a single quenchant (air or liquid) and cooling to near room temperature (usually slightly above) without interruption.

- Use of two quenching media on a timed basis, such as a partial quench in water, followed by an oil; quench or quenching in molten salt, followed by finishing in air.
- Isothermal quenching, that is, cooling to some predetermined temperature, then holding at this temperature until the phase transformation is completed.
- Spray quenching, which usually uses an aqueous medium. This technique is most frequently used for quenching of induction- or flame-hardened parts, although it can be used for quenching of furnace-heated workpieces. Spray quenching offers the advantage of instantaneous control in timing as well as pressure regulation.

10.2.1 Interrupted Quenching Techniques

Interrupted quenching, also known as timed quenching, or delayed quenching, covers several different techniques which are described separately in the paragraphs which follow.

The various forms of delayed quenching, and their degrees of refinement, as used in current practice, delayed quenching is almost as old as the general quenching process. At one time, most tool makers performed their own heat treatments—such as they were. Despite their relatively crude equipment and near total lack of technical knowledge, they discovered certain principles that are a part of present day practice. For example, the old time tool makers and blacksmiths knew how to harden tools by heating them in forge furnaces and by immersing them in water. However, cracking was common. They soon learned that if the tools were immersed in water for some short length of time, after they were cooled below red heat but not nearly to room temperature, then allowed to complete their cooling in still air, or immersed in oil, the tendency to cracking was minimized without sacrifice of hardness. Even though these craftsmen had never heard of a time-temperature-transformation (TTT) curve, they were, unknowingly quenching in accordance with TTT curve practice. The drastic water quench simply cooled the tools quickly below the nose of the "S" curve, an area in which rapid quenching/cooling is not necessary, because the beginning of the transformation is prolonged. This delayed, or double-medium, quenching is still in practice. For the most part, however, more sophisticated approaches are now used. Some of these modern techniques are described below.

10.2.2 Rinse Quenching

Parts made from highly alloyed tool or martensitic stainless steels usually are air hardening; that is, they can be fully hardened by cooling in still air from their austenitizing temperature, which for these steels is usually 980 to 1260 °C (1800 to 2300 °F). In many instances, a barium-type molten salt is the austenitizing medium. Under these conditions, if the workpieces are simply removed from the austenitizing salt and allowed to cool in air, the residual salt that adheres to the surfaces is extremely difficult to remove. To prevent this, parts are quenched in an intermediate salt bath of carbonate/chloride-type salts, usually held at 595 to 650 °C (1100 to 1200 °F). This interrupted bath dissolves the high-temperature salt from the surfaces. As a rule, parts are held in the quench just long enough to equalize the temperature and remove all of the high-temperature salt, after which parts are hardened in still air, but when sufficient cooling rates are not provided by still air the more sophisticated methods of gas quenching are preferable.

10.2.3 Austempering

Another commercially important interrupted quenching technique is known as austempering. A thorough knowledge of transformation characteristics for a specific steel (TTT curves) can result in a number of different, but reproducible, results. It is obvious that as transformation from the austenite is allowed to complete isothermally at a preselected temperature below Ae_3 a number of microstructures and, in turn, hardness values can be obtained. In general, the lower the transformation temperature, the higher the hardness, extending to the zone where martensite is formed. This phenomenon can lead to the formation of several specific microstructures or acceptable mixed structures. The most common structure is produced by isothermal transformation at some temperature below the "nose" of the "S" curve and is well known as austempering (see Fig. 10.2.1).

Austempering is the isothermal transformation of a ferrous alloy at a temperature below that of pearlite formation and above that of martensite formation. Steel is austempered by being:

- Heated to a temperature within the austenitizing range (usually 788 to 871 °C, or 1450 to 1600 °F)
- Quenched in a bath maintained at a constant temperature usually in the range of 260 to 400 °C (500 to 750 °F)
- Allowed to transform isothermally to bainite in this bath
- Cooled to room temperature, usually in still air.

The principal purpose of austempering is to obtain increased ductility or notch toughness at a given high hardness. For true austempering, the metal must (a) be cooled from the austenitizing temperature to the temperature of the austempering bath fast enough so that no transformation of austenite occurs during cooling, and (b) be held at bath temperature long enough to ensure complete transformation of austenite to bainite.

Austempering is used an economical approach for heat treating many hardware items. Steel springs (mainly small sizes) are among the most notable items produced by austempering.

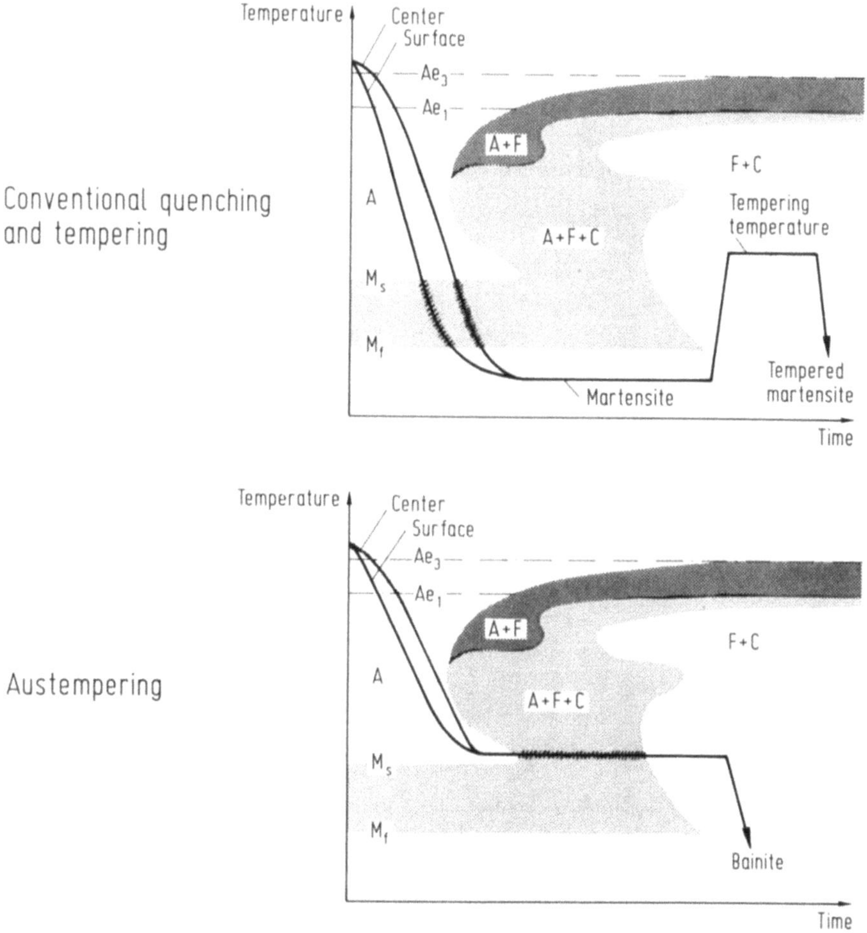

Fig. 10.2.1. Schematic comparison of time-temperature-transformation cycles for conventional quenching and tempering and for austempering. Source: Metals Handbook, Heat Treating, Cleaning and Finishing, Vol. 2, 8th ed., ASM, 1964, p. 56

Systems for isothermal quenching such as the austempering process are generally considered as quenching furnaces rather than as quenching tanks. Isothermal quench furnaces are designed to operate over a temperature range of 177 to 400 °C (350 to 750 °F) and primarily are intended for cooling rather than heating. Salt mixtures of nitrate–nitrite are required for the lower temperature range. Nitrates are used for 315 °C (600 °F) and higher. Size is unlimited and ranges from 0.08 m^3 (3 ft^3) for small bath applications to 9 to 12 m^3 (30 to 40 ft) long with immersed conveyors for high production. A typical isothermal quench furnace is shown in Fig. 10.2.2.

The molten salt is held in a steel pot that may have external fins for heat extraction. An air chamber between the pot and casing permits circulation of

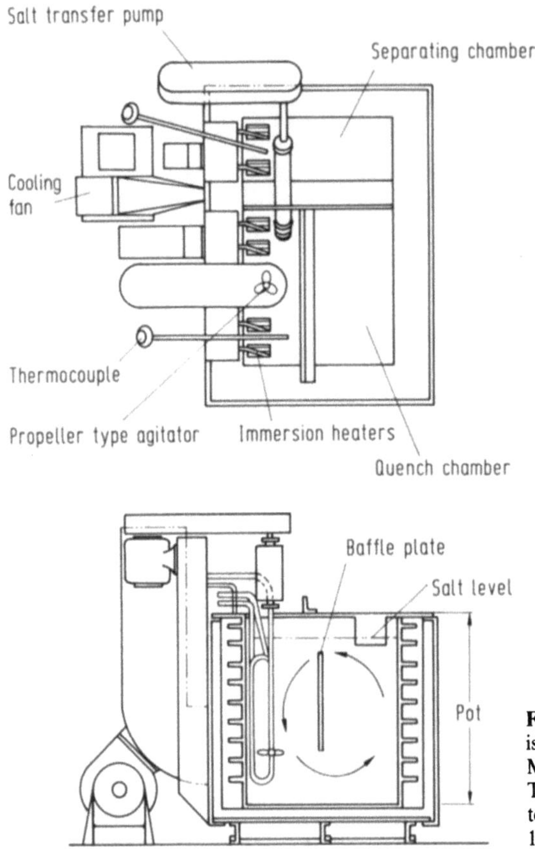

Fig. 10.2.2. Typical construction of an isothermal quench furnace. Source: MEI Course 9, Maintenance of Heat Treating Equipment, Lesson 4, "Maintenance of Salt Bath Equipment," ASM, 1979, p. 3

air for forced cooling. The external casing is insulated for the operator's comfort. The bath usually is heated by internal electric resistance elements, although gas-fired immersion tubes occasionally are used. Air filters are recommended if the environment is dusty. Immersed water-cooled coils are used to cool the salt, but extreme caution is required because a cracked coil could cause a steam explosion or an eruption of salt from the pot. Clean, high-pressure air and aspirated water can be safely introduced through immersed tubes for heat extraction.

Temperature control of $+/-2.5\,K$ easily is obtained using commonly available temperature control instruments. Over-temperature protection is provided with a temperature controller or a separate excess temperature-indicating instrument. The electrical circuits usually are designed to actuate the cooling systems when needed, and the excess temperature instrument actuates a signal, such as a warning light or an alarm.

Agitation of the salt is provided by propeller mixers, centrifugal pumps, or air bubblers. Belt-driven propeller-type agitation is most common. Belt drives require periodic inspection for tightness and replacement of cracked or frayed belts. Water-cooled propeller shaft bearings require lubrication and inspection to ensure that adequate cooling is maintained.

Additional quenching severity can be obtained by the addition of water to a nitrate–nitrite salt. The water is added to the bath by directing a stream of water through a conical spray nozzle into the surface of the salt or into the vortex created by a mechanical mixer. A shroud should surrounded the nozzle to prevent spattering of the water and salt. The mechanical agitation of the bath intermixes the water and salt. Because water continuously evaporates from the surface of the bath, the rate of evaporation increases during the quenching of hot workpieces. Therefore, it is necessary to add water periodically to maintain water concentration and a uniform quenching severity. The amount of water to be added varies with the operating temperature of the salt. The recommended concentrations are 204 °C (400 °F). 0.5 to 2% water; 260 °C (500 °F), 0.5 to 1% water; 315 °C (600 °F), 0.25 to 0.5% water; and 371 °C (700 °F), 0.25% water.

10.2.4 Martempering

Martempering is a form of delayed, or interrupted, quenching. It is not closely related to austempering, because martempering is not an isothermal process. Martempering is sometimes called marquenching, which is a more appropriate term, because martempering implies a tempering operation, which is not true. Martempered parts should be subsequently tempered in the same manner as parts quenched to or near room temperature by cooling in oil, water, or other quenching medium.

Martempering of steel consists of (a) quenching from the austenitizing temperature into hot oil or molten salt at a temperature in the upper part of (or slightly above) the martensite range, (b) holding in the quenching medium until the temperature throughout the steel is substantially uniform, and then (c) cooling (usually in air) at a moderate rate, to prevent any greater difference in temperature between the outside and the center of the section. The formation of martensite occurs fairly uniformly throughout the workpiece during cooling to room temperature, thereby avoiding the formation of excessive amounts of residual stress.

After martempered parts have cooled to room temperature, they are tempered in the same manner as though they had been conventionally quenched. The time–temperature relations of martempering and tempering are compared with those of conventional quenching and tempering in the left and center diagrams of Fig. 10.2.3. Because the final phase of cooling during martempering is relatively slow, heavy and light sections transform from the surface to the center in about the same time. Thus, martempering minimizes or eliminates the distortion resulting from unequal transformation rates normally occurring in conventional quenching.

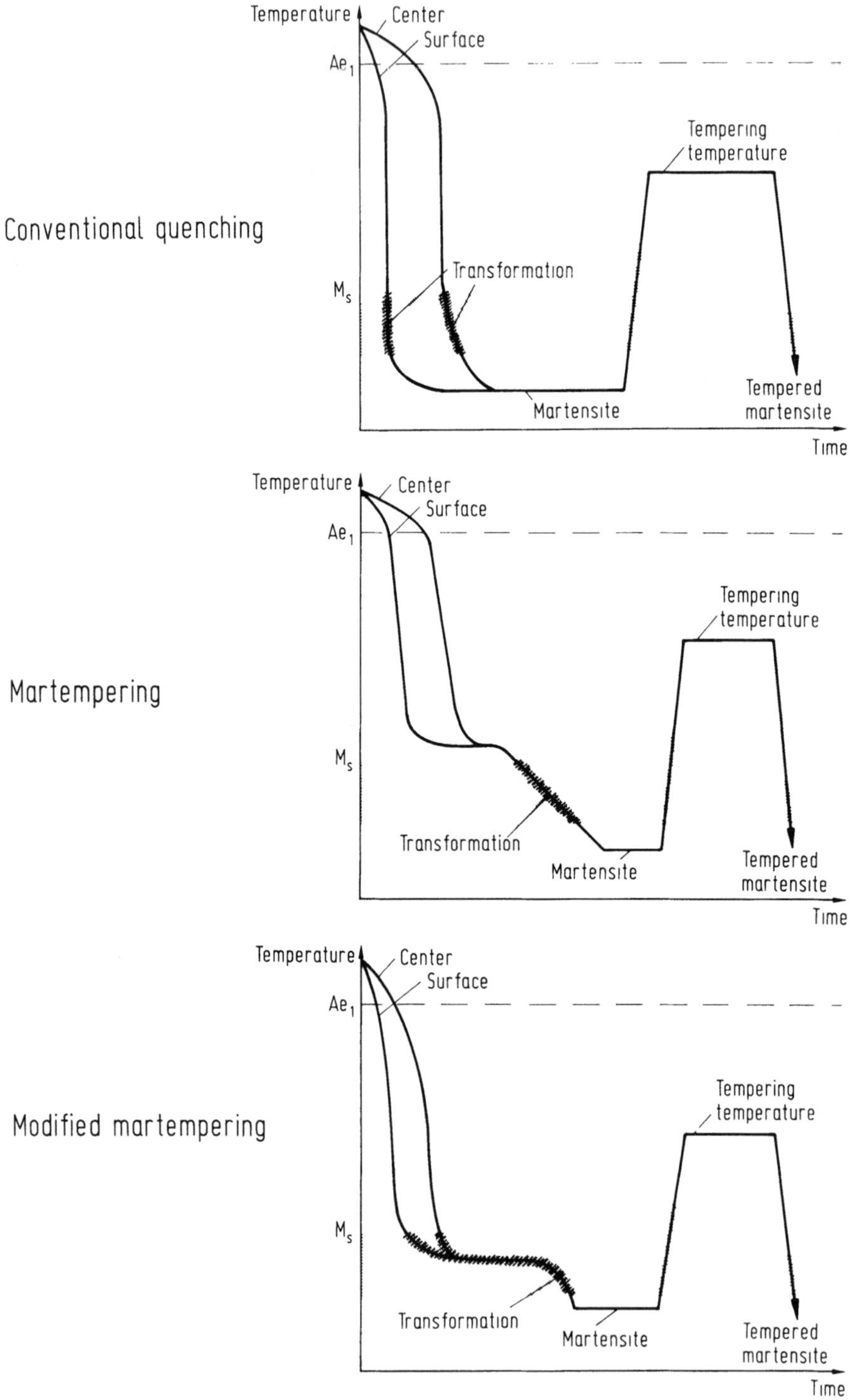
Conventional quenching
Temperature
Center
Surface
Ae1
Tempering temperature
Ms
Transformation
Martensite
Tempered martensite
Time
Martempering
Temperature
Center
Surface
Ae1
Tempering temperature
Ms
Transformation
Martensite
Tempered martensite
Time
Modified martempering
Temperature
Center
Surface
Ae1
Tempering temperature
Ms
Transformation
Martensite
Tempered martensite
Time

Martempering is used primarily to minimize distortion. When the hardenability of the steel is adequate, martempering provides better dimensional control than conventional quenching. This is primarily because fewer harmful residual stresses are developed in martempering than in conventional quenching, due to the fact that the greatest thermal variations occur while the steel is in the relatively plastic austenitic condition, and because final transformation and thermal changes occur throughout the part at approximately the same time.

In many instances, martempering has eliminated the need for quenching fixtures that were required for minimizing distortion during conventional quenching and has thus reduced the cost of tooling and handling. However, when changing from conventional quenching to martempering, it may be necessary to study the dimensional variations in individual parts before establishing preheat treatment dimensions.

Another virtue of martempering is that, in many instances, it is possible to incorporate a straightening operation before the part has hardened. As shown in Fig. 10.2.3 (center), the part, when quenched to a temperature just above M_s, has become equalized in temperature (from its surface to center), but is has not yet transformed to martensite. At this time, a rapid straightening operation can be used that minimizes total distortion. After straightening, the part is replaced into the quenching medium and allowed to transform.

Modified Martempering differs from "standard" martempering only in that the temperature of the quenching bath is lower—ranging from just below the M_s point to about 95 °C (200 °F) (Fig. 10.2.3 right). Faster cooling rates are obtained with modified martempering than in the standard process. This is important for steels of lower hardenability that require faster cooling to harden to sufficient depth. Thus, modified martempering is applicable to a greater range of steel compositions than is the standard process.

Another advantage of modified martempering is that the lower temperature of the quenching medium often permits the use of simpler quenching equipment. Also, because oil is invariably the quenchant used for the modified process, existing equipment may be converted at a low cost.

In general, uniform and reproducible results are obtained by both the standard and modified practices.

10.2.5 Gas and Fog Quenching

Gas Quenching is used to provide a cooling rate that is faster than the cooling rate obtained in still air, but slower than that obtained in oil. In gas quenching, the austenitized workpiece is placed directly into the gas quenching zone or

◄

Fig. 10.2.3. Time-temperature-transformation diagrams with superimposed cooling curves showing quenching and tempering (left) conventional process, (center) martempering, and (right) modified martempering. Source: Metals Handbook, Heat Treating, Vol. 4, 9th ed., ASM, 1981, p. 86

chamber, and heat is rapidly extracted from the metal by a fast-moving stream of gas. The cooling rate of the metal is related to the surface area and mass of the part and to the type, velocity and pressure of the cooling gas (see Fig. 10.2.4). The cooling rate can be adjusted and controlled by altering the last three of these variables, thus providing a degree of flexibility that can be obtained with very few other quenching media.

During the quenching period, large volumes of relatively cold gases are directed through nozzles or vanes to achieve the impingement of high-velocity gas on the surface of the work load. After absorbing heat from the material being processed, the gases are cooled by being passed over water-cooled or refrigerated coils. Recirculating fans return the chilled gases to the high-velocity nozzles, through which they are again directed at the work to absorb more heat.

Various gases, ranging from air to complex mixtures, may be used for cooling, depending on processes requirements. Protective atmospheres commonly are used, not only to produce bright work, but also to increase the heat transfer rate between the gas and the work. This procedure also increases the rate of cooling.

Gas quenching systems may be designed for either batch or continuous processing. The units are gastight to prevent infiltration of air or loss of gas pressure.

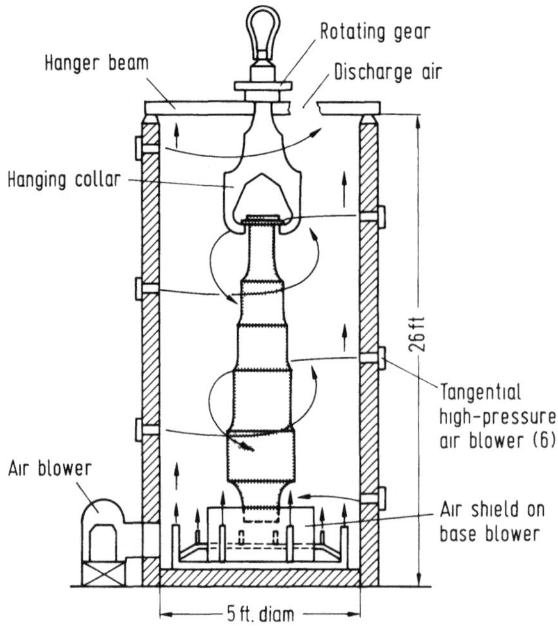

Fig. 10.2.4. "Wing tunnel" used in gas quenching large turbine-rotor forgings. Source: Metals Handbook, Heat Treating, Vol. 4, 9th ed., ASM, 1981, p. 46

Gas quenching also has been used to develop the desired mechanical properties in other steel products, such as thin sheet and small-diameter wire.

Gas quenching commonly is used in the hardening of large forgings to obtain more uniform cooling of heavy sections and to prevent cracking and thermal instability. Generally, as the size of the forging increases, it becomes more difficult to secure reasonable uniformity of properties throughout heavy sections. Assuming that the steel selected possesses adequate hardenability, it may still be necessary to gas quench to obtain desired properties. Gas quenching also will produce uniform mechanical properties in large parts of intricate shape and varying section thickness. For these parts, fixturing usually is required during the quenching cycle.

For example, large turbine-rotor forgings are gas quenched in the "wind tunnel" apparatus shown schematically in Fig. 10.2.4. As each forging is rotated, six tangential high-pressure blowers create a convection current of air in the tunnel. A large blower, located at the base of the rotor and suitably deflected from the end of the shaft, forces a constant stream of air over the surfaces of the forging and out of the top of the tunnel. This method of quenching has been used to process hundreds of forgings, yielding acceptable mechanical properties from surface to center with virtually no distortion or cracking.

Many highly alloyed tool steels (depending greatly on section thickness as well as on their composition) can be fully hardened in still air, but when section thicknesses increase, and sufficient cooling rates are not provided by still air, the more sophisticated techniques of gas quenching may be required, especially when the workpieces are austenitized in a vacuum furnace.

For example, tool steels of types A2 and T1, in the form of solid blocks 50 by 100 by 100 mm (2 by 4 by 4 in.), were successfully austenitized and gas quenched with nitrogen in a vacuum furnace. A sealed fan in the roof of the furnace chamber chilled the nitrogen by forcing it over the water-cooled walls of the chamber; the cooled gas was then admitted to the chamber at 69 kPa (10 psig) and directed at the steel.

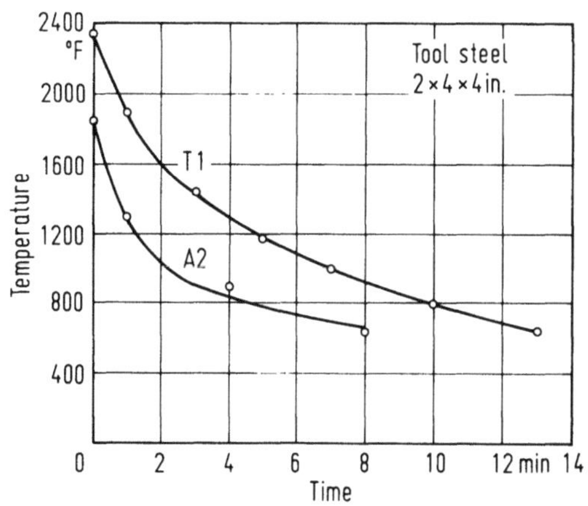

Fig. 10.2.5. Surface cooling curves for types T1 and A2 tool steels quenched from austenitizing temperatures by cooled nitrogen in a vacuum furnace. Source: MEI Course 6, Heat Processing Technology, Lesson 15, "Quenching Media and Equipment," ASM, 1977, p. 21

As indicated in Fig. 10.2.5, the A2 steel cooled from 1010 to 345 °C (1850 to 650 °F) in 8 min, and the T1 steel from 1290 to 345 °C (2350 °F) in 13 min. These rates were sufficient for maximum hardness.

For Quenching

In fog quenching, heat is rapidly extracted from the metal by a fast-moving stream of gas that contains water droplets. The cooling capacity of the "fog" is derived from both the absorption of heat by the gas and the heat of vaporization of the water. The addition of water droplets (or fog) to an air stream can increase its cooling capacity by a substantial amount. Fog quenching is most effective in the lower temperature ranges.

An outgrowth of gas quenching, fog quenching normally is substituted for liquid quenching in an effort to minimize distortion. Although less severe than a liquid quench, fog quenching is more severe than a plain gas quench of equivalent velocity. In general, the types of parts that are fog quenched are similar to those that are considered suitable for gas quenching, but for situations in which the cooling rates provided by conventional gas quenching were insufficient.

10.2.6 Press and Cold Die Quenching

There are many round, flat, or cylindrical parts that distort to an unacceptable degree when subjected to conventional immersion quenching. Automotive ring gears are notable examples. Under such conditions, it is necessary to resort to special techniques. It must be emphasized, however, that the equipment for those special techniques is expensive and production rates are slow. Consequently, the resulting cost of heat treatment is relatively high. Therefore, use of these techniques should be considered only when minimal distortion is mandatory. Press quenching is closely related to intense quenching, which is discussed elsewhere in this book.

Press Quenching

Probably the most widely used special technique is press quenching. To realize the maximum advantages of controlled quenching and to minimize distortion caused by the heating cycle, press quenching dies must be made to provide the proper oil flow and hold critical dimensions of the part being quenched. In press quenching, the die contacts the heated part, and the pressure of the press aligns the part mechanically. This occurs before quenching begins, while the part is hot and plastic. The machine and dies then force the quenching medium into contact with the part in a controlled manner.

The quantity of quenchant and its rate of flow are controlled by the press. The distribution of the flow is controlled by the dies. The speed of the quench depends on the quenching medium and the rate of oil flow. The rate of cooling can be controlled by adjusting the rate of flow of oil through the die.

Design of the quenching dies is based on definite concepts, and the basic equipment has been standardized. Although some basic types of dies are available for certain typical shapes of workpieces, additional die parts are required to adapt and accommodate a specific part. The manufacturer should be consulted in regard to specific die requirements where it is intended to take advantage of controlled quenching.

Quenching to a tolerance of 0.025 to 0.050 mm (0.001 to 0.002 in.) for roundness and flatness is common practice for parts such as ring gears when proper equipment and correctly designed dies are used in quenching accurately and uniformly machined gears. The most important factor in all die design is to correlate the die contact surfaces with the surfaces of the machine workpiece before it is heat treated.

Quenching Presses. These machines have a reservoir and recirculating pump system with valve and time controls, so that at all stages of the quenching cycle—the volumes and rates of flow through the die—are automatically changed to preset values to suite the requirements of the part. The built-in pumping system and reservoir reduce the external supply of quenching medium required to approximately 132 l/min (35 gal/min) while providing controlled flow through the die of as much as 852 l/min (225 gal/min) or more, depending on the size of the machine. The external supply of quenching medium needed is determined by the heat absorption required and by the desired temperature rise of the medium. Because the die passages direct all the quenching fluid into close proximity to the workpiece, the supply from external sources need not be increased to allow for wastage and unutilized overflow.

Many presses are equipped so that the rate of oil flow can be changed during the quenching cycle. This practice merely takes advantage of the time-temperature-transformation (TTT) characteristics. That is, the oil first flows at a high rate (more than 757 l/min or 200 gal/min) so as to cool the workpiece below the nose of the TTT curve as fast as possible. The flow is then retarded (as low as 132 l/min, or 35 gal/min) while transformation to martensite occurs below 260 °C (500 °F). If the press is being used in high production, oil flow may be increased when martensite transformation is nearly complete, to finish the cycle more quickly and make the press available for the next part.

Types of Presses. Presses based on the "rapid release" or "pulsing" principle are available in manual or fully automatic units. In such a press, the workpiece is placed on the lower die in the usual manner. "Pulsing" is developed by the intermittent release of the upper die pressure applied to the part, which allows the part to "breathe" during normal thermal contraction, by eliminating die contact friction. This portion of the cycle requires approximately 20 s, during which time pulsing occurs at a frequency of about once every 2 s. Maximum flow of oil is maintained during pulsing. In the manual machine, the flow of oil is then decreased and finally increased to hasten the operation, as mentioned above.

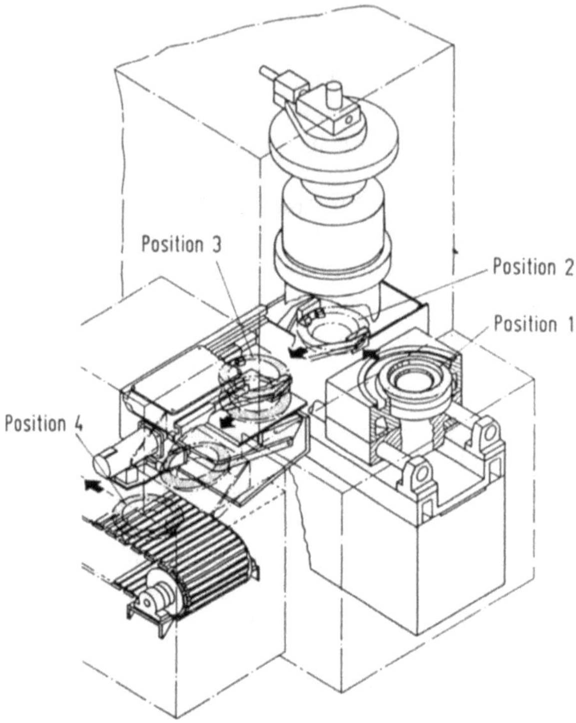

Fig. 10.2.6. Cutaway view of a four-position, automatic gear-quenching press using the pulsing principle. Source: MEI Course 6, Heat Processing Technology, Lesson 15, "Quenching Media and Equipment," ASM, 1977, p. 34

Automatic machines designed for relatively high production (Fig. 10.2.6), also utilize the pulsing principle. In this specific process, the hot gear is placed on the extended lower die at the front of the machine (position 1). The lower die and workpiece then move into the center of the quench chamber (position 2). The upper mechanism (containing the pressure rings and expanding cone) moves down to contact both the workpiece and the expanding mechanism in the lower die, after which the door to the quenching chamber closes. Pulsing and flow of oil (about 757 l/min, or 200 gal/min) then take place for about 20 s. Following this predetermined interval, pulsing stops and the upper die lifts. The door to the auxiliary quench chamber opens, allowing it to fill with oil. The unloader jaws then transfer the workpiece from the lower die to the auxiliary quench chamber (position 3) for further cooling. The door then closes, sealing the auxiliary quench chamber.

The oil in the initial quench chamber drains immediately, and the lower die moves to the loading chamber, ready for the next workpiece. While the next workpiece is being quenched, the part in the auxiliary quench chamber is discharged, still under oil, to the conveyor tank (position 4) and is cooled further as it is conveyed out of the oil. Quenching of ring gears at a rate as fast as two per minute has been accomplished by this method, which might be considered very good when alternatives (usually more finishing stock and/or some form of straightening) are examined. However, this operation would be considered slow, tedious, and expensive compared to normal mass-production quenching out of a furnace.

Cold, Dry-Die Quenching

Commonly used quenching media usually are liquid or gas, thin flat disks, or workpieces with intricate designs cannot be quenched in conventional medium without unacceptable distortion.

One means of solving such problems is to utilize various forms of cold, flat, or shaped dies, depending on the shape of the part being quenched. If only a few parts are involved, the dies will not require special cooling, provided they are allowed to cool before another part is quenched. The dies usually are used in some sort of press located close to the austenitizing operation.

As an example of dry die quenching, relatively large diameter, 3.2 mm (1/8 in.) thick disks made from AISI 1095 steel were austenitized at 788 °C (1450 °F), removed from the furnace one at a time, and quickly dropped into a lower die made from copper (for high heat conductivity) and cooled by means of water ducts. As the workpiece was dropped onto the lower die, the press was activated and the upper die (also made from copper and water cooled) came down solidly onto the workpiece and held it while it cooled very quickly, simulating the speed of a water quench. The disks were hardened with no distortion, but the operation was slow and tedious.

Almost any form of dry die quenching is slow and expensive, and it should not be considered for conventional applications. There are, however, special applications, where the cost can be justified. Gray cast iron, which is an excellent heat sink because of a large amount of Kish graphite, is an excellent material for dry die quench fixtures.

10.2.7 Self Quenching

Although there are sometimes exceptions, as a rule, self quenching is applicable only to induction or flame hardened parts, as well to other local surface hardening methods based on rapid heating by high energy density (implies-hardening, laser-hardening and electron beam hardening). These techniques of heating are about the only procedures whereby extremely wide variations in temperature can be achieved within a very short space or depth.

After all, the basic requirement for hardening of steel is a rate of cooling which exceeds the critical cooling rate. The steel does not know whether this rapid cooling was achieved by "dry or wet" methods, and it does not matter. In many instances, the surface of the workpiece is heated so rapidly by induction that the major part of the total mass is "stone cold" when the surface is heated to—say 870 °C (1600 °F). Thus, the underlying material acts as a significant heat sink which often cools the heated zone at a rate sufficient to develop full hardness. In other instances, the cold portion of the workpiece may not be sufficient to provide all of the required cooling. Under such circumstances, a bit of added liquid quenchant may be used, such as a short-time spray.

10.3 Computer Controlled Spray Cooling

P. Archambault and F. Moreaux

From a practical point of view, the classical immersion quenching techniques do not allow a real control of the cooling processes which are mainly related to the nature and temperature of the part and of the quenchant. Indeed, no correction can be applied throughout the duration of the cooling operation and so the variation of the extracted heat flux density cannot be controlled.

This is not the case for spray cooling. As it has been previously seen, the cooling efficiency of the water spraying process depends on the characteristics of the spray and so, on the water and air supply pressures. These pressures can be varied with the aid of servovalves to provide a continuous variation of the density of the spray and of the velocity of the water droplets. However, in order to real time control the cooling process, one has to verify the continuity of the cooling phenomena in response to air and water pressures variations.

This appears on Fig. 10.3.1 on which several experimental cooling laws, obtained for the same cylindrical sample with different but constant pressures, are reported. Natural convection and cold water immersion cooling laws are also presented [1]. The shaded area between the curves covering cooling with air only and cooling with air and water reveals the overlap between the one and two-phase cooling regimes which means that a continuous variation of the air and water pressures can provide a continuous variation of the extracted heat flux density. This is an important point in view of real time controlling the cooling process. Then, by adjusting the supply pressures ranges, one can theoretically reach any instantaneous cooling rate between the slowest (natural convection) and the fastest (cold water quenching) cooling. Of course this is an ideal situation and experimental parameters can limit the versatility of the cooling process.

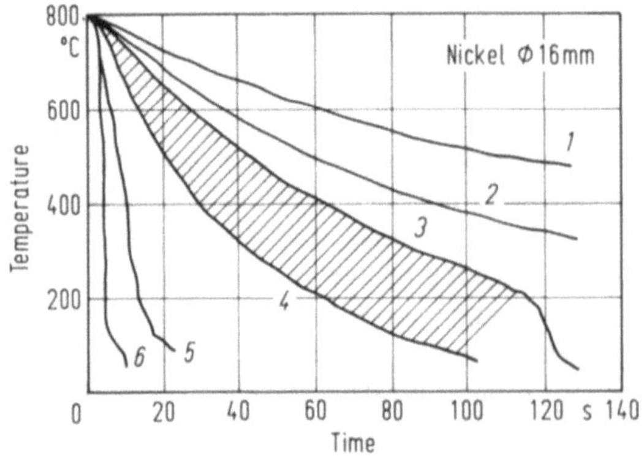

Fig. 10.3.1. Identification procedure: Control of the overlap between air only and air + nul pressure water spraying regimes. *1* natural convection; *2* Pair = 0.05 MPa (air only); *3* Pair = 0.05 MPa – Pwater = 0 MPa; *4* Pair = 0.6 MPa (air only); *5* Pair = 0.6 MPa – Pwater = 0 MPa; *6* cold water (immersion)

Fig. 10.3.2. Example of a two-phase jet nozzle

Choice of the Spray Nozzles

Depending on application and on cooling rates to be reached, one will use a one or two-phase jet nozzles. The shape of the jet (linear, square or round section are available) will be adapted to the geometry of the part to be cooled. However, the jet angle must not be significantly altered by pressure supply variations. In the case of two-phase jets, the water generally enters the jet axially and is projected by air centrifugation. Then the waterflow can be varied by changing either air or water pressure. The air and water supply pressure ranges have to be considered by taking into account the desired maximum cooling capability of the system under development. Realistic orders of magnitude are 0–0.5 MPa for the air and 0–0.05 MPa for the water. The maximum air and water flows needed by the choosen nozzle must be known in order to design the general supply circuit which must be able to provide the maximum flows for N nozzles (N being the number of nozzles of the designed system) [2]. Figure 10.3.2 shows an example of nozzle already in use which is equipped with lateral small nozzles for modifying the cross section of the jet.

Disposition of the Nozzles

The number and the positionning of the nozzles mainly depend on the shape of the part to be heat treated and on the desired cooling efficiency. An important parameter is the covering of the surface of the part which must be as homogeneous as possible in order to provide an homogeneous cooling of the surface. The difficulty here is to control the overlap on the surface between adjacent jets to minimize cooling heterogeneities due to the fact that the profile of the cooling efficiency in the cross section of the jet is not flat and not controlled during the real time process. In fact, one has to find a compromise between the covering of the surface of the part, the distance of the nozzle from the surface and the cooling effectiveness. Of course this compromise conditions the number of nozzles to be used [3]. No precise rule can be applied here, each positionning being specific to the cooling problem under consideration.

Characteristics of the Servovalves

The air and water supply servovalves play an important role in the system. Because pressures are adjusted automatically, servovalves must have special features such as a well-defined servocontrol behaviour and a fast reaction time. The shape of the clapper valve must be adjusted in order to obtain a pressure variation as linear as possible with the analog command signal level and the possible flow through the valve must allow the maximum flow needed by the whole spraying system. On the other hand the reaction time from closed to opened position must be as short as possible (typically about 0.5 s).

Identification of the Spraying System

Once the spraying system has been designed, it is necessary to identify its behaviour for different air and water supply conditions. This can be performed by using a testing specimen heated at a high temperature (constant parameter) and then spray cooled with well defined air and water pressures. Temperature evolutions are registered (thermocouple or infra red measurement) for each constant pressure condition. This leads to a classical diagram (temperature versus time) with as many cooling curves as pressure supply conditions [2]. However, for a real time purpose, one prefers the kind of diagram shown in Fig. 10.3.3 on which the heat flux density is reported versus the surface temperature. The interest of such a representation is, on a first approximation, to be shape independent which is quite convenient for a real time process. On this example three groups of curves are presented depending on the air and water supply conditions (air only, nul water pressure with varying air pressure, maximum air pressure with varying water pressure). Each curve of each group is parametered by a constant pressure value. It has to be noticed here that the experimental curves have been linearized in the three cooling regimes (non wetting, wetting and convection) to make their use easier in the real time algorithms.

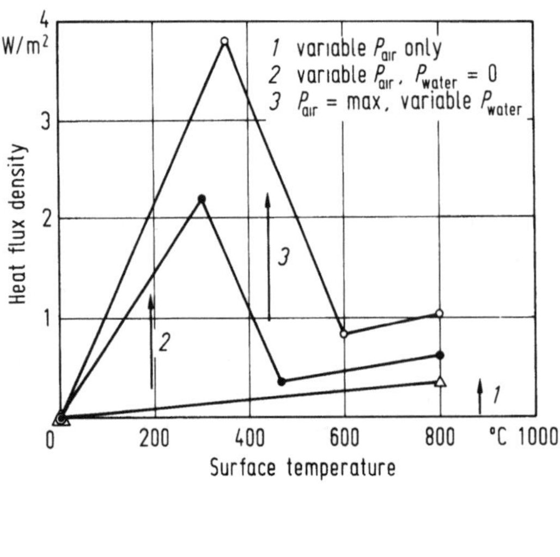

Fig. 10.3.3. Identification procedure: Heat flux density versus surface temperature. Each domain *1*, *2*, *3* contains several curves obtained for different but constant air and water supply conditions [3]

One can see that the heat flux density variations with the surface temperature are non linear, which makes real time guidance difficult and requires an accurate identification of the processes. However it is possible to bypass such a difficulty by introducing special corrective parameters in the regulation algorithms for taking into account experimental variations like dimensional changes of the heat treated parts.

Regulation Algorithms

Different algorithms are needed to control the whole cooling process. They concern (i) the identification of the spraying system, (ii) the servovalves, (iii) the regulation process itself.

(i) the identification algorithms are determined with the linearized curves of Fig. 10.3.3 in each spraying regime (for specific air and water pressure conditions) as $P_i = f(q_s, T_s)$ where T_s is the temperature of the surface of the part and q_s the heat flux density extracted from the part. P_i is the air or water servovalve output pressure depending on the considered spraying regime i.e. P_i is either a constant or a variable output parameter. The general expression is $P_i = a_0 + a_1X^1 + a_2X^2 + \ldots a_nX^n$ where X depends on q_s and T_s. So each spraying regime is identified with a specific P_i expression [3].

(ii) the air and water servovalves are identified by using experimental results which lead to a polynomial expression $U = g(P)$ where U is the analog signal level to be supplied to the servovalve in order to obtain the desired output pressure P.

(iii) the regulation algorithm is a classical one. It is expressed as a function of the measured temperature at time t and of the desired temperature at time $t + \Delta t$ and uses a P.I.D. scheme. An intermediate variable is then defined as:

$$V_i = K_p * \Delta T + K_i * (\Delta T * \Delta t) + K_d * (\Delta T / \Delta t)$$

where K_p, K_i, K_d are the proportional, integral and derivative coefficients and with $\Delta T = (T\text{measured} - T\text{setpoint})$.

In the preceding algorithms one can introduce corrective coefficients to accommodate for experimental variations. Indeed, from one case to another, significant modifications (for example changing the part or the desired cooling law) can alter the response of the regulation process. Moreover, for technical reasons due, for instance, to particular variations of the cooling rates, one can need weighting separately the response of the regulation process in the different cooling regimes. Therefore it becomes necessary to modify the algorithms in order to get a satisfying regulation. As the determination of these algorithms is a rather long process, one prefers introducing weighting coefficients in a set of formulations determined in a standard configuration.

Regulation Loop

The regulation loop involves several steps. First of all one has to input a set of data which concern the geometry of the part, the alloy to be heat treated, the desired cooling law (setpoint) and the regulation coefficients. The time step of the regulation process must be determined as low as possible in order to allow the whole calculations of the regulation loop to be performed [3]. It depends on the computer which is used for this process and on the writing of the software. Reasonable values for the time step are around 50 ms. A few parameters must be used to control the functioning of the spray cooling system hardware and a start signal must indicate the beginning of the cooling operation in order to synchronize the regulation process. The regulation loop is then schematically divided as follows:

—surface temperature measurement at time t (T_m),
—temperature interpolation on the setpoint curve at time $t + \Delta t$ (T_{sp}),
—calculation of the heat flux density q_s to be extracted between t and $t + \Delta t$ with P.I.D. correction,
—determination of the spraying regime to be considered. This is the longest process in the regulation loop as it involves numerous tests to determine the spraying regime in which such an heat flux density q_s can be extracted at temperature T_m. The air and water pressures are then calculated by using the specific P_i expression (see (i) above),

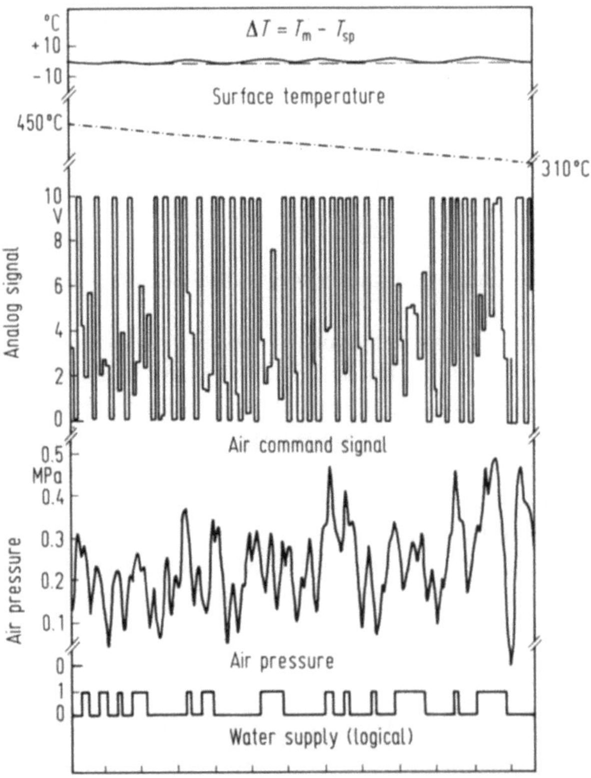

Fig. 10.3.4. Temperature and air and water command signals variations during a linear cooling regulation (time step = 0.1 s). In this example the water pressure is constant (0 MPa), so the water supply is a logical parameter.

—calculation of the servovalves command signals (see (ii) above),
—output of the command signals,
—time incrementation for next step regulation.

During or at the end of the cooling, one can output the evolution of the surface temperature and of the analog command signals on a paper recorder or on a plotter. An example of this appears on Fig. 10.3.4. These curves are very helpful for adjusting the corrective parameters introduced in the preceding algorithms and for controlling the effectiveness of the cooling regulation process.

Examples of Real Time Controlled Quench Coolings

Until now, only few examples of real time controlled coolings are available; as this is a new technique, they mainly concern cylindrical parts on a laboratory scale. The first example shown on Fig. 10.3.5 deals with the case of perlitic transformation. One can see that the real time cooling control allows the extraction of the heat flux due to the transformation. Then, the perlite which is isothermally formed (curve 3) is very thin and it has been verified that the mechanical

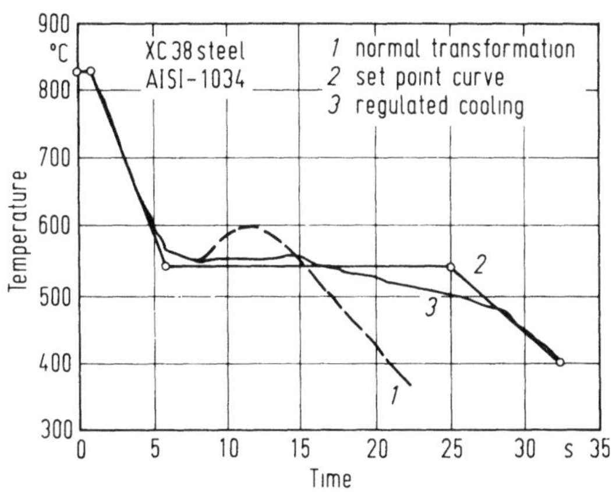

Fig. 10.3.5. XC38 (AISI 1034) cylinder ($\phi 16 - h = 48$ mm). Normal and real time controlled isothermal perlitic transformation

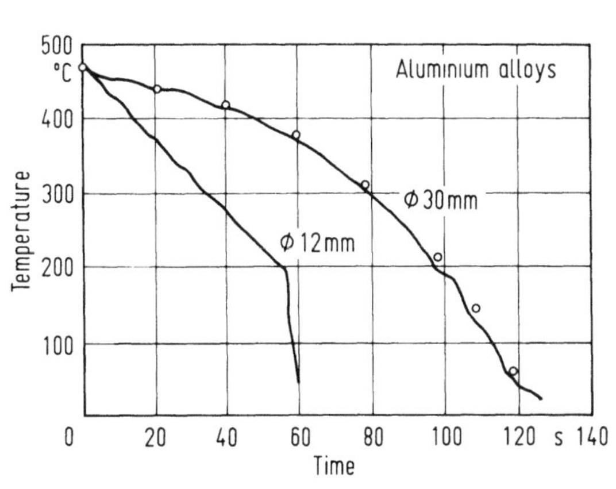

Fig. 10.3.6. Accelerated and linear quench coolings for aluminium alloys. The linear coling is interrupted at 200 °C

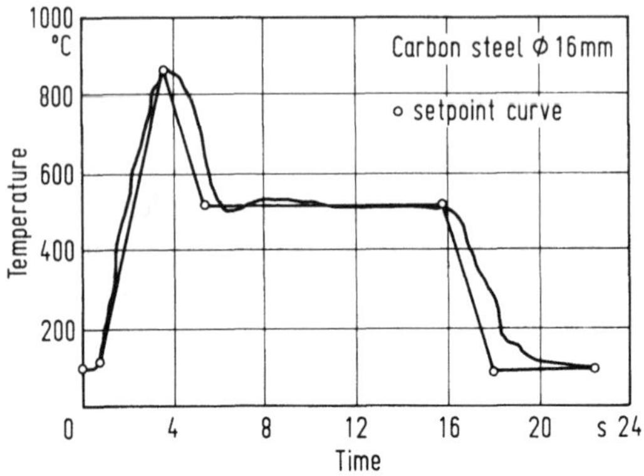

Fig. 10.3.7. Surface heat treatment of a carbon steel. Induction heating and spray cooling are coupled in a fully automated procedure

characteristics are higher than in the case of the non isothermal transformation (curve 1) [4].

The next example (Fig. 10.3.6) deals with aluminium alloys for which linear and accelerated coolings were needed for metallurgical and mechanical reasons (solid solution decomposition and residual stresses control) [5]. It has to be noticed that this kind of cooling curves cannot be realized with the aid of classical immersion quenching techniques.

Other Developments

The versatility of such a real time controlled cooling system allows the coupling between heating and cooling operations to provide a fully automated heat treating process. For such a configuration the response time of the heating system must be as low as possible and the transfer of the part between the heating and cooling zones synchronized by the computer. As an example this has been done with induction heating which can be real time controlled by using similar identification and regulation algorithms [6]. Figure 10.3.7 presents a temperature evolution obtained in the case of surface heat treating. One can see that the attainable thermal kinetic can be very high which allows a precise control of the phase transformations. The deviations are only due here to the high desired cooling rates but, in that particular case, had no significant influence on the metallurgical results [7]. The other interest of coupling the heating and cooling processes is to allow an on-line heat treating of the part on the production line. Such researches are on the way and could lead to a substantial reducing in the fabrication cost of the parts under consideration.

References

1. Didier G, Archambault P, Chevrier JC, Moreaux F (1981) Utilisation d' un miniordinateur pour le pilotage en temps réel de refroidissements par pulvérisation. Mémoires Scientifiques, Revue de Métallurgie, June, pp 293–302

2. Bieth M, Moreaux F, Beck G (1976) Caractéristiques du transfert thermique entre un solide à haute température et un jet diphasique. Entropie, 71, pp 42–49
3. Didier G, Archambault P, Moreaux F (1983) Automatisation d'un système de refroidissement par pulvérisation pneumatique. Revue Générale de Thermique, 256, April, pp 333–339
4. Denis S, Basso C, Fernandes FMB, Simon A (1986) Contribution des contraintes internes d'origine thermique dans le calcul de l'avancement des transformations de phase en refroidissement continu d'un acier XC80. Mémoires Scientifiques, Revue de Métallurgie, October, pp 533–342
5. Archambault P, Chevrier JC, Beck G, Bouvaist J (1980) Contribution to the optimisation of the 7075 heat treatment. Materials Science and Engineering, 43, pp 1–6
6. El Fenne M, Archambault P, Moreaux F, Pourprix Y (1987) Trempe par pulvérisation d'eau après chauffage superficiel par induction. Cas d'un acier XC42. Traitement Thermique, 207, pp 31–34
7. Pierronnet M, Archambault P, Moreaux F, Clement B, Beauget M (1988) Comportement métallurgique et mécanique d'aciers trempés par induction après cémentation. Traitement Thermique, 222, pp 35–41

10.4 Intensive Steel Quenching Methods

NIKOLAI I. KOBASKO

10.4.1 New Methods for Quenching Alloyed Steels Based on the Heat Exchange Intensification

Parts of machine and tools of alloy steels are usually quenched in slowly cooling media: mineral oils, aqueous polymeric solutions, fluidized beds in air-water flow, etc.

Accelerated cooling is connected with a risk of quench cracks formation that arise under high thermal and structural stresses.

Moreover, during the accelerated cooling, increased buckling of the quenched parts was observed which convinced the heat tresters that intensification of cooling with phase changes ($\gamma \rightarrow M$) is dangerous. That is why all the efforts were directed to selection of a slowly quenching medium.

On the basis of numerous experimental studies, it has been established [1, 2] that with the cooling rate increase within the martensite decomposition range, the probability of quench cracks formation grows up to a maximum value and then lowers to zero (Fig. 10.4.1). The cooling rate plotted on the axis of abscissae was defined at the specimen center at temperature of 300 °C.

Similar regularity was also observed by the authors of Ref. [3] who tested splined surface specimens. The specimens were produced from alloy steels and cooled in the water flow being adjusted. In the intensive water flow (the cooling velocity increase) the quench cracks disappeared.

Substantiation of the regularity obtained is of a great practical and scientific importance. With the aim of this, the calculation was fulfilled of the thermal and stressed-strained states of steel parts on the basis of finite element methods [4]. For mathematical simulation of the process of part quenching, a mathematical model was used that included a non-linear equation of nonstationary heat

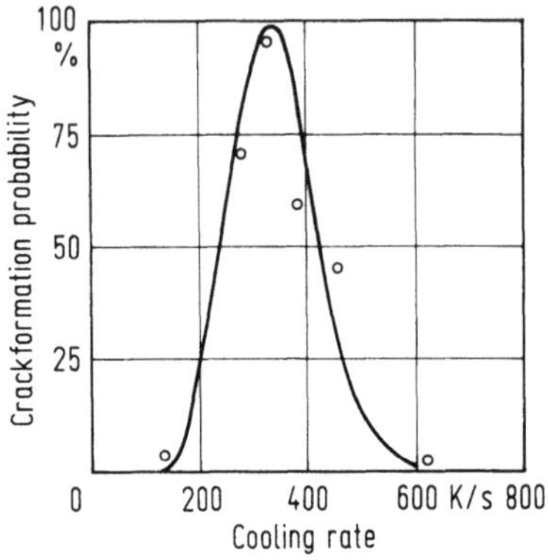

Fig. 10.4.1. Effect of cooling rate within the martensite formation range on crack formation in cylindrical specimens 6 mm in diameter made of steel 40X*

*40X = 41Cr4 (DIN)

conduction, equations of the theory of thermoelastic-plastic flow with kinematic strengthening under the appropriate boundary conditions.

At each time and space step, the calculation results were compared with the thermokinetic diagram of the supercooled austenite transformation, and new thermophysical and mechanical characteristics for the next step were chosen depending on the structural components.

The principal block-diagram of the computing complex developed at the Institute of Strength Problems with the participation of the Institute of Engineering Thermophysics, both of the Ukrainian SSR Academy of Sciences is shown in Fig. 10.4.2 [4]. The calculation results are: temperature field, material phase composition, migration of points in the volume which is calculated, components of stress and strain tensors, intensities of stress and strain, and the field of the safety factor it means a relation of stresses by which the material will be destroyed. These values are presented in the form of tables and isometric lines that allow to observe the kinetics of phase changes in the process of heating and cooling.

Using the potentialities of the computing complex mentioned above, the current and residual stresses were determined depending on the cooling intensity for cylindrical specimens made of different steel qualities. Similar calculations were fulfilled for quenching the parts of complex configuration.

The most important results used at developing the intensive methods of quenching were as follows.

It has been stated [5] that with the quench intensity increase, the residual stresses grow at first, then lower, with further increase of Biot number and pass over to compression stresses (Fig. 10.4.3).

The dependence of residual stresses on the cooling rate of the specimen core at temperature of 300 °C is readily represented.

It appeared that the maximum probability of quench cracks formation and the maximum tensile stresses that depend on the specimen cooling rate coincided [6].

The absent of quench cracks at quenching alloyed steels under intensive heat transfer can be explained by high compression stresses arising at the surface of the parts being quenched.

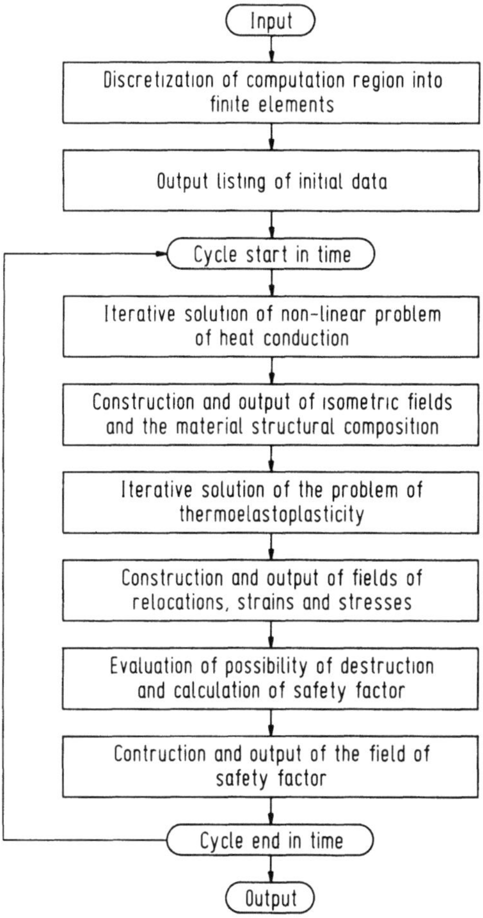

Fig. 10.4.2. Block-diagram of the computing complex

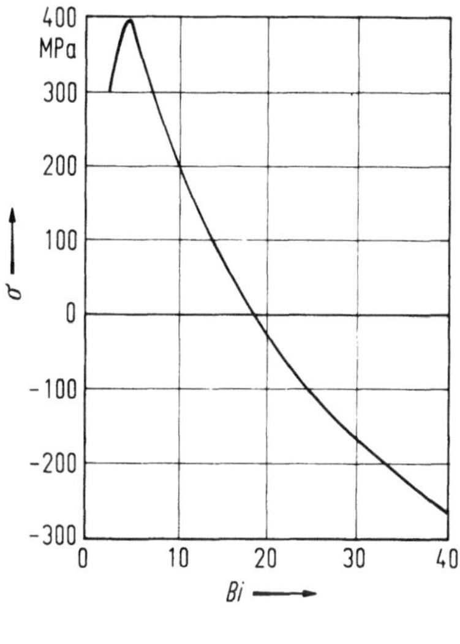

Fig. 10.4.3. Residual circumferential stresses at the surface of a cylindrical specimen versus Biot number

The mechanism of arising high compression stresses in the process of intensive heat transfer is described in Refs. [5, 6, 7].

The results obtained were confirmed experimentally by measuring residual stresses at the surface of quenched parts (specimens) with the help of X-ray crystallography technique [8].

The intricate character of residual stresses dependence on the cooling rate can be explained by "superplasticity" and variation of the phase specific volume at phase changes [9]. Under the conditions of high-forced heat transfer ($Bi \to \infty$) the part surface layer is cooled down initially to the ambient temperature, while the core temperature remains practically constant.

In the process of cooling, the surface layers should compress. However this process is hampered by a heated and expanded core. That is why, compression is balanced out by the surface layer expansion at the moment of superplasticity. The higher the temperature gradient and the part initial temperature are, the greater the surface layer expansion is. At further cooling, the core is compressed due to which the surface layer begins to shrink towards the centre and compression stresses arise in it. At the core cooling, transformation of austenite to martensite takes place. The martensite specific volume is higher than that of austenite. For this reason, the core swelling is taking place that causes the surface layer extension at moderate cooling.

At $Bi \to \infty$ the surface layer is stretched to a maximum and therefore despite swelling it cannot completely occupy an additional volume formed due to the external layer extension. It is just under the conditions of high-forced heat transfer that compression stresses occur in the surface layer. More detailed information about the calculation results can be found in Refs. [9, 10, 11].

In order to confirm the above statements, one can refer to Ref. [12] where a method of IQ-quenching steel parts by using intensive jet cooling is described. The author points out that the method of intensive cooling is applied to superficial hardening of small parts (shafts, axes, pinions, etc.) made of alloy steels. Here a very high intensity of cooling is achieved that allows to obtain a 100 percent martensite structure in the outer layer and high residual compression stresses. It should be noted that under the conditions of very intensive cooling, the strain decrease is observed. While treating the parts of complex configuration, it is necessary to use several combined jets in order to prevent the steam jacket formation. A disadvantage of this method is a high cost of the equipment.

10.4.1.1 Methods of Quenching Alloy Steel Parts

In order to substantiate the method of quenching let us consider the character of changes in the current stress at the surface of the parts being quenched depending on various intensities of cooling. It has been stated that in the course of time, small tensile stresses occur at first at the surface of the specimen to be quenched, and then during the martensite layer formation these stresses transform into the compression ones that reach their maximum at a certain moment of time and then decrease (Fig. 10.4.4). The current stresses become the residual ones that can be either tensile or compression depending on the cooling intensity (Fig. 10.4.3). The maximum tensile stresses correspond to the maximum compression stresses in the control layer of the part being quenched (Fig. 10.4.5).

The mechanism of the current stress formation is as follows. When the part is completely in the austenite state, there arise tensile stresses that pass over to the compression ones in the process of the martensite phase formation and due to increase of its specific volume. The larger part of austenite is transformed to martensite and the larger the martensite layer is, the higher the compression stress. The situation retains until a thick enough martensite crust is formed resembling a rigid vessel that still contains the supercooled austenite in the supercooled phase. Further advance of martensite inside the part causes the effect of water freezing in a glass vessel [9]. Due to the core volume increase at martensite transformations, either decrease of compression stresses in the surface layers or destruction of the external layer will take place if the phase specific change is large enough and the external layer is insufficiently stretched and strong. Under such conditions, compression stresses in the surface layer change over to stressed states that cause destruction of the surface layer.

Reduction of compression sresses at further advance of martensite into the part being quenched is caused by the parting action that is attributed to variation in the phase specific volume in the core.

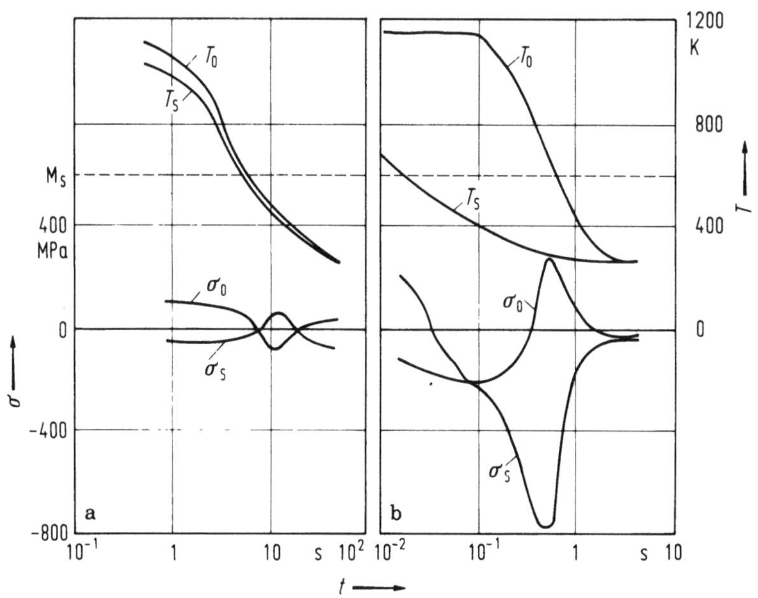

Fig. 10.4.4 a, b. Change of current circumferential stresses at the surface (σ_s) and in the center (σ_0) of the cylindrical specimen being cooled under various heat transfer conditions [10]. **a** Slow cooling; **b** Intensive cooling

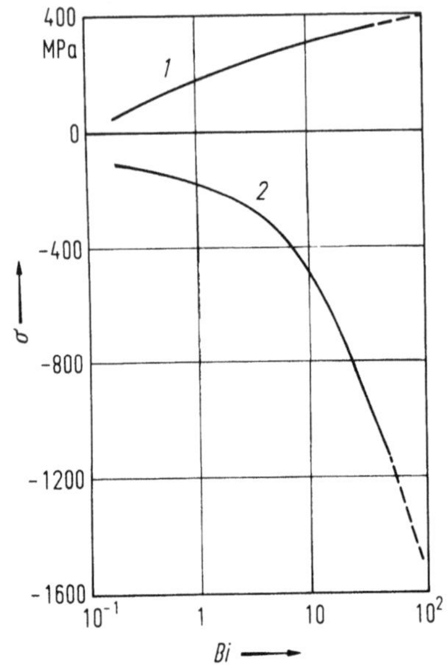

Fig. 10.4.5. Maximum circumferential compression stresses at the surface (*2*) and tensile stresses in the center (*1*) of the cylindrical specimen 6 mm in diameter versus Biot number

If the process of intensive cooling is stopped at the moment of achieving the maximum compression stresses, and isothermal holding is realized at the temperature of the onset of martensite transformation (M_S) then the martensite phase advance would cease and sufficiently high compression stresses can be fixed. They would slightly decrease due to the isothermal holding at which stress relaxation takes place.

An optimal depth of the quenched layer that depends on part dimensions corresponds to maximum compression stresses.

Using the calculation methodics developed and the possibilities of the computing complex [4, 5] it is easy to find with the help of an electronic computer the time of achieving the maximum for bodies of arbitrary axisymetric form being quenched under various heat transfer conditions.

The degree of intensive cooling can be characterized by Bi_v number or by Kondratjev number Kn. There is a universal interconnection between these numbers

$$Kn = \psi \cdot Bi_v = \frac{Bi_v}{\sqrt{Bi_v^2 + 1.437\, Bi_v + 1}} \tag{10.4.1}$$

that is valid for bodies of various configurations.

The author of the well known manual "Theory of Heat Conduction" A.V. Lykov has called equation (1) an important relation of the theory of regular conditions [13]. Criterion $Kn = \psi\, Bi_v$ is the main value determining the heat transfer mechanism of the body. It was named Kondratjev number (criterion) in honour of the outstanding thermophysicist G.M. Kondratjev.

It appeared that the curves $Kn = f(Bi_v)$ for geometrically different bodies (sphere, parallelipiped, cylinder, etc.) were located so close to each other that practically all the family could be replaced by a single averaged curve [13].

The parameter criterion ψ characterizing the temperature field non-uniformity is equal to the ratio of the body surface excess temperature to the mean excess temperature over the body volume. If the temperature distribution across the body is uniform ($Bi_v \to 0$) than $\psi = 1$. The higher is the temperature non-uniformity the less ψ is. At $\psi = 0$, the temperature distribution non-uniformity is the highest ($Bi \to \infty$, while $T \to T_m$).

Thus, Kondratjev number characterizes not only the temperature field non-uniformity but also the intensity of interaction between the body surface and the environment.

Kondratjev number is the most generalizing and the most universal value which may serve to describe the cooling conditions under which compression stresses occur at the surface of various bodies.

For rather high compression stresses to occur at the surface of the part being quenched it is sufficient to meet the following condition:

$$0.8 \leqq Kn \leqq I.$$

Thus the essence of the new method is that alloy and high-alloy steel parts are cooled under the conditions of high intensive heat transfer ($Kn \geqq 0.8$) up to the moment of reaching maximum compression stresses at the surface with the following isothermal holding under temperature M_s [14]. A year later-similar quenching method was proposed in Japan [15].

In accordance with the method described, alloyed steel parts are quenched in such a way that a very hard surface layer of the given depth and an arbitrarily hard matrix are obtained. An example of such method realization is given below. An alloy steel specimen containing (in %) 0.65–0.85 C; 0.23–0.32 Si; 0.4–0.9 Mn; 2 Ni; 0.5–1.5 Cr; 0.1–0.2 Mo is heated up to 800–850 °C and spray quenched with water fed under pressure of 0.4 – 0.6 MPa during 0.2–0.8 s. The specimen is subject further to isothermal heating at 150–250 °C during 10–50 minutes [15].

It is obvious that the spray quenching under high pressures provides intensive cooling ($Kn > 0.8$) that is completed when a certain depth of the composition quenched layer is achieved.

For the steel composition cited, the temperature of the martensite transformation onset is within the range of 150–200 °C.

The isothermal holding time at this temperature (about 10–15 minutes) is chosen from thermokinetic diagrams of super cooled austenite dissociation in such a way that to provide this dissociation into intermediate components in the part central layers.

The analysis of the methods described shows that various authors have come independently to an identic conclusion that is a rather pleasant coincidence because it testifies to urgency and authenticity of the technology being studied.

10.4.1.2 Steel Quenching Method Based on the Mechanism of Non-Stationary Nucleate Boiling

The method is based on using the regularities of non-stationary nucleate boiling that consists in the fact that at nucleate boiling the surface temperature of the part being quenched is maintained for a long time at the level of T_s. [16].

Based on this regularity, the surface temperature of the part being quenched with the help of various additives to water or under a certain pressure in the process of nucleate boiling in maintained at such a level that not more than 25–30 percent of martensite can be formed in the supercooled austenite. When the nucleate boiling is over the parts are discharged and held in air until the temperature smoothing over their cross-section. Then the process of cooling goes on in such a way that the part surface cooling rate is within 30–600 K/s [17].

This method allows to prevent the possibility of quench cracks formation and to improve the material strength properties and plastic characteristics.

At the first stage, a very intensive cooling takes place due to the growth and departure of numerous bubbles that are formed at the part surface during non-stationary nucleate boiling. (When this method is used the period of film boiling is to be reduced to a minimum.) In the period of nucleate boiling the highest thermal stresses are achieved, that is why the delay in martensite transformations is needed.

The analysis of thermokinetic diagrams of structural steels has shown that for the most of materials about 25–30% of martensite is formed in the supercooled austenite at the temperature within the limits of $M_s - (50\text{–}200\,°C)$.

If in the supercooled austenite there is less than 30% of newly formed martensite then the material is still plastic and the possibility of quench crack formation is absent. When the parts are discharged and hold in air the newly formed martensite is selftempered in the surfaces layer due to which its strength properties are improved and stresses decrease. Thus the isothermal holding under the boiling temperature of the liquid medium that corresponds to $M_s -$ (250–200 °C) temperature and holding in air prevent quench crack formation that arise under high current and residual stresses.

When the nucleate boiling and holding in air are over the process of cooling the part being quenched goes on again at high-speed cooling within the martensite range below [18]: in agitated water flow, in criogenic medium oscillating with a resonance frequency, etc.

The surface cooling rate should exceed a certain value beyond which an additional material strengthening is observed.

For high-alloy medium carbon steels of 40XH2MA (0.37–0.44 C; 0.5–0.8 Mn; 0.17–0.37 Si; 0.6–0.9 Cr; 1.25–1.65 Ni; 0.15–0.25 Mo) type this rate lies within the limits of 60–80 K/s, while for the steel grade 45 (0.42–0.5 C; 0.17–0.37 Si; 0.5–0.8 Mn) it is higher than 300 K/s.

For alloy steels with carbon content up to 0.6%, for example 60C2A (0.58–0.63 C; 0.6–0.9 Mn; 1.6–2 Si; <0.3 Cr) the critical cooling rate lowers to 30 °K/s. This technology is protected by the USSR certificate of authorship [17].

Let us compare the method described with that of high-carbon steel quenching that is patented in the USA [18].

The essence of invention is as follows. A steel specimen is heated up to the temperature of austenization for carbides to be solved, then cooled down to the temperature below that of the onset of martensite transformations (100–280 °C), held at this temperature for the period from 5 minutes to 1 hour during which 20–35% of martensite is formed. Afterwards the specimen is tempered within the range of 205–260 °C for 10–30 minutes and then cooled down to the room temperature or to a negative temperature for the rest of austenite to be transformed into martensite. Now the tempering of the newly formed martensite is carried out. The method of quenching proposed ensures the absense of microcracks in the quenched steel and improves the material strength characteristics.

Thus the comparison of the two quench methods has revealed that they have very much in common.

The isothermal holding temperature is maintained lower to provide formation of not more than 35% of martensite in the supercooled austenite.

—Tempering or self-tempering takes place at the second stage.
—Intensive cooling down to the room temperature or to negative temperatures takes place at the final stage.

In the case under study, the rate of the part surface cooling at the final stage is higher than the critical cooling rate within the martensite range.

The first method refers to the steels with carbon content to 0.6%, the second one—to high-carbon steels.

The priority of the first method refers to 1978 [17], and of the second one—to 1983 [18].

The both methods are widely used at quenching structural and instrument steels.

10.4.1.3 Application of New Methods for Quenching Parts of Complex Configuration

In the process of quenching parts of complex configuration, it is not always possible to supercool the austenite uniformly over the whole surface to the temperature of onset of martensite transformations (M_s) as the latter begin first of all in a thinner cross-section of the part. For this reason it is sometimes advisible to carry out the retarded cooling of the parts of complexes configuration, so that a uniform temperature field could be observed over the whole cross-section. When M_s temperature is reached at all the points of the part, holding in air is carried out and then a very intensive and uniform cooling over the whole surface of the part being quenched is created. Some of the experimental results obtained on the basis of the numerical calculation using the workstation electronic computer are given below. The calculation was fulfilled with the help of the computing complex described above. The process of quenching the bearing rings of complex configuration made of X15 (0.95–1.05 C; 0.2–0.4 Mn; 0.17–0.37 Si; 1.3–1.65 Cr) steel and carbonised plugs of X CH steel (0.11–0.16 C; 0.4–0.7 Si; 0.9–1.5 Mn; 0.5–0.8 Cr) was simulated. The part shape and dimensions are shown in Fig. 10.4.6. It should be noted that the plug is carbonised only from inside (Fig. 10.4.6b).

At the quenching process mathematical modelling, different intensities of cooling within the martensite range were set by giving different values of coefficients of heat transfer α.

It appears that at retarded and very intensive cooling within the martensite range, the conicity of the rings being quenched is the least, while at moderate intensity of cooling it is higher (Table 10.4.1).

Races are cooled in two steps. The first step is mineral oil cooling until the minimum conicity is achieved. At the second step, different cooling rates are preset.

Minimum values of conicity at the maximum cooling rate at the second step can be explained by the fact that martensite transformations at a very high cooling rate are completed over the whole surface almost simultaneously and thus form a hard crust that hampers further deformation of the part due to phase changes inside the rings.

Maximum compression and tensile residual stresses observed over the ring cross-section are given in Table 10.4.1. While quenching in the circulating water and in the shower, the residual stresses are low since the condition of $Bi \rightarrow 18$

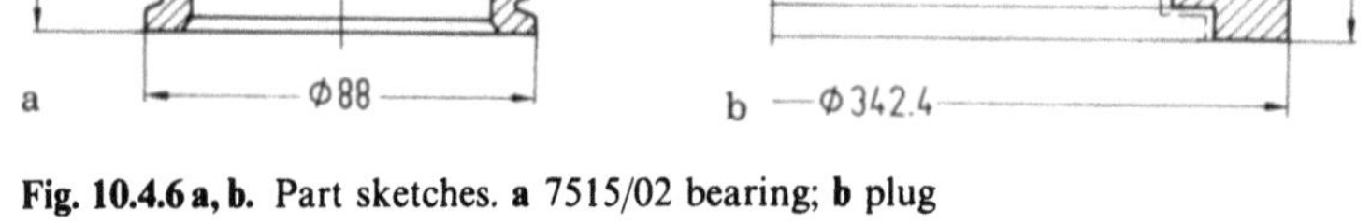

Fig. 10.4.6 a, b. Part sketches. **a** 7515/02 bearing; **b** plug

Table 10.4.1. Residual stresses and bearing ring conicity depending on the cooling conditions within the martensite range [4]

Method of cooling	Stresses, MPa			Conicity
	σ_{22}	σ_{33}	σ_i	
In air	−90 + 130	−130 + 165	70	0.044
In oil	−110 + 130	−150 + 180	180	0.045
In circulating water	−75 + 200	−180 + 230	240	0.051
Under intensive shower	−70 + 130	−100 + 180	170	0.005

is satisfied (Fig. 10.4.3). The coning decrease at intensive shower cooling can be explained by simultaneous formation, over the whole surface, of the martensite layer that fixes the ring initial form. Under the moderate cooling rates, a thin cross-section of the ring is cooled at first in which martensite transformations begin that cause displacements and coning formation.

By adjusting the intensity of cooling, one can control the value of residual stresses and reduce the strain to the minimum (Fig. 10.4.5) [11].

Figure 10.4.7 shows the calculation results on residual stresses in the carbonised plug quenched under the conditions of intensive cooling. As it is seen from the figure, the maximum compression stresses arise in the carbonised layer that are offset by small tensile stresses in the layer without carbonising. Thus intensification of cooling within the martensite range causes lowering the strain and creating high compression residual stresses to prevent quench crack formation that agrees well with the experimental data [12]. This conclusion is confirmed by numerous calculations fulfilled by the methodics described in Ref. [19]. In this work a criterion of failure of quenching is described with the help of which

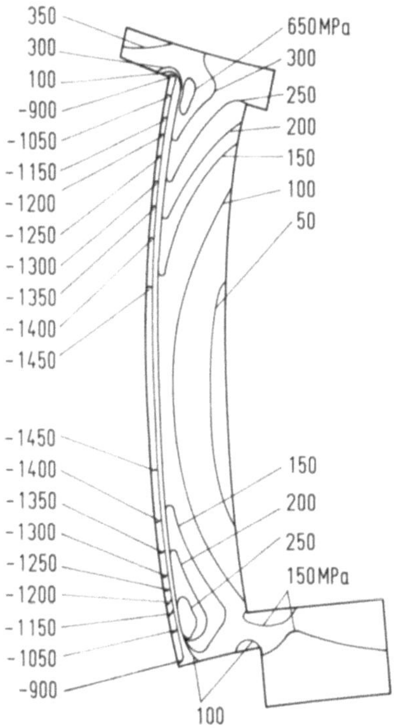

Fig. 10.4.7. Residual circumferential stresses in the case hardened plug. (Deformation is deliberately enlarged 25 times)

one can evaluate the probability of the parts failure under the intensive cooling. Generalized Pisarenko–Lebedev criterion was chosen as a criterion of failure [19].

In the course of calculation at each time step, superposition of the cooling path on the thermokinetic diagram was carried out in order to determine the material structural composition, and the field of the safety coefficient $k(x_i, t)$ was constructed.

Safety factor $k = f(x_i, t)$ is the function of coordinates x_i and time t and is determined by the ratio between the values of the ultimately admissible stress and the equivalent stress. The equivalent stress is a certain generalized value which is calculated on the basis of one of the theories of strength.

Destruction occurs when the equivalent stress reaches an admissible value, i.e. at $k \leqq 1$.

Similar results were obtained at modelling the process of two-stage quenching of the carburized bush.

With the help of the calculation methodics described, it can be shown that under the conditions of intensive cooling that ceases at the moment of reaching the maximum compression stresses at the surface, quench cracks are absent even at quenching the bearing rings made of X15 steel (0.95–1.05 C; 0.2–0.5 Mn; 0.17–0.37 Si; 1.3–1.65 Cr; < 0.3 Ni) (Fig. 10.4.7). It is known that at water quenching the bearing rings, cleavages and cracks are observed. Therefore the calculation methodics was tested at quenching the bearing rings. Figure 10.4.8 shows the fields of isothermal lines of circumferential stresses (a), and the safety factor (b) at the moment of the bearing ring failure. The most dangerous zone is indicated by the isothermal line in Fig. 10.4.8 where the safety factor equals unity. The character of changes in the principal stress σ_1, stress intensity σ_i, equivalent

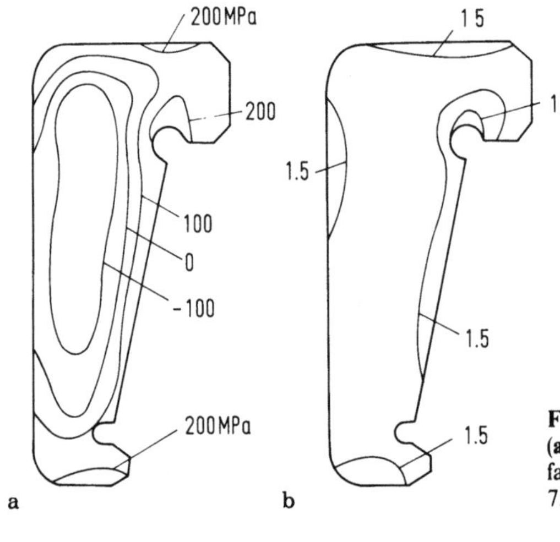

Fig. 10.4.8 a, b. Fields of isometric lines of (**a**) circumferential stresses and (**b**) safety factor at the moment of failure onset of 7515/02 bearing ring [17]

stress σ_{eq} and tensile strength σ_{ten} with time in the center of the most dangerous zone is shown in Fig. 10.4.9. In 4 seconds σ_{eq} reaches the value of σ_{ten} and, just at this moment the material failure is taking place.

The calculated data on the time of failure coincided well with the experimental study results that were obtained using the noise level transducer. The time from the cooling beginning till failure recorded by the noise level transducer constituted 4.3 s. The location of the crack generaton coincided as well (Fig. 10.4.10).

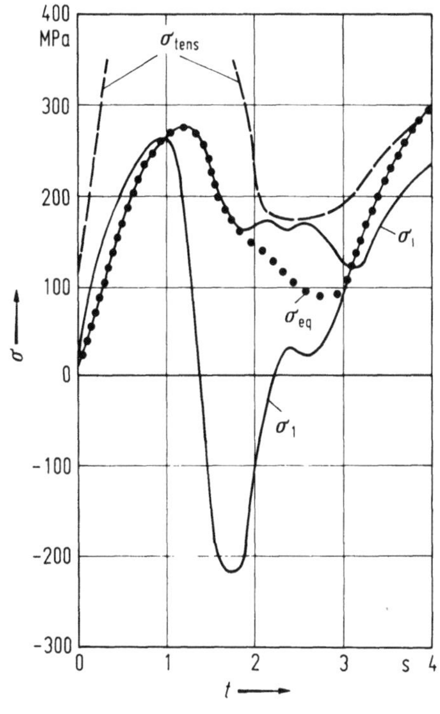

Fig. 10.4.9. Change of principal stress σ_1, stress intensity σ_i, equivalent stress σ_{eq}, tensile strength σ_{tens} with time [17]

Fig. 10.4.10. Character of 7515/02 bearing ring failure at water quenching [17]

It should be mentioned that tensile stresses are acting in the surface layers during the ring water quenching, while in the inner layers compression stresses prevail. At quenching under the conditions of intensive heat transfer the situation is converse—compression stresses are observed in the surface layers and tensile stresses—in inner layers. The latter are less dangerous under new technological conditions due to the viscous core. Thus, application of intensive methods of quenching for alloy steel parts is substantiated by numerous experimental and theoretical results.

10.4.2 Reasoning for A Promotion of the Reliability of Parts of Machines and Tools Which were Hardened with Intensive Quenching Methods

Increase of the service life of machine parts and tools quenched under intensive heat transfer can be explained by two factors: creation of high compression stresses at the part surface and improvement of the material mechanical properties. The effect of the hardening treatment, that cause high compression stresses at the part surface, on the part service life as compared to the known methods is discussed in Ref. [6].

It is evident from this work that at various treatments under compression stresses it is possible to increase the safe life of the part almost by an order of magnitude. The cause of increasing the life of various parts are high surface compression stresses that broaden the limits of endurance σ_W [6, 20].

It is stated in Ref. [20] that following the surface plastic strain and the chemical-thermal treatment, the compression stresses at the part surface reach 700–1200 MPa. The same relates to the thermal treatment procedure. As it is possible to create high enough compression stresses at the part surface by intensifying the cooling process, hence it is possible to increase substantially their safe life.

The research carried out at the Institute of Engineering Thermophysics in cooperation with the automobile plant "AvtoKRAZ" has shown that the service life of semi-axes made of 40 steel (0.37–0.45 C; 0.17–0.37 Si; 0.5–0.8 Mn; <0.25 Cr; <0.25 Ni) (at quenching under the condition $Bi \rightarrow \infty$) during the intermittent tests is an order of magnitude higher than that of the same semi-axes made of 40XH2MA steel (0.37–0.44 C; 0.5–0.8 Mn; 0.17–0.37 Si; 0.6–0.9 Cr; 1.25–1.65 Ni; 0.15–0.25 Mo) quenched under the bulk heating in oil [21].

Apart from the creating high compression stresses at the surface of the parts being quenched, the cooling intensification within the martensite range increases the material strength properties and improves its plastic characteristics. The matter is that for each steel grade there exists a certain cooling rate within the martensite range beyond which such hardening can be observed. This idea was formulated for the first time in Ref. [22] and confirmed experimentally by further investigations [21].

The experimental rig at which the tests were carried out to verify the intensive cooling effect within the martensite range is shown schematically in Fig. 10.4.11. The specimen quenching was realized first in high-concentrated electrolytes ($CaCl_2$).

Then after the nucleate boiling has ended the specimens were placed into a vibrating tank filled with liquid nitrogen where the cooling process within the martensite range was forced.

At resonance vibrations of the tank with the liquid containing bubbles, an unusual phenomenon of anomalous heat and mass transfer processes is observed [23, 24]. The bubbles are desconding down to the bottom rather than floating at the surface forming a pulsating cluster of bubbles (Fig. 10.4.11b) that gives rise to heat transfer processes intensification. This phenomenon can be attributed to the fact that the stable pulsating cluster of bubbles transforms the energy of the external stationary source into the individual bubble energy discrete in space. At the same time, at the bubble surface there arise high velocities and accelerations as well as local processes pressures which leads to wave generation and the developed turbulization [24]. All these factors promote maximum intensification of heat and mass transfer processes.

The effect discovered by us was used to intensify the process of cooling at the second stage.

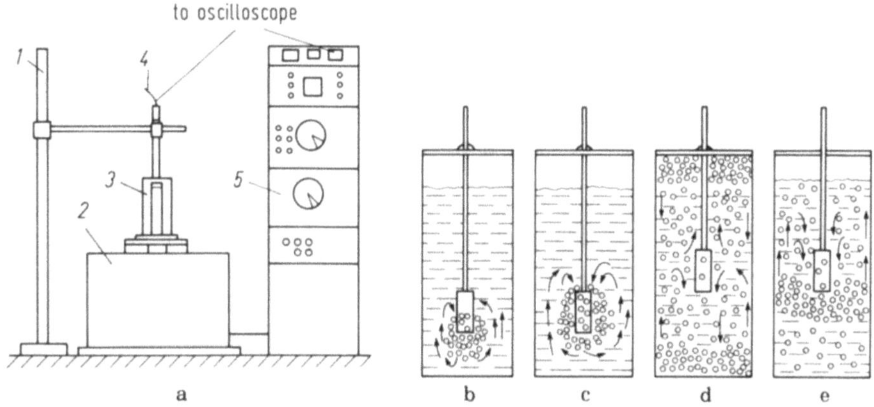

Fig. 10.4.11 a–e. Experimental set-up to test the effect of steel specimen intensive cooling within the martensite range **a** general view; **b**, **c**, **d**, **e** schemes of quench medium motion at partial (**b**) and complete (**c**) contact of bubble swarms with heated sample as well as under conditions of vibrational mixing in water (**d**) and in liquid nitrogen (**e**). *1* bracket; *2* electrodynamic vibrator; *3* working reservoir with liquid nitrogen; *4* thermocouple; *5* control panel

The durability of the non-stationary nucleate boiling is determined from the equation of non-stationary heat conduction:

$$c\rho\frac{\partial T}{\partial \tau} = \mathrm{div}\,[\lambda \text{ grad } T] \tag{10.4.2}$$

under the boundary conditions that allow for the process of nucleate boiling, i.e.

$$\left[\frac{\partial T}{\partial r} + \frac{\beta^{\mathrm{m}}}{\lambda}(T - T_{\mathrm{s}})^{\mathrm{m}}\right]_{r=R} = 0 \tag{10.4.3}$$

and under the initial conditions

$$T(\tau, 0) = T_0. \tag{10.4.4}$$

The moment of the nucleate boiling completion was determined from the equality of specific heat fluxes at single-phase convection and nucleate boiling, i.e. $q_{\mathrm{f}} = q_{\mathrm{con}}$.

By solving Eqs. (10.4.2)–(10.4.4), an approximate analytical relation was obtained to determine the durability of the non-stationary nucleate boiling

$$t_{\mathrm{rf}} = \left[\Omega + f \ln\frac{\vartheta_{\mathrm{I}}}{\vartheta_{\mathrm{II}}}\right]\frac{K}{a}; \tag{10.4.5}$$

where

$$v_{\mathrm{I}} = \frac{1}{\beta}\left[\frac{2\lambda(\vartheta_0 - \vartheta_{\mathrm{I}})}{R}\right]^{0.3}; \tag{10.4.5}$$

$$v_{\mathrm{II}} = \frac{1}{\beta}\left[\alpha_{\mathrm{con}}(\vartheta_{\mathrm{II}} + \vartheta_{\mathrm{u.h.}})\right]^{0.3}.$$

$c\rho$ is specific heat capacity; ρ is density; λ is thermal conductivity; m, β, Ω, f are constants; T_{s} is the quenching medium saturation temperature; T_0 is the heated part initial temperature; q_{f} and q_{con} are specific heat fluxes at nucleate boiling and single-phase convection; t_{rf} is durability of non-stationary nucleate boiling; K is Kondratjev form factor of the body; α is the material thermal diffusivity; α_{con} is heat transfer coefficient at single-phase convection; $v_0 = T_0 - T_{\mathrm{s}}$; $v_{\mathrm{u.h.}} = T_{\mathrm{s}} - T_{\mathrm{m}}$; T_{m} quenching medium temperature; R is the characteristic body dimension.

Relation (10.4.5) agrees well with the experimental data and is used in calculating the durability of the first-stage cooling at implementation of the second method of quenching.

The specimens quenched under different conditions were subject to tempering and polishing with the subsequent testing for their mechanical properties. Y 7A (0.66–0.73 C; 0.17–0.18 Mn; 0.17–0.33 Si; <0.2 Cr; <0.2 Ni) steel specimens were tempered at 360–370 °C during 2 hours, 60C2A (0.58–0.63 C; 0.6–0.9 Mn; 1.6–2 Si; <0.3 Cr) steel specimens—at 460 °C with the same holding time. The experimental results of the tests carried out as described above are shown in Table 10.4.2. It is evident that if the process of the specimen cooling down in the concentrated solution goes on up to the room temperature, then the main part of austenite transformation into martensite takes place in the solution principally at the first stage. Therefore there is no sense in its further cooling under vibration as the effect of additional strengthening is absent in this case.

The best results were obtained with the specimen cooling down at the first stage during is that corresponds to the durability of the nucleate boiling. This speaks to the fact that the main part of austenite transformation into martensite has taken place under intensive heat transfer in the process of cooling in the liquid nitrogen at vibrational agitation. For this reason, the best mechanical properties were achieved under these conditions.

The material additional strengthening is conditioned by the fact that in the supercooled austenite, martensite plates are formed with a high specific volume. As a consequence at intensive cooling the deformation of austenite takes place between the martensite plates having been formed. The higher the cooling rate within the martensite range, the more is austensite deformed. It is known that at low-temperature thermo-mechanical treatment, the material hardening is taking place together with the improvement of its plastic properties. Obviously the same effect can be observed at cooling intensification during quenching [21].

The cooling intensification within the martensite range leads to creation of high density of dislocations that are "frozen" at the material cooling to a low

Table 10.4.2. Steel mechanical properties at step quenching (in high-concentrated solution $CaCl_2$-liquid nitrogen) depending on holding time at the first stage (specimens 5 mm in diameter)

Method of cooling	Time of the first stage of cooling, s	Steel grade	Strengh, MPa		%	
			R_m	$R_{\rho 0.2}$	A	Z
Step quenching in 50% $CaCl_2$ solution with transport into	1	y7A*	1610	1570	7.9	31
		60C2A*	1960	1740	4	22
liquid nitrogen		y7A	1510	1450	6.1	22
with vibrational	2	60C2A	1750	1630	7	51
agitation	3	60C2A	1670	1590	7	33
Quenching in 50% $CaCl_2$ solution to complete cooling	∞	y7A	1460	1370	7.6	22
		60C2A	1430	1260	8.2	26

* y7A = (0.65–0.74 C; 0.15–0.35 Si; 0.15–0.30 Mn; <0.15 Cr; <0.20 Ni; <0.20 Cu) in %
60C2A = (0.58–0.63C; 1.6–2.0 Si; 0.6–0.9 Mn; <0.3 Cr; <0.25 Ni; <0.20 Cu) in %

temperature having no time to accumulate near the grain boundary where the nuclei of future microcracks are formed [21]. In the process of low temperature holding, the dislocations are evidently fixed by carbon atoms and some other elements. It is noted in Ref. [25] that the part cooling down to the liquid nitrogen temperature and long holding under this temperature facilitates increasing the material wear resistance.

In the CIS, intensive quenching of alloy and high-alloy steels in water electrolytic solutions of optimum concentration and in intensive water flows is widely used. It results in the increase of the material strength properties by 20–30% as compared to oil quenching. At the same time, the impact toughness increases two- or three-fold. Here sharp decrease of the threshold cold capacity and improvement of the parameters that characterize the material fracture toughness are taking place.

Thus, the cooling intensification increases the material mechanical properties and creates high compression stresses at the surface which provides high safe life of machine parts and tools.

10.4.3 Practical use of New Quenching Methods and Perspective of Their Wide Application in Industry, Based on the Development of New Equipment

Metallurgists of "ЗИЛ" Industrial Corporation (Moscow) were the first who have paid special attention to intensive methods of quenching alloyed steels, tested them and introduced into production [26]. It was just at this plant where the technology of quenching semi-axes in the intensive water flow [3] was applied that allowed to replace alloyed steels by cheaper materials.

Automatic lines for quenching motor car semi-axes were developed as applied to "ГАЗ" motor lorry [27].

The block diagram of automatic device for quenching motor lorry semi-axes in the water flow under excessive pressure is shown in Fig. 10.4.12.

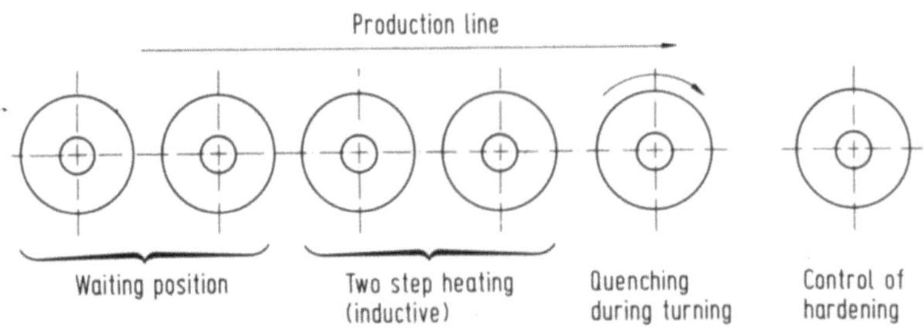

Fig. 10.4.12. Automatic device for quenching of long semi-axes in water flow [27]—schematic

A two-step heating is applied. At the first step, semi-axes are heated up to 600–700 °C, at the second step the temperature of quenching is reached. At the third step, semi-axes are quenched in the fast water flow, and at the fourth one the quality control of semi-axes being quenched is carried out. Following this step and self-tempering, the semi-axes are finally couled in the bath [18].

From the above description it is seen that the set-up allows to use new quenching methods. Indeed at the third step, the process of intensive cooling in the quench chamber can be readily stopped at the moment of achieving the maximum compression stresses at the semi-axes surface after which it is possible to carry out the semi-axis holding at M_s temperature. The second method is easily realized as well due to the presence of the bath 18 for semi-axis final cooling. In this case after the nucleate boiling in the quench chamber is ended, the process of intensive cooling is completed and it is possible to carry out holding in air and intensive cooling in the bath [18].

The technology was further developed in [28]. The authors have connected the quench chamber intended for semi-axis intensive cooling to the transducers that analyzed the processes of nucleate and film boiling and transformation of austenite into martensite that in their turn were connected to actuating mechanisms of water supply to the chamber and semi-axes charge and discharge (Fig. 10.4.13).

If it is needed to realize the first method of quenching at which the optimum depth of quenching is achieved, then transducer 6 is the principal one that records the magnetic phase increase. When the maximum compression stresses are achieved at the surface of the part to be quenched, then the optimum depth of the quenched layer is reached that corresponds to a certain value of the magnetic phase recorded by transducer 6. In this case transducer 5 controls the vapor film presence. When there is no vapor film then the water flow velocity should be increased up to a value at which the developed film boiling is reduced to minimum and disappears at all.

An original procedure of determining the vapor film presence suitable for these purposes has been developed in Ref. [29].

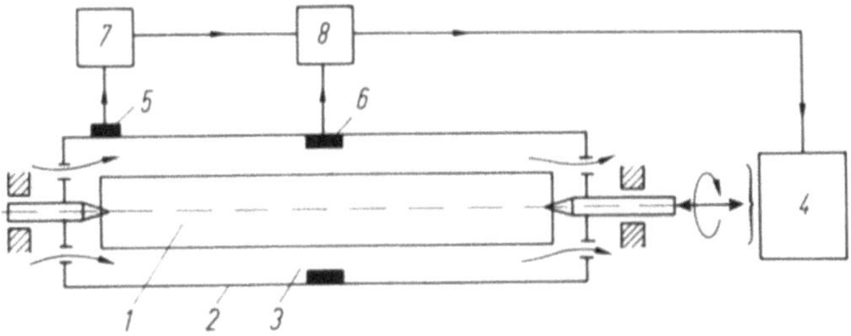

Fig. 10.4.13. *1* semi-axes; *2* quenching chamber; *3* pressurised water flow; *4* mechanical drive for the semi-axis; *5* sensor for analysing the process of nucleate and film boiling; *6* sensor for analysing the portion of transformed structures by changing the ferromagnetic state; *7* electronic device (amplifier and microprocessor); *8* amplifier

If the second method is used then transducer 5 is the principal one that records non-stationary nucleate boiling. When this boiling is completed the process of intensive cooling is ended by means of transducer 6. In this case the chamber pressure is to be maintained such that not more than 35% of martensite forms in the supercooled austenite during the nucleate boiling.

It should be noted that actuating mechanism 8 includes a microprocessor that processes the transducer signals and produces instructions for drive 4.

In this case the application of the quenching methods proposed is shown by an example of using the directed water flows.

However it is rather difficult to provide the quench medium circulation in the closed tank at tool quenching with cold application. That is why a simple device was proposed [21] that allowed to intensify the tool quenching process in the media cooled to the temperature below the room one (Fig. 10.4.14). Using a vibrating device (a nozzle with a variable through-section) creates flooded or free streams cooled down to low temperatures of the quench medium.

This set-up allows to use successfully the quench methods proposed.

According to the first method, the tool is heated up to the austenitization temperature and placed into a quench medium cool to low temperatures. The tool is held there to the moment of achieving the maximum compression stresses at the surface after which the isothermal holding at M_s temperature begins.

While realizing the second method, a refrigeration unit with the quench medium is used for the cooling intensification at the final stage.

The examples cited above indicate that the quench methods proposed can be realized by using various media.

It should be noted that the authors of [30] have proved the possibility to control the surface temperature of the parts being quenched by the preset law with the help of the electronic computer.

Determination of the optimum depth of the quenched layer is of a certain practical interest because it provides the maximum stresses at the surface of the parts being quenched. The methodics of computational determination of

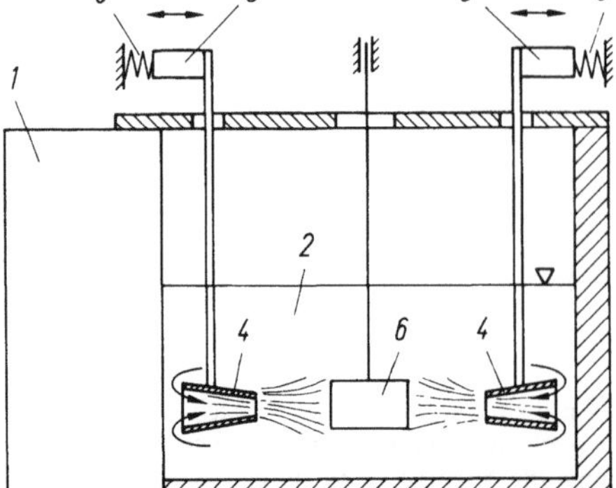

Fig. 10.4.14. Device for quenching tools in cryogenic media: *1* cooler; *2* quenching medium; *3* actuator of mechanical vibration; *4* nozzle with variable cross-section; *5* damper; *6* part (specimen)

the cylindrical steel hardenability developed in [31] is of special importance. The experimental equipment and the electronic computer that were used by the author of [32] can also be of use for these purposes.

Using the computer controlled water-air flow [26] or water quenching under pressure (Fig. 10.4.15) the part surface temperature is maintained at the first stage of cooling at such a level that not more than 25–30% of martensite may be formed in the supercooled austenite. As soon as the process of water-air cooling or nucleate boiling (Fig. 10.4.15) ends, the parts are maintained at the temperature of 250 °C during 10 minutes after which they are placed into the tank within cryogenic medium (Fig. 10.4.14) where they are cooled under the intensive heat exchange conditions down to the bath temperature; afterwards the parts are subject to retempering.

In conclusion it should be noted that intensive methods of quenching are rather promising as they substantially increase the service life and reliability of parts of machine and tools. This is confirmed by a rather interesting and important work [33] which demonstrates the advantages of intense quenching that provides increase in the part service life under cyclic loads by almost one order of magnitude. The author of this work points out the possibility of replacing alloyed steels by usual carbon steel which is economically efficient.

However for these methods to be widely used in industry it is necessary to create modern equipment with an appropriate hardware and to carry out further investigation into the physical phenomena taking place at steel quenching under the intensive cooling.

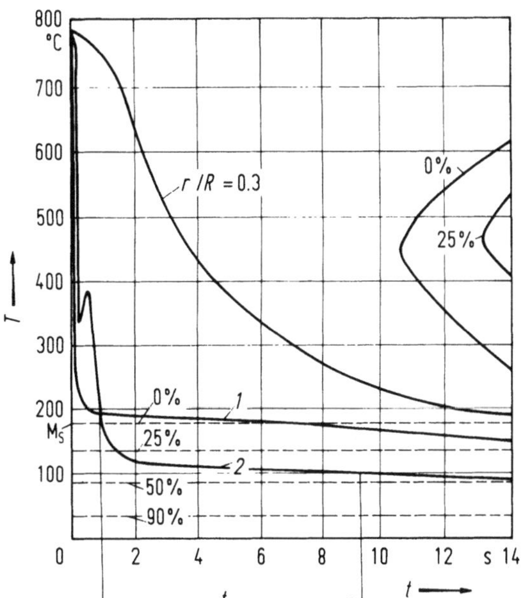

Fig. 10.4.15. Pressure adjustment of the cylindrical part (ϕ20 mm) surface temperature at water cooling at 20 °C. *1* 0.1 MPa; *2* 0.7 MPa

References

1. Kobasko NI, Prokhorenko NI (1964) Cooling rate effect of quenching on crack formation in 45 steel—Metallovedenie i termicheskaya obrabotka metallov (MiTOM). No. 2, pp. 53–54
2. Kobasko NI (1970) Crack formation at steel—quenching MiTOM. No. 11, pp. 5–6
3. Bogatyrev JuM Shepelyakovsky KZ Shklyarov IN (1967) Cooling rate effect on crack formation at steel quenching—MiTOM. No. 4, pp. 15–22
4. Kobasko NI, Morganyuk VS (1981) Study of thermal and stressed states of machine-building parts in the process of heat treatment. Kiev: Znanije Ukr. SSR, pp. 16
5. Kobasko NI, Morganyuk VS (1983) Study of thermal and stressed-strained state at heat treatment of power plant industry parts. -Kiev: Znanije Ukr. SSR, pp. 16
6. Ganiev RF, Kobasko NI, Frolov KV (1987) On principally new ways of increasing metal part service life—Doklady Academii Nauk (DAN) USSR. Vol. 194, No. 6, pp. 1364–1473
7. Kobasko NI (1986) Increase of service life of machine parts and tools by means of cooling intensification at quenching—MiTOM. No. 10, pp. 47–52
8. Kobasko NI, Nikolin BI, Drachinskaya AG (1987) Increase of service life of machine parts and tools by creating high compression stresses in them—Izvestija VUZov (Machinostrojenie). No. 10, pp. 157–157
9. Kobasko NI (1989) Increase of steel part service life and reliability by using new methods of quenching—Metallovedenie i termicheskaya obrabotka metallov. No. 9, pp. 7–14
10. Kobasko NI, Morganyuk VS (1985) Numerical study of phase changes, current and residual stresses at quenching parts of complex configuration—4th Int. Congr. Heat Treatment Mater. Berlin. June, pp. 466–486
11. Kobasko NI (1988) On the possibility of controlling residual stresses by changing the cooling properties of quench media—Metody povyshenija konstruktivnoi prochnosti metallicheskikh materialov. -M.: Znanije RSFSR, pp. 79–85
12. Ovaku Sigeo (1987) Intensive cooling—Kinzoku Metals & Technol. Vol. 57, No. 3, pp. 48–49
13. Lykov AV (967) Theory of Heat Conduction. -M.: Vysshaya Shkola, p. 560
14. Kobasko NI (1988) Method of part quenching made of high-alloyed Steels—Inventor's certificate 1215361 (USSR).—Bulletin of inventions. No. 12. Applied 13.04.1983. No. 3579858 (02–22).
15. Naito Takeshi. Method of steel quenching. Application 61-48514 (Japan), 16.08.1984, No. 59 -170039
16. Kobasko NI (1968) Thermal processes at quenching—MiTOM. -No. 3. pp. 2–6
17. Kobasko NI (1988) Method of steel quenching—Inventor's certificate No. 797243 (USSR).—Bulletin of inventions No. 7, Applied 13.02.1978, No. 2578660 (02–22), C 21 1/56
18. Lyman John R. High carbon steel microcrecking control during hardening—Pat. 4523965 (US). 07.03.1983. 472656
19. Morganyuk VS, Kobasko NI, Kharchenko VK (1982) On the possibility of predicting quench cracks—Problemy prochnosti. No. 9, pp. 63–68
20. Troshchenko VT (1980) Metal strength under variable loads. -Kiev: Naukova Dumka, pp. 206
21. Kobasko NI (1980) Steel quenching in liquid media under pressure. -Kiev: Naukova Dumka, pp. 206
22. Kobasko NI (1967) Effect of structural and thermal stresses on crack formation at steel quenching—In book: All-Union conference on increasing output and economical efficiency of heating furnaces. Dnepropetrovsk. December 1967: Abstracts of papers, Dnepropetrovsk, pp. 26–27
23. Ganiev RF, Kobasko NI, Kulik VV et al. (1980) Vibration phenomena in multiphase media and their application in technology. Kiev: Tekhnika, p. 144
24. Ganiev RF, Dolinsky AA, Kobasko NI, Frolov KV (1987) Phenomenon of anomalous mechanism of heat and mass transfer processes in multiphase media.—Diklady Academii Nauk SSSR, Vol. 294, No. 3, pp. 560–563.
25. Nakajama Khisakhoko, Tsunatsu Eidzi, Sakamoto Massaki, Akasi Tsunema (1985) Deep-free treatment in liquified nitrogen—Tumandzo to netsu syori: Cast. Forg. and Heat Treat. Vol. 38, No. 10, p. 25.
26. Correspondence readers' conference at Likhachev plant—MiTOM. 1965, No. 6, p. 72
27. Natanzon EI, Rwmyankd LS (1976) Simultaneous quenching of truck semi-axes—Avtomobilnaya promyshlennost. No. 10, pp. 33–35

28. Kobasko NI, Lisovoy VA, Khalatov AA (1988) Method of controlling quenching process—Inventor's certificate (USSR). —Bulletin of inventions No. 18. Applied 10.11.1986, No. 4145886 (31-02)
29. Tensi HM, Künzel Th, Stitzelbergen F (1987) Benetzungskinetik als wichtige Kenngrösse für die Härtung beim Tauchkühlen—Härt. Techn. Mitt. -Bd. 42. Nr.3. -S. 125–131
30. Archambault P, Didier G, Moreaux F, Beck G. (1984) Computer controlled spray quenching—Metal Progr. Vol. 126, No. 5, pp. 67–72
31. Liščić B, Filetin T (1987) Computer-aided determination of the process parameters for hardening and tempering structural steels—Heat Treat. Metals. Vol. 14, No. 3, pp. 62–66
32. Liščić B, Filetin T (1985) Mikrokomputer-gestütztes Sonde-Ekalon-Verfahren zur Messung der realen Abschreckintensität und zur Vorausbestimmung der Härteverteilung am Rundquerschnitt beim Härten—Int. Congr. Heat Treatment Mater. Berlin. June, S. 1266–1285
33. Kern Roy F (1986) Intense quenching—Heat Treating. No. 9, pp. 19–23

11 Prediction of Hardness Profile in Workpiece Based on Characteristic Cooling Parameters and Material Behaviour During Cooling

11.1 Prediction of Hardening Behaviour from Wetting Kinematics

HANS M. TENSI

11.1.1 Possibilities and Limits to Predict the Hardening Behaviour

Predictions of material strength are an important technological challenge but extremely complicated and complex. When defining the strength properties obtained through cooling processes, researchers today tend to confine themselves to the hardness value. Determining the hardness of a component is simple and thus economical, and hardness measurement in particular allows the high resolution description of the spatial strength distribution on the surface as well as on the cross-section of quenched components.

As was shown in Chap. 1, the variations in the strength values of steels are determined over a very wide range by thermal and athermal microstructural transformations of the austenite. Listed in the order of their formation with increasing rate of cooling from the high-temperature phase of the steel, the different microstructures are:

Ferrite/pearlite; pearlite; bainite; martensite with residual austenite.

These microstructures have quite different strength values (with Ck 45, for example, the hardness values range from approx. Vickers hardness 200 to 800). With certain cooling rates, the structures can occur in combinations. Different microstructural combinations in the quenched volume can, however, have the same (!) hardness values, as described by relation (1), which, strictly speaking, is non-reversible:

Microstructure and Microstructural Combinations	⇨ ⇽?-	Hardness/ Strength	(11.1.1)

A clearly reversible physical correspondence only exists between the course of cooling and microstructural formation, as described by relation (2):

$T(t)$ Curve during Quenching	⇨ ⇦	Microstructure and Microstructural Combinations	(11.1.2)

As a result, unequivocal prediction of the attained strength from the course of cooling is only possible using the following chain:

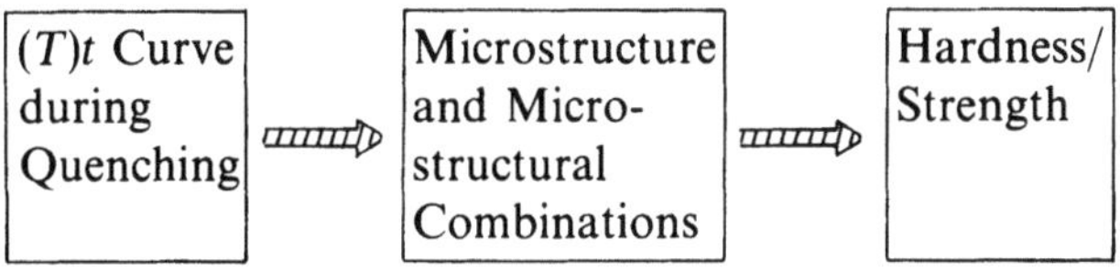

(11.1.3)

The prediction of strength properties from the course of quenching therefore would always have to be preceded by intermediate quantification of the transformation structure. The relation

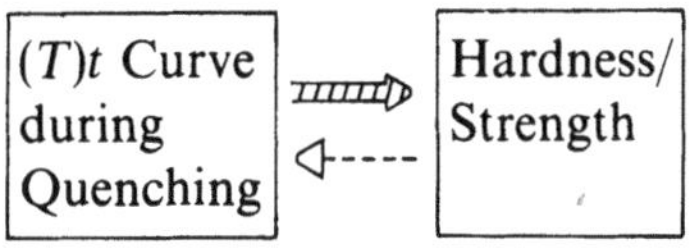

(11.1.4)

is highly dependent upon the steel composition with respect to its range of reversibility. In functional theory, this relation is termed "non-injective".

In spite of these fundamental difficulties, fairly accurate strength predictions can be made using the simplified path of assigning a hardness to the cooling curves[1-5]).

Such basic material-related information can be extracted from CCT diagramms (i.e. continuous cooling transformation diagrams), from the hardness profiles in JOMINY samples and from corresponding curves for immersion-cooled (usually cylindrical) parts.

It is, of course, crucial for the applicability of such fundamental experiments that all metallurgical and thermal boundary conditions are kept as similar as possible when predicting a component's hardening behaviour.

As a *metallographic prerequisite* for this, matching of

—material composition,
—initial microstructure prior to austenitizing, and
—austenitizing temperature and time

for component volume and test samples is required.

It is also desirable that *thermal cooling parameters* such as

—surface condition of the component,
—chemical and physical properties of the quenchant,
—heat conductivity coefficients (values dependent upon temperature and structure), and
—cooling conditions, i.e. quenchant's temperature and agitation rate

agree to the greatest possible degree.

Differences between component and sample regarding the ratio volume/surface area are taken into account by computing the time-dependent temperature distribution in the component.

In case of "non-Newtonian cooling"[1] it is essential to consider the wetting process to allow reliable calculation of the temperature distribution in the component during cooling [2, 3].

11.1.2 Influence of Wetting on the Temperature Distribution During Quenching

The progression of wetting in the case of "non-Newtonian wetting" of an immersion-cooled steel cylinder is shown in Fig. 5.3 (page 96). The time required for completely wetting the cylinder of 25 mm dia. with a length of 100 mm was approx. 20 seconds. Since the heat transfer coefficients before and behind the wetting front differ by up to two orders of magnitude (Fig. 5.5, page 99), it becomes clear that the wetting front's rate of progression on the surface of the immersion-cooled part strongly affects the time-dependent temperature distribution of the sample during quenching.

Temperature/time profiles for different distances z from the lower end of the cylindrical steel sample (dia 25 mm × 100 mm) were calculated to illustrate the influence of wetting front migration from the bottom to the top of the sample, exerting variations of local cooling in the sample surface (Fig. 11.1.1). The steep slopes of the five temperature curves mark the wetting front's progression over the different points on the sample (with distances z from sample

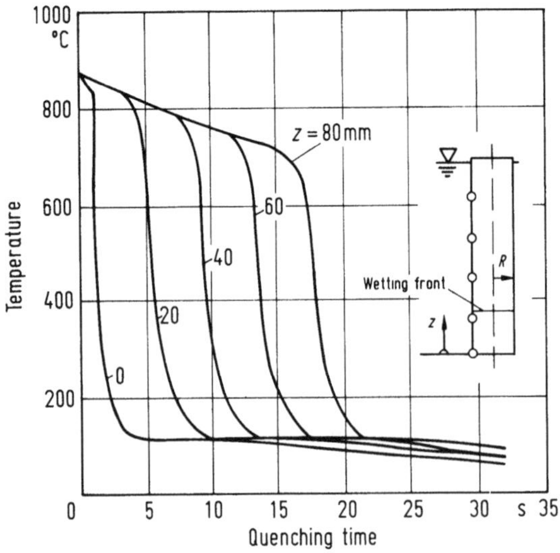

Fig. 11.1.1. Temperature drop on the surface of a cylindrical steel part (dia 25 mm × 100 mm) at various distances z from the lower end of the sample (0 mm < z < 80 mm) during slow progression of wetting (max. wetting time approx. 20 seconds). Cf. Fig. **5.3**. References to Fig. **5.4a**

[1] There are varying heat transfer coefficients for different surface ranges of a component at the same point of time

bottom = 0 to 80 mm). The time for the ascension of the wetting front from the lower end of the sample ($z = 0$) to location $z = 80$ is in this example about 16 seconds (according to Fig. 5.3). In the case of a very short wetting time, i.e., when the transition from vapour blanket to nucleate boiling occurs like an "explosion" (cf. Fig. 5.4a, p. 97), the temperature/time curves for the five locations on the sample surface would be virtually congruent.

When the temperature change is measured in the sample's centre of mass, such a contrast in the course of wetting is hardly recognizable. This is illustrated by Fig. 11.1.2, in which the temperature curves for slow wetting (curve a; migration velocity of wetting front 4.8 mm/s) and for sudden ("explosive") wetting (curve b; migration velocity of wetting edge unmeasurably fast) are compared [4].

When calculating the time-dependent temperature distributions in the two cylindrical samples it becomes apparent that they differ considerably for a slow and sudden wetting (Fig. 11.1.3) even if the sample centre cools in almost the same way (cf. with Fig. 11.1.2). The temperature profiles were calculated with the differential method using the heat transfer coefficients for vapour blanket, nucleate boiling and convective heat transfer[1] and the measured wetting front velocities.

Figure 11.1.3 shows the isotherms present after 5, 10 and 15 s cooling time. During slow wetting, high axial temperature gradients are produced (detail a), while axial temperature gradients are almost equal to zero in case of sudden wetting (detail b). In contrast, there is only a radial temperature gradient at all levels of the sample (with the exception of the lower foot) in detail b. The extremely differing temperature distributions in the samples at every point of

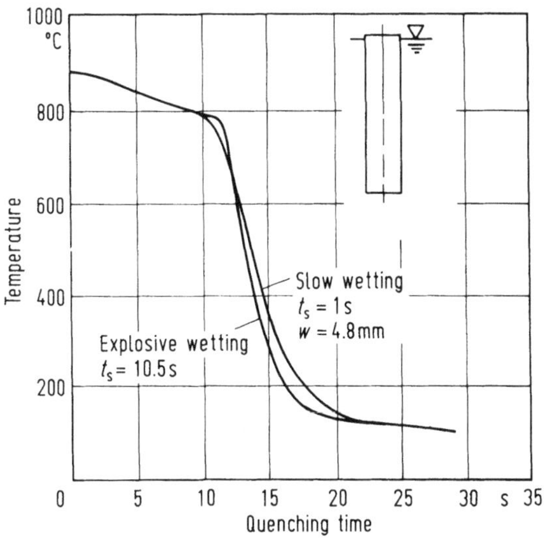

Fig. 11.1.2. Comparison of cooling rates in the centre of cylindrical steel samples (dia 25 mm × 100 mm) during slow wetting (as in Fig. **11.1.1**) and during sudden ("explosive"). wetting

[1] $\alpha_{FS} \approx 250\ W/(m^2\,K)$; $\alpha_{BS} \approx 15\,000\ W/(m^2\,K)$; $\alpha_{CONV} \approx 700\ W/(m^2\,K)$

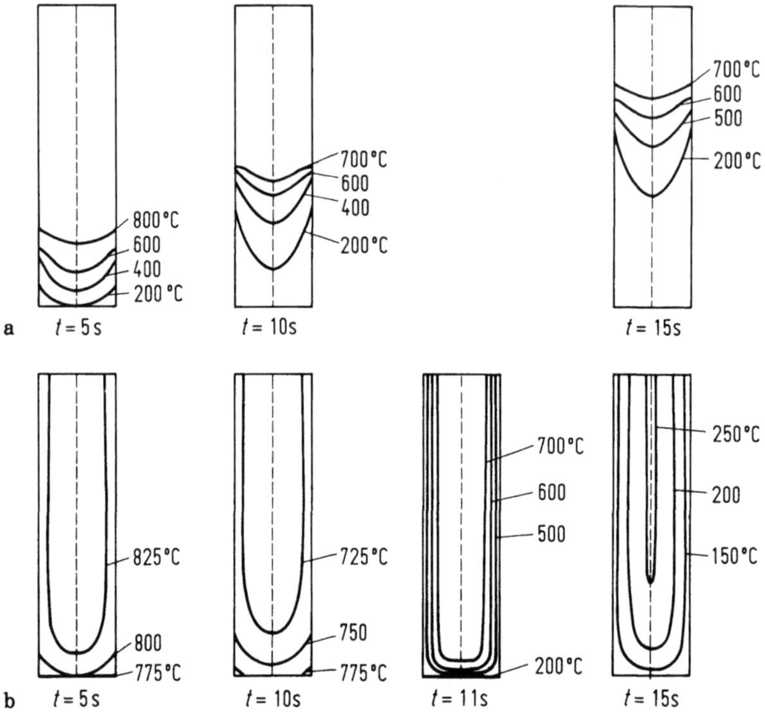

Fig. 11.1.3 a, b. Time-dependent temperature distribution during cooling of cylindrical steel samples (dia 25 mm × 100 mm) in case of slow and sudden wetting. **a** Slow wetting: Speed of the wetting edge (migrating from bottom to top) approx. 4.8 mm/s; **b** Sudden wetting

cooling (see, for example, details a and b at 5, 10 or 15 s) indicate that the hardness distributions in the two sample bodies must be extremely different as well: While the slowly wetted sample will have a distinct axial hardness gradient, the axial hardness gradient in the suddenly wetted sample will be practically zero (with the exception of a small lower range ($0 < z < 10$ mm)).

11.1.3 Prediction of Hardening Behaviour Using the Wetting Kinematics

When steel parts of identical material, charge and initial microstructure but different shapes with similar module are quenched after identical austenitizing, the locations exhibiting the same cooling curves will have identical hardness and the same microstructure or microstructural combinations. Since the heat transfer coefficients during the phases: vapour blanket, nucleate boiling, and convection vary only slightly in identical or comparable liquid quenchants, the wetting time describes the thermal cycle on the surface with adequate precision. Therefore the same hardness is to be expected for locations with the same wetting time.

11.1.3.1 Calculation of the Surface Hardness

A wetting time of more than 25 seconds was observed during immersion cooling of a cylindrical sample (dia 25 mm × 100 mm; material 100 Cr 6) in distilled water at a bath temperature of 50 °C. In this time, the wetting front migrates at an almost constant speed from the bottom to the top of the sample. The wetting time t_B increases towards the top of the sample in accordance with the progression of wetting. Figure 11.1.4 shows the wetting time related to the distance z covered from the sample bottom. In addition, the hardness profile was measured along several mantle lines and has also been mapped in relation to distance z. Relation (5) can be derived from these measurements:

$$\boxed{\text{Wetting Time } t_B} \rightleftarrows \boxed{\text{Hardness}} \qquad (11.1.5)$$

This relation is used as a "calibration curve" for hardness prediction (Fig. 11.1.5 [2, 4]). It allows the reliable prediction of surface hardness values attained under varying courses of wetting resulting, for example, from changes in bath convection or bath temperature.

In the following diagrams, the surface hardness distributions calculated from the calibration curve (Fig. 11.1.5) and the measured wetting time are compared to the measured hardness values (Figs. 11.1.6 and 11.1.7). Both figures illustrate the reliability of such calibration curves when the quenchant is modified physically (in this case with respect to the temperature) or through chemical additives (15% polymer in aqueous solution).

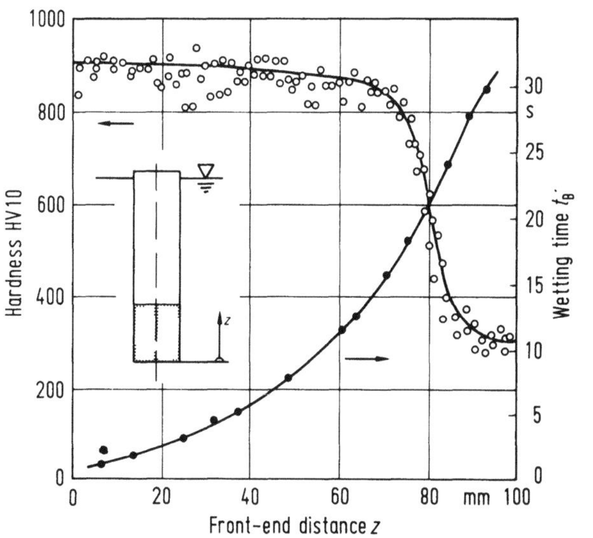

Fig. 11.1.4. Wetting time t_B and surface hardness HV (V.P.N.) as a function of the distance from the front face of the cylindrical sample (100 Cr 6; dia 25 mm × 100 mm) after immersion cooling in distilled water at 50 °C; no forced convection. Austenitizing of samples at 860 °C, 15 minutes

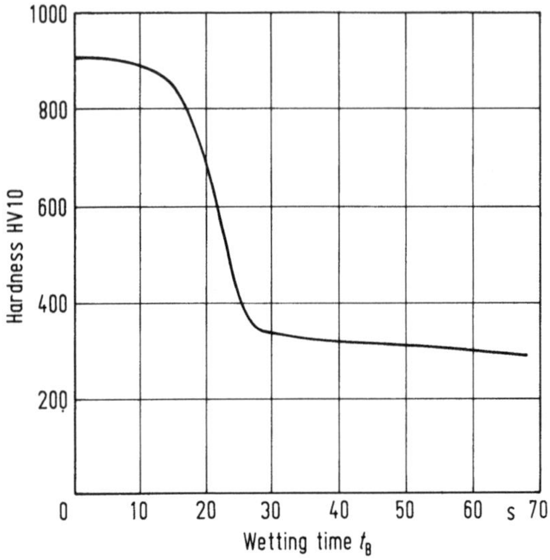

Fig. 11.1.5. Calibration curve for a 100 Cr 6 steel; Dependency of surface hardness on wetting time t_B for an initial state of the material metallographically defined by alloy composition/batch and austenitizing (derived from Fig. **11.1.4**)

In Fig. 11.1.6, still the same quenchant ("distilled water") was used, but at a temperature of 80 °C. Due to the high temperature of the quenching bath, the maximum wetting time measured at the top of the sample was more than 60 seconds. The hardness profile (continuous line) along the mantle line was calculated from measured wetting times and the calibration curve. The measured

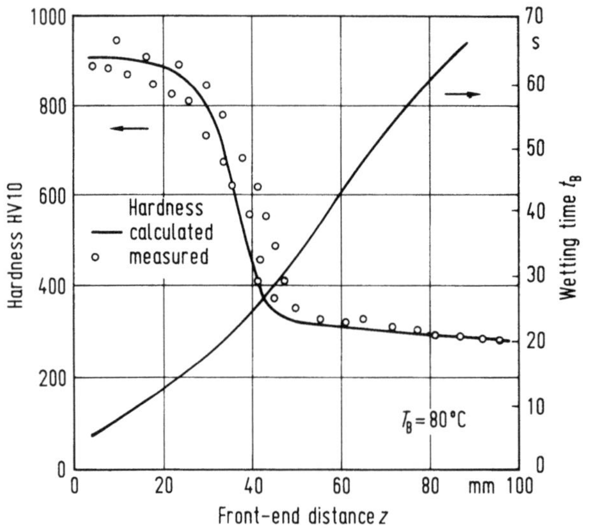

Fig. 11.1.6. Surface hardness profile calculated from the measured wetting time t_B and the specific calibration curve for the material, related to the distance from the lower end of the sample and compared to the measured hardness profile. Sample: 100 Cr 6, dia 25 mm × 100 mm, austenitizing according to the calibration curve. Bath: Distilled water, $T_B = 80\,°C$

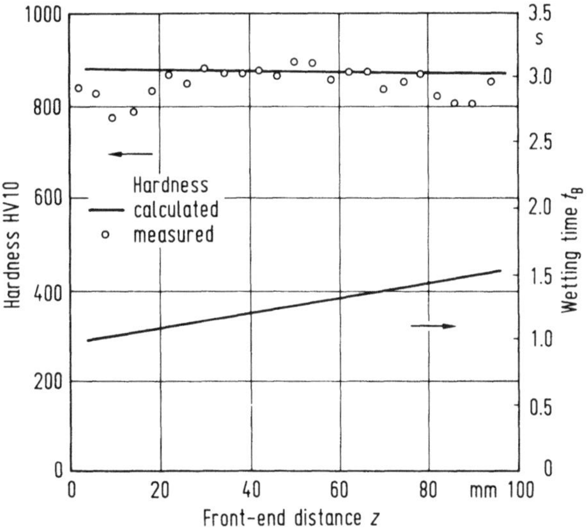

Fig. 11.1.7. Surface hardness profile calculated from the measured wetting time t_B and the specific calibration curve for the material, related to the distance from the lower end of the sample and compared to the measured hardness profile. Sample: 100 Cr 6, dia 25 mm × 100 mm, $T_A = 860\,°C$, $t_B = 15$ min. Bath: Polymer solution, $T_B = 35\,°C$, no forced convection; the calculation of the surface hardness was carried out using a calibration curve specifically produced for this material

values (dots) indicate congruent curves for calculated hardness and measured hardness.

In Fig. 11.1.7, the chemical composition of the quenching bath was changed by adding 15% polymer to the water. The start of wetting from the sample bottom is delayed by approx. 1 second, but the wetting front then proceeds to the top in only 0.5 seconds (approx.), as is shown by the wetting time measured over distance z. Thus the maximum wetting time is 1.5 seconds. The hardness distribution along the mantle line was calculated from the measured wetting times by means of the calibration curves (Fig. 11.1.5) and is also shown as a continuous line in the diagram; again the measured hardness values (dots) largely correspond to the calculated hardness curve.

The impact of the material condition prior to immersion cooling on the shape of the calibration curve is illustrated in Fig. 11.1.8, which shows the differing calibration curves obtained for 100 Cr 6 steel after varying austenitizing times. The samples were austenitized 10 and 15 minutes respectively at the same austenitizing temperature and from the same initial state. Austenitizing time $t_A = 10$ min resulted in calibration curve a. Austenitizing time $t_A = 15$ min caused the hardness drop to shift towards a shorter wetting time and a smoother slope (calibration curve b). The 5 additional minutes of austenitizing result in a larger portion of dissolved carbon in the austenite; the lower martensite starting temperature associated with this causes the hardness to drop in curve b after a shorter wetting time than in curve a.

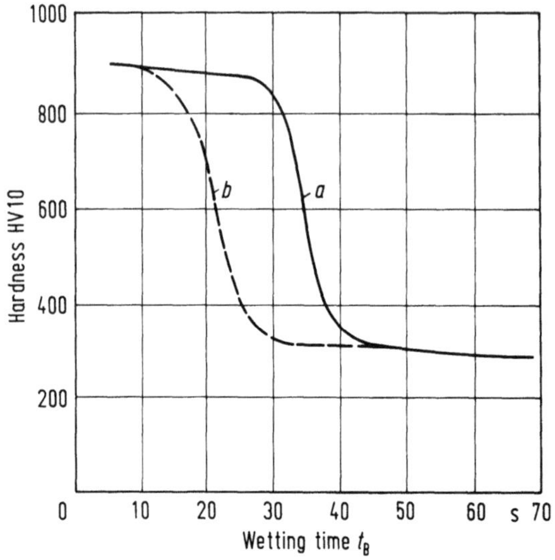

Fig. 11.1.8. Influence of austenitizing conditions (here austenitizing time t_A) on the calibration curve of a 100 Cr 6 steel (produced during immersion cooling in distilled water at 80 °C). *a* Austenitizing time 10 min at 860 °C; *b* Austenitizing time 15 min at 860 °C

The diagram shown in Fig. 11.1.9 further illuminates the prediction of surface hardness from measured time t_B and calibration curve: Wetting time t_B^* is measured at a surface point z^* of the immersion-cooled workpiece (these two values cause dot 2). The measurement of t_B is done by recording the course of wetting on the workpiece surface by optical means, or by measuring the change in conductance between the workpiece and a counterelectrode, or by means of

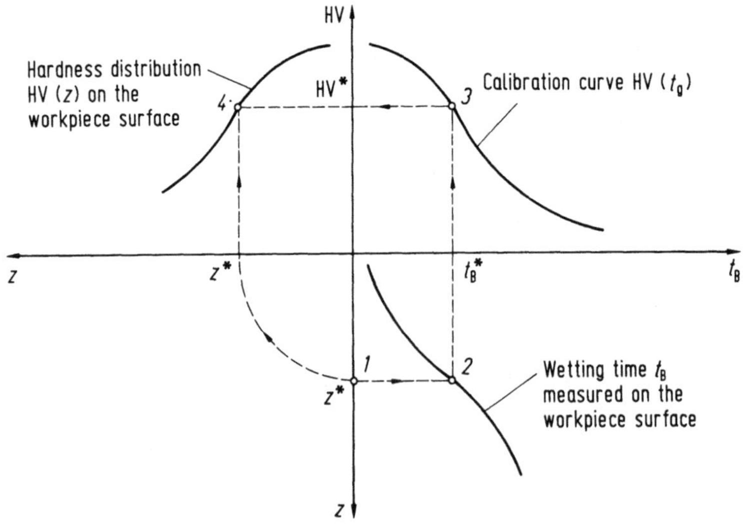

Fig. 11.1.9. Determination of the surface hardness distribution $HV(z^*)$ of a steel sample after quenching, using the calibration curve $HV(t_w)$, and the measured wetting time on the immersion cooled workpiece surface $t^*(z^*)$. Steps of evaluation: 1→2→3→4 and 1→4

thermocouples placed inside the workpiece close to its surface (cf. Chap. 8.1.2). The hardness value HV* attained on the workpiece surface is obtained through the measured wetting time via the calibration curve (dot 3 in Fig. 11.1.9). The hardness value is correlated to the actual point z^* on the workpiece surface. z^* and HV* yield coordinate 4 (cf. Fig. 11.1.9). Using a large number of auxiliary points (t_B, z), the entire hardness profile HV (z) on the workpiece surface can be ascertained.

11.1.3.2 Calculation of the Hardness Distribution in Cross-Section of Cylindrical Specimens

Since the hardness profile on the surface of components can be very reliably ascertained based on the measured wetting kinematics during immersion cooling, the use of surface hardness for predicting the distribution of strength throughout the component volume is a logical option. If the wetting kinematic is known, it is possible to calculate the cooling curve in the workpiece volume during immersion cooling even in case of "non-Newtonian heat transfer", and to predict microstructural transformation and hardness distribution on this basis. The use of the heat flow equivalent H as per Grossmann and Assimov [6] is suitable for cylindrical parts as shown in Chap. 8.2.1.

In addition to the calibration curve indicating the relation between surface hardness and wetting time, a Jominy hardness curve has to be recorded for the same type of steel (under the same austenitizing conditions). The surface hardness at location z^* of the immersion-cooled component whose radial hardness distribution is to be ascertained is assigned to a corresponding hardness on the Jominy hardness curve (Fig. 11.1.10). Note the different ways of cooling in the Jominy sample and in the immersion-cooled cylindrical sample: While there is axial and radial heat flow (i.e., $dT/dz \neq 0$ and $dT/dr \neq 0$) in the cylindrical body in case of "non-Newtonian cooling", the Jominy sample has an almost one-dimensional heat flow in the x-axis with no radial heat flow $(dT/dr \approx 0)$.

In this model (see Fig. 11.1.10), the temperature profile at the distance x^* of the Jominy sample is related to location z^* on the surface of the immersion-cooled cylindrical workpiece. This correlation is assumed to be permissible since cooling of the two locations in the Jominy sample and in the immersed part occurred in such a way that similar hardness could result. Since the cooling of the Jominy specimen is approximately in accordance with Newton's Law of Cooling[1], a heat flow equivalent H as per Grossmann can be obtained for the cylinder slice with thickness Δz of the immersion-cooled component. H corresponds to a virtual heat transfer coefficient α. This coefficient is virtual because Newtonian cooling[1] is assumed for ascertaining the value H, although it is usually not present in an immersed part.

[1] The heat transfer on the quenched end remains approximately constant throughout cooling

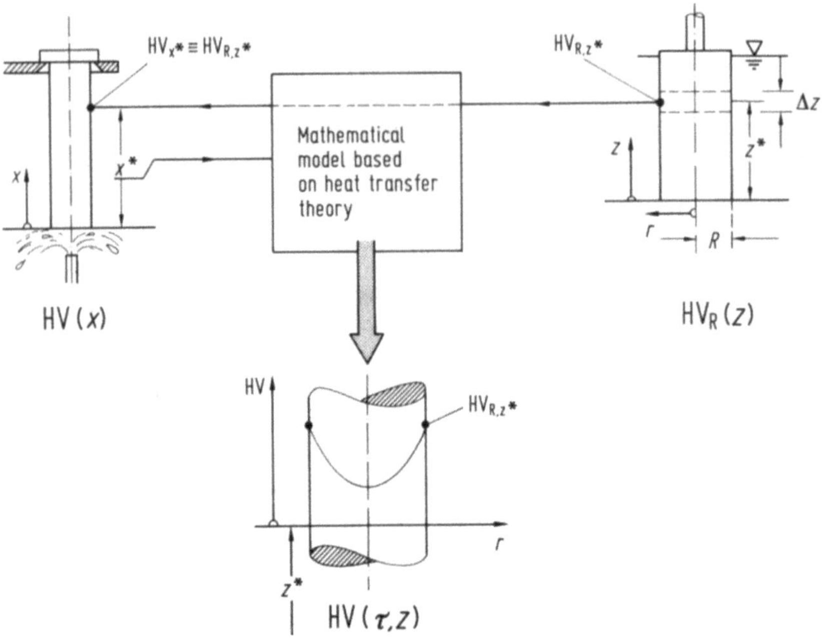

Fig. 11.1.10. Scheme for the prediction of the radial hardness profile in immersion-cooled cylindrical samples on the basis of wetting kinematics (cf. calibration curve), Jominy hardness and H value as per Grossmann and Assimov

The value H comes from the non-steady-state heat conduction equation for cylindrical objects and was used instead of the Biot number by Grossmann and Assimov [7]. The following equation applies:

$$H \times D = h \times R = \mathrm{Bi} = \alpha/\lambda \times R \qquad (11.1.6)$$

where

D: Diameter of the body in m; $R = D/2$
H: Heat flow equivalent in 1/m; $H = h/2$
α: Heat transfer coefficient in $W/(m^2 \times K)$
λ: Heat conductivity in $W/(m \times K)$

If, to simplify matters, a constant and temperature-independent thermal conductivity coefficient λ is assumed, dimension H then depends solely upon the heat transfer coefficient on the surface α und thus on the cooling performance of the quenching fluid.

The value H can then be used to calculate the times when the so-called "half temperature" $T_H(T/T_A = 0.5)$ occurs at different radius locations r/R of the cylinder slice in question. H is derived from the non-steady-state heat conduction equation. Comparison with the Jominy temperature curves for the same half temperature times yields the corresponding "quenched end distances" x^*. Based on the Jominy hardness curve, a specific hardness value is assigned to each distance x^*. If these hardness values are transferred to the immersion-cooled part, the radial hardness profile at cylinder's distance z^* is ascertained.

Since slow transition from vapour blanket to nucleate boiling causes great axial variations in cooling, the corresponding value H for each distance z must be calculated for predicting hardness throughout the cylinder volume. If thickness Δz of the fictitious cylinder slice at location z^* is small enough, it is assumed that the difference in the axial heat flows at locations $(z^* + \Delta z/2)$ and $(z^* - \Delta z/2)$ can be considered zero. The calculation steps are schematically shown in Fig. 11.1.10.

Diagrams as per Crafts and Lamont can be used for implementing the described hardness prediction method (cf. Fig. 11.1.11). The diagrams correlate heat flow equivalent H, workpiece diameter D and Jominy distance $\times$ and were produced for different ratios $r/R (0 < r/R < 1.0)$.

As an example, Table 11.1.1 lists the calculated heat flow equivalents H for different distances z from the lower end of a cylindrical 100Cr6 part. The quenchant in this case was distilled water at 50 °C.

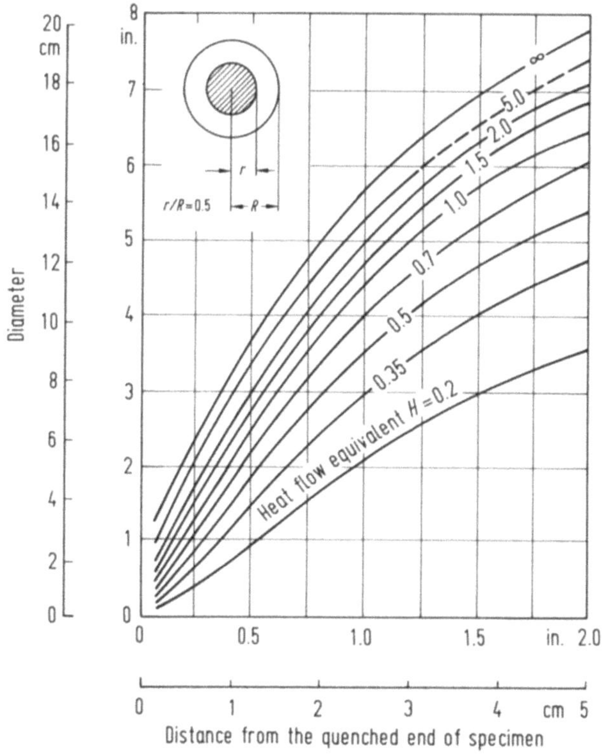

Fig. 11.1.11. Example of a Crafts-Lamont diagram for radius ratio $r/R = 0.5$; relationship between distance x from the quenched end of Jominy specimen and component diameter D for different H values

Table 11.1.1. Calculated Grossmann numbers H in different distances z from the lower end of an immersion cooled 100 Cr 6 sample; quenchant: distilled water of 30 °C

Distance of the lower end in mm	25	50	70	80	90
Heat flow equivalent in inch^{-1}	1.70	1.27	0.57	0.35	<0.2

Figure 11.1.12 depicts radial hardness profiles predicted according to the above described approach (continuous lines) and actually measured hardness distributions (dots) for three arbitrarily chosen cross sections ($z = 25, 80$, and 90 mm). The wetting edge reached location $z = 25$ mm after approx. 5 s. According to the calibration curve, this particular wetting time corresponds to a surface hardness of 900 V.P.N.. The same hardness is found at a distance of 1 mm from the quenched end of a Jominy sample (cf. Fig. 11.1.10). Using the Crafts–Lamont diagram for the surface ($r/R = 1$) the H-value for location 25 mm can be obtained. This H-value can be used to calculate the hardness in the cylinder slice at a distance $z = 25$ mm e.g. for radius ratio $r/R = 0.5$. Using this H value, more equidistant Jominy distances x_i for as many radius ratios as possible can be calculated from the Crafts–Lamont diagrams [8]. Thus the radial hardness

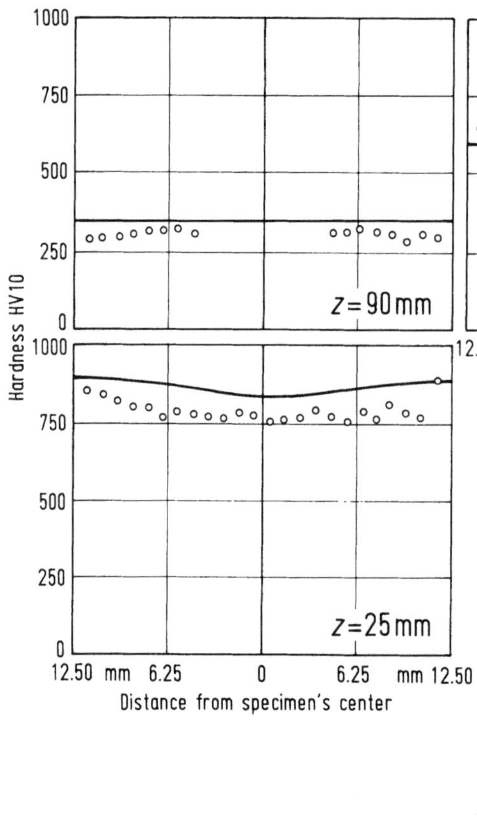

Fig. 11.1.12. Radial hardness distributions predicted (continuous line) and measured (dots) in a cylindrical part (here dia 25 mm × 100 mm) at different distances z. Material and austenitizing: 100 Cr 6, T_A = 860 °C, t_A = 15 min. Quenchant: Water at T_B = 50 °C, no forced convection

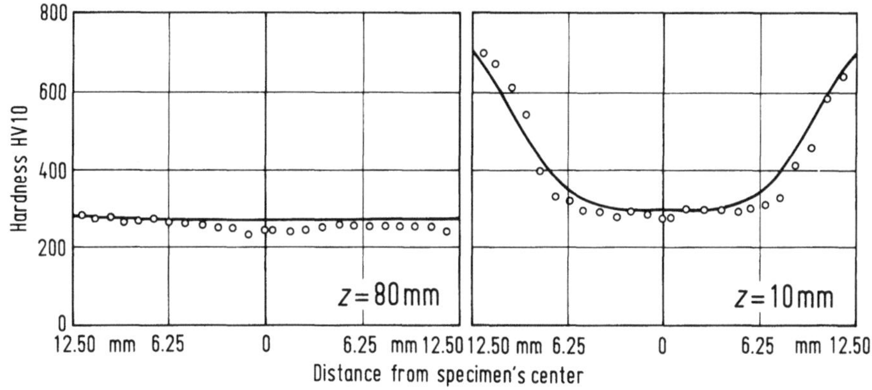

Fig. 11.1.13. Radial hardness distributions predicted (continuous line) and measured (dots) in a cylindrical part (here dia 25 mm × 100 mm) at different distances *z*. Material and austenitizing: C45, $T_A = 880\,°C$, $t_A = 3$ min, Quenchant: Water 35 °C

profile at distance $z = 25$ mm of the cylindrical part can be predicted. This method is to be repeated in an analog way for any further distances *z*. The limit for the use of these Crafts–Lamont diagrams is reached at a distance of $z = 90$ mm from the lower end of the part for the physical dimensions applicable here, since the *H* values are only considered to a minimum of $0.2\,\text{inch}^{-1}$. For the lowest *H* value of this evaluation, the calculated hardness profile is determined as a straight horizontal line.

The radial hardness profiles in steels whose transformation behaviour is easily affected by changes in the cooling rate vary greatly with increasing distance from the lower end of the part, i.e., with increasing wetting time t_B, as can be seen in Fig. 11.1.13. During immersion cooling of a C 45 part (dia 25 mm × 10 mm), the radial hardness curve after quenching is extraordinarily steep at a distance $z = 10$ mm from the bottom surface, while hardness is constant throughout the cross section at distance $z = 80$ mm. Again, the hardness prediction was made using the described method, and represented by curves. The predicted hardness and the actual hardness (dots) of this steel, which responds sensitively to cooling rate variations, are in agreement as well.

The described method allows the prediction of the surface hardness and hardness distribution over the cross section of cylindrical parts when the course of wetting is known. The immersed object is for this purpose assumed to undergo Newtonian cooling, even if the heat transfer coefficients during the different phases of boiling have greatly differing values.

11.1.3.3 Calculation of the Hardness Distribution in Specimens of Optional Geometries

The temperature/time profile for each point of not simply rotation-symmetrical or even of non-symmetrical objects can be calculated when the physical cooling conditions stated in Chap. 11.1.3 are known. The time-dependent temperature distribution in parts of virtually any shape can be calculated by means of calculation methods such as the "differential method" or the "Finite Element Method", with computing requirements being considerably increased by complicated work geometries.

Isothermal time-temperature-transformation diagrams (TTT diagrams) specific to materials are often employed when generally available materials-science information on the transformation behaviour of austenite is also to be used [9]. It must be emphasized, however, that the temperature profiles described here are basically continuous, and not "isothermal" coolings.

A frequently used method of transformation or hardness prediction is as follows [10–14]:

The transformed microstructure $W_{P/B}$ during a constant temperature is to be calculated using an exponential function according nucleation and nuclear growth, corresponding.

$$W_{P/B} = 1 - \exp(-b \cdot (t/t_0)^n \qquad (11.1.7)$$

where

$W_{P/B}$: Volume fraction of transformed Austenite in Perlite or Bainit
b: Coefficient of the kinetic of Austenite transformation
n: Exponent of the kinetic of Austenite transformation
t: time of transformation
t_0: reference time of transformation

The truth of Eq. (11.1.7) relative to the thermal activated transformation characteristics of Austenite in Perlite or Bainit was determined experimentally [11, 12]. The values of "b" and "n" are dependent from the transformation temperature. It is obvious that the values are additionally varying for Perlite- and Bainit-transformations.

Having a temperature below the martensite start temperature the athermic activated transformation of austenite in martensite is described by equations like 8a and b.:

$$W_M = 1 - \exp\{c(M_s - T)^m\} \qquad (11.1.8a)$$

or

$$W_M = 1 - \{(T - M_f)/(M_s - M_f)\}^m \qquad (11.1.8b)$$

where

c: Coefficient of Austenite transformation
m: Exponent of Austenite transformation kinetic
W_M: Volume fraction of transformed Austenite in Martensite
M_f: Martensite finish temperature
M_s: Martensite start temperature
T: Temperature of transformation ($T \leq M_s$)

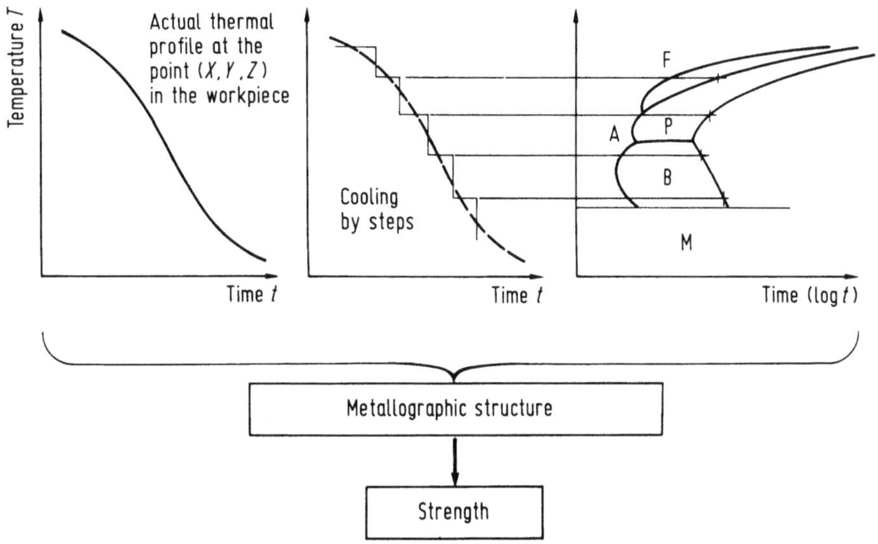

Fig. 11.1.14. Scheme for microstructure and strength prediction from cooling curve and an isothermal TTT diagram

The corresponding isothermal TTT-diagram is fed into a computer as true to its shape as possible and provided with any supplementary quantitative information on transformed structure and hardness values. The intermediate values corresponding to the individual temperatures are determined by interpolation (straight-line or according to metallurgial findings).

If the microstructural transformation or the hardness value at an arbitrary point in the immersion-cooled part are now to be predicted, the monotonous temperature drop for this location is measured or calculated using the described methods. To make use of an isothermal TTT-diagram for continuous cooling, the monotonous temperature drop is split into a step function, according to Fig. 11.1.14. It is assumed that the individual horizontal (isothermal) sections in the step function produce such a transformation as occurs at the individual temperatures in the isothermal TTT diagram. By an iteration of the transformation steps until the end of continuous cooling the integral transformation behaviour is derived (Fig. 11.1.14).

To explain the iteration, the steps of transformation at the time "i" and "i + 1" using Eq. 11.1.7 are shown in Fig. 11.1.15. At the isothermal step T_i there was a volume of Austenite $(1 - W_{i-1})$ available for transformation (Fig. 11.1.15b). By the function in Fig. 11.1.15b we obtain with W_{i-1} the fictive point of time $t_{i;fict.}$. During the time Δt_i we get an additional transformation of microstructure $(W_i - W_{i-1})$. The total transformed microstructure volume W_i produces for the next temperature step T_{i+1} the fictive start point $t_{i+1;fict.}$ for the next iteration (Fig. 11.1.15c). The strength values achieved through cooling can then be esti-

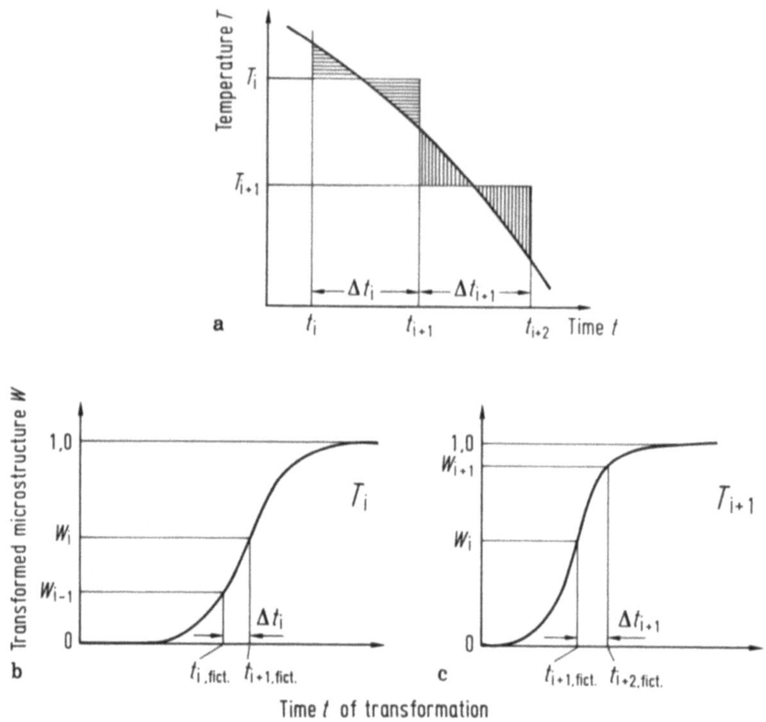

Fig. 11.1.15 a–c. Transformed microstructure W during the iteration steps "i" and "$i+1$" (having the temperature T_i and T_{i+1}) to be calculated using the Eq. 7 (schematic)

mated (with some restrictions) (cf. Fig. 11.1.14) if the non injective relation between curve $T(t)$ and strength (cf. relation 4) is taken into account.

This method offers the advantage of using existing materials science information on transformation behaviour, but it also has considerable drawbacks: The TTT diagram usually fails to give a sufficiently reliable description of a given steel's transformation behaviour. The reasons for this are deviations in alloy composition (within the standard range) and, to a considerable degree, variations in the initial microstructure prior to austenitizing. Austenitizing temperature and time can, of course, be precisely synchronized with the requirements of the TTT diagramm. These purely metallographic differences alone can produce significant divergences between the actual transformation behaviour and the one described in the TTT diagram.

Another drawback is the fact that further deviations can occur during the transfer of TTT curves into a computer; these deviations will result in a further misrepresentation of transformation behaviour. In this case point-by-point representation of state field boundary lines might be preferable to the generally used representation by adapted functions. Moreover, many isothermal TTT diagrams are incomplete and/or unsuited for computer-aided information processing (e.g., adjacent transition fields are often not separated; thus pearlite or even ferrite transition fields merge into bainite fields without boundaries). An additional important issue is the exact extent to which it is permissible to apply the microstructural transformations occurring during isothermal holding to a continuous transformation divided into steps. Finally it must be noted that the time/temperature diagrams primarily describe the transformation behaviour of the structure on the basis of a given course of cooling, and not the hardness.

Conclusions

The time dependent temperature distribution can be calculated with a high reliability, using all the cooling parameters, especially those of the wetting kinematic. For these temperature drops the metallographic structures can be precalculated with the help of isothermal TTT-diagrams at every point in any given geometry. By predicting the surface-hardness and -structure from the wetting time (using a calibration curve) the above mentioned method can be extended and checked. For simple rotation-symmetrical geometries the calculated surface hardness can be used together with the modified H-value to predict the distribution of hardness on the cross-section with a sufficient accuracy.

In spite of the numerous limitations mentioned, this and similar methods offer an adequate option for the provision of valuable basic information on the transformed structure and, with some reservations, the strength values of a component.

References

1. Steffen E (1986) "Untersuchung von Abkühlmedien auf Polymerbasis für die Stahlhärtung", (140 Seiten) Doctoral Thesis, Technical University Munich
2. Künzel Th (1986) "Einfluß der Wiederbenetzung auf die allotrope Modifikationsänderung tauchgekühlter Körper in wässrigen Kunstsofflösungen", (138 Seiten) Doctoral Thesis, Technical University Munich
3. Stitzelberger-Jakob P (1981) "Härtevorherbestimmung mit Hilfe des Benetzungsablaufes beim Tauchkühlen von Stählen", (160 Seiten) Doctoral Thesis, Technical University Munich
4. Tensi HM, Stitzelberger-Jakob P, Stich A (1990) Steuern und Berechnen der Abkühlung zylinderförmiger Körper und Vorherbestimmung der Härteverteilung, HTM 45 145–153
5. Tensi HM, Stich A (1990) Effect of Wetting Process on Cooling Behaviour and Hardness Distribution in Immersion-cooled Workpieces, VII. Int. Congress for the Heat Treatment of Metals, Vol. V: 136–150
6. Asimow M, Craig WF, Großmann MA (1941) Correlation between Jominy Test and Quenched Round Bars, SAE Trans., 283–292
7. Großmann MA, Asimow M, Urban SF (1936) Relation to Quenching and some Quantitative Data in: Hardenability of Alloyed Steels, American Society for Metals, Cleveland 124–196
8. Crafts W, Lamont JL (1954) Härtbarkeit und Auswahl von Stählen, Springer Verlag Berlin/Göttingen/Heidelberg, (261 Seiten), 66–74
9. Atlas zur Wärmebehandlung der Stähle Published by MPI für Eisenforschung GmbH in cooperation with VDEh; Vol. I: Rose A, Peter W, Straßburg W, Radenacher L, Düsseldorf 1954–58. Vol. II: Rose A, Hougardy HP, Düsseldorf, 1972
10. Tzitzelkov I (1973) "Eine mathematische Methode zur Beschreibung des Umwandlungsverhaltens eutektoider Stähle" Doctoral Thesis, RWTH-Aachen
11. Tzitzelkov I, Hougardy HP, Rose HP (1974) Mathematische Beschreibung des ZTU-Schaubildes für isotherme Umwandlung und kontinuierliche Abkühlung, Arch. f. Eisenhüttenwes., 45: 525–532

12. Hougardy HP, Yamazahi K (1986) An Improved Calculation of the Transformation of Steels, Steel Research, 57: 466–471
13. Tanaka K, Iwasaki R, Nagaki S (1984) On the TTT- and CCT-diagrams of steels: A phenomenological approach to transformation kinetics. Ingenieur-Archiv 54: 81–90
14. Umemeto M, Nishioka N, Tamura I (1981) Prediction of Hardenability from Isothermal Transformation Diagrams Heat Treating 2: 130–138

List of Symbols

Bi:	Biot number ($=\alpha \times 1/\lambda_s$)
b; *c*:	Coefficient of the kinetik of austenite transformation
n:	Exponent of the kinetic of austenite transformation
H:	Heat flow equivalent as per Grossmann and Assimov in inch^{-1} or (m^{-1})
h:	$H \times 2$ in inch^{-1} or (m^{-1})
*HV**:	Hardness *HV*10 in V.P.N. at a defined point
M_f:	Martensite finish temperature
M_s:	Martensite start temperature
R:	Outer radius of sample work piece in mm
r:	Radius in mm
T:	Temperature in °C
T_A:	Austenitizing temperature in °C
T_B:	Bath temperature in °C
T_H:	"Half temperature" $=0.5 * T_A$ in °C
t:	Time in *s*
t_A:	Austenitizing time in min
t_B:	Wetting time in *s*
$W_{P/B/M}$:	Volume fraction of Austenite transformed in Perlite/Bainite/Martensite
x:	Distance from quenched end of Jominy specimen in mm
z:	Distance from lower end of cylindrical samples or work pieces in mm
α_{CONV}:	Heat transfer coefficient during convection in Wm^{-2}K^{-1}
α_{FB}:	Heat transfer coefficient during vapour blanket in Wm^{-2}K^{-1}
α_{NB}:	Heat transfer coefficient during nucleate boiling in Wm^{-2}K^{-1}
λ_s:	Thermal conductivity in sample in Wm^{-1}K^{-1}.

11.2 Predetermination of Hardness Results

B. Liščić

11.2.1 The QTA-Method

Introduction

Contrary to the "*H*"-model developed by Grossmann who had used the Newton's law of cooling (i.e. a constant α-value), Wünning tried to take into account film-boiling and nucleate boiling effects (with different α-values) and to describe the cooling curves by three model parameters corresponding to: the vapour blanket stage, the boiling stage and the convection stage [1], [2].

As it is seen from Fig. 11.2.1.1 using the QTA-3 model, a real cooling curve showing the characteristic knee during the boiling stage, can be well described by the model parameters Q, T and A the meaning of which is the following:

—"Q" is the average heat flux density in MW/m^2 on the surface of the workpiece—until it cools down to 500 °C. It depends on: the agitation, on the transition temperature ϑ_L, the undercooling, and the viscosity of the liquid.

Corresponding values for water are between 0.3 and 3 MW/m^2, and for oils from 0.3 to 1 MW/m^2.

To determine the Q-value in each particular case, a standard steel probe with known hardenability is used.

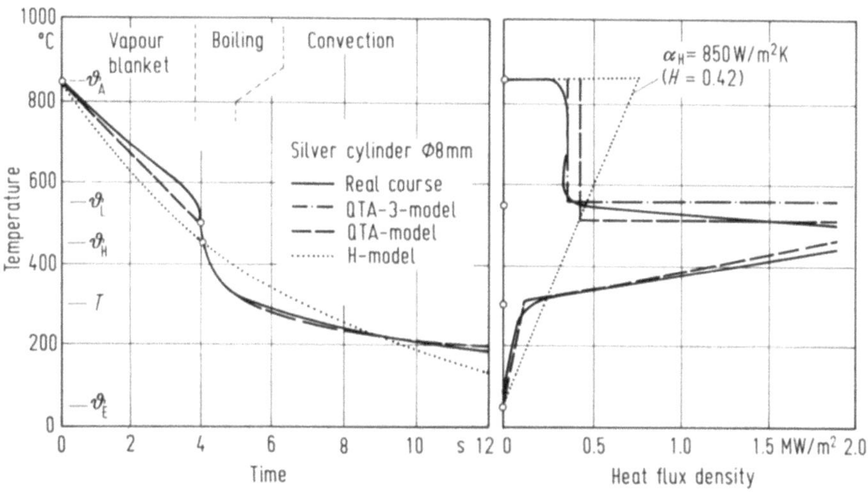

Fig. 11.2.1.1 a, b. The real course of quenching compared with different models, valid for quenching a 8 mm Dia silver Cylinder in oil. **a** Cooling curves; **b** Heat flux density v.s. temperature

—"T" is the temperature in °C at which the steep decline of the heat flux density in the boiling stage becomes zero (see extrapolation in Fig. 11.2.1.1b).

This temperature is very near (hardly above) the boiling point i.e. above the temperature at which the boiling starts. This T-value is a property of the quenchant and does not depend on cooling conditions. This means, it can be determined from the relevant laboratory test.

—"A" is the average heat transfer coefficient (α_B) in W/m^2K during the convection stage. It depends on: agitation and viscosity of the liquid as well as on the dimensions of the workpieces. It can be determined approximately by calculation using the heat transfer laws.

Heat Transfer and Temperature Course in the Vapour Blanket Stage

The heat transfer during the vapour blanket stage (provided the undercooling of the quenchant is sufficient), is assumed to be dependent only on: flow and temperature boundary layer within the fluid. The author assumes that the heat flux density (q_F) depends in this case on: undercooling, on the heat conductivity of the fluid and on the flow-conditions (agitation rate, viscosity, dimension), but neither on the surface (wall) temperature nor on the properties of the vapour blanket itself. As a proof for that the author quotes the usually linear cooling course in the vapour blanket stage.

At a certain temperature ϑ_L—corresponding to the Leidenfrost point, the vapour blanket passes over to the boiling (nucleate boiling) stage. The supposition of an average constant heat flux density (q_F) during the vapour blanket stage comes very near to the actual relations. The temperature course on the surface and in the core, depending on the cylinder diameter can be taken from Fig. 11.2.1.2. The decrease of temperature can be calculated, related to the heat flux density (using the second boundary condition: $q = \text{const}$), according to the following formula:

$$\frac{2\cdot\lambda}{q\cdot D}(\vartheta_A - \vartheta) = F_2(X, F_0) \tag{11.2.1}$$

For the practice it is interesting to know the duration of the vapour blanket. This duration is longer, proportionally to: $(\vartheta_A - \vartheta_L)\cdot\lambda/q_F$, and to the diameter, but only upto a certain max. value, that is generally valid for all bodies:

$$t_{F\,max} = \pi/4\lambda c\rho(\vartheta_A - \vartheta_L)^2/q_F^2 \tag{11.2.2}$$

Heat Transfer and Temperature Course in the Boiling Stage

In the boiling stage the heat flux density, in case of oil quenching, increases to values more than $2\,MW/m^2$ because of the effect of boiling bubbles, without any agitation from the outside. Cooling rates of more than 200 K/s on the surface are produced, so that the heat transfer in the boiling stage can be calculated using the first boundary condition: $\vartheta_R = \text{constant}$, (related to T), according to the formula:

$$\theta = F_1(X, F_0) \tag{11.2.3}$$

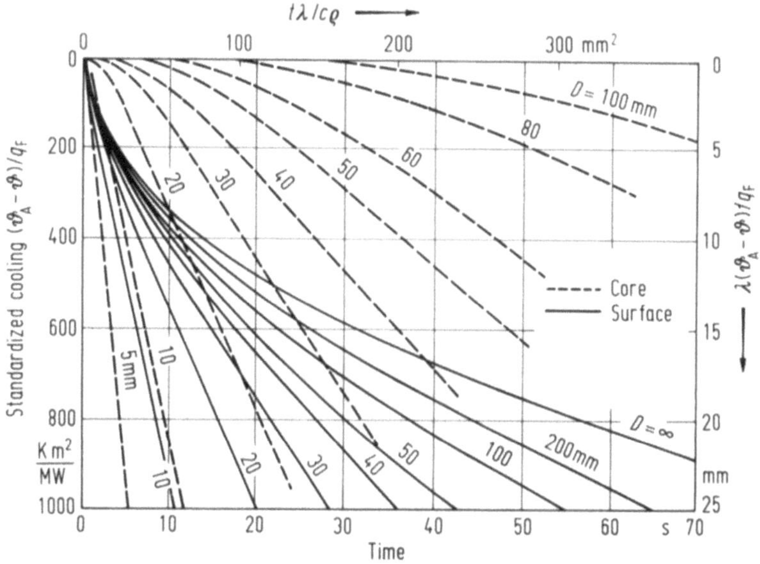

Fig. 11.2.1.2. Cooling of the surface area and core of a steel cylinder during the vapour blanket stage; $l > 4\ D$; $\lambda = 25$ W/mK; $c \cdot \rho = 4,5$ MJ/m³K $\cdot\ \vartheta =$ actual temperature

The temperature course calculated according to (11.2.3) is valid for the core of the workpiece. The surface temperature would drop instantaneously, according to the first boundary condition, to the final value T.

The transition from the boiling stage to the convection stage occurs when the temperature gradients on the surface become equal.

According to the QTA-Method T is the temperature at which the steep declining heat flux density curve during boiling stage would intersect (if extrapolated) the temperature axis (see Fig. 11.2.1.1). This temperature is nearly the same as the boiling start temperature of the quenchant. It is a property of the quenching medium itself. It is not influenced by other conditions as e.g. the agitation, bath temperature, surface condition of the workpiece etc. The only factor which can influence this temperature is its dependence on pressure.

Other authors have reported also about the important influence of the temperature T on the course of the quenching process.

Heat Transfer and Temperature Course in the Convection Stage

In the convection stage we strive to a possibly small heat transfer in order to equalize the temperature differences within the workpiece before the martensite formation starts. A sharp transition from the boiling stage to the convection stage is desired.

In the upper part of the convection stage, one can find (even for low agitation <0.2 m/s) predominantly the heat transfer realized by natural convection with a turbulent boundary layer. In this case one can use the following dimensionless numbers to calculate the heat transfer characteristic values:

$$N_u = 0.13\ (\mathrm{Gr}\cdot\mathrm{Pr})\ (11.2.4),\ \text{valid for } \mathrm{Gr}\cdot\mathrm{Pr} > 10^8$$

$$N_u = \frac{\alpha l}{a}$$

$$Gr = \frac{\rho - \rho_0}{\rho} \cdot \frac{g l^3}{v^2}$$

$$Pr = \frac{v}{a}$$

Figure 11.2.1.3 shows the results of such calculation, i.e. the heat transfer coefficient in the convection stage, $\alpha_B = $ "A". Beside the surface temperature of the workpiece, the biggest influence on the heat transfer has the viscosity of the oil.

In case of higher agitation, the heat transfer laws for coercive convection should be used.

The "Q"-Probe—Relation Between "Q"-Value and Hardness

While the "T"-value can be taken from a relevant labor-test for oil in question and the "A" value can be calculated from its physical properties, the "Q" value must be measured under workshop conditions.

The author of the QTA method supposes that it is difficult to measure and record the cooling curves in the quenching tank, therefore an indirect method for evaluation of "Q" was suggested. This method is based on the known relation

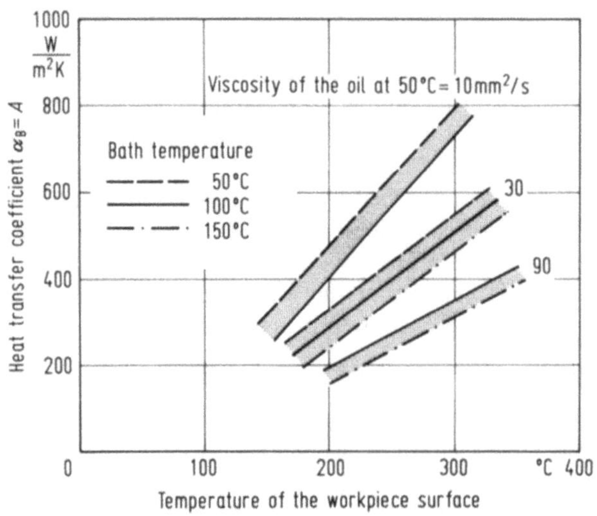

Fig. 11.2.1.3. Heat transfer in oil baths during the convection stage at free convection with turbulent boundary layer

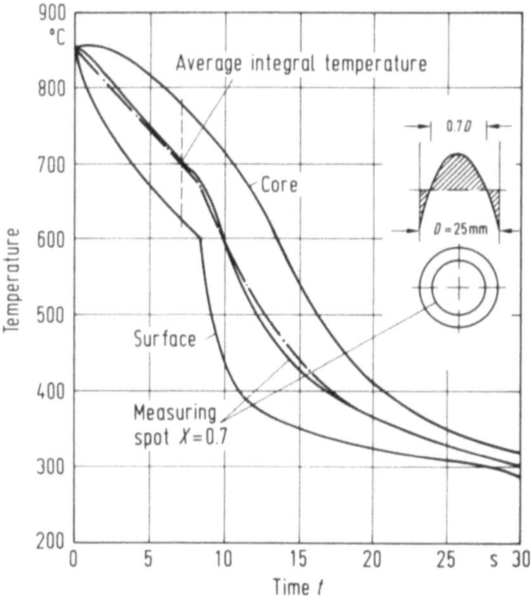

Fig. 11.2.1.4. Q-probe; course of the integral average temperature, and the temperature at $X = 0.7 \cdot D$

between the cooling rate and resulting hardness, when quenching a certain steel. By using a calibration curve, not only a relative value can be obtained, but also the heat flux density in MW/m^2 can be defined.

Figure 11.2.1.4 shows the Q-probe. It is a 25–30 mm Dia × 100 mm long cylinder made of a nonalloyed or low alloyed steel, which should be quenched together with the whole batch of workpieces under workshop conditions.

Although the temperature field during quenching, within this probe is distorded and changes very quickly, the calculations have shown, that the *integral average temperature* is following the temperature at $X = 0.7 \cdot D$. For this temperature and for a thermodynamically thin body, the following formula is valid:

$$q = c\rho V \frac{d\vartheta}{dt}. \tag{11.2.5}$$

According to this formula Q is inversely proportional to the cooling time upto 500 °C.

For a certain steel grade a series of calibration tests should be done by measuring the temperature course at the point $X = 0.7 \cdot D$ for different values of Q. The "Q" value can be determined then by measuring the hardness at the point $0.7 \cdot D$, as it can be seem from Fig. 11.2.1.5. The Q-probe takes into account the surface condition (oxidation, roughness) of the real workpieces, as well as the flow conditions in the quenching bath, because it is heated and quenched together with the batch of the workpieces.

If the relation is known between different Q-values and the resulting hardness (in form of the calibration curve) it is possible by only one hardness

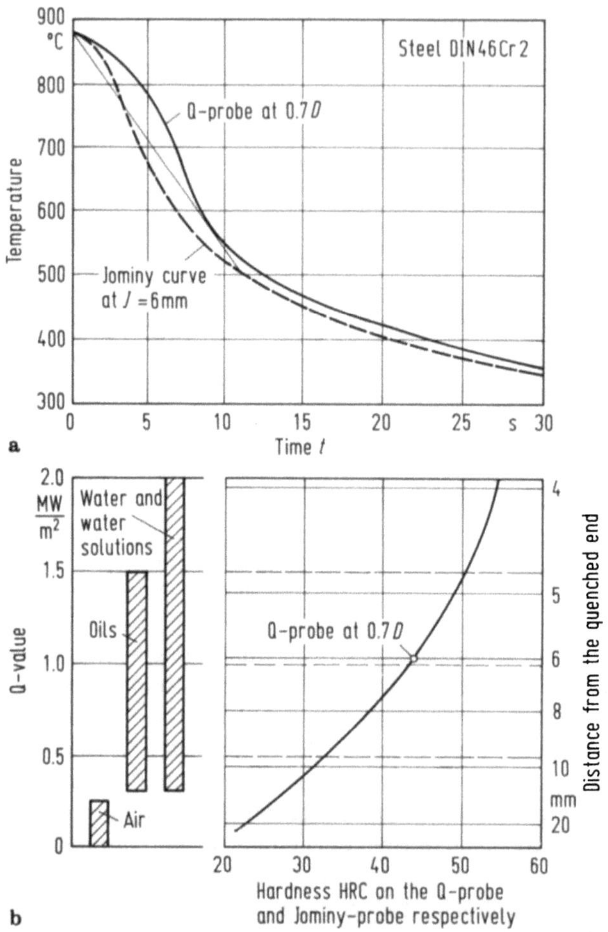

Fig. 11.2.1.5 a, b. Cooling curves at spots of equal hardness (44 HRC) for the Q-probe at 0.7 *D*, as well as for Jominy specimen at equivalent Jominy distance $J = 6$ mm (**a**). Calibration curve for the Q-probe (**b**)

measurement to define the Q-value. Figure 11.2.1.5a shows the cooling curve at $X = 0.7 \cdot D$ of the Q-probe section. After quenching a hardness of 44 HRC was measured in this case at the same point. The same hardness is found at a distance of 6 mm from the quenched end of the relevant Jominy hardenability curve, showing that in both cases more or less the same cooling laws were followed. The dashed curve in Fig. 11.2.1.5a represents the calculated cooling curve for 6 mm distance on the Jominy curve compared to the measured cooling curve at $0.7 \cdot D$ of the Q-probe (full line). This proves that using the hardness as parameter, Q-value can also be related to the Jominy distance.

E.G. the point showing 44 HRC on the calibration curve (see Fig. 11.2.1.5b), corresponds to a Q-value of 1.05 MW/m^2 which can be calculated from the average cooling rate during the vapour blanket stage (33.3 K/s—the tangent at the cooling curve—see Fig. 11.2.1.5a), and the volume specific heat capacity of 4.5 MJ/m^3K—related to the surface of the probe.

The "Q"-value, as it can be seen from Fig. 11.2.1.5b, is inversely proportional to the square of the Jominy distance (distance from the quenched end). The correlation between the heat flux density as a function of the hardness—gives the necessary calibration curve and can be used for determination of the "Q"-value from the hardness measured in a distance of $0.7 \cdot D$ from the core using a standard steel grade (in this case DIN – 46Cr2).

It should be noted, that practical measurements have shown scattered values when measuring the hardness at point $0.7\,D$. This can be attributed to segregation within the material itself, and presents one of the obstacles for achieving good reproducibility.

Implementation of the QTA-Model

The temperature profiles can be calculated and approximately graphically presented as it could be seen from Fig. 11.2.1.6 in which a comparison between Grossmann's "H"-model, and QTA-model is given. Contrary to the "H"-model in QTA-model the effect of cooling with vaporisation (in the boiling stage) is evident through the small distance λ/α_K.

A computer programme was developed in order to calculate the temperatures at the surface and in the core of round bars at different times from the immersion. The imput data related to the steel grade, bar diameter and to the quenchant characteristics—can be changed.

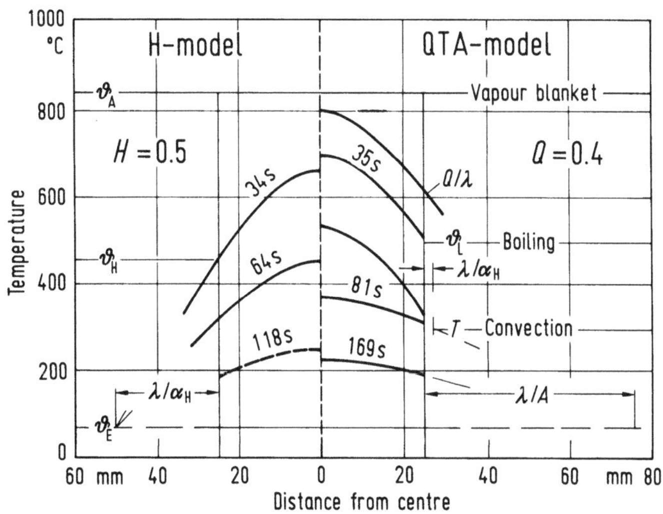

Fig. 11.2.1.6. Graphical determination of the temperature profiles by using the model-parameters: Q, T and A

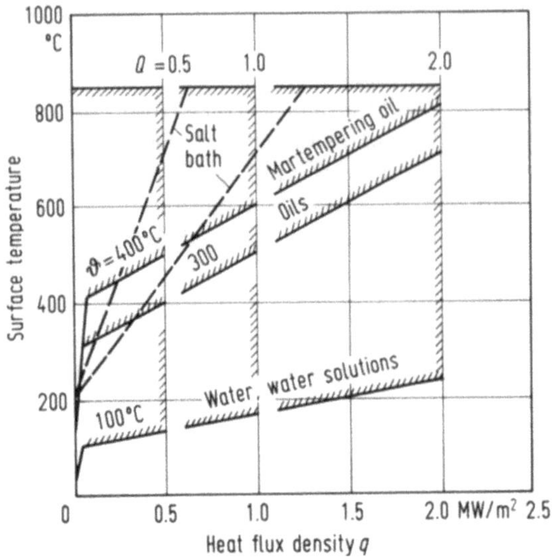

Fig. 11.2.1.7. Heat-Flux density diagram for quenching in water, oils and salt-baths

Figure 11.2.1.7 shows the heat flux density diagram in which effects of the most important quenchants are characterized, according to the model by parameters "Q" and "T". The steep fall ends in temperature T according to which one can put in order different quenchants. Using these model-data the critical cooling time from 800 °C to 500 °C was calculated for the surface area and for the core of steel cylinders as a function of their diameter.

Figure 11.2.1.8 shows this correlations for water, oils and salt-bath. The upper abscissa in diagrams shown in Fig. 11.2.1.8 gives the equivalent Jominy distances for the same cooling times from 800 °C to 500 °C. In the upper left diagram on Fig. 11.2.1.8—valid for water solutions big influence of the parameters "Q" can be seen.

The main aim of the diagram shown in Fig. 11.2.1.8 is the possibility to quantitatively evaluate the effects of changes of the parameters "Q" and "T".

With the relevant Q-value, and physical data of the used quenchant i.e. the boiling temperature (T-value) and the heat transfer coefficient in the convection stage (A-value), it is possible to calculate cooling curves in different points of a workpiece's cross-section which has been quenched under the same conditions as the standard Q-probe.

For workpiece shapes other than cylinder the calculated cooling time can be multiplied approximately by a shape coefficient k:

$$k = \frac{\text{Volume} \cdot \text{Diameter (resp. thickness)}}{\text{Surface}};$$

for cylinder $k = 1.0$
for plate $k = 2.0$
for sphere $k = 0.67$

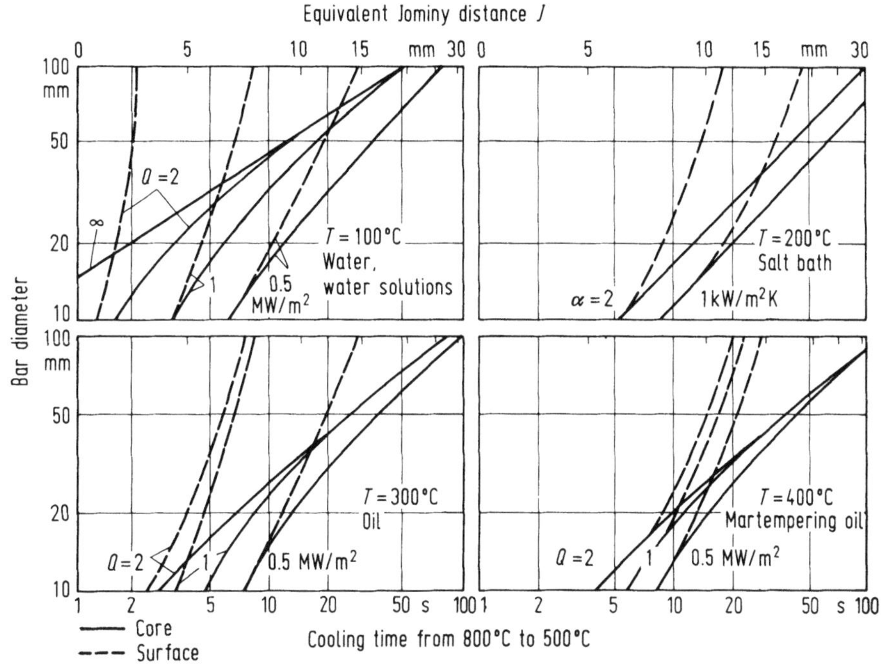

Fig. 11.2.1.8. Critical cooling time (800 → 500 °C) for core and surface areas of round bars when quenching them in water, oils, and salt-baths, depending on "Q" and "T"

Supposing that the relevant calibration curve is available, the QTA-method can be used for the following:

—The cooling conditions on different places within a batch can be quantitatively evaluated, and corresponding measures for changing the cooling conditions can be studied.
—The necessary cooling conditions for reaching certain hardness values at a certain workpiece (kind of quenchants, agitation, bath temperature) can be determined.
—Temporary influences of the workshop conditions on the quenching behaviour of a quenchant can be studied and the workshop condition can be regularly checked.
—Quenchants can be mutually compared in regard to their effects, and selected correspondingly.

As described above it is quite possible, by using the mentioned parameters of the QTA-method, to calculate the temperature course in different points of a workpiece taking into account specific effects which take place during quenching in vapourizable quenchants.

It should be noted, however, that the relation between the hardness measured at the point 0.7·D of the Q-probe and the quenching conditions is not necessarily unambiguous because the hardness depends not only on the cooling conditions but also on many microstructural aspects (e.g. segregations etc.) of the material itself which have not been taken into account. Therefore the prediction of hardness distribution using this method may be uncertain.

List of Symbols

ϑ_A:	Austenitization temperature °C
ϑ_L:	Transition-temperature: vapour blanket/boiling stage °C (Leidenfrost point)
$\vartheta_s = T$:	Boiling temperature of the quenchant (Beginning) °C
ϑ_E:	Temperature of the quenching bath °C
ϑ:	Actual temperature °C
ϑ_R:	Temperature on the surface of the workpiece °C
ϑ_M:	Temperature in the core of the workpiece °C
$\frac{d\vartheta}{dt}$:	Cooling rate K/s
D:	Diameter of the workpiece m
l:	Length of the workpiece m
V:	Volume m^3
t:	time s
t_F:	Duration of the vapour blanket stage s
λ:	Heat conductivity of the body material W/m·K
ρ:	Density kg/m^3
$c\rho$:	Volume-related heat capacity $MJ/m^3 \cdot K$
$a = \frac{\lambda}{c\rho}$:	Thermal diffusivity m^2/h
g:	gravitation m/s^2
v:	kinematic viscosity m^2/s
ϑ_H:	Half-temperature of the Grossmann's "H"-model $= (\vartheta_A - \vartheta_E)/2 + \vartheta_E$ °C
q:	heat flux density MW/m^2
Q:	Heat flux density on the surface from ϑ_A to 500 °C
q_F:	Heat flux density on the surface in the vapour blanket stage MW/m^2
α:	Heat transfer coefficient $W/m^2 K$
α_K:	Heat transfer coefficient in the boiling stage $W/m^2 K$
$\alpha_B = A$:	Heat transfer coefficient in the convection stage $W/m^2 K$
α_H:	Heat transfer coefficient of the "H"-model from ϑ_A to ϑ_H $W/m^2 K$
θ:	Standardized temperature ratio between 1 and 0; $\theta = (\vartheta - \vartheta_x)/(\vartheta_0 - \vartheta_x)$
X:	Distance from the core, related to $D/2$; $X = 2x/D$
F_0:	Fourier number $= 4\left(\frac{\lambda}{c\rho}\right)t/D^2$

References

1. Wünning J (1981) Berechnung und Steuerung des Temperatur-verlaufs beim Abschrecken von Stahl in Wasser und Öl, HTM, 36: 5, S. 231–241.
2. Wünning J, Liedtke D (1983) Versuche zum Ermitteln der Wärmestromdichte beim Abschrecken von Stahl in flüssigen Abschreckmitteln nach der QTA-Methode, HTM, 38: 4, 5 149–155

11.2.2 Relations Between Cooling Curves and Hardness Distribution (after K.E. Thelning)

B. Liščić

Introduction

As coordinator of Task 4 of the IFHT technical committee "Scientific and Technological Aspects of Quenching", Karl–Erik Thelning has conducted during 1984/85, a round-robin investigation aimed at elucidating the correlation of the measured quenching intensity of oils with the hardness distribution in quenched parts [1].

The test material consisted of steel bars having diameters: 10, 16, 25 and 50 mm and lengths three times their respective diameter made out of four steel grades. All the steels were supplied by one source. Nine different kinds of quenching oils were used which characteristics were determined by means of cooling curves and cooling rate curves measured and recorded by means of an Inconel 600 probe 12.5 mm diameter by 60 mm length having the thermocouple located at its geometric centre. The cooling capacity of each oil has been classed according to its capacity for producing high hardness and a good degree of through hardening on the cross-section of steel specimens.

The original terms of reference for Task 4 were: bringing together the cooling curves of the quenching media employed and results of hardening tests.

In a wider perspective these investigations aimed at making it possible to deduce from a cooling curve the suitability of the cooling medium under consideration to be used for hardening different steel grades in various dimensions. In any case it ought at least to let the best-suited cooling medium be chosen from among different cooling media available for the steel grade and dimension in question.

The following designations were applied to different kinds of quenching oils used: A, C, E, F, M, U, V, P, 36—the physical properties of which are given in Table 11.2.2.1.

The four steel grades used were chosen according to a scale of ascending hardenabilities. The lowest-hardenability steel should through—harden in the smallest dimension (10 mm dia); the highest-hardenability steel should through—harden in the largest dimension (50 mm dia).

Table 11.2.2.1.

Oil	A	C	E	F	M	U	V	P	36
Density g/cm³	0.884		0.871		0.895	0.856	0.887	0.875	0.862
Viscosity 40 °C	24.1	32	33	22	62.5	15.3		22	23
cSt 100 °C	4.5	5.4		4.1	8.4	3.5		4.1	4.5
Flash Point °C	196	222	228	164	227	190	207	200	190

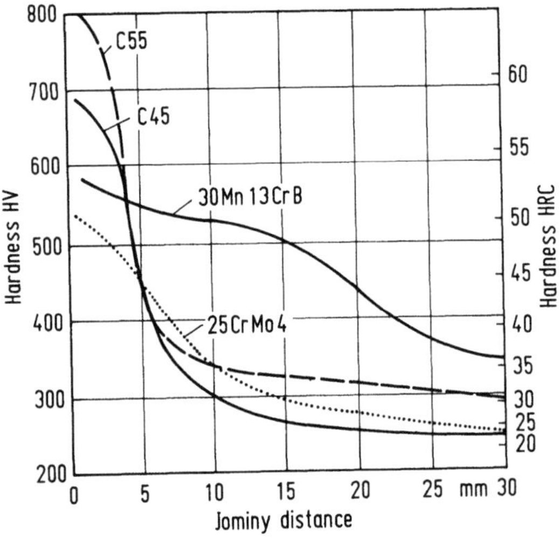

Fig. 11.2.2.1. Jominy hardenability curves for steel grades used in the investigation

Table 11.2.2.2 contains the composition of the steels that refers to the heat analysis. Steels have been designated according to DIN and their respective hardenabilities according to the Grossmann's critical diameter D_i. Figure 11.2.2.1 shows the relevant Jominy hardenability curves for the steels in question.

Hardness Measurement After Hardening

Each specimen was sectioned across half its length and Rockwell C (HRC) or Vickers (HV) hardness measurements were taken transversely along a diameter. The mean values were plotted in the diagrams. From each participating institution,

Table 11.2.2.2. Chemical composition and hardenability

Steel grade	C	Si	Mn	P	S	Cu	Cr	Ni	Mo	B_{sol}	Di mm
C45	0.46	0.25	0.72	0.016	0.033	0.22	0.12	0.13	0.01		31.5
C55	0.55	0.29	0.77	0.022	0.022	0.18	0.13	0.07	0.01		37.0
25 CrMo 4	0.25	0.24	0.61	0.015	0.028	0.19	1.02	0.12	0.17		76.2
30 Mn 13 CrB	0.32	0.25	1.23	0.015	0.011	0.26	0.52	0.11	0.02	0.003	122.4[a]

[a] Including B. Boronfactor 1.77

Table 11.2.2.3. Dimensions suitable for hardening in oil

Steel grade	ϕ 10 × 30	ϕ 16 × 48	ϕ 25 × 75	ϕ 50 × 150 mm
C45	x	x		
C55	x	x		
25 CrMo 4		x	x	
30 Mn 13 CrB			x	x

first the results of the tests to determine the cooling capacities of the oils used were given, as temperature–cooling time (Fig. 11.2.2.2) and temperature–cooling rate (Fig. 11.2.2.3) diagrams presented here only from one participating institution for oils C, E, F. Further, an account is given of the hardening process, specifying which steel grades and bar dimensions were hardened and which oils were used, respectively. Finally, the results of the hardness measurements were presented as depth-of-hardness diagrams for each steel grade, specimens' dimension and oils used as can be seen as an example in Figs. 11.2.2.4–11.2.2.7. This was followed by a preliminary discussion concerning the ranking order of the oils as regards their capacity to produce high as-quenched hardness and satisfactory through-hardening of the test bars.

Reckoned from the capacity of the quenching media to produce high hardness in the steel specimens, the ranking order of the oils in this case predominantly was: F, E, C.

For the main tests the steel test pieces were heated in a protective atmosphere, but also without protective atmosphere. When quenching the test pieces

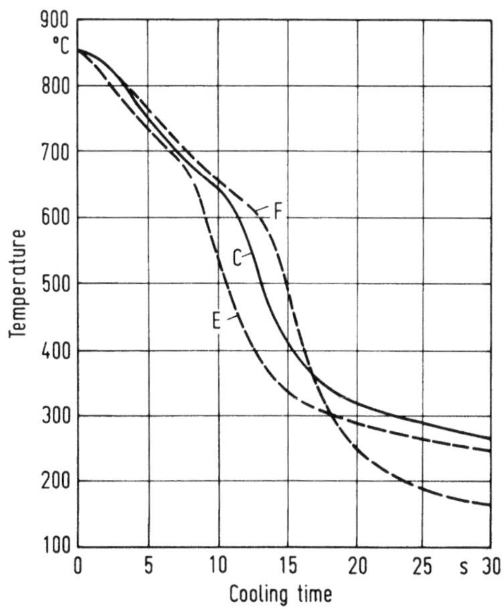

Fig. 11.2.2.2. Temperature-cooling time curves for oils *C*, *E*, *F* measured and recorded by the Inconel 600 testing probe ϕ12.5 × 60 mm

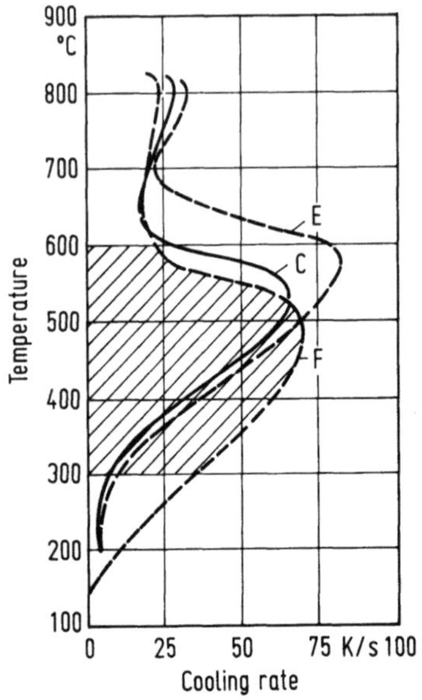

Fig. 11.2.2.3. Temperature-cooling rate curves for oils *C*. *E*, *F* measured and recorded by the Inconel 600 testing probe $\phi 12.5 \times 60\,mm$

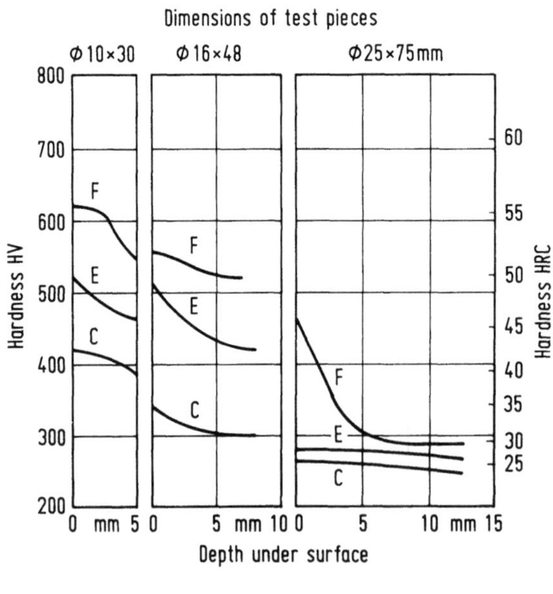

Fig. 11.2.2.4. Steel C 45 quenched without protective atmosphere in different oils of 50 °C with agitation rate 0.4 m/s

the oil bath was agitated, but additional tests were carried out without agitating the oil bath.

In this way the influence of the surface condition as well as the influence of agitation were investigated. Based on this investigation the following was found:

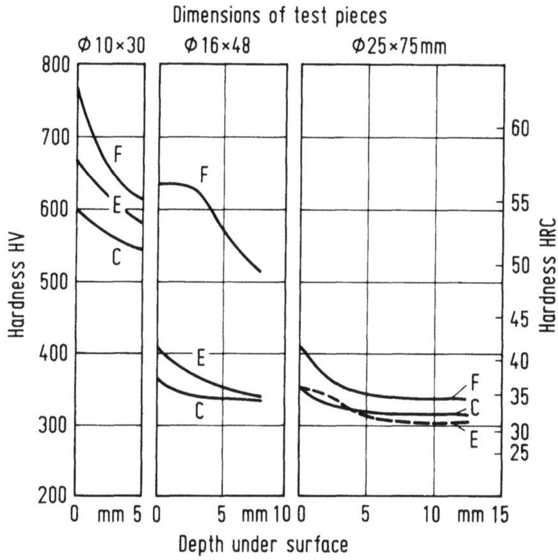

Fig. 11.2.2.5. Steel C 55 quenched without protective atmosphere in different oils of 50 °C with agitation rate 0.4 m/s

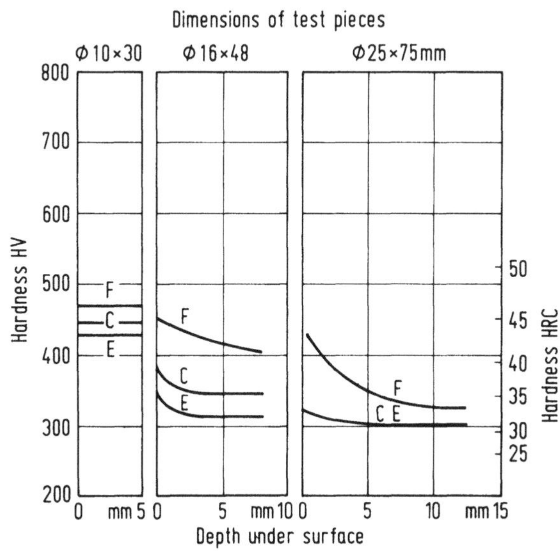

Fig. 11.2.2.6. Steel 25 Cr Mo 4 quenched without protective atmosphere in different oils of 50 °C with agitation rate 0.4 m/s

The ranking order of the oils with respect to their capacity to give high as-quenched hardness in the steel test bars is approximately the same whether the bars were heated with or without protection against oxidation.

As was expected, agitation during quenching had a manifest influence on the increase in hardness. Specimens quenched with agitation gave always higher hardness values and better through hardening than some specimens quenched

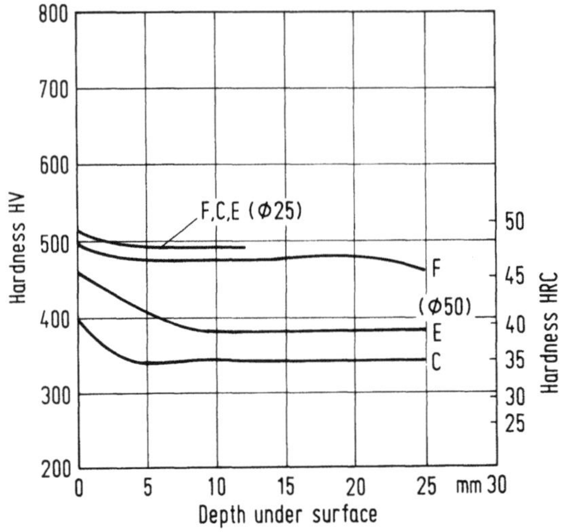

Fig. 11.2.2.7. Steel 30 Mn 13 CrB quenched without protective atmosphere in different oils of 50 °C with agitation rate 0.4 m/s

without agitation. With regard to the alloy steel test pieces of the same dimensions, the influence of agitation is probably not so marked and most likely it decreases as the cooling capacity of the oils increases. It appeared for both of testpiece dimensions, that the ranking order of the decreasing cooling capacity of the oils used as measured by the as-quenched hardness, was the following:

—Protective atmosphere-quenched with agitation
—Protective atmosphere-quenched without agitation
—No protective atmosphere-quenched with agitation
—No protective atmosphere-quenched without agitation.

The hardness difference between the superior and the inferior hardening method alternatives was about 400 HV for the smaller test pieces and fully 200 HV for the larger ones.

Criteria for Estimating the Cooling Capacity of an Oil When Used for Quench-Hardening of Steel

It is an almost universal assumption that when a steel is being quench-hardened its rate of cooling at some specific temperature, for example at: 700 °C, 600 °C or 500 °C is crucial for the hardness obtained. This conception suggest that the cooling rate curves obtained for different quenching oils in a standardized laboratory test procedure may be used to judge the hardness distribution on the cross section of real parts after quenching. Unfortunately it is at once apparent by looking for example at Fig. 11.2.2.3, in connection with Figs. 11.2.2.4–11.2.2.7, that this conception is not a general truth since it is not applicable either to different oils or to different steel grades.

From this investigation, according to Thelning, it was not possible to find a really reliable method of classifying all the oils investigated solely by judging

Table 11.2.2.4. Cooling capacity of the oils, expressed as area under cooling curves in cm^2

Oil group	1				2			3	
Oil	36	A	F	M	P	E	U	V	C
Area, cm^2	40	33	32	31	29	29	27	25	24

The oils in Group 1 may be classed as accelerated oils, those in Group 3 as conventional quenching oils and those in Group 2 as intermediate quenching oils

from the cooling curves obtained, and from the capacity of the oils to produce high as-quenched hardnesses and satisfactory through-hardening in the steels. However, a reasonably good overall understanding of the hardening result may be obtained by studying the shape of the cooling curves and the *area of the surfaces* obtained by the cooling-rate curve, the temperature axis and the two temperature horizontals that enclose the diffusion-dependent transformation range (see e.g. the dashed area under cooling curve for oil F in Fig. 11.2.2.3). It was found that the temperature interval—for the temperature horizontals 600 °C to 300 °C yielded the best results.

Following this concept all kinds of oils used in this investigation have been classified into 3 groups—see Table 11.2.2.4.

The above described investigation shows, that a direct correlation between cooling curves and hardness distribution on the cross-section doesn't yield satisfactory results.

References

1. Thelning KE (1986) "Correlation of the Measured Quenching Intensity with the Hardness Distribution in Quenched Parts", Proceedings of the 5th International Congress on Heat Treatment of Materials, Budapest, Volume III, p. 1737–1759
2. Thelning KE (1983) "New Aspects on the Appraisal of the Cooling Process During Hardening of Steel", Scand. Journ. of Metallurgy 12, 189–194

11.2.3 IVF Method for Classification of Quenching Oils

B. Liščić

At the Swedish Institute of Production Engineering Research (IVF) in Göteborg practically every commercially available oil in Sweden was tested on its cooling characteristics [1] using a standardized test method [2], and complementary hardening tests have been carried out. The oils were tested using Inconel 600 cylinder of ϕ 12 × 60 mm with the thermocouple in its geometric center and plotting the temperature vs. time and temperature vs. cooling rate curves.

Systematic hardening tests with different oils have been carried out using unalloyed steel cylinders of 10 mm and 16 mm diameters, made of SAE 1045 steel grade. For the austenitization protective atmosphere was used, and quenching was performed in stationary oil at 50 °C.

This investigation has made it possible to group the hardening performance of different quenching oils in a consistent and appropriate manner.

The criteria for this grouping were (Fig. 11.2.3.1): the transition temperature between the film boiling and the nucleate boiling phase (T_{VP}), the transition temperature between the nucleate boiling phase and the convection phase (T_{CP}), and the cooling rate in the critical ferrite and pearlite nose temperature range (CR). A simple formula was developed on the basis of these results, allowing the hardening power of a particular oil for a given grade of steel to be calculated.

For unalloyed steels the empirical formula is as follows:

$$HP = 91.5 + 1.34 T_{VP} + 10.88 CR - 3.85 T_{CP},$$

where:

HP = the hardening power of the quenching oil
T_{VP} = the transition temperature between the vapour phase and the boiling phase, in °C
CR = the cooling rate over the temperature range 500–600 °C, in °K/s
T_{CP} = the transition temperature between the boiling phase and the convection phase, in °C.

The formula has been developed in order to facilitate selection of a suitable quenching oil for a particular application, based on the ability of the quenchant

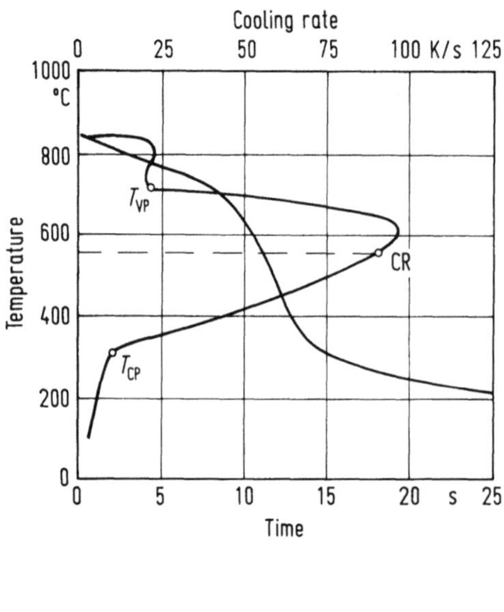

Fig. 11.2.3.1. Definition of the points that have been used as criteria to characterize a quenching oil

Table 11.2.3.1.

	Name of oil	Trans. temp. vapour/boiling phase °C	Cooling rate at 500–600 °C K/s	Trans. temp. boiling/conv. phase °C	HP value approx.
High	Superquench R	730	100	280	1218
	Durixol 4	720	98	290	1145
	Voluta F	650	78	230	1090
	Isomax 166	720	95	300	1036
	QH 10	720	95	320	982
	Frenso 36	710	86	330	927
	Quendila A25	675	83	330	864
	Belini FS	735	87	340	800
	Mobiltherm F	670	76	300	754
	Quenchtex C	680	78	330	691
	ESL 145	730	76	360	645
Hardness	Thermola 22	715	70	360	595
	Iloquench 32	730	70	370	530
	Durixol W72	760	67	370	482
	QH 25	725	77	370	427
	QH 40	730	70	380	373
	Hardolja 1423	675	65	390	312
	ESL 85	735	63	390	260
	Hardolja 1397	685	65	390	205
	Quench 2	650	62	370	155
	Quench 3	655	63	370	100
	Voluta C	655	56	400	45
Low	Iloquench 5	685	56	420	10

to harden steel. It is intended only for the classification of one oil relative to another; because of the empirical character of HP value there is no direct relationship, between HP value and the Grossmann's factor "H".

The coefficients in the formula have been chosen so that the range of *HP* values for the oil considered is about 10–1000.

The cooling characteristics of a quenching oil depend on: the base oil used, the refining process and the additives employed to increase cooling capacity or reduce oxidation, or improve washability. There is considerable variation between oils, with cooling rates ranging from slow to fast.

Table 11.2.3.1 shows numerical values of the three characteristic points for the oils investigated, and corresponding calculated *HP* value.

The results obtained by using the above mentioned formula have been matched to a straight line by the method of least squares.

Of the oils tested by IVF, the "Superquench R" has the highest calculated *HP* value, and also gives the highest hardness. "Iloquench 5" has the lowest *HP* value, and gives the lowest hardness. It can be seen from Table 11.2.3.1., that oils with lower cooling rates generally have a higher transition temperature between the boiling phase and the convection phase. Both of these factors contribute to reduced hardening power. This is related to the nature of the oil base, its refining and additives.

For the user who plans to change from one type of quenching oil to another, if the hardening performance of the oils is unknown, the user will have to determine the cooling rates vs temperature curves from which the three characteristic points can be determined.

Hardening power can then be calculated from the above given formula, after which the oils can be compared.

References

1. Segerberg S (1988) "Classification of Quench Oils: a Method of Comparison".—Heat Treating/ Dec p. 30–33
2. Industrial quenching oils—Determination of cooling characteristics—Laboratory test method. Draft International Standard ISO/DIS 9950

11.2.4 Prediction of Hardness Values Based on Cooling Parameter

B. Liščić

(The cooling parameter in this text has been marked as "λ", which should not be mixed with the thermal conductivity λ[W/Km])

Continuous cooling transformation (CCT) diagrams as shown in Fig. 11.2.4.1 for a certain steel composition may give a lot of information about the transformation kinetic, microstructure and hardness expected when quenching or cooling this steel from the austenitizing temperature. For the practical application, however, the relation between the transformation behaviour of the steel from one side and dimensions of the workpiece as well as quenching intensity of the quenching medium from the other side, cannot be taken straight from such a diagram. This problem arises because each cooling curve in a CCT diagram may be attributed to a certain point of the workpiece cross-section of different dimensions quenched in different quenching media.

One way for solving this problem was the introduction of the "cooling parameter" λ and the corresponding "cooling time diagram" [1] [2].

Every cooling curve in a CCT diagram corresponding to the cooling law in a certain point of a workpiece can be defined by the time necessary to cool this point from 800 °C to 500 °C, or by the cooling parameter λ which is:

$$\lambda = \frac{\text{cooling time from 800 °C to 500 °C}}{100} \text{ in seconds}$$

$$\lambda = \frac{t_{500\,°C} - t_{800\,°C}}{100} \text{ s}$$

%C	%Si	%Mn	%P	%S	%Cr	%Mo	%Ni
0.43	1.29	0.77	0.012	0.014	0.63	0 16	0.33

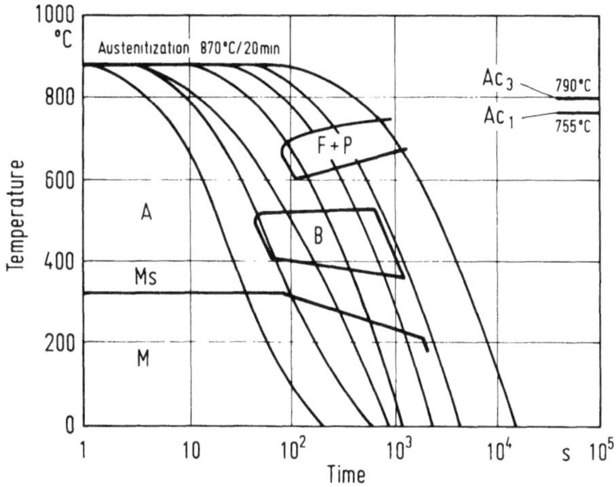

	Cooling time from 800°C to 500°C in s						
	14	37	78	190	390	680	2200
Martensite in %	100	97	45	25	10	-	-
Bainite in %	-	3	55	72	65	-	-
Ferrite in %	-	-	-	3	10	15	17
Pearlite in %	-	-	-	-	15	85	83
Microhardness HV10	689	642	503	440	378	274	254

Fig. 11.2.4.1. Continuous cooling transformation (CCT) diagram for the steel of specified chemical composition

One λ-value may correspond to the cooling laws in different points of workpiece of different dimensions quenched in different quenching media. Figure 11.2.4.2 shows some examples for the cooling curve with $\lambda = 1.0$.

The cooling time diagram (Fig. 11.2.4.3, right hand side), differs from the CCT diagram (Fig. 11.2.4.3, left hand side) thereby that on the abscissa, instead of time from the beginning of cooling, it is plotted the time necessary in each case of cooling to pass through the temperature region from 800 °C to 500 °C. This time interval is called: "cooling time from 800 °C to 500 °C". Instead of this cooling time from 800 °C to 500 °C, the value of cooling parameter λ may be plotted on the abscissa. In this way to every cooling curve in the CCT diagram, corresponds a vertical line in the cooling time diagram.

The cooling time diagram itself consists of two parts: the bigger upper part corresponds exactly to the time-temperature field of the CCT diagram, having the temperature scale on the left ordinate, and hardness scales (both in HV 10 and HRC units) on the right ordinate.

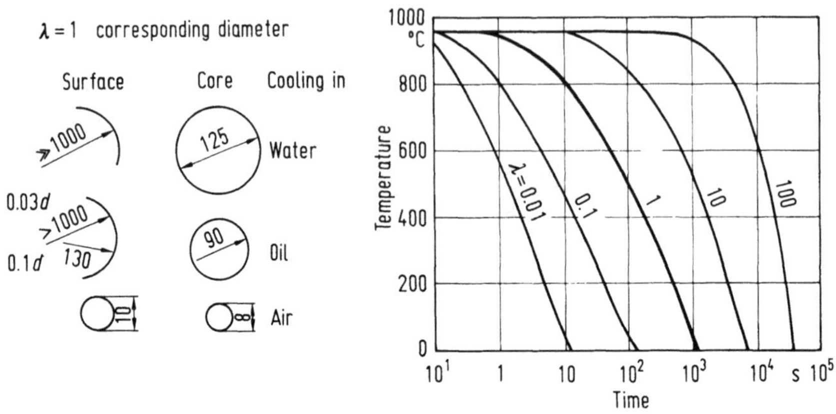

Fig. 11.2.4.2. Relation among the cooling parameter λ, bar diameter and quenching medium

The abscissa of the upper margin of the diagram represents the λ-value, and the lower abscissa represents the cooling time from 800 °C to 500 °C in seconds. Both of these abscissa are valid also for the smaller auxiliary diagram in the lower part.

This auxiliary diagram, which has been taken from [3], has on the ordinate a scale for bar diameter ranging from 10^1 to 10^3 mm. The curves in this diagram valid for the surface, 1 and 1a, and the core 2 of the bar when quenched in water or oil or cooled in air give the relation between the bar diameter and the cooling time from 800 °C to 500 °C or the λ-value respectively.

Below the abscissa of this auxiliary diagram, — see 3 — distances from the quenched end of the Jominy test bar in mm have been marked. This enables a direct comparison between the CCT diagram and a Jominy hardenability curve as well as to judge the depth of hardening for different bar diameters (quenched or cooled in different media) from the Jominy hardenability curve.

Besides the cooling time diagram another diagram, derived also from CCT diagram, becomes more and more used. This is the diagram presenting the amount of microstructural phases (Fig. 11.2.4.4). This diagram differs from the cooling time diagram thereby that on the left ordinate instead of a temperature scale a scale in percentage from 0 to 100% for the amount of different microstructural phases has been presented.

It should be pointed out that by using this diagram particular phases of the microstructure have to be summarized so that always 100% of the microstructure constitutents has to be found when reading this diagram along a vertical. For every cooling curve in the CCT diagram corresponds a vertical also in this diagram.

The smaller auxiliary diagram is here used in the same way as with the cooling time diagram.

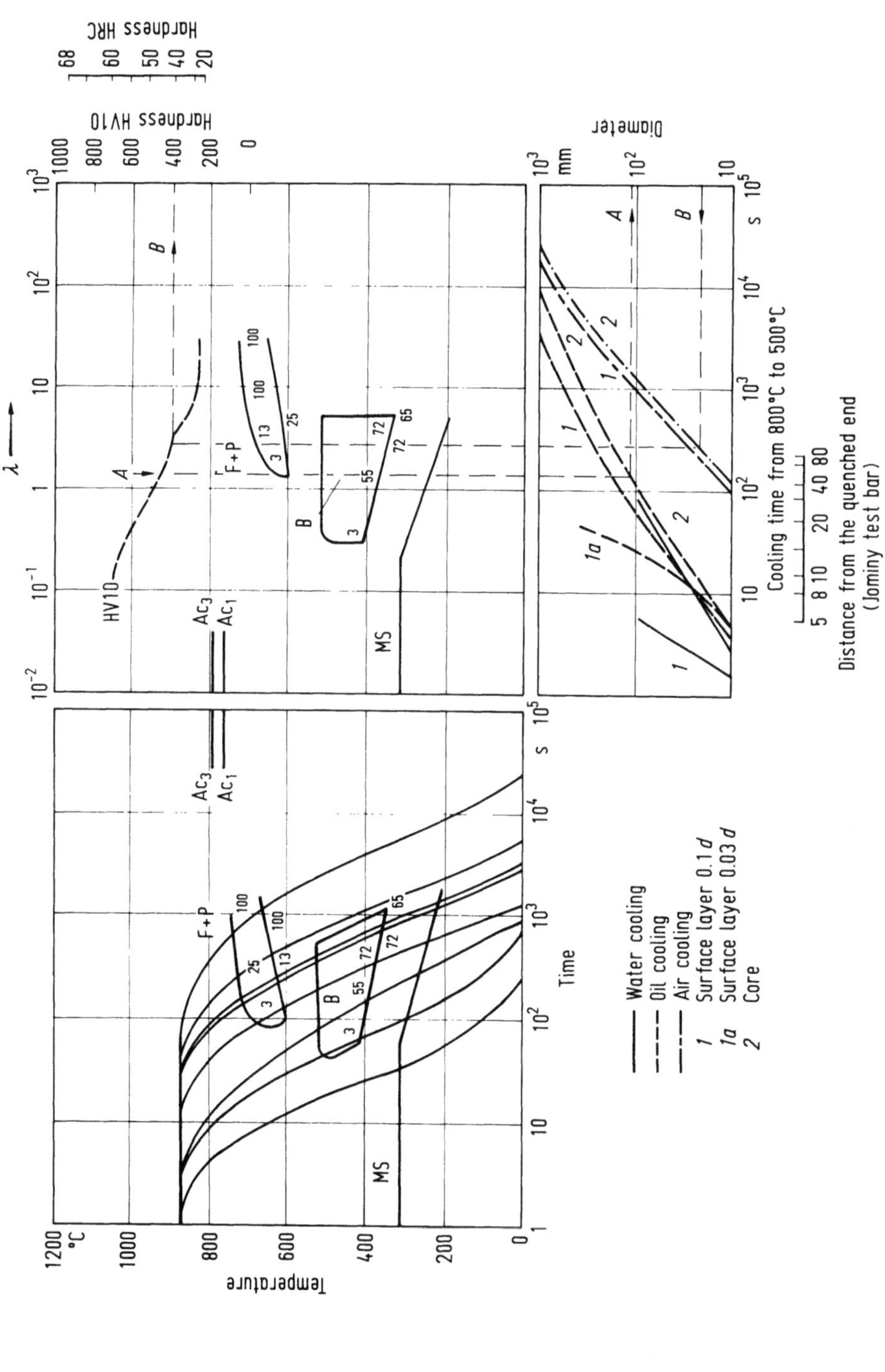

Fig. 11.2.4.3. CCT diagram and cooling time diagram for the steel 40 CrNiMo 21 (DIN)

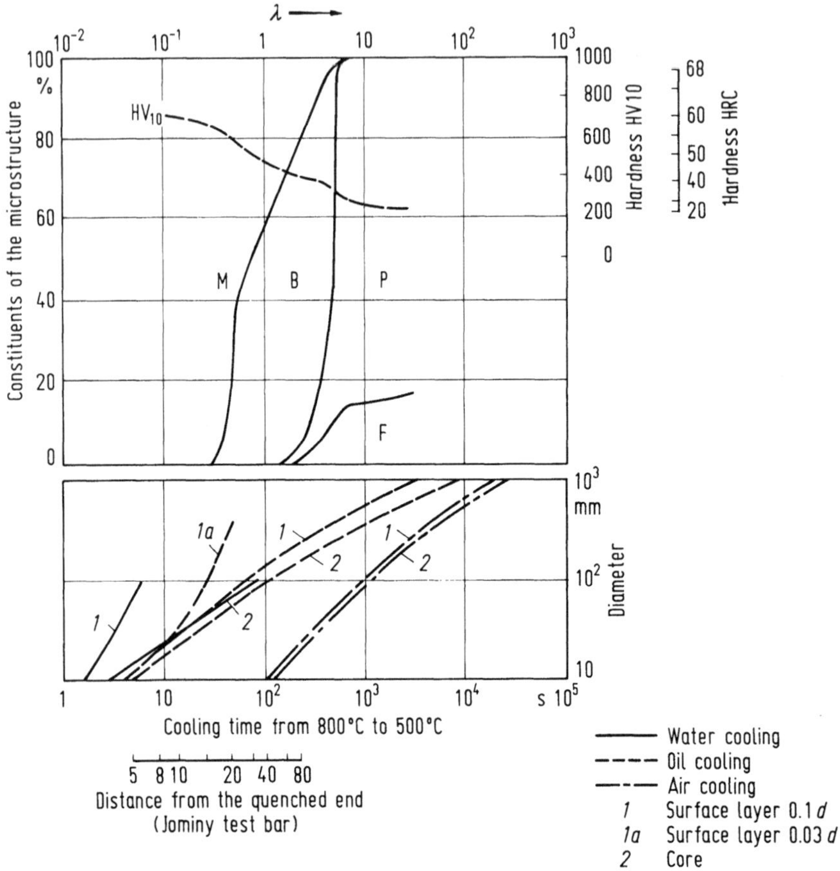

Fig. 11.2.4.4. Diagram presenting the amount of microstructural phases for the steel 40 CrNiMo 21 (DIN). M = martensite; B = bainite; P = pearlite; F = ferrite

Practical Application of Described Diagrams

a) Finding out of the Critical Bar Diameter

Contrary to a CCT diagram the cooling time diagram (Fig. 11.2.4.3) enables easily to find out e.g. up to which bar diameter the steel 40 Cr Ni Mo 21 (DIN) quenched in oil can be hardened in the core itself without having any ferrite in the microstructure. Following the line A from the upper to the lower part of the cooling time diagram (Fig. 11.2.4.3) the result obtained in this case reads 130 mm.

Fig. 11.2.4.5. Cooling time diagrams for tool steels: 100 MnCrW 4 (DIN), and X155 CrVMo 12 1 (DIN)

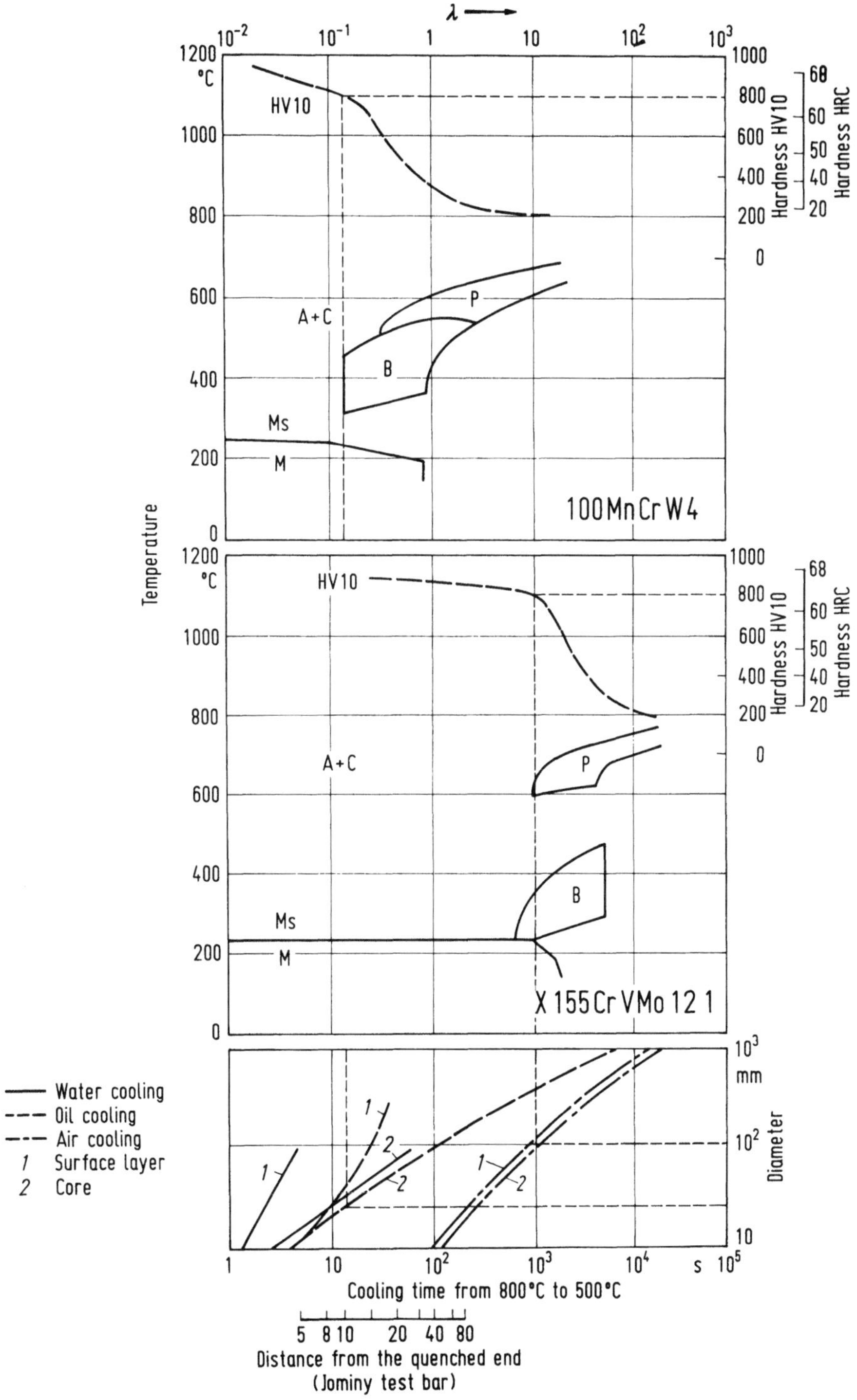
λ
100MnCrW4
X 155 Cr VMo 12 1
HV10
A+C
P
B
Ms
M
Temperature
°C
Hardness HV10
Hardness HRC
Diameter
mm
Water cooling
Oil cooling
Air cooling
1 Surface layer
2 Core
Cooling time from 800°C to 500°C
s
Distance from the quenched end
(Jominy test bar)

Following the line B from the lower to the upper part of the same diagram one can immediately find out that in the core of a round bar having diameter 20 mm, made of the same steel, when cooled in air, —a hardness of about 400 HV 10 or 40 HRC has to be expected.

In Fig. 11.2.4.5 two similar examples have been shown. The first one in the upper part of the diagram shows that for the steel 100 MnCrW 4 (DIN), a round bar of about 20 mm diameter only can be hardened by quenching it in oil without having any bainite in its core.

The second example, in the lower part of this diagram, shows that for the steel X 155 CrVMo 12 1 (DIN) round bars up to 100 mm diameter can be hardened, when cooled in air, without having any pearlite in the core.

The same results could be obtained by using the diagram which presents the amount of microstructural phases for the relevant steel grade.

b) Establishing of the Quenching/Cooling Intensity for a Specified Quenching/Cooling Medium

In the auxiliary diagrams in the lower part of the cooling time diagram as well as of the diagram presenting the amount of microstructural phases, the plotted curves represent cases only when usual quenching/cooling media (water, oil and air) are used. In case another quenching/cooling medium should be used, and relevant curves plotted in this auxiliary diagram, the procedure is as follows:

- Round bars having different diameters should be quenched/cooled in the relevant medium (taking care that austenitizing conditions are always the same), and temperature should be measured near the surface as well as in the core, during cooling.
- From the time-temperature curves measured the cooling time from 800 °C to 500 °C has to be taken and λ values calculated.
- The calculated λ-values should be plotted at relevant diameters in the auxiliary diagram.

Figure 11.2.4.6 shows e.g. some curves in the auxiliary diagram that are valid for cooling in vacuum, as well as for cooling in a stream of nitrogen pressurized to 1, 3 and 5 bar.

In the cooling time diagram in Fig. 11.2.4.6 the regions of structure transformation for two tool-steel grades: X 210 Cr 12 (DIN) and X 155 CrVMo 12 1 (DIN), have been plotted. It can be seen from this diagram, that the steel grade X 210 Cr 12 when cooled in air can be hardened in its core without having any pearlite only for bars upto 10 mm diameter, whereas the steel grade X 155 CrVMo 12 1 under the same cooling conditions may be hardened in its core without having pearlite upto 50 mm bar diameters.

When e.g. a round bar of 80 mm diameter should be hardened in a vacuum furnace, without having any pearlite in its core for the steel grade X 210 Cr 12 a nitrogen stream pressurized to 5 bar is necessary, while for the steel grade X 155 CrVMo 12 1, for the same bar diameter, a nitrogen stream of only 1 bar is sufficient.

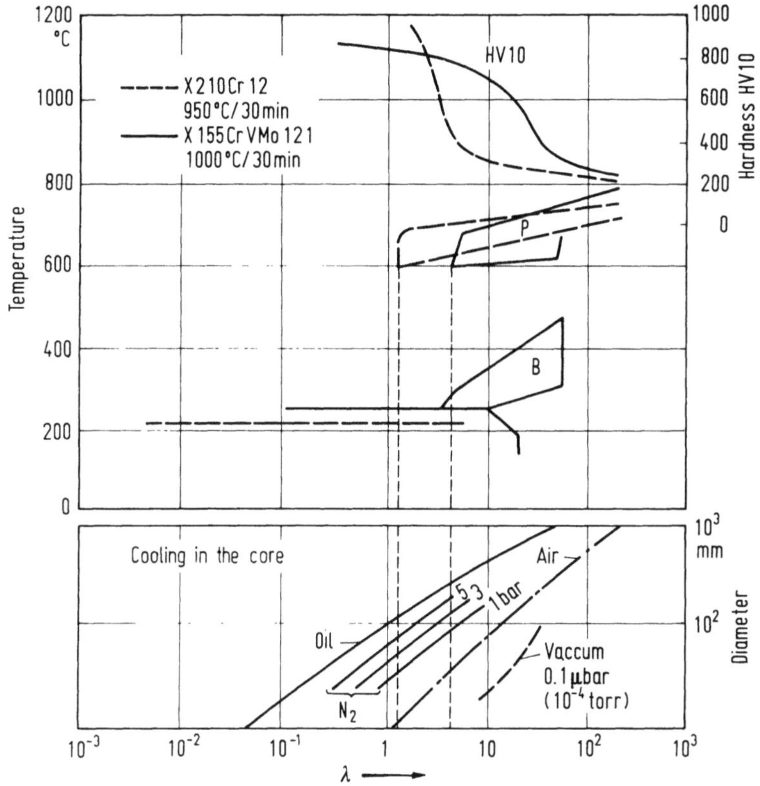

Fig. 11.2.4.6. Cooling time diagram for the steels: X 210 Cr 12 (DIN) and X 155 CrVMo 12 1 (DIN) with curves plotted in auxiliary diagram also for cooling in vacuum and in a stream of pressurized nitrogen respectively

The following objection should be made to the method described:

Usually the cooling curves in the CCT-diagrams are prepared on the basis of linear cooling (according to Newton's law) which differs from the actual cooling of metallic components in vapourable liquid quenchants.

References

1. Kulmburg A (1976) "Die Anwendung der ZTU-Schaubilder in der Härtereitechnik"—Österr. Ing. Zeitschrift 19 S 185–191
2. Kulmburg A. "Die praxisnahe Darstellung von Umwandlungsvorgängen"—Mitteilung aus den Forschungsanstalten der Vereinigten Edelstahlwerke, Kapfenberg (Austria)
3. Wever F, Rose A, Peter W, Strassburg FW, Rademacher L (1954) Atlas zur Wärmebehandlung der Stähle, Band 1. Düsseldorf-Verlag Stahleisen

11.2.5 Method Cetim[1] for Prediction of Hardening Power of Quenching Oils

B. Liščić

Bases of the Method

The method developed at Cetim [1] is based on the supposition that if two specimens made of the same steel grade, have the same hardness, after being quenched, they have been submitted to identical cooling processes, following the same law of cooling. If e.g. after quenching (Fig. 11.2.5.1) the same hardness values are found in the point A of the cross-section of a round bar of a certain diameter D, as on the Jominy specimen in the distance J_1 from the quenched end (the same is valid for the points B and J_2 as well as C and J_3), we suppose that the law of cooling in correspondent points was the same.

It should be noticed, that this supposition is not entirely correct, but could be taken only as an approximation. This is so because, as it is known, the heat flux, when quenching a cylindrical specimen in oil, is not precisely the same as on the quenched end of a Jominy specimen.

Based on the above concept, the experimental procedure to be performed when a quenching oil has to be judged on its hardening power, is the following:

- Series of round bars of different diameters, made of the same grade of steel, are to be quenched (under same conditions) in the oil in question.
- Hardness values have to be measured, after hardening, in points $R/2$ (or in center) of the cross section in mid length of the bars (Fig. 11.2.5.2a).

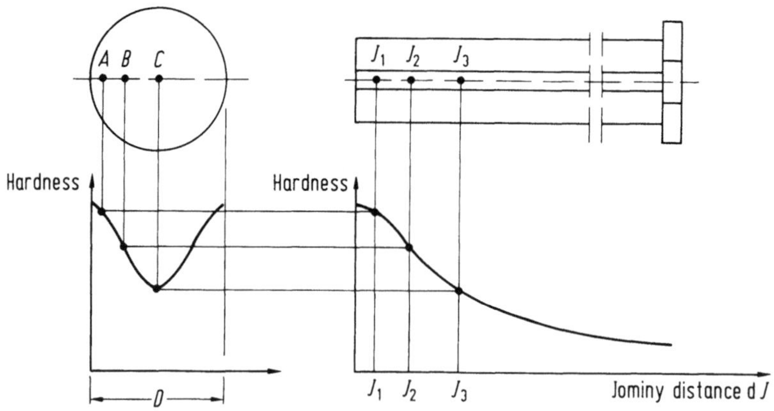

Fig. 11.2.5.1. Points A, B and C respectively represent the same hardness values, being cooled with the same cooling law as the points J_1, J_2 and J_3 respectively

[1] Cetim stands for: Centre Technique des Industries Mechaniques 52, avenue Félix-Louat, 60300 Senlis, France

—Taking the Jominy curve of the relevant steel grade,—to find out the equivalent Jominy distances on the abscissa (J_1, J_2, J_3) respectively which give the same hardness values as measured on the cross-section of each bar (D_1, D_2, D_3)—Fig. 11.2.5.2b.

—Finally this enables to establish a correlation between different bar diameters and equivalent Jominy distances (valid for point $R/2$ where the hardness was measured).

As it could be seen from Fig. 11.2.5.2c this correlation is a straight line following the equation:

$$D = KJ,$$

where the straight line gradient "K" characterizes the hardening power of the quenching oil tested.

Actually the result obtained by using the above described experimental procedure characterizes the hardening power in the quenching tank on the whole, taking into account also the temperature of the oil as well as its degree of agitation.

The experiments at Cetim were performed under following conditions:

Round bars used:	D 20, 30, 40 and 50 mm (Length = 3 × D)
Steel grade used:	38 C 4 (AFNOR) or AISI 5140
Austenitization:	850 °C in protective atmosphere of $N_2 + 5\%$ H_2

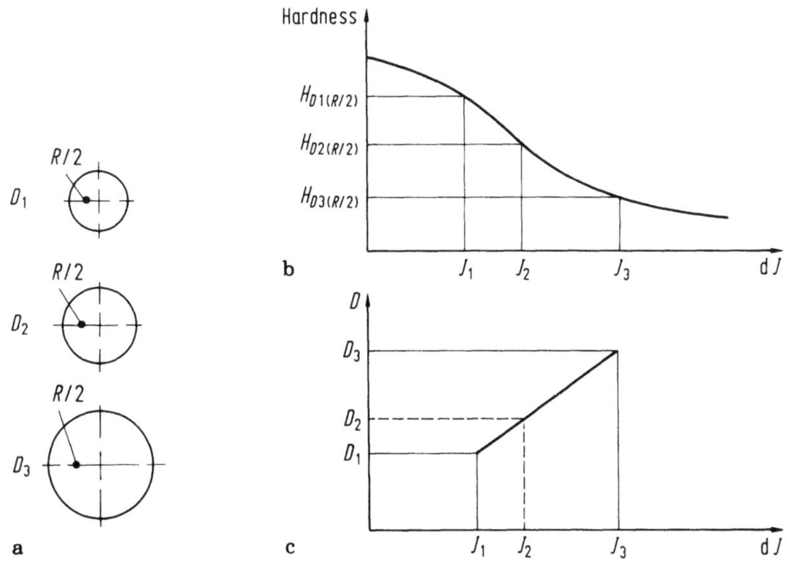

Fig. 11.2.5.2 a–c. Procedure for determination of the hardening power "K" of a quenching oil

Volume of the quenching tank:	60 l
Oil temperatures used:	50 °C and 150 °C
Agitation:	zero or violent.

The authors of this method themselves indicate that the method is applicable for bar diameters of less than 50 mm.

Estimation of the Hardening Power of Oils Based on Their Temperature vs. Cooling Rate Curves

First of all a clear distinction should be noted between the "cooling capacity" and the "hardening power" of a quenchant. The "cooling capacity" (French expression: "pouvoir refroidissant") of a quenching fluid represents its ability to extract heat from the quenched specimen more or less rapidly during cooling from the austenitizing temperature to the room temperature.

In cases of vaporable fluids it characterizes the heat exchange between the quenched specimen and the quenching liquid during the successive: vapour blanket, boiling and convection stages of cooling. It is usually measured by small cylindrical specimens—using standardized methods, and represented by temperature vs. cooling rate diagrams as could be seen on Fig. 11.2.5.3 where such diagrams have been plotted for 8 different kinds of oils.

Although such diagrams enable to characterize the cooling capacity of an oil, very precisely, it is delicately to use them in practice because we don't have an enough accurate method which would allow us to foresee metallurgical incidence from the cooling capacity of an oil.

The "hardening power" (French expression: "pouvoir trempant") of a quenching fluid represents its behaviour to support formation of martensite within a steel specimen during quenching. It is necessary to perform quenching tests with steel specimens to combine the relative influence of different cooling stages on the decomposition of the austenite. The resultant parameter from such cooling tests is the hardness of the quenched steel specimen which indicates precisely enough the degree of hardening in the point considered.

The cooling capacity as well as the hardening power of a quenching fluid depend not only on the specific properties of the quenchant, but also on the quenching conditions i.e. degree of agitation and quenchant temperature.

Authors of the method Cetim have made a valuable contribution to the study of the relation between the cooling capacity and the hardening power of quenching oils for defined degree of their agitation, by utilizing the information from the temperature vs. cooling rate curves.

As it could be seen from Fig. 11.2.5.3 the boiling stage of each temperature vs. cooling rate curve (during which the cooling rate maximum is achieved), could be defined by the temperature interval between T_1 and T_2 where:

T_1 = transition temperature between vapour blanket stage and boiling stage
T_2 = transition temperature between the boiling stage and convection stage

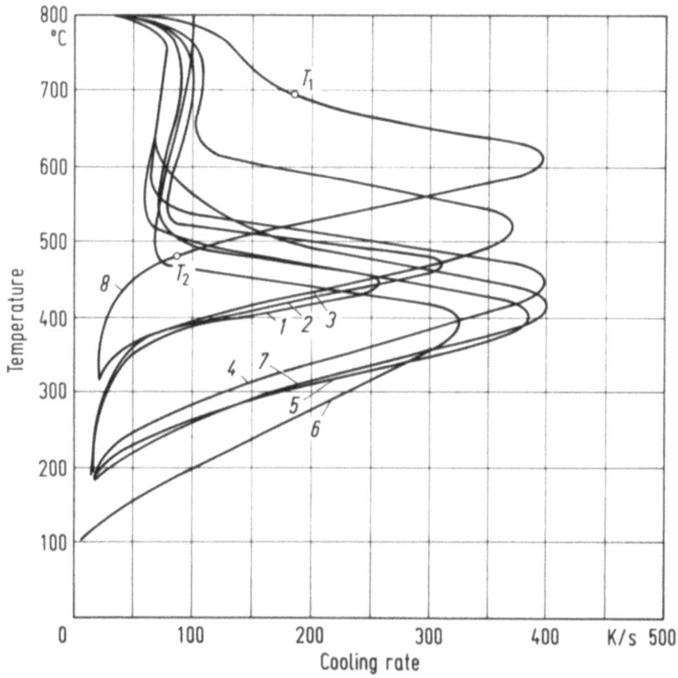

Fig. 11.2.5.3. Temperature vs. cooling rate curves for 8 different kinds of oils tested without agitation

T_1 and T_2 have been determined according to the method of the French association ATTT, described in [2].

Figure 11.2.5.4 gives the temperatures T_1 on the ordinate and T_2 on the abscissa for the same 8 kinds of oils which temperature vs. cooling rate curves have been plotted in Fig. 11.2.5.3. Numbers marked with • inside that diagram represent the values of "K_0" i.e. the relative values of hardening power (without agitation) obtained from metallurgical tests. The following can be concluded from Fig. 11.2.5.4:

— For identical values of T_1 (cases of oils No. 7 and 3 as well as No. 5 and 1) the oil having lower temperature T_2 has a higher value of the hardening power.
— For identical, but relatively low values of T_2 (cases of oils No. 7 and 5), the oil having higher temperature T_1 has a higher value of the hardening power; on the contrary at similar but high values of T_2 (cases of oils No. 3, 2 and 1), the value of T_1 has a weak influence on the hardening power which value passes from 1.86 to 2.16 when T_1 drops from 600 °C to 500 °C.
— Pairs of different values T_1 and T_2 (cases of oils No. 4 and 6 or 3 and 2) lead to very similar values of hardening power.

It seems to be possible to link together those points by curves called "curves of the equal hardening power" (see in Fig. 11.2.5.4).

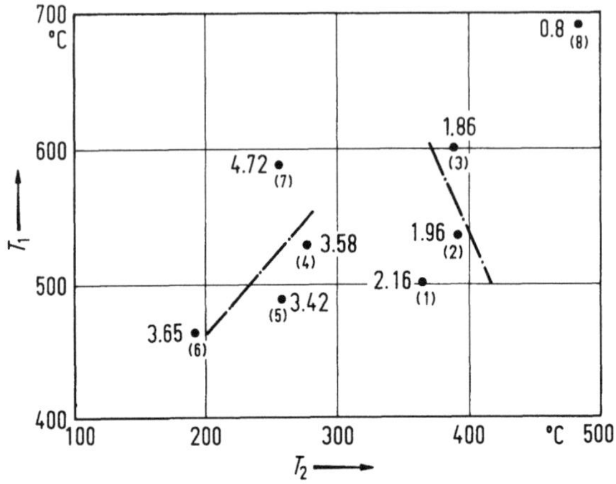

Fig. 11.2.5.4 Values of "K_0" as a function of T_1 and T_2 for 8 different kinds of oils tested without agitation • values of "K_0"; (8) numbers of different oils

As conclusion about the relation between the cooling capacity of an oil and its hardening power in first approximation the following can be deducted:

The hardening power of an oil (taking into account its cooling capacity based only on the oil characteristics i.e. excluding the influence of agitation) depends on the temperature zone in which boiling stage occurs. If the maximum of the cooling rate during boiling is at higher temperatures, the hardening power has a tendency to be low, and vice versa. (see the curves 8 and 6 on Fig. 11.2.5.3, and lines 8 and 6 on Fig. 11.2.5.8).

There have also been established two mathematical models which link together the:

K_0 = hardening power of an oil, without agitation, or
K_{95} = hardening power of an oil, with intensive agitation

to temperatures T_1 and T_2 of the same oil taken from its temperature vs. cooling rate curve. These equations read:

$$\log K_0 = A + B \log T_1 + C \log T_2 + D \log T_1 \cdot \log T_2$$
$$\log K_{95} = A' + B' \log T_1 + C' \log T_2 + D' \log T_1 \cdot T_2,$$

where $A, B, C, D, A', B', C', D'$, are constants.

Both of this equations are purely empirical. Using this formulae, the following objection has to be made:

The calculated values of "K_0" or "K_{95}" depend heavily on how accurately the temperatures T_1 and T_2 have been defined (this is more evident when the mentioned formulae are written in exponential form). Because the precise determination of T_1 and T_2 from the cooling rate vs. temperature curves is difficult, —there is possibility of making substantial errors when calculating "K_0" or "K_{95}".

Using these mathematical models it was possible to design a nomogram as shown in Fig. 11.2.5.5. This nomogram enables to determine the values of

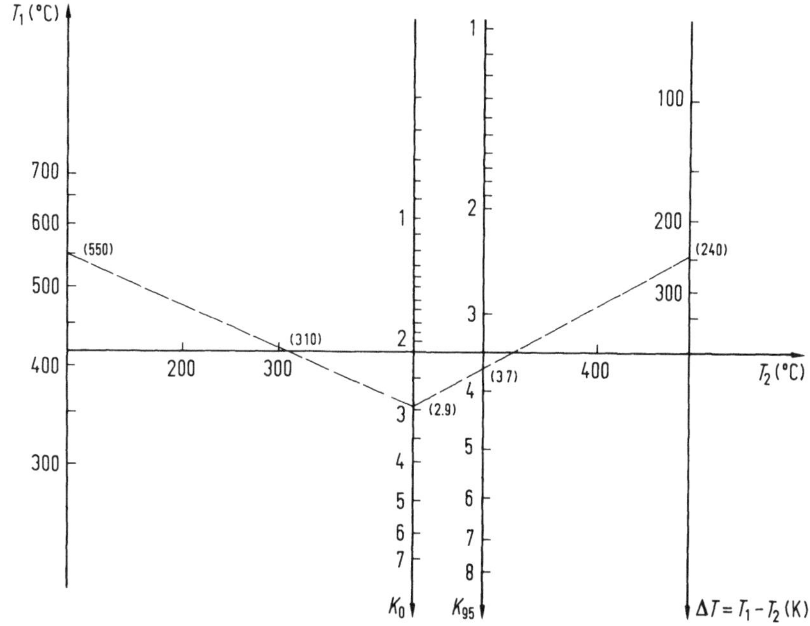

Fig. 11.2.5.5. Nomogram for determination of the values "K_0" and "K_{95}" of an oil by means of its temperatures T_1 and T_2

"K_0" and "K_{95}" of a quenching oil by means of its temperatures T_1 and T_2. As an example:

From the temperature vs. cooling rate curve of a quenching oil we read:

$$\left.\begin{array}{l} T_1 = 550\,^{\circ}\mathrm{C} \\ T_2 = 310\,^{\circ}\mathrm{C}, \end{array}\right\} \quad \text{accordingly} \quad \Delta T = 240\,^{\circ}\mathrm{C}$$

From these values, using the nomogram in Fig. 11.2.5.5 we can determine the values of hardening power K_0 and K_{95} for points at mid radius of the cylinder as:

$$K_0 = 2{,}9; \quad K_{95} = 3.7$$

In this way the values for K_0 and K_{95} have been determined for 122 samples of different quenching oils for which corresponding temperature vs. cooling rate curves have been plotted. The resulting curves of the equal hardening power are shown in Fig. 11.2.5.6.

Practical Application of the Method Cetim

In order to obtain the required hardness after quenching, in particular point of a steel bar cross section having a defined diameter D, we take first from the relevant Jominy-curve the equivalent Jominy distance which corresponds to

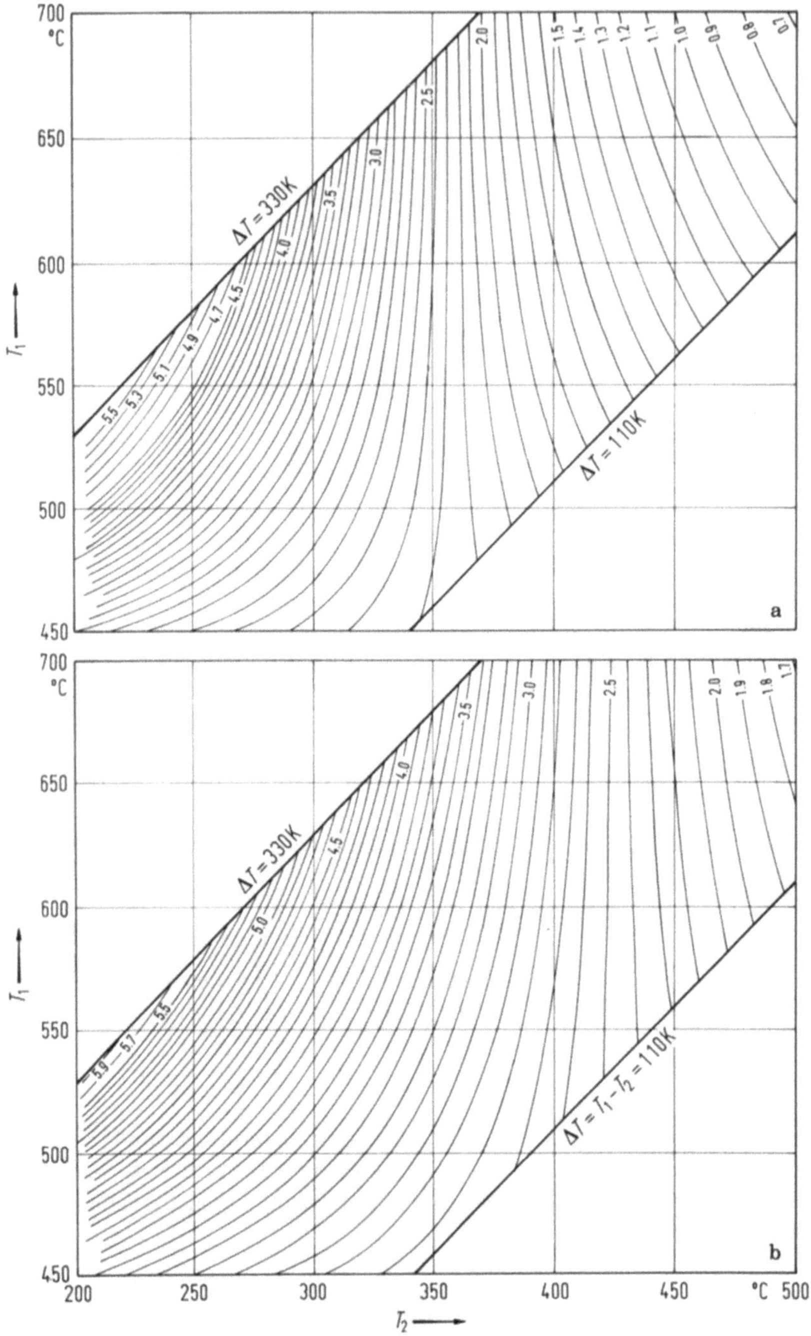

Fig. 11.2.5.6 a, b. Curves of the equal hardening power of different quenching oils **a** without agitation; **b** with intensive agitation

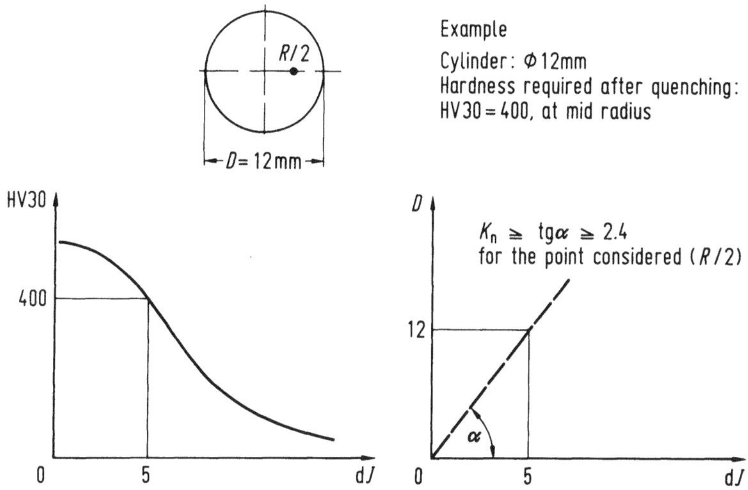

Fig. 11.2.5.7. Principle of determination of the necessary hardening power "K_n"

the required hardness—see Fig. 11.2.5.7. In this case a hardness of $HV_{30} = 400$ is required at mid radius of a cylinder of 12 mm diameter. The equivalent Jominy distance is 5 mm. By bringing together the diameter of the cylinder and the equivalent Jominy distance (see the diagram on the right side of Fig. 11.2.5.7) we obtain the "*necessary* hardening power" K_n by the following equation:

$$K_n \geqq tg\,\alpha \geqq \frac{12}{5} \geqq 2.4 \quad \text{(for the point } R/2\text{)}$$

Now the necessary hardening power K_n should be compared to the values K_0 or K_{95} (in case we take also into consideration an intensive agitation which exists in the quenching tank).

If the necessary hardening power K_n is low, almost all quenching oils will be satisfying, but if K_n is high, only oils having high values of the hardening power can be used. In case the necessary hardening power K_n is higher than K_0 or K_{95} values of all oils, these means that the required hardness cannot be attained by quenching in oil. The solution has to be found than in selecting either a steel grade having higher hardenability, or another more severe quenching medium (water).

Figure 11.2.5.8 shows two diagrams representing the relations $D = KJ$ for 8 different kinds of oils which temperature—vs. cooling rate curves have been plotted in Fig. 11.2.5.3. Hardness was measured at mid radius of cylinders 10–50 mm Dia made of steel grade 38 C 4 (AFNOR), or AISI 5140.

The left hand side diagram on Fig. 11.2.5.8 is for still oils (without agitation) and the right hand side diagram is for same oils with intensive agitation.

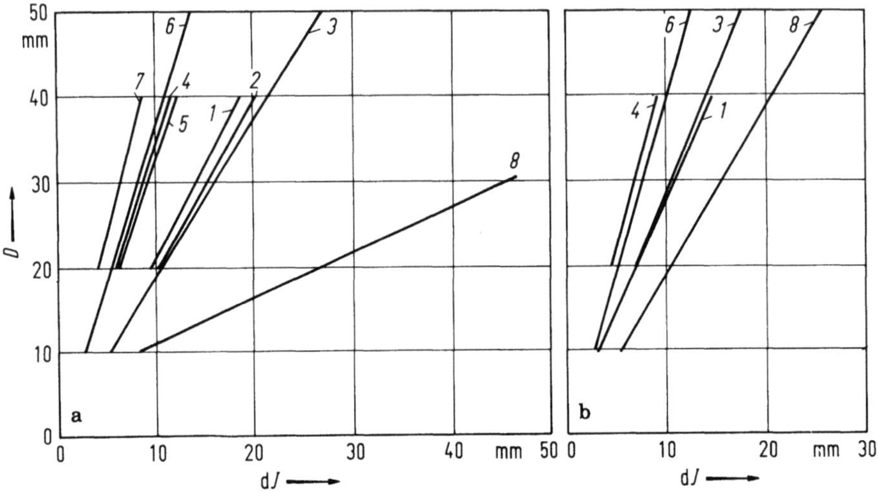

Fig. 11.2.5.8 a, b. Relation $D = KJ$ for R/2 of cylinders 10–50 mm Dia made of steel grade 38 C 4 (AFNOR) or AISI 5140 and quenched in 8 different kinds of oils (which temperature-cooling rate curves are plotted in Fig. **11.2.5.3**) **a** quenched in still oil (without agitation); **b** quenched with intensive agitation

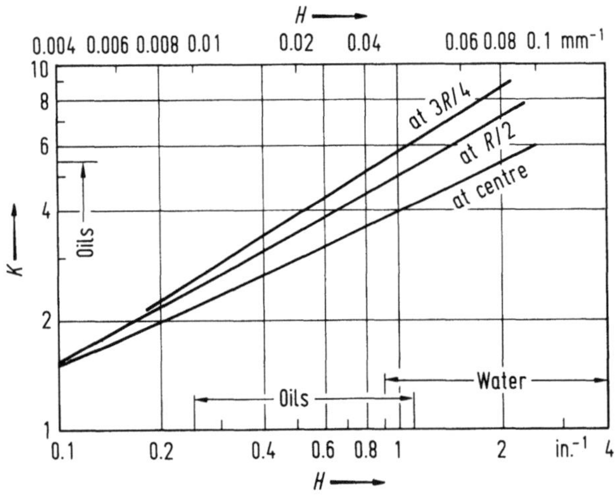

Fig. 11.2.5.9. Relation between the quenching severity "H" according to Grossmann and the hardening power "K" for 3/4 R; 1/2 R; and center of round bars

Figure 11.2.5.9 shows the relation between the quenching severity "H" according to Grossmann and the hardening power "K" according to method Cetim for 3/4 R; 1/2 R and center of round bars.

It should be noticed that this is a relation between an empirical value ($K = \frac{D}{J}$) and a value (H) which has a physical background!

References

1. Mansion J, Bournicon C, Sayettat C (1978) "Les huiles de trempe", CETIM
2. ..."Drasticite, des huiles de trempe. Essai au capteur d'argent", Association Technique de Traitement Thermique, Paris 1986

11.2.6 Calculation of Mechanical Properties According to Blondeu, Maynier, Dollet and Vieillard-Baron

T. Filetin

Introduction

Mechanical properties of steel are a basic function of the chemical composition and microstructure, that is, a consequence of the metallurgical process and heat treatment. The establishment of correlation between composition, microstructure and properties of steel has always presented a challenge to investigators. Today with the help of computer technique, these tasks have become easier to deal with and more attractive to solve.

Creusot-Loire system of the mentioned authors [1], [2], [3], is one of the practically checked methods [4] for prediction of hardness, tensile strength and yield strength via chemical composition as well as the type and share of structural phases as a consequence of different cooling regimes at a heat treatment process.

Calculation of Cooling Rates

According to the basic flow diagram (Fig. 11.2.6.1) for the system of prediction, it is essential to establish the *intensity of cooling* of a certain product in specific cooling conditions, as well as the *critical cooling rate* from the chemical composition in CCT diagram. By overlaying these cooling intensities, quenched structure can be obtained. Mechanical properties of a steel product can be derived from the hardness of basic phases—martensite, bainite and ferrite-pearlite, both before and after tempering. For large components there is an adequate relationship between the equivalent workpiece diameter (D) and *the cooling rate* (v) which is given by the following empirical formula [4]:

$$\log D = a - b \cdot \log v \qquad (11.2.6.1)$$

D = equivalent diameter, mm
v = cooling rate at 700 °C, K/h
a, b = regression parameters for a given quenchant

These coefficients are defined experimentally, separately for each medium and quenching equipment, dimensions (D) and cooling rates (v) being varied.

In order to obtain empirical equations connecting the chemical composition and the cooling rate, numerous data from CCT diagram have been analyzed

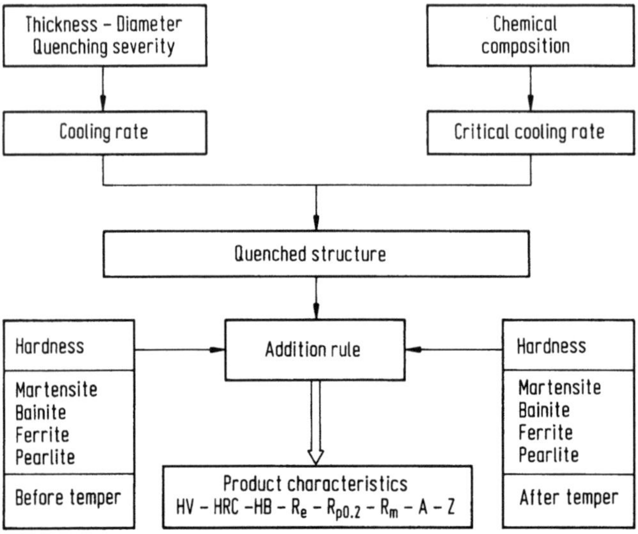

Fig. 11.2.6.1. Flow diagram of the prediction system [2]

%C	%Si	%Mn	%P	%S	%Cr	%Cu	%Mo	%Ni	%V
0.35	0.23	0.65	0.026	0.013	1.1	0.18	0.05	0.23	< 0.01

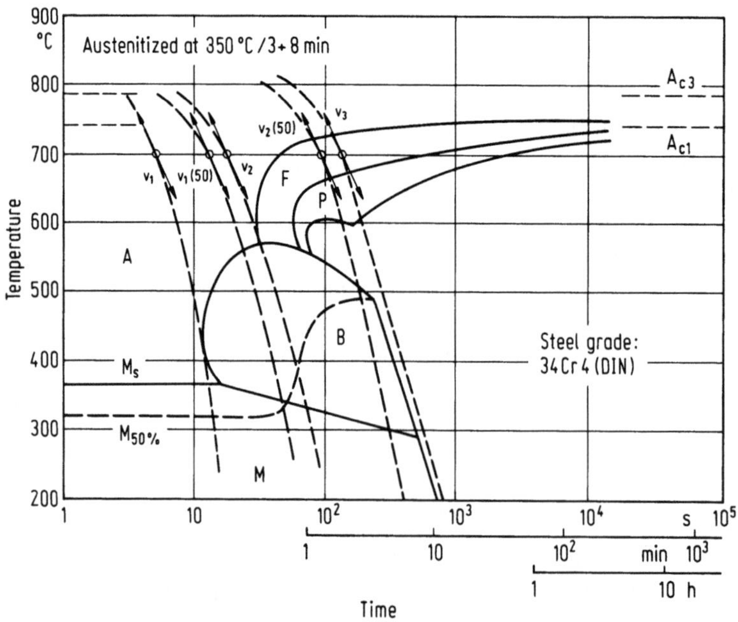

Fig. 11.2.6.2. CCT diagram for steel 34Cr4 (DIN)

statistically. If we consider, for instance, a CCT diagram of a certain type of steel (Fig. 11.2.6.2), we shall notice three kinds of basic structures: martensite (M), bainite (B) and ferrite-pearlite (F + P). These structures can be achieved through different cooling rates: v_1, v_2 and v_3. Other cooling rates worth mentioning are:

$v_1(50)$ for obtaining 50% M and 50% B
$v_1(90)$ for obtaining 90% M and 10% B
$v_1(10)$ for obtaining 10% M and 90% B
$v_2(50)$ for obtaining 50% B and 50% F + P
$v_2(90)$ for obtaining 90% B and 10% F + P
$v_3(90)$ for obtaining 10% B and 90% F + P

Regression equations for the basic critical—relevant cooling rates were obtained by the statistical analysis:

$$\left.\begin{array}{rll}
100\%\,\mathrm{M}: & \log v_1 & = 9.81 - 4.62\mathrm{C} - 1.05\mathrm{Mn} - 0.54\mathrm{Ni} - 0.5\mathrm{Cr} - 0.66\mathrm{Mo} - 0.00183\mathrm{P_A} \\
90\%\,\mathrm{M} + 10\%\,\mathrm{B}: & \log v_1(90) & = 8.76 - 4.04\mathrm{C} - 0.96\mathrm{Mn} - 0.49\mathrm{Ni} - 0.58\mathrm{Cr} - 0.97\mathrm{Mo} - 0.0010\mathrm{P_A} \\
50\%\,\mathrm{M} + 50\%\,\mathrm{B}: & \log v_1(50) & = 8.50 - 4.13\mathrm{C} - 0.86\mathrm{Mn} - 0.57\mathrm{Ni} - 0.41\mathrm{Cr} - 0.94\mathrm{Mo} - 0.0012\mathrm{P_A} \\
100\%\,\mathrm{B}: & \log v_2 & = 10.17 - 3.8\mathrm{C} - 1.07\mathrm{Mn} - 0.7\mathrm{Ni} - 0.57\mathrm{Cr} - 1.58\mathrm{Mo} - 0.0032\mathrm{P_A} \\
90\%\,\mathrm{B} + 10\%\,\mathrm{P+F}: & \log v_2(90) & = 10.55 - 3.65\mathrm{C} - 1.08\mathrm{Mn} - 0.77\mathrm{Ni} - 0.61\mathrm{Cr} - 1.49\mathrm{Mo} - 0.004\mathrm{P_A} \\
50\%\,\mathrm{B} + 50\%\,\mathrm{P+F}: & \log v_2(50) & = 8.74 - 2.23\mathrm{C} - 0.86\mathrm{Mn} - 0.56\mathrm{Ni} - 0.59\mathrm{Cr} - 1.6\mathrm{Mo} - 0.0032\mathrm{P_A} \\
10\%\,\mathrm{B} + 90\%\,\mathrm{P+F}: & \log v_3(90) & = 7.51 - 1.38\mathrm{C} - 0.35\mathrm{Mn} - 0.93\mathrm{Ni} - 0.11\mathrm{Cr} - 2.31\mathrm{Mo} - 0.0033\mathrm{P_A} \\
100\%\,\mathrm{P+F}: & \log v_3 & = 6.36 - 0.43\mathrm{C} - 0.49\mathrm{Mn} - 0.78\mathrm{Ni} - 0.27\mathrm{Cr} - 0.38\mathrm{Mo} - 2\sqrt{\mathrm{Mo}} - 0.0019\mathrm{P_A}
\end{array}\right\} \quad (11.2.6.2)$$

where P_A—Austenitizing parameter at the temperature T_A in K is:

$$P_A = \left(\frac{1}{T_A} - \frac{4.6}{\Delta H}\log\frac{t_A}{t_0}\right)^{-1}, K \qquad (11.2.6.3)$$

t_A/t_0 = holding time comparison ratio;
t_A = austenitizing time, h,
t_0 = unit of time = 1 h
ΔH = activation energy of the process, J/mol
ΔH = 577778 J/mol for steels with Al and
ΔH = 460548 J/mol for steels without Al.

This is the parameter which defines the equivalence of time and temperature of those processes where diffusion is present.

By calculation of these cooling rates and their interpolation into CCT diagrams the relative portion of structural phases in cooling of the real dimension workpieces can be established.

Mechanical properties

From the obtained empirical equations for cooling rate it has been deduced that the hardness of basic phases, namely martensite, bainite and ferrite-perlite is highly dependent on chemical composition and a corresponding cooling rate.

On the basis of this precept and with the help of the statistical analysis of data obtained through the research of a great number of industrial steels, formulae for the calculation of hardness of basic phases have been derived:

a) Hardness before tempering—quenched or slowly cooled condition:

$$HV_M = 127 + 942C + 27Si + 11Mn + 8Ni + 16Cr + 21 \log v_r \quad (11.2.6.4)$$

$$HV_B = -323 + 185C + 330Si + 153Mn + 65Ni + 144Cr + 191Mo + \log v_r(89 + 53C - 55Si - 22Mn - 10Ni - 20Cr - 33Mo) \quad (11.2.6.5)$$

$$HV_{P+F} = 42 + 223C + 53Si + 30Mn + 12.6Ni + 7Cr + 19Mo + \log v_r \cdot (10 - 19Si - 4Ni + 8Cr + 130V) \quad (11.2.6.6)$$

v_r is the actual cooling rate

Hardness of mixed structures can be calculated according to the following summation rule:

$$HV_{mix} = \frac{\%\alpha}{100} HV_\alpha + \frac{\%\beta}{100} HV_\beta \quad (12.2.6.7)$$

where

$\%\alpha$ = percent of the first phase (α)
$\%\beta$ = percent of the second phase (β)
HV_α = hardness of the first phase
HV_β = hardness of the second phase.

b) Hardness after tempering. Tempered martensite:

$$HV = -74 - 434C - 368Si + 15Mn + 37Ni + 17Cr - 335Mo - 2235V + \frac{10^3}{P}(260 + 616C + 321Si - 21Mn - 35Ni - 11Cr + 352Mo + 2345V) \quad (11.2.6.8)$$

Bainite:

$$HV = 262 + 163C - 349Si - 64Mn - 6Ni - 186Cr - 458Mo - 857V + \frac{10^3}{P}(-149 + 43C + 336Si + 79Mn + 16Ni + 196Cr + 498Mo + 1094V) \quad (11.2.6.9)$$

where $\frac{10^3}{P} = 1.365 - (0.205C + 0.233Mo + 0.135V)$

These equations are valid for the range of compositions:

$0.1\% < C < 0.5\%$ (weight percent)
$Si < 1\%; Mn < 2\%$
$Ni < 4\%; Cr < 3\%; Mo < 1\%$
$V < 0.2\%; Cu < 0.5\%$
$Mn + Ni + Cr + Mo < 5\%$
$0.01\% < Al < 0.05\%$

The range of heat treatment parameters are:

Austenizing: $800\,°C < T_A < 1100\,°C$, 1 hr
Tempering: 1 hr, $500\,°C < T_T < 700\,°C$, 200 hr

Applications

These formulae can be inserted into a wide range of calculations of mechanical properties (hardness, yield strength, tensile strength), for instance:

—hardness in heat affected zones of welded joints obtained on the RPI Gleeble machine for simulation of short termal cycles,
—corresponding hardness in CCT diagram,
—prediction of hardness at Jominy curves,
—properties of as-rolled steel product-plates, bars and forging components for nuclear applications.

References

1. Maynier Ph, Dollet P, Bastien P (1970) "Influence des éléments d' alliage sur la trempabilite des aciers alliés", Rev. Met., 4, p. 343–352
2. Maynier Ph, Jungmann B, Dollet J (1978) "Creusot–Loire System for the Prediction of the Mechanical Properties of Low Alloy Steel Products", Hardenability Concepts with Applications to Steel, Met. Society of AIME p. 518–545
3. Maynier Ph, Dollet J, Bastien P (1978) "Prediction of Microstructure via empirical Formulae based on CCT Diagrams" Hardenability Concepts with Applications to Steel", Met. Society of AIME, p. 163–178
4. Shakeshaft M (1988) "A Mathematical Framework for the economic Selection of Carburizing Steel", World Materials Congress, Chicago p. 131–135

11.2.7 Own Data Bank for Quenching Intensities, Jominy Hardenability and Hardness Distribution on Test Specimens

T. FILETIN

Introduction

Every heat treatment shop deals with a range of quenching facilities and quenching conditions. In the workshop practice we can realize many different quenching conditions by changing any of the quenching parameters. These are: the kind of quenchant, its concentration (e.g. of a polymer solution), agitation rate and direction, pressure of media with spray quenching, bath temperature etc. In fact, there are many combinations of quenching conditions having different quenching intensities. To find out which combination gives optimum properties in a certain case and to reproduce exactly the hardening results, the following are required: measuring, recording, evaluation, storing and comparing of the real quenching intensities.

In addition, several different steel grades and heats are used in every production. In order to find out their hardenability, the representative Jominy curves should be stored for each steel grade. Other steel characteristics should also be collected and stored in the own files.

In order to predict the hardness distribution across round bars of different diameters (see next paragraph) after quenching under specified conditions, hardness along the cross section of a hardened test specimen has to be measured and values for characteristic points of the cross-section stored into the files.

The files of measured data for quenching intensities, steel characteristics and hardness values on test specimens constitute the basis of an own data bank. Such data files can be additionally extended by storing data about new quenching conditions and/or steel grads and/or the hardness values measured on test specimens.

An adequate software has been developed in order to establish and gradually enlarge the individual data bank. The software supports: recording, evaluation, storing and retrieval process, comparison and numeric or graphic presentation of measured, stored and predicted data (Fig. 11.2.7.1). The system has been developed and can be implemented, partially or completely, on standard personal computer (PC) having necessary additional hardware (Liščić/Nanmac probe and *A/D* convertors).

Data Stored in Files (See Fig. 11.2.7.1)

File (F1)—Steels

This file contains data about steel grades available for the application in the own heat treatment shop. These are:

—Steel grade, heat number
—Chemical composition—%C and other elements in wt.%

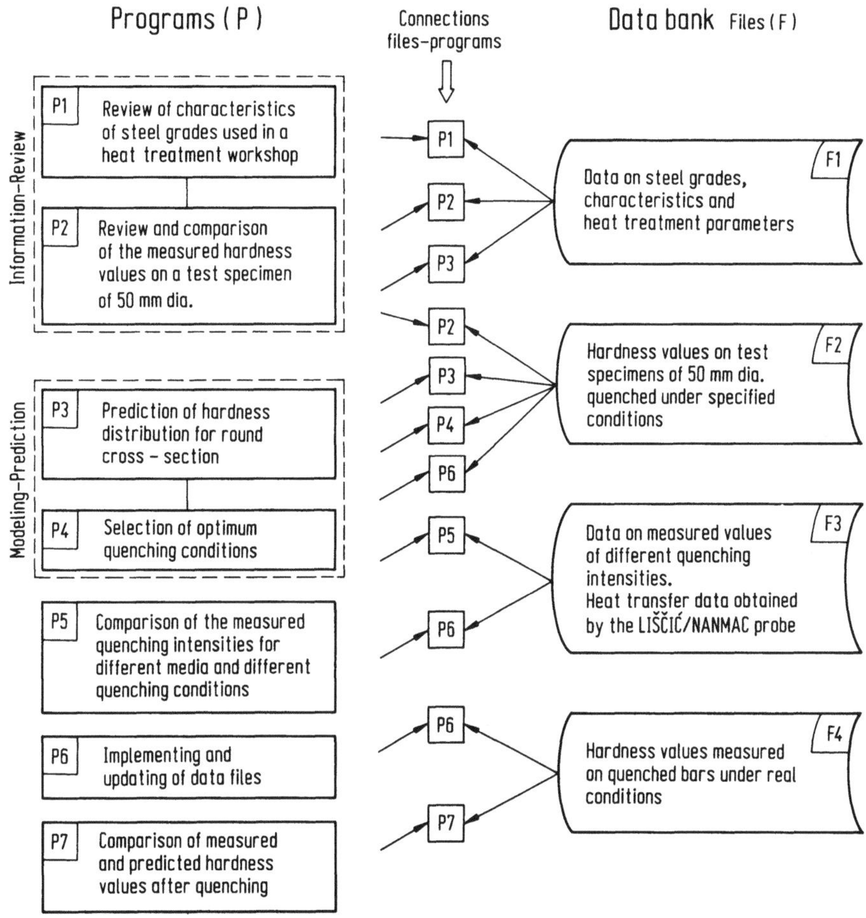

Fig. 11.2.7.1. The system of programs for the use of the stored data on quenching conditions; modeling of hardness distribution and selection of the desired quenching parameters

—Austenitizing temperature T_A, °C
—A_{c3} and A_{c1} temperature, °C
—Martensite start temperature M_s, °C
—Hardness values from the Jominy curve—measured or calculated from the chemical composition (equations from Just [5])
—Other data on steel relevant for the heat treatment process.

File (F2)—Specimens

The main data are hardness values measured in five characteristic points on the cross section of the test specimen made of the corresponding steel grade and quenched under specified conditions. The specimen should be of the same dimensions (50 mm Dia × 200 mm) and of the same surface conditions as the probe (Liščić/Nanmac). In hardening of the test specimen, special care should be taken to provide for the equal heating and austenitizing process as well as the quenching conditions for both the probe and the test specimen. After quenching, the specimen should be carefully cut at mid length, the cross-section ground, and the hardness measured along the three radii at 120°. The average values for the five characteristic points on the cross section:—surface, 3/4*R*, 1/2*R*, 1/4*R* and the center, are then stored in the file. Besides, the following data are stored:

—steel grade, heat number
—kind of quenching medium
—agitation rate, m/s
—pressure, bar
—concentration (e.g. of a polymer solution) of the quenching medium
—quenching bath temperature, °C
—*H*-value according to Grossmann
—date of measuring and other identification data

File (F3)—Quenching Intensities:

Quenching intensity measurements should be performed by the above mentioned probe see paragraph 8.4 and in every case the quenching conditions should be exactly specified. For every quenching process with exactly specified quenching parameters, the following functions have to be recorded:

a) Temperature (T) versus time (t):

$$T_s = f(T), T_{1.5} = f(t), T_c = f(t)$$

s = Surface, 1,5 = Distance of 1.5 mm below the surface
c = Center

The following should be calculated and stored from these values:

b) Heat flux density (q):

$$q = f(t), q = f(T_s), q = f(T_c)$$

c) Temperature difference between surface and center versus time

$$\Delta T = T_c - T_s = f(t)$$

The quenching intensity (I) is defined and expressed by the four characteristic values as it is described in paragraph 8.5.

In addition, the following two values can be evaluated:

a) Maximum temperature difference between the center and the surface: $\Delta T_{max} = (T_c - T_s)_{max}$, °C, with stating of the cooling time at which maximum temperature difference occurs. This time results from the course of basic recorded functions: $T_s = f(t)$ and $T_c = f(t)$

b) The surface temperature at which the maximum heat flux density occurs:

$T_s q_{max}$, °C.

Different quenching media are compared from the point of view of heat extraction intensities via these two indicators. The higher ΔT_{max}, the higher heat extraction intensities are to be expected in a certain medium but also higher thermal stresses will occur. The other indicator, $T_s q_{max}$, together with the M_s temperature for the respective steel grade, suggests whether a superimposing of thermal and structural internal stresses is going to occur. Namely, the lower the $T_s q_{max}$, and especially if it drops below M_s, the greater the risk of superimposing of high thermal and structural stresses.

These evaluated data are also stored in the files for each measurement.

The established data bank for steel grades of interest and quenching conditions available, completed with hardness values of test specimens, enables a quick computer search for finding the optimum quenching conditions when a certain hardness value in a specified point of the bar cross section is required (see Chap. 11.2.8).

Software possibilities

The package of developed programs offers a flexible possibility of an interactive choice of the working mode, retrieval of specific file, as well as the graphic and the tabular presentation of data, functions and the modeled results.

Figure 11.2.7.1 shows two basic program possibilities which can be devided into two groups with respect to the typical requirements of the use of data from data files:

a) Information, review and comparison:
 —review of characteristics and properties of steels commonly used in the production incorporating heat treatment (program P1 according to Fig. 11.2.7.1).
 —review and comparison of measured hardness values on test specimens made from different steel grades and quenched in different media under different quenching conditions. There are two possibilities of comparison of the hardness values stored:- achievable hardness values of different steel grades under the equal quenching conditions, and
 —achievable hardness values of the same type of steel grade under different quenching conditions (program P2 according to Fig. 11.2.7.1).
 —Additionally, for each quenching/cooling conditions which have been measured and recorded by the Liščić/Nanmac probe and then stored into the computer memory,—one can compare (in grafic or tabular form) all the functions representing the quenching/cooling intensity, by using the Program P5.

b) Prediction of hardness following quenching for the round cross-section of different dimensions, and of different steel grades quenched under different cooling conditions (Program P3 according to Fig. 11.2.7.1).

Program no. 4. is used for the selection of the optimum cooling conditions which would provide for the desired properties in a certain cross section spot of a defined steel grade.

Should one wish to compare the hardness distribution obtained by modeling with the hardness distribution measured after quenching, files F4 and Program P7 (Fig. 11.2.7.1) should be used. This enables to compare the real hardness distribution on the cross section of quenched round bars with predicted values, giving the possibility to correct the model for prediction and expanding the knowledge in the own data-base system.

Table 11.2.7.1. Comparison of hardness distribution on test specimens made of three different steel-grades quenched under the same conditions

```
        INPUT DATA:
        ***********

    QUENCHING CONDITION :
         -QUENCHANT   : OIL UTO-2
         -TEMPERATURE: 20 °C
         -AGITATION   : 1 M/S
         -H-VALUE     : 0.41

    STEEL GRADES FOR COMPARISON:
    ============================

  I Č.4830 (SAE 6150H);B.NO.88199
 II Č.4732 (SAE 4140H);B.NO.73456
III Č.4731 (SAE 4135H);B.NO.45107

HARDNESS DISTRIBUTION MEASURED
ON THE TEST SPECIMEN 50 MM DIA
*******************************
           STEEL GRADE:
      I          II          III

SURFACE=59.0    56.0       51.0, HRC
3/4 R  =58.6    52.0       48.9
1/2 R  =58.7    51.0       45.5
1/4 R  =57.8    50.0       42.0
CENTER =56.9    50.0       38.0
```

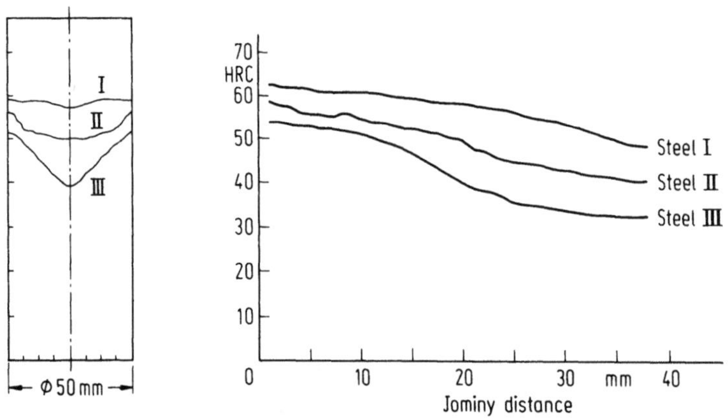

Tables 11.2.7.1 and 11.2.7.2 give two examples of several possibilities for comparison of stored data by means of the Program P2. The first example (Table 11.2.7.1) shows a comparison of measured hardness distribution on a 50 mm Dia test specimen for three different steel grades quenched in oil of 20 °C and 1 m/s agitation rate (H = 0.41), together with corresponding Jominy curves.

For the same dimensions of round bars quenched under equal cooling conditions a dominant influence of steel hardenability on the hardness distribution, as it is well known, can be seen. The second example (Table 11.2.7.2) shows a comparison of measured hardness values on a 50 mm Dia test specimen quenched under different quenching conditions i.e. in a mineral oil of 20 °C and

Table 11.2.7.2. Comparison of hardness distribution on a test specimen made of the same steel-grade quenched in two different media

```
          INPUT DATA:
          ***********

STEEL GRADE :Č.4732 (SAE 4140H)B.NO.73456

          QUENCHING CONDITION FOR COMPARISON:
          ===================================

                       I            II

QUENCHANT:       SALT BATH    OIL HALENOL

TEMPERATURE:       200 °C         20 °C
AGITATION:         0.6 M/S        1 M/S
H-VALUE     :      0.40           0.30

       HARDNESS DISTRIBUTION MEASURED
       ON THE TEST SPECIMEN 50 MM DIA
       ******************************
             QUENCHING CONDITION
                I         II
       SURFACE=55.0      56.0, HRC
       3/4 R   =49.5     52.5
       1/2 R   =49.0     52.0
       1/4 R   =48.0     51.0
       CENTER  =47.0     51.0
```

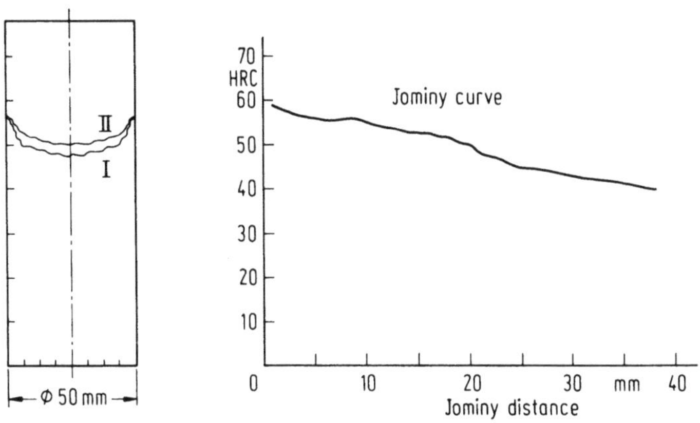

1 m/s agitation rate and a salt bath of 200 °C with 0.6 m/s agitation rate respectively.

Although the Grossmann "H" value for this salt bath was higher (H = 0.4) then for this mineral oil (H = 0.3), higher hardness values have been obtained after quenching round bars of 50 mm Dia in oil then after quenching them in salt bath. This again is a proof that Grossman's "H" value is not in all cases proportional to the depth of hardening achieved.

The concept of the own data bank will be of greater value for a practical application in those cases where the data for a great number of different steel grades and quenching conditions are stored.

11.2.8 Computer-Aided Prediction of Hardness Profile Upon Quenching Using the Own Data Bank

T. Filetin

Prediction of Hardness Distribution on Round Bars' Cross-Sections

The previous chapter has set forth the concept and the content of the data bank and corresponding programs which together make the basis for the further development and application of the method of prediction of hardness distribution after hardening, as well as the selection of the optimal quenching/cooling parameters [1, 2, 3].

The method of computer aided prediction of hardness distribution after quenching is valid for bar diameters from 20 to 90 mm.

Figure 11.2.8.1 gives the flow diagram of the program, required input data and necessary calculations that must be performed step by step as follows (see also the scheme in Fig. 11.2.8.2) [1].

1. Specify the steel grade and quenching conditions.
2. Harden a test specimen of the same steel grade by quenching it under specified conditions.
3. Measure the hardness (HRC) across the test specimen cross section as described in Chap. 11.2.7.
4. Input into the computer the hardness values for five characteristic points on the test specimen's cross section—S = surface, 3/4R, 1/2R, 1/4R, C = center.

   If files already contain hardness values which are results of previous measurements using the same cooling conditions, then the second and the third step are passed over. In that case the hardness values corresponding to the defined combination: steel grade—cooling conditions, are simply retrieved from the file. If the testing that has been carried out included a new steel grade and/or different cooling parameters, the data obtained are

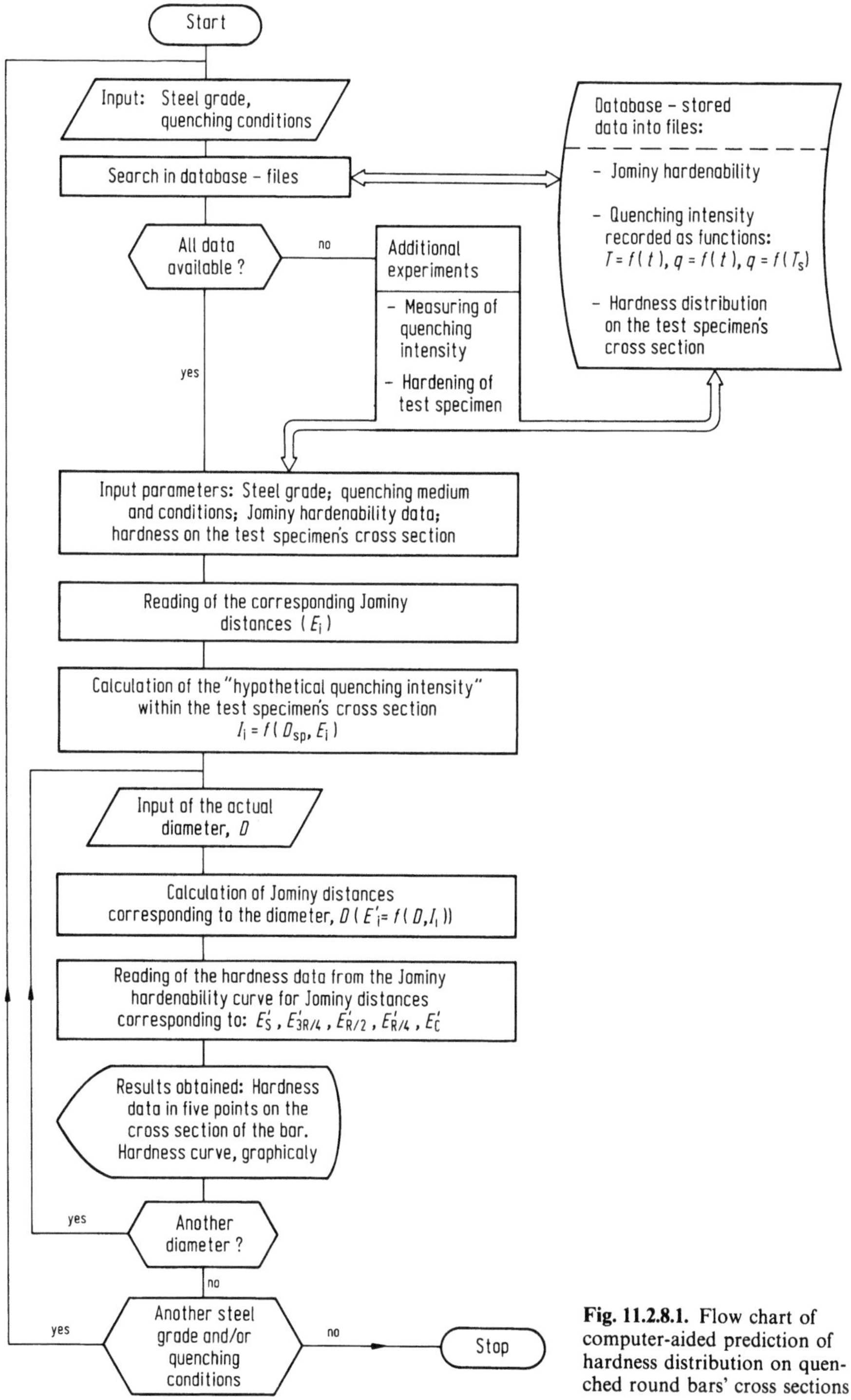

Fig. 11.2.8.1. Flow chart of computer-aided prediction of hardness distribution on quenched round bars' cross sections

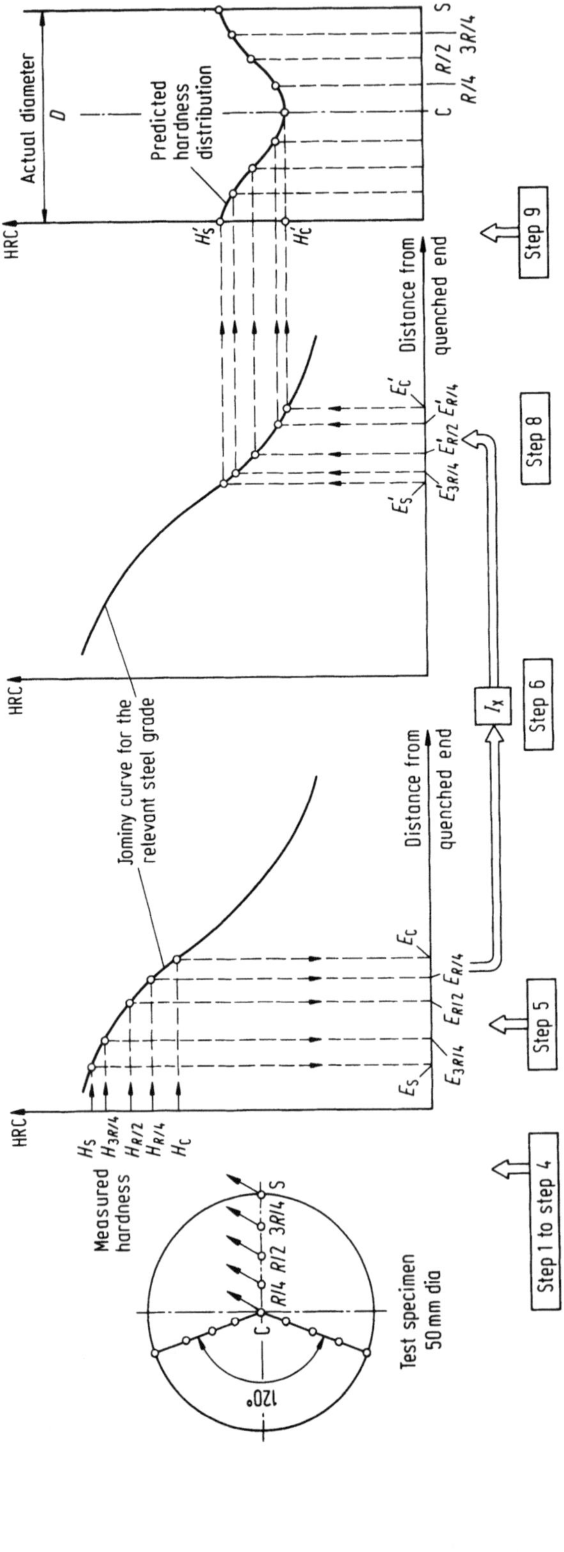

Fig. 11.2.8.2. Stepwise scheme of the process of prediction of hardness distribution

stored in the file F2 (see Chap. 11.2.7). In such a way the own funds containing experimental data and knowledge for future modelling, are increased.

5. By means of stored Jominy hardenability data for the relevant steel grade, read the *equidistant points on the Jominy curve* (E_s; $E_{3/4R}$; $E_{1/2R}$; $E_{1/4R}$; E_C), which have the same hardness as measured at the relevant points on the test specimen's cross section.
6. Calculate the hypothetical quenching intensity I_x at each of the above—mentioned characteristic points on the test specimen's cross section by means of regression Eqs. (11.2.8.2) through (11.2.8.6) based on known values for E_s; $E_{3/4R}$; $E_{1/2R}$; $E_{1/4R}$; E_c, and the test specimen's diameter D_{sp}. Equations (11.2.8.2) through (11.2.8.6), which combine the equidistant points on the Jominy curve, the test specimen diameter D_{sp} and the hypothetical quenching intensity I_x, have been derived from the regression analysis of a series of Crafts-Lamont diagrams [4.7].

   This analysis is based on the relations from Just [5] for the surface and the center of a cylinder that generally can be written as:

$$E_i = A \frac{D^{B_1}}{I^{B_2}} \tag{11.2.8.1}$$

where

E_i = corresponding equidistant point on the Jominy curve
A, B_1, B_2 = regression coefficients
D = cylinder (bar) diameter
I = quenching intensity ("*H*" according to Grossmann)

By means of regression analysis, the following relations have been obtained:

$$E_s = \frac{D^{0.718}}{5.11 \times I^{1.28}} \tag{11.2.8.2}$$

$$E_{3/4R} = \frac{D^{1.05}}{8.62 \times I^{0.668}} \tag{11.2.8.3}$$

$$E_{1/2R} = \frac{D^{1.16}}{9.45 \times I^{0.51}} \tag{11.2.8.4}$$

$$E_{1/4R} = \frac{D^{1.14}}{7.7 \times I^{0.44}} \tag{11.2.8.5}$$

$$E_c = \frac{D^{1.18}}{8.29 \times I^{0.44}} \tag{11.2.8.6}$$

These equations are valid for $20 < D < 90$ mm, $1 < E < 40$ mm, and $0.2 < I < 2.0$. It is presupposed that at every point of the cylinder's cross section there exist different cooling conditions that have been taken into account through the hypothetical quenching intensity I_x. The value of the hypothetical quenching intensity can be calculated for every specific point from Eq. (11.2.8.2) through (6). The following equations have been deduced:

$$I_s = \left(\frac{D^{0.718}}{5.11 \times E_s}\right)^{0.78} \tag{11.2.8.7}$$

$$I_{3/4R} = \left(\frac{D^{1.05}}{8.62 \times E_{3/4R}}\right)^{1.495} \tag{11.2.8.8}$$

$$I_{1/2R} = \left(\frac{D^{1.16}}{9.45 \times E_{1/2R}}\right)^{1.495} \tag{11.2.8.9}$$

$$I_{1/4R} = \left(\frac{D^{1.14}}{7.7 \times E_{1/4R}}\right)^{2.27} \tag{11.2.8.10}$$

$$I_c = \left(\frac{D^{1.18}}{8.29 \times E_c}\right)^{2.27} \tag{11.2.8.11}$$

7. Enter the actual bar diameter D for which the predicted hardness distribution is desired.
8. Calculate the *equidistant Jominy distances* (E'_s; $E'_{3/4R}$; $E'_{1/2R}$; $E'_{1/4R}$; E'_c) according to Eqs. (11.2.8.2) to (11.2.8.6) which correspond to the actual bar diameter D and to the calculated quenching intensities: I_s; $I_{3/4R}$; $I_{1/2R}$; $I_{1/4R}$; I_c.
9. Read the hardness values H'_s; $H'_{3/4R}$; $H'_{1/2R}$; $H'_{1/4R}$; H'_c from the relevant Jominy curve that belong to the calculated Jominy distances: E'_s; $E'_{3/4R}$; $E'_{1/2R}$; $E'_{1/4R}$; E'_c, and plot the hardness distribution curve over the cross section at the chosen actual diameter D.

Figure 11.2.8.3 gives an example of computer aided prediction of hardness distribution for 30 and 70 mm dia bars made of Č.4732—(SAE 4140 H) quenched in oil at 20 °C without agitation

Figure 11.2.8.4 shows an example of computerized prediction of hardness distribution compared with measured values obtained upon quenching a round bar 35mm dia made of SAE 1043 (with low hardenability) in a salt bath of 200 °C without agitation.

Comparison with Experimentally Obtained Results

Figure 11.2.8.5 gives comparison of the hardness distribution across round bars measured after experimental quenching and obtained by computer-aided prediction. Three different steel grades, four different bar diameters and four different quenching conditions have been used in this comparison.

In order to ascertain the precision of the hardness distribution prediction, the prediction was, in some cases, additionally performed by the Gerber–Wyss method [6]. In this case, the quenching intensity "H" was established experimentally, according to Grossmann; for the oil used at 20 °C with an agitation rate of 1 m/s, $H = 0.417$; for water at 20 °C the agitation rate was 1 m/s; $H = 1.38$.

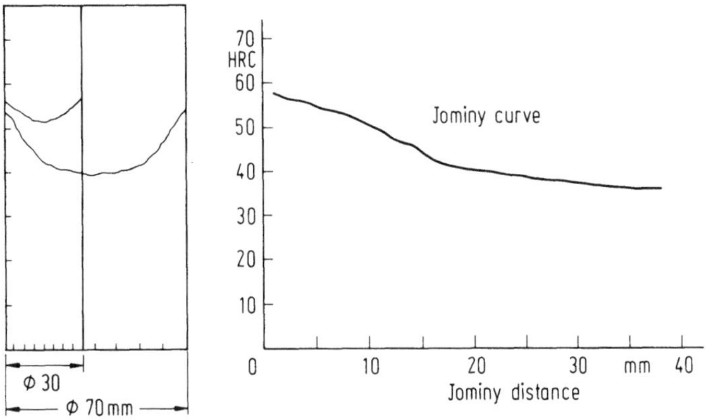


```
PREDICTION OF HARDNESS DISTRIBUTION
-----------------------------------

      INPUT DATA:
      -----------

  STEEL GRADE:Č.4732 (SAE 4140H);B.NO.43111
  QUENCHING CONDITIONS:OIL-UTO-2;20 °C;OM/S

    DIAMETER FOR HARDENING,MM:30
    ----------------------------
  RESULTS OF COMPUTER AIDED PREDICTION:
  *************************************

    CALCULATED HARDNESS:
    --------------------
         DIAMETER=30MM
         -------------

         SURFACE,HRC.......=55.3
         3/4 RADIUS........=54.3
         1/2 RADIUS........=53
         1/4 RADIUS........=51.5
         CENTER............=51.1

  GRAPHIC PRESENTATION(YES=1,NO=0)
         ANOTHER DIAMETER (YES=1,NO=0)

    DIAMETER FOR HARDENING,MM:70

         DIAMETER=70
         -----------
         SURFACE,HRC.......=53.1
         3/4 RADIUS........=46.4
         1/2 RADIUS........=40.7
         1/4 RADIUS........=39.6
         CENTER............=39

  GRAPHIC PRESENTATION(YES=1,NO=0)
         ANOTHER DIAMETER (YES=1,NO=0)
```

Fig. 11.2.8.3. An example of computer-aided prediction of hardness distribution for quenched round bars of 30 and 70 mm diam; steel grade SAE 4140H

```
     QUENCHED ROUND BAR SECTIONS

PREDICTION OF HARDNESS DISTRIBUTION
----------------------------------------

          INPUT DATA:
          -----------

  STEEL GRADE:Č 1530-SAE 1043

  OUENCHANT:SALT BATH
  TEMPERATURE:200 °C
  AGITATION:0 M/S
  ------------------------

    DIAMETER FOR HARDENING,MM:35
    -------------------------

  RESULTS OF COMPUTER AIDED PREDICTION:
  ***********************************

    CALCULATED HARDNESS,HRC:
    -------------------------

          SURFACE,HRC........=30
          3/4 RADIUS.........=24
          1/2 RADIUS.........=22.8
          1/4 RADIUS.........=21
          CENTER.............=21.3

  GRAHIC PRESENTATION(YES=1,NO=0) ?
```

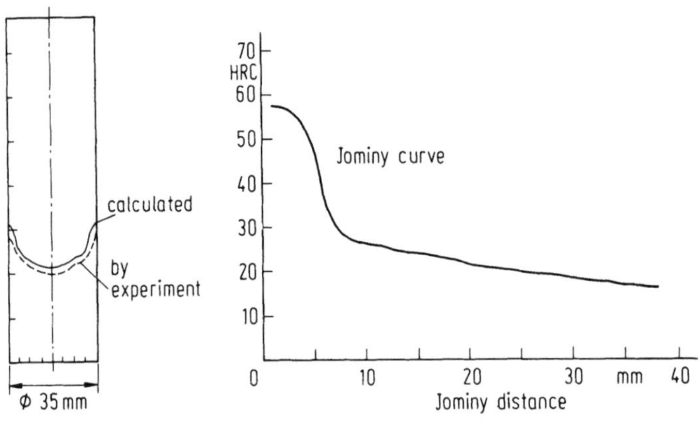

Fig. 11.2.8.4. An example of computer aided prediction of hardness distribution compared with experimentally obtained values, for a round bar of 35 mm Dia, steel Č. 1530—(SAE 1043), quenched in hot salt bath of 200 °C, without agitation

From examples 2, 3, 5 and 6 on the Fig. 11.2.8.5 it can be seen that the computer-aided prediction gives a better fit to the experimentally obtained results.

So far the experiments have shown a better fitting and concordance of experimental and modelled values of hardness in steel grades of better hardenability, while less reliable results of hardness predictions are obtained in

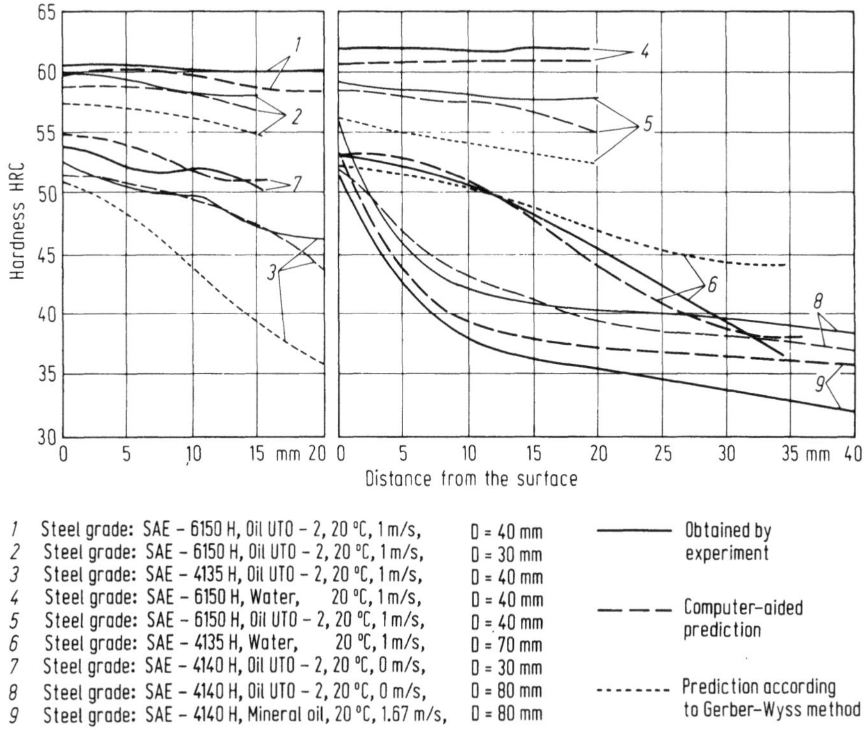

Fig. 11.2.8.5. Comparison of the hardness distribution on round bars' sections of different diameters measured after experiments and obtained by computer-aided prediction as well as by prediction according to the Gerber-Wyss method

steel grades of lower hardenability—(unalloyed steels). That is why this method requires additional testing for steel grades with Jominy curves having sheer decrease.

There is also a need of developing a procedure of hardness prediction for other typical forms of cross sections of steel products: cubes, parallelepipeds, plates, and tubes. The method should also be extended to bar diameters bigger than 100 mm.

Selection of Optimum Quenching Conditions

The use of the described relations and the stored data also makes it possible to find out the optimum quenching conditions when a certain hardness distribution (or a hardness value) on a specified bar cross section of a known steel grade is required.

The procedure for the selection of quenching conditions has been described using the example on Fig. 11.2.8.6. It consists of the following procedure:

First, the required hardness after quenching (H_q) in the particular point on the cross section of the workpiece must be defined—here 51 HRC at 3/4R.

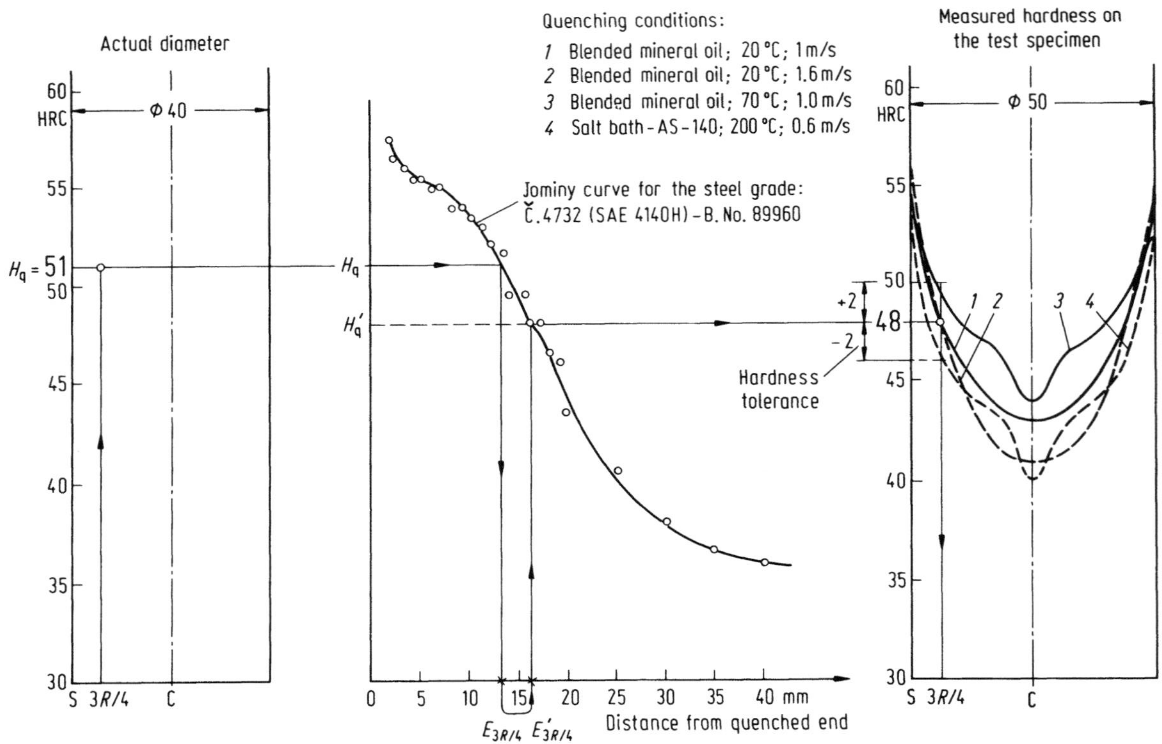

Fig. 11.2.8.6. An example of computer-aided selection of quenching conditions

By using the stored hardenability curve for this steel—here SAE 4140 H—the next step is establishing the Jominy distance $E_{3/4R}$ (Eq. 11.2.8.3) at which the same hardness H_q is obtained.

By means of the values of $E_{3/4R}$ and diameter D, the "hypothetical" quenching intensity factor $I_{3/4R}$ can be calculated according to the empirical, Eq. (11.2.8.8) derived from the regression analysis. The same equation applies to the test specimen of 50 mm dia and can be written as:

$$I_{3/4R} = \left(\frac{7.05}{E_{3/4R}}\right)^{1.495} \tag{11.2.8.12}$$

By substituting the calculated value of $I_{3/4R}$ and $D = 50$ mm, the Jominy distance $E'_{3/4R}$ corresponding to the 3/4R of the specimen's cross section can be obtained:

$$E'_{3/4R} = \frac{10.85}{0.668} I_{3/4R} \tag{11.2.8.13}$$

Further reference to the Jominy curve (Fig. 11.2.8.6) shows that hardness at the distance $E'_{3/4R}$ is H'_q (here 48 HRC), i.e. the same quenching conditions to produce a hardness value of $H_q = 51$ HRC at 3/4R of a 40 mm dia bar, yield a hardness H'_q of 48 HRC at the 3/4R of the 50 mm dia specimen.

Now the computer searches through all the stored hardness distribution curves of specimens made of the same steel grade for those quenching conditions by means of which the nearest hardness H'_q has been obtained (tolerance is ± 2 HRC). If, in such a computer search, the same value of hardness is obtained

Table 11.2.8.1. An example of the computer selected quenching conditions

```
INPUT DATA:
***********

     -STEEL GRADE  :   Č.4732(SAE 4140H)
     -BATCH NO.    :   89960
     -DIAMETER FOR
      HARDENING    :  [40 MM]
     ------------------------------
 REQUIRED HARDNESS:   [51 HRC]
 CRITICAL POINT ON THE
 CROSS SECTION ?:2

      1.SURFACE
     [2. 3/4 R]
      3. 1/2 R
      4. 1/4 R
      5. CENTER

    SELECTED QUENCHING CONDITIONS:
    ******************************
  MEDIA         ΔHRC    TEMPERATURE    AGITATION     H-VALUE
============================================================
1.OIL-KALENOL    0          20 °C       1.0 M/S        0.25
2.OIL-KALENOL    0.5        20 °C       1.6 M/S        0.30
3.OIL UTO-2      1.5        70 °C       1.0 M/S        0.22
•4.SALT  BATH    2.0       200 °C       0.6 M/S        0.41

ADDITIONAL INFORMATION ABOUT THESE CONDITIONS (Y/N)?:
```

by using two or more different quenching conditions, then those conditions which give the smallest value of residual stresses should be selected.

In this case the maximum temperature difference ΔT_{max} between center and the surface should be of utmost importance for the final decision. Table 11.2.8.1 gives an example (the same as given in Fig. 11.2.8.6) of the input requirements and the print-out of selected quenching conditions.

Hardness of 51 HRC at the $3/4R$ of a 40 mm dia bar may be obtained by quenching it in four different conditions. Certainly, the greater the number of recorded quenching intensities and relevant hardness distributions stored, the more precise and reliable selection of optimum conditions in the actual case is.

The method requires certainly more experiments in order to determine its reproducibility and applicability when applied to other steel grades an other quenching/cooling conditions.

References

1. Liščić B, Filetin T (1988) "Computer-aided Evaluation of Quenching Intensity and Prediction of Hardness Distribution", J. Heat Treatment, Vol. 5 No. 2, pp. 115–124
2. Liščić B, Filetin T (1987) "Computer-aided Determination of the Process Parameters for Hardening and Tempering Structural Steels", Heat Treatment of Metals 3, pp. 62–66.
3. Liščić B, Filetin T (1986) "Computergestützte Auswahl der Abschreckbedingungen", V. Int. Congress on Heat Treatment of Materials, Budapest, pp. 1760–1767
4. Crafts W, Lamont JI (1954) "Härtbarkeit und Auswahl von Stählen", Springer Verlag
5. Just E (1976) "Vergüten–Werkstoffbeeinflussung durch Härten", VDI-Berichte No. 256 pp. 125–140
6. Gerber W, Wyss U (1948) "Die Härtbarkeit und Vergütbarkeit der Stähle", Von Roll Mitt. 7, No. 2/3, Juni/Sept. pp. 13–49
7. Asimov M, Craig WF, Grossmann MA (1941) "Correlation between Jominy Test and Quenched Rounds", Trans. SAE Vol. 49, pp. 283–293

11.2.9 Prediction of Structural Constituents and Hardness Values upon Quenching by Using the CCT Diagrams

B. Liščić

A continuous-cooling-transformation (CCT) diagram, see Fig. 11.2.9.1, may be read only along the cooling curves drawn on it. As it could be seen from Fig. 11.2.9.1 only in case when the cooling rate is higher than critical one, a single phase structure (martensite) will be formed. In all other cases of cooling a mix of multi-phase structure will be formed. With continuous cooling e.g. there is no possibility to obtain structure that would consist only of bainite. Another important remark when reading a CCT diagram is that the size of an

%C	%Si	%Mn	%P	%S	%Cr	%Y	Method of melting	Grain size (ASTM) Mc Quaid-Ehn
0.38	1.05	1.14	0.041	0.019	0.23	0.02	b. S.-M.	4 and 5

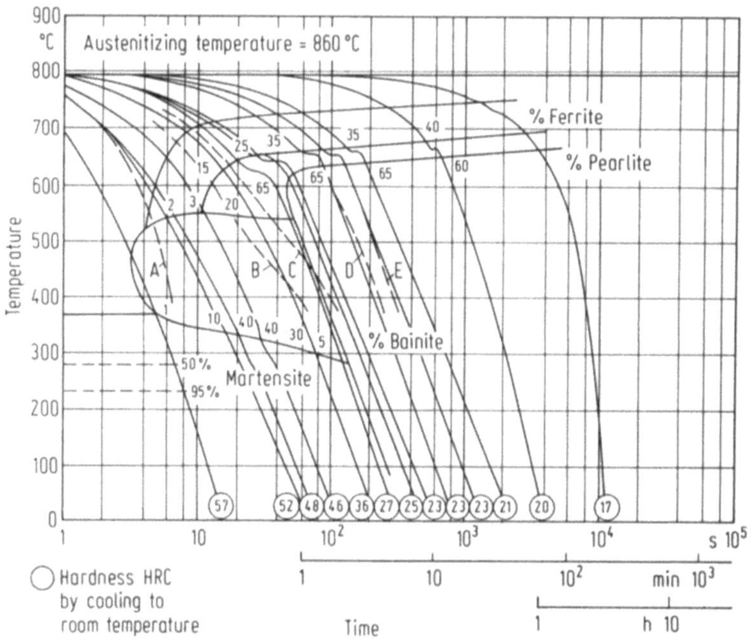

Jominy distances A = 2 mm 56 HRC, B = 10 mm 38 HRC,
C = 20 mm 28 HRC, D = 40 mm 24 HRC, E = 65 mm 22 HRC

Fig. 11.2.9.1. Continuous-cooling-transformation (CCT) diagram (of German origin) for the steel grade DIN-37 MnSi 5—Valid exactly for the heat of the given composition

area of one phase (obtained by crossing of a cooling curve) does not necessarily represent the amount of transformed phase. More information can be obtained from a CCT diagram if the amount of the transformed phase, in relevant areas, are marked with a number which denotes the percentage of the phase transformed. This is mostly the case with CCT-diagrams of German origin (see Fig. 11.2.9.1). The cooling curves in those diagrams have been obtained by cooling small specimens of 4.5 mm Dia in a dilatometer. The number on the end of each cooling curve denotes the hardness value in HV or HRC, obtained after quenching.

In CCT diagrams of American origin, see Fig. 11.2.9.2, the cooling curves superimposed on it correspond to the cooling law in different distances from the quenched end of the Jominy test specimen. To predict the hardness after quenching using those diagrams, one has to use the relevant Jominy hardenability curve.

There are two most important facts, when using a CCT diagram for prediction of structural constituents and hardness values upon quenching, on which

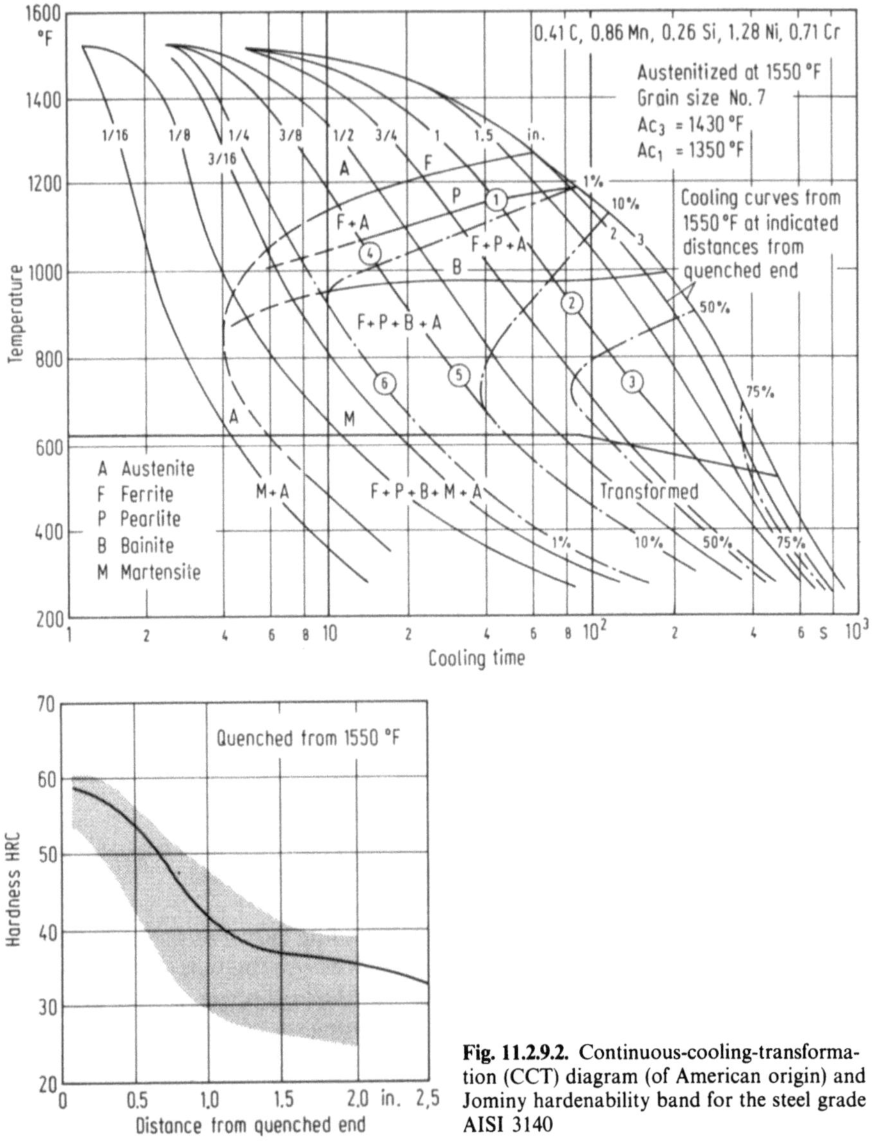

Fig. 11.2.9.2. Continuous-cooling-transformation (CCT) diagram (of American origin) and Jominy hardenability band for the steel grade AISI 3140

the exactness of such a prediction depends:

a) Each CCT diagram is valid exactly only for the chemical composition of specimens used for its construction. The chemical composition, within a steel grade however, differs from heat to heat. Because we don't have the relevant CCT-diagram for each heat, in order to take into account the differences in the chemical composition within a steel grade, usually the CCT diagram is constructed using specimens from that heat which Jominy hardenability curve corresponds best to the average curve of the relevant hardenability band.

b) Each CCT diagram describes only those transformations of the structure which occur along the cooling curves of specimens used for its construction. This means that a CCT diagram is valid only for cooling conditions under which it was constructed. The cooling law of above mentioned specimens of 4.5 mm Dia, that were used in a dilatometer for construction of the CCT diagram can, according to [1], be described sufficiently accurate by the following exponential function:

$$T = T_0 \cdot e^{-\alpha t},$$

where T_0 is the austenitization temperature; α is the heat transfer coefficient and t is time. Because each CCT-diagram is valid exactly for certain austenitization conditions, the austenitizing temperature and time (or austenite grain size) are usually indicated on it.

The exactness of predicted structural constituents and hardness values upon quenching depends on how far the cooling law in real workpieces corresponds to the cooling curves of specimens drawn in each CCT diagram.

In order to clearly establish the relations between the cooling curves drawn in a CCT diagram, and cooling curves obtained in different points of the cross-section of real workpieces, experiments have been conducted [1] using cylinders of different diameters. The length of cylinders was always 6 × Dia in order to avoid the influence of the heat transfer through the cylinder's ends.

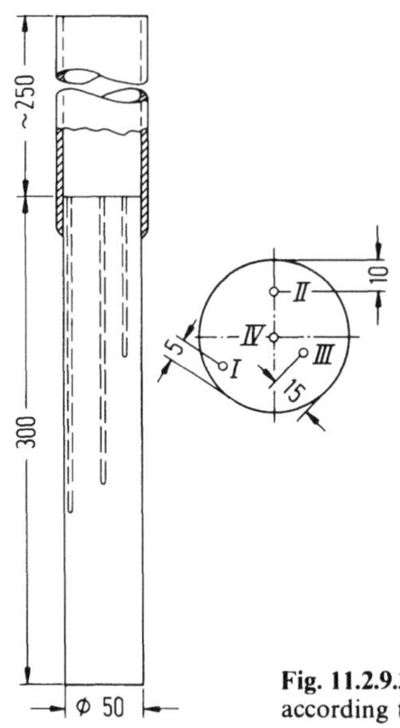

Fig. 11.2.9.3. Instrumented workpiece of 50 mm Dia used in experiments, according to [1]

The cylindrical workpieces have been instrumented with thermocouples, as shown in Fig. 11.2.9.3, and the cooling curves (temperature-time course) were measured in points: 5, 10, 15 and 25 mm below the surface. Quenching was performed from the austenitizing temperature 860 °C in water, oil and air. The measured cooling curves have been plotted in temperature-time diagrams—Figs. 11.2.9.4 and 11.2.9.5—which temperature and time axis have the same scale as the relevant CCT diagram (see Fig. 11.2.9.1). Comparing the cooling curves in different points of the cross-section on Figs. 11.2.9.4 and 11.2.9.5 for quenching in water, quenching in oil and cooling in air, one can see how the temperature gradient between core and surface becomes smaller when the quenching intensity decreases. The temperature difference e.g. between points I and IV (i.e. 5 and 25 mm below the surface) after 10 seconds from the beginning of quenching is: 180 °C for water-quenching, 70 °C for oil-quenching and less than 10 °C for air cooling. In case of water quenching the heat transfer (heat extraction) on the surface is big compared to heat conduction within the workpiece. In case of air cooling the heat transfer on the surface is small compared to the heat conduction within the workpiece.

The shape of the cooling curves measured in different points of the cross-section of a 50 mm Dia cylindrical workpiece agrees quite well with the shape of cooling curves obtained with small dilatometric specimens used for construction of the CCT diagram, as it can be seen comparing Figs. 11.2.9.4 and

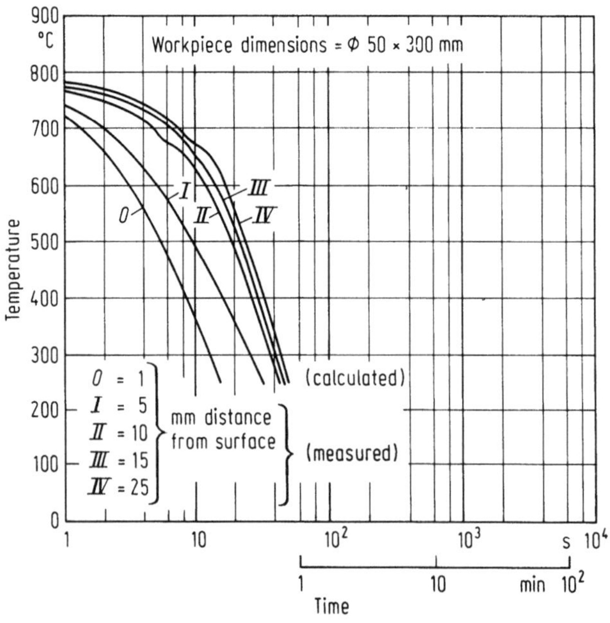

Fig. 11.2.9.4. Cooling curves in different points of the cross-section of a cylindrical workpiece of 50 mm Dia (steel grade DIN 37 MnSi 5), quenched in water

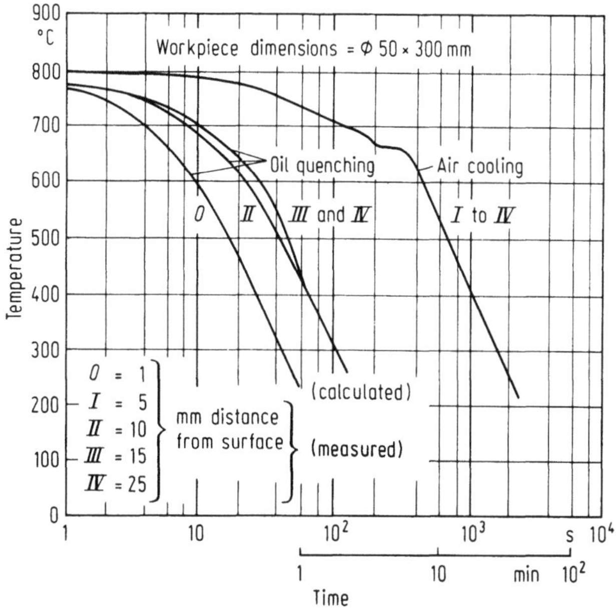

Fig. 11.2.9.5. Cooling curves in different points of the cross-section of a cylindrical workpiece of 50 mm Dia (steel grade DIN 37 MnSi 5), quenched in oil or cooled in air, respectively

11.2.9.5 with Fig. 11.2.9.1. This conclusion is valid not only for rapid cooling in water but also for slow cooling in air. In the point I (see Fig. 11.2.9.4) when quenching in water, in 2 seconds a temperature of 700 °C, in 10 seconds a temperature of 500 °C and in 25 seconds a temperature of 300 °C was achieved. The corresponding cooling curve on the CCT diagram (see Fig. 11.2.9.1) obtained with small dilatometric specimen (marked on its end with hardness of 48 HRC) achieves in 2 seconds 700 °C, in 8.5 seconds 500 °C and in 22 seconds 300 °C. This means that the point I (5 mm below the surface) of a cylindrical workpiece having 50 mm Dia, when quenched in water cools approximately according to the same exponential cooling law as does the small dilatometric specimen which attains a hardness of 48 HRC. It means that the same hardness, and same structure constituents have to be expected after quenching in point I also.

For cooling in air the shape of the cooling curve is such that it fits exactly between the cooling curves marked with 21 and 20 HRC (see Fig. 11.2.9.1). Not only the first part of the cooling curve is compatible but also the effect of the recalescence, due to pearlite transformation, occurs in the same way and at the same temperature. (As it can be seen from Fig. 11.2.9.5 there is no difference in cooling curves between surface and core, when cooled in air).

The shape of cooling curves in points II, III and IV i.e. 10, 15, and 25 mm below the surface, when quenching in water, is a little steeper in the lower temperature region (below 550 °C) than the corresponding cooling curves in the

CCT diagram. In this case a difference in the shape of cooling curve for surface and core respectively, for bigger cross sections, starts to be noticeable.

From the above described experiments the following conclusions can be drawn:

—Each quenching/cooling process results in specific structural constituents and hardness value as indicated by the cooling curves of small dilatometric specimens in the CCT diagram.

—If the cooling curve in a definite point of the workpiece's cross-section is sufficiently similar to a cooling curve of small dilatometric specimen used to construct the CCT diagram, —equal structural constituents and hardness value can be expected after quenching as indicated on the end of the relevant cooling curve in the CCT diagram. This applies generally to workpieces of different shape and dimensions.

—It appears that the main problem in practical use of CCT diagrams for prediction of structural constituents and hardness upon quenching is to exactly establish the cooling curve in the specified point of the workpiece's cross-section and find out if it is sufficiently similar to a cooling curve in the relevant CCT-diagram.

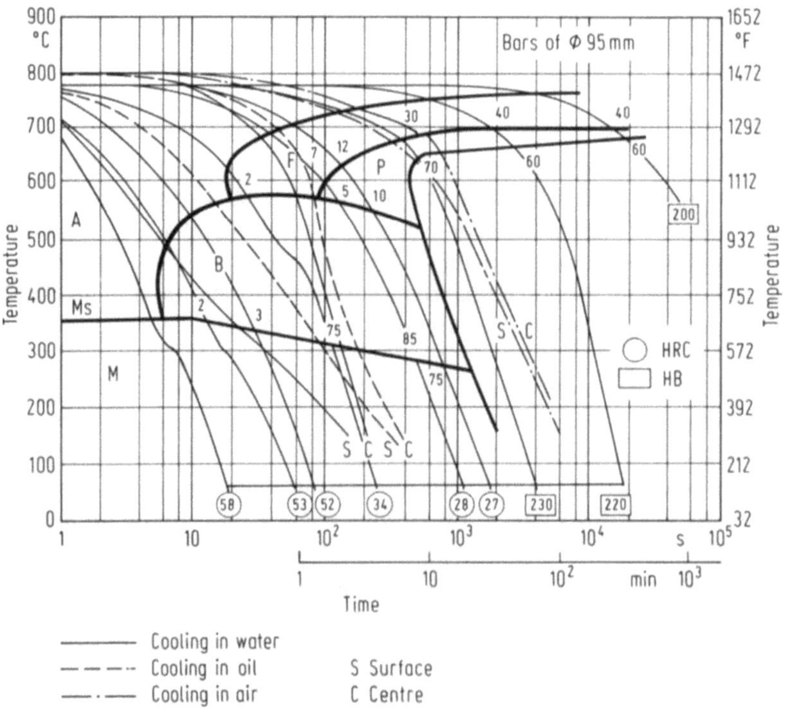

Fig. 11.2.9.6. Cooling curves for round bars of 95 mm Dia superimposed on the CCT diagram of the steel grade DIN 42 CrMo 4 or AISI 4140

There are two ways for establishing a cooling curve in a certain point of the cross-section of a workpiece:
—either experimentally by measuring the temperature-time course for definite quenching conditions by a thermoelement
—or by calculating the cooling curve in the specified point from another cooling curve within same cross-section obtained by measurement for same quenching conditions. This method is first of all applicable to axially symmetrical cross-sections with one dimensional heat flow (cylinders). Computer-software for such calculations already exist.

Ones having the necessary cooling curve plotted in the same scale as the CCT diagram, it can be superimposed on the relevant CCT-diagram as shown on Fig. 11.2.9.6.

The exactness of prediction from a CCT diagram decreases with increasing dimension of the workpiece's cross-section. According to [3] it seems that sufficiently exact predictions are possible by using CCT diagrams for round bars of 100 mm Dia when quenching in oil, and 150 mm Dia when quenching in water.

—The heat conductivity of steel may also influence the exactness of the prediction. When a cooling curve was established for a certain steel grade and

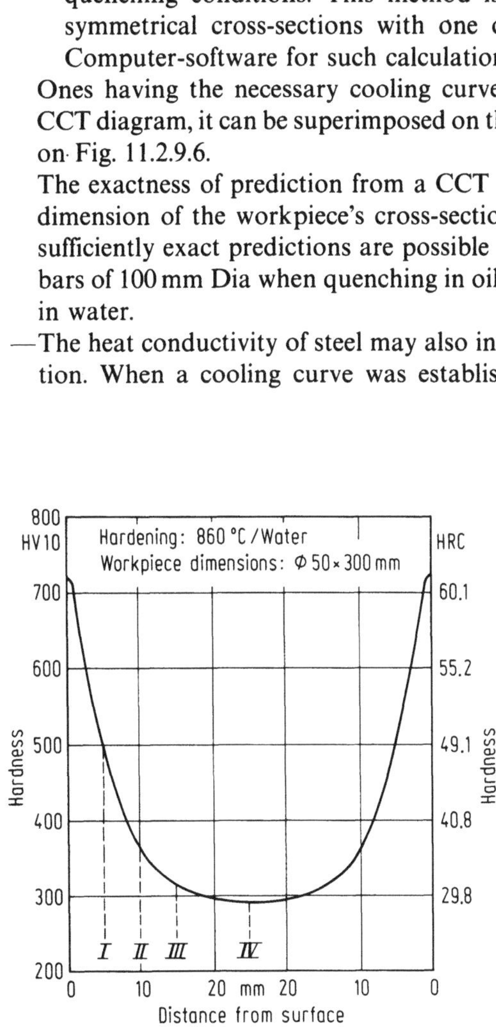

Points	I	II	III	IV
Hardness in HRC	49	38	31	29
Ferrite (%)	2	10	15	25
Pearlite (%)	0	5	20	25
Bainite (%)	20	30	30	20
Martensite (%)	~78	~55	~35	~30

Fig. 11.2.9.7. Hardness distribution and structural constituents upon quenching a round bar of 50 mm Dia in water. Steel grade: DIN—37 MnSi 5

should be transfered to another steel grade it is supposed that the heat conductivity of both steel grades are similar.
Fortunately for most of structural non alloyed and low alloyed steel grades this is more or less the case at least in the temperature region between 800 and 500 °C.

The above described relations between the cooling curves and relevant structural constituents and hardness are valid inversely too. It is namely possible from the structure and hardness in definite point of the cross-section upon quenching, to reconstruct the relevant cooling curve. From the change in structure and hardness in different points of the cross-section, different cooling curves can be determined, using the relevant CCT diagram. Figure 11.2.9.7 shows the hardness distribution upon quenching in water of a round bar of 50 mm Dia. The table below gives the structural constituents and hardness

%C	%Si	%Mn	%P	%S	%Cr	%Y	Method of melting	Grain size (ASTM) Mc Quaid-Ehn
0.38	1.05	1.14	0.041	0.019	0.23	0.02	b. S.-M.	4 and 5

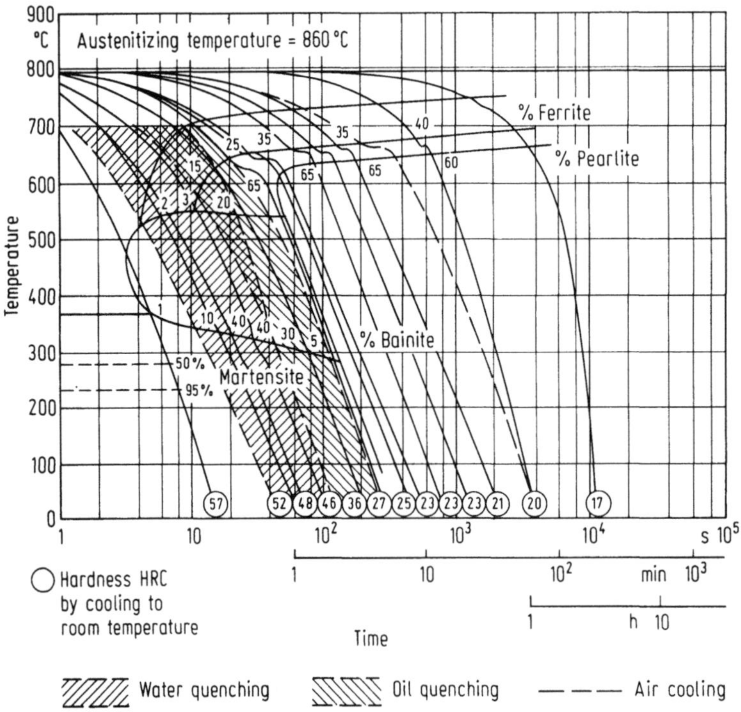

Fig. 11.2.9.8. CCT diagram of the steel grade DIN—37 MnSi 5 with the areas of reconstructed cooling curves in different points of a 50 mm Dia round bar when quenched in water, oil and air

values in points I, II, III and IV (i.e. 5, 10, 15, and 25 mm below the surface). The cooling curve e.g. that corresponds to the hardness value and structural constituents in point I is the curve in Fig. 11.2.9.1 marked on its end with 48 HRC, and this cooling curve is sufficiently similar to the cooling curve for point I in Fig. 11.2.9.4 which was exprimentally obtained. As an example of this procedure Fig. 11.2.9.8 shows a CCT diagram with areas of cooling curves through the whole cross-section of a 50 mm Dia round bars quenched in water, oil and air, which have been reconstructed from relevant structural constituents and hardness values.

If instead of dashed areas the cooling curves for specific points were drawn on a transparent paper, that could be superimposed on the CCT diagrams of different steel grades, one could select the most suitable steel grade that will upon quenching give the required structure and hardness.

%C	%Si	%Mn	%P	%S	%Cr	%Y	Method of melting	Grain size (ASTM) Mc Quaid-Ehn
0.38	1.05	1.14	0.041	0.019	0.23	0.02	b. S.-M.	4 and 5

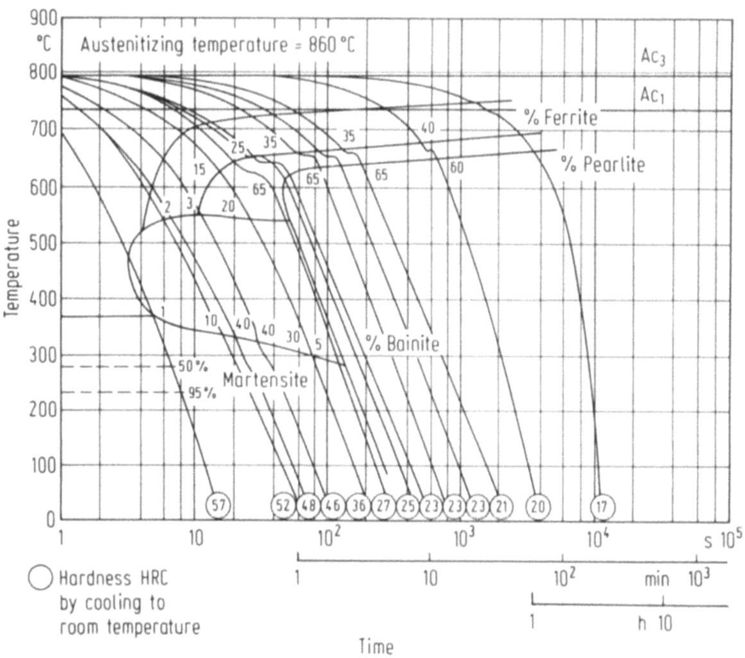

Fig. 11.2.9.9. CCT diagram of the steel grade DIN—37 MnSi 5 with superimposed cooling curves in points of different distances from the quenched end of a Jominy test specimen

In order to establish the relation between a CCT diagram and the cooling law in points of different distances from the quenched end of a Jominy test specimen, cooling curves have been measured during the Jominy test at distances of: 2, 10, 20, 40 and 65 mm from the quenched end. The obtained cooling curves have been superimposed on a CCT diagram of the steel grade DIN-37 MnSi5 (see Fig. 11.2.9.9). In this way, by using the adequate CCT diagram it is possible to find out the transformation behaviour in every point along the Jominy test specimen and to predict there the structural constituents.

It can be seen from Fig. 11.2.9.9 that the cooling curves of the Jominy test specimen are not so compatible with the cooling curves in the CCT diagram as cooling curves measured in different points of a round bar cross-section, especially for distances less than 20 mm from the quenched end.

In spite of these deviations there is a useable correspondence between the hardness values on the Jominy test specimen and the hardness values in the CCT diagram. It is therefore possible to superimposee the cooling curves for different distances from the quenched end of a Jominy test specimen (as Fig. 11.2.9.2 shows), but one should be aware of the fact that the accuracy of prediction in this way is less than in case of direct relation between the cooling curves of a CCT diagram and actual cooling curves in a definite point of a round bar cross-section.

References

1. Rose A, Strassburg W (1953) "Anwendung des Zeit-Temperatur-Umwandlungs Schaubildes für kontinuierliche Abkühlung auf Fragen der Wärmebehandlung", Archiv für das Eisenhüttenwesen, 24. Jahrgang, Heft 11/12-Nov/Dec. S. 1–10
2. Wever F, Rose A (1954) "Zur Frage der Wärmebehandlung der Stähle auf Grund ihrer Zeit-Temperatur-Umwandlungs-Schaubilder", Stahl und Eisen 74. Jahrgang, Heft 12, S. 749–759
3. Peter W, Hassdenteufel H (1967) "Aussagefähigkeit der Stirnabschreckprüfung und des Zeit-Temperatur-Umwandlungsschaubildes für das Ergebnis der Härtung von Rundstäben", Stahl und Eisen 87, Nr. 8, S. 455–457
4. Atlas of Isothermal Transformation and Cooling Transformation Diagrams, American Society for Metals, Metals Park, Ohio (1977)

Subject Index

MIX
Papier aus verantwortungsvollen Quellen
Paper from responsible sources
FSC® C105338

If you have any concerns about our products, you can contact us on
ProductSafety@springernature.com

In case Publisher is established outside the EU, the EU authorized representative is:
Springer Nature Customer Service Center GmbH
Europaplatz 3, 69115 Heidelberg, Germany

Printed by Libri Plureos GmbH
in Hamburg, Germany